Antioxidants in Animal Feed

Antioxidants in Animal Feed

Editors

Min Xue
Junmin Zhang
Zhenyu Du
Jie Wang
Wei Si

MDPI • Basel • Beijing • Wuhan • Barcelona • Belgrade • Manchester • Tokyo • Cluj • Tianjin

Editors

Min Xue
Chinese Academy of Agricultural Sciences
China

Junmin Zhang
Institute of Animal Sciences of Chinese Academy of Agricultural Sciences
China

Zhenyu Du
East China Normal University
China

Jie Wang
Chinese Academy of Agricultural Sciences
China

Wei Si
Institute of Animal Sciences of Chinese Academy of Agricultural Sciences
China

Editorial Office
MDPI
St. Alban-Anlage 66
4052 Basel, Switzerland

This is a reprint of articles from the Special Issue published online in the open access journal *Antioxidants* (ISSN 2076-3921) (available at: https://www.mdpi.com/journal/antioxidants/special_issues/Antioxidants_Animal_Feed).

For citation purposes, cite each article independently as indicated on the article page online and as indicated below:

LastName, A.A.; LastName, B.B.; LastName, C.C. Article Title. *Journal Name* **Year**, *Volume Number*, Page Range.

ISBN 978-3-0365-5471-6 (Hbk)
ISBN 978-3-0365-5472-3 (PDF)

Contents

antioxidants

Editorial

Antioxidants in Animal Feed

Jie Wang [1,†], Wei Si [2,†], Zhenyu Du [3], Junmin Zhang [2] and Min Xue [1,*]

1 National Aquafeed Safety Assessment Center, Institute of Feed Research, Chinese Academy of Agricultural Sciences, Beijing 100081, China
2 State Key Laboratory of Animal Nutrition, Institute of Animal Sciences of Chinese Academy of Agricultural Sciences, Beijing 100193, China
3 School of Life Sciences, East China Normal University, Shanghai 200241, China
* Correspondence: xuemin@caas.cn
† These authors contributed equally to this work.

Citation: Wang, J.; Si, W.; Du, Z.; Zhang, J.; Xue, M. Antioxidants in Animal Feed. *Antioxidants* **2022**, *11*, 1760. https://doi.org/10.3390/antiox11091760

Received: 17 August 2022
Accepted: 2 September 2022
Published: 6 September 2022

Production animals are often exposed to several oxidative stress conditions, including, but not limited to, heavy metals, alternative protein sources, environmental stress, disease, high densities, as well as handling, which may suppress growth performance, animal health and production, subsequently impacting economic feasibility. Promising research results have revealed that the administration of natural or synthetic antioxidants in the diet would be an important nutritional strategy to mitigate the negative influence induced by oxidative stress conditions. The Special Issue "Antioxidants in animal feed" has been conceived to set out the knowledge on the effects of dietary antioxidants on host health and performance of production animals, including livestock, poultry and fish. It provides various nutritional approaches to improve antioxidant capacity and benefit host health in animal production. Here, we offer an overview of the contents of this Special Issue, which collects 17 original articles.

For livestock and poultry, oxidative stress could affect ovarian function. Wang et al. found that oxidative stress could decrease the laying performance, ovarian function and influence gut microbiota and body metabolites in the layer model [1]. They then explored the role of melatonin on ovary oxidative stress, suggesting melatonin could exert an amelioration in ovary oxidative stress through the SIRT1-P53/FoxO1 pathway. Melatonin is considered as a bio-antioxidant. Peng et al. evaluated the impacts of dietary melatonin supplementation during pregnancy on reproductive performance, maternal–placental–fetal redox status, placental inflammatory response and mitochondrial function [2]. They concluded that melatonin supplementation during gestation could improve maternal–placental–fetal redox status and reproductive performance by ameliorating placental antioxidant status, inflammatory response and mitochondrial dysfunction. The work from Xu et al. focused on the potential effects of adding acidifiers to drinking water [3]. The results showed that supplementing drinking water with an acidifier has potential as an antioxidant, which was reflected in improvements in growth performance, immunity, antioxidant capacity and intestinal flora. The study by Liu et al. determined the effects and mechanisms of increased consumption of methionine by sows and piglets on the capacity of the progeny to counteract lipopolysaccharide (LPS) challenge-induced injury in the liver and spleen of piglets [4]. The results showed that dietary methionine supplementation alleviated liver and spleen damage that was induced by the LPS challenge. In addition, the results indicated that beneficial effects of dietary methionine were potentially due to the increased antioxidant capacity and inhibition of the TLR4 and NOD signaling pathway.

Various studies focused on the use of antioxidants in ruminates to improve health, performance and product quality. Wang et al. explored the effects of L-glutamine (L-Gln) on calves during weaning [5]. They found that a dietary lower-level L-Gln supplementation (1 and 2% of DMI) had higher average daily gain, glutathione peroxidase and IgG concentration and villus height/crypt depth of the duodenum and jejunum, as well as lower cortisol, haptoglobin and interleukin-8 concentration in weaned calves. These results provided evidence that the addition of L-Gln in the diet improved the negative effects of sudden weaning in calves. A study from Ma et al. investigated the effects of rumen-protected glucose (RPG) on the hepatic oxidative/antioxidative status and protein profile [6]. They showed that RPG supplementation reduced insulin sensitivity but increased the liver triglyceride concentration and the oxidative stress in early postpartum cows, which may indicate an increased risk of liver metabolic disorders caused by RPG supplementation. Wang et al. found that a plant-protein-based milk replacer had a negative effect on calves' liver function, immunity and antioxidant capacity [7]. In addition, transcriptome analysis revealed that energy metabolism, immune function and mineral metabolism showed differences during the pre-weaning period, while during the post-weaning period, osteoclast differentiation and metabolic pathways showed a difference. Kong et al. studied the potential of *Acremonium terricola* culture (ATC) or ATC as a new feed additive in dairy cow feeding [8]. The results showed that ATC improved milk yield and milk protein yield. Furthermore, the improvement in milk yield was likely related to improved immune function and antioxidant capacity.

Regarding fish production, various studies show how to mitigate the oxidative stress caused by unconditional production conditions using micronutrients. In particular, Wu et al. explored the effects of vitamin A on the muscle quality, nutritional quality and antioxidative ability of grass carp [9]. The results highlighted that dietary Vitamin A could improve flesh quality by increasing antioxidant capacity via the Nrf2/Keap1 signaling pathway. Similarly, dietary vitamin C can attenuate oxidative damage, inflammation and acute hypoxia-induced apoptosis in gibel carp via the Nrf2/Keap1 signaling pathway [10]. These findings further suggest that vitamins A and C, as essential micronutrients, could be powerful antioxidants in the diet to regulate antioxidant capacity via certain potential signaling pathways, such as Nrf2/Keap1. Further, the work of Xu et al. focused on docosahexaenoic acid (DHA), as a nutritional modulator, to alleviate palmitic-acid-induced inflammation of macrophages via the TLR22-MAPK-PPAR/Nrf2 pathway in large yellow croakers, thereby improving the utilization rate of palm oil in aquafeed [11]. Another study found that dietary glutamine could regulate immune and antioxidant capacity to protect against *Flavobacterium columnare* infection in yellow catfish [12]. The authors suggest their study firstly demonstrated the regulatory roles of glutamine in the fish immune and antioxidant system and reported its inhibitory effects on fish apoptosis and autophagy during pathogenic infection. Furthermore, Shi et al. found that oxidized-fish-oil diets can cause negative physiological health effects in channel catfish, while adding taurine can increase growth and antioxidant ability, reduce lipid deposition and improve intestinal health [13]. These findings could advance the understanding of the molecular mechanism of oxidative stress and provide nutritional mitigative strategies via supplemented micronutrients in aquaculture production.

Special attention has also been paid to plant extracts, functional components or alternative protein sources. Starch is necessary as a binder and sweller during extrusion processing of pelleted aquatic feeds. However, carnivorous fish, for example, largemouth bass, fed excess starch may induce metabolic liver disease [14]. Liang et al. found that largemouth bass fed high-starch feed induced oxidative stress and lipid metabolic disorder, while dietary olive extract could improve antioxidant capacity, anti-inflammatory responses and lipid metabolism, but could not completely repair high-starch-diet-induced lipid metabolic disorder [15]. Xu et al. also reported that probiotic *Lactobacillus plantarum* MR1 ameliorated high-carbohydrate-diet-induced hepatic lipid accumulation and oxidative stress by increasing the circulating uridine [16]. Additionally, conventional soybean meal, replacing fishmeal protein in aquatic feed, could adversely influence the growth performance and health of the host. Wang et al. showed that the mixture of plant extracts (thymol and carvacrol) and chelated trace elements (Cu, Mn and Zn) in the diet could mitigate soymeal-induced adverse effects on growth and disease resistance through the improvement in antioxidant capacity and regulation of gut microbiota [17]. Interestingly,

Arthrospira platensis, a blue-green alga, could activate the antioxidant response and alleviate oxidative stress and pigmentation disorder induced by air exposure in yellow catfish [18].

All the research articles in this Special Issue show that establishing a better understanding of oxidative stress is of pivotal importance in production animals. The variety of subjects treated proves that this is a complex and multifaceted topic, on which researchers are working from different viewpoints and perspectives. We thank all the authors for their contributions. We hope that this Special Issue will encourage more scientists to move forward on the path to increasing knowledge on the effect of natural or synthetic antioxidants on the growth and health of production animals.

Funding: This research received no external funding.

Conflicts of Interest: The authors declare no conflict of interest.

References

1. Wang, J.; Jia, R.; Gong, H.; Celi, P.; Zhuo, Y.; Ding, X.; Bai, S.; Zeng, Q.; Yin, H.; Xu, S.; et al. The Effect of Oxidative Stress on the Chicken Ovary: Involvement of Microbiota and Melatonin Interventions. *Antioxidants* **2021**, *10*, 1422. [CrossRef] [PubMed]
2. Peng, X.; Cai, X.; Li, J.; Huang, Y.; Liu, H.; He, J.; Fang, Z.; Feng, B.; Tang, J.; Lin, Y.; et al. Effects of Melatonin Supplementation during Pregnancy on Reproductive Performance, Maternal-Placental-Fetal Redox Status, and Placental Mitochondrial Function in a Sow Model. *Antioxidants* **2021**, *10*, 1867. [CrossRef] [PubMed]
3. Xu, Q.L.; Liu, C.; Mo, X.J.; Chen, M.; Zhao, X.L.; Liu, M.Z.; Wang, S.B.; Zhou, B.; Zhao, C.X. Drinking Water Supplemented with Acidifiers Improves the Growth Performance of Weaned Pigs and Potentially Regulates Antioxidant Capacity, Immunity, and Gastrointestinal Microbiota Diversity. *Antioxidants* **2022**, *11*, 809. [CrossRef]
4. Liu, M.; Zhang, Y.; Cao, K.X.; Yang, R.G.; Xu, B.Y.; Zhang, W.P.; Batonon-Alavo, D.I.; Zhang, S.J.; Sun, L.H. Increased Ingestion of Hydroxy-Methionine by Both Sows and Piglets Improves the Ability of the Progeny to Counteract LPS-Induced Hepatic and Splenic Injury with Potential Regulation of TLR4 and NOD Signaling. *Antioxidants* **2022**, *11*, 321. [CrossRef] [PubMed]
5. Wang, S.; Wang, F.; Kong, F.; Cao, Z.; Wang, W.; Yang, H.; Wang, Y.; Bi, Y.; Li, S. Effect of Supplementing Different Levels of L-Glutamine on Holstein Calves during Weaning. *Antioxidants* **2022**, *11*, 542. [CrossRef] [PubMed]
6. Ma, Z.; Fang, L.; Ungerfeld, E.; Li, X.; Zhou, C.; Tan, Z.; Jiang, L.; Han, X. Supplementation of Rumen-Protected Glucose Increased the Risk of Disturbance of Hepatic Metabolism in Early Postpartum Holstein Cows. *Antioxidants* **2022**, *11*, 469. [CrossRef] [PubMed]
7. Wang, S.; Hu, F.; Diao, Q.; Li, S.; Tu, Y.; Bi, Y. Comparison of Growth Performance, Immunity, Antioxidant Capacity, and Liver Transcriptome of Calves between Whole Milk and Plant Protein-Based Milk Replacer under the Same Energy and Protein Levels. *Antioxidants* **2022**, *11*, 270. [CrossRef] [PubMed]
8. Kong, F.; Zhang, Y.; Wang, S.; Cao, Z.; Liu, Y.; Zhang, Z.; Wang, W.; Lu, N.; Li, S. Acremonium terricola Culture's Dose-Response Effects on Lactational Performance, Antioxidant Capacity, and Ruminal Characteristics in Holstein Dairy Cows. *Antioxidants* **2022**, *11*, 175. [CrossRef] [PubMed]
9. Wu, P.; Zhang, L.; Jiang, W.; Liu, Y.; Jiang, J.; Kuang, S.; Li, S.; Tang, L.; Tang, W.; Zhou, X.; et al. Dietary Vitamin A Improved the Flesh Quality of Grass Carp (*Ctenopharyngodon idella*) in Relation to the Enhanced Antioxidant Capacity through Nrf2/Keap 1a Signaling Pathway. *Antioxidants* **2022**, *11*, 148. [CrossRef] [PubMed]
10. Wu, L.; Xu, W.; Li, H.; Dong, B.; Geng, H.; Jin, J.; Han, D.; Liu, H.; Zhu, X.; Yang, Y.; et al. Vitamin C Attenuates Oxidative Stress, Inflammation, and Apoptosis Induced by Acute Hypoxia through the Nrf2/Keap1 Signaling Pathway in Gibel Carp (*Carassius gibelio*). *Antioxidants* **2022**, *11*, 935. [CrossRef] [PubMed]
11. Xu, D.; Cui, K.; Li, Q.; Zhu, S.; Zhang, J.; Gao, S.; Hao, T.; Mai, K.; Ai, Q. Docosahexaenoic Acid Alleviates Palmitic Acid-Induced Inflammation of Macrophages via TLR22-MAPK-PPARgamma/Nrf2 Pathway in Large Yellow Croaker (*Larimichthys crocea*). *Antioxidants* **2022**, *11*, 682. [CrossRef] [PubMed]
12. Jiao, C.; Zou, J.; Chen, Z.; Zheng, F.; Xu, Z.; Lin, Y.H.; Wang, Q. Dietary Glutamine Inclusion Regulates Immune and Antioxidant System, as Well as Programmed Cell Death in Fish to Protect against Flavobacterium columnare Infection. *Antioxidants* **2021**, *11*, 44. [CrossRef] [PubMed]
13. Shi, Y.; Hu, Y.; Wang, Z.; Zhou, J.; Zhang, J.; Zhong, H.; Fu, G.; Zhong, L. The Protective Effect of Taurine on Oxidized Fish-Oil-Induced Liver Oxidative Stress and Intestinal Barrier-Function Impairment in Juvenile Ictalurus punctatus. *Antioxidants* **2021**, *10*, 1690. [CrossRef] [PubMed]
14. Chen, P.; Zhu, Y.P.; Wu, X.F.; Gu, X.; Xue, M.; Liang, X.F. Metabolic adaptation to high starch diet in largemouth bass (*Micropterus salmoides*) was associated with the restoration of metabolic functions via inflammation, bile acid synthesis and energy metabolism. *Br. J. Nutr.* **2022**, 1–38. [CrossRef] [PubMed]
15. Liang, X.; Chen, P.; Wu, X.; Xing, S.; Morais, S.; He, M.; Gu, X.; Xue, M. Effects of High Starch and Supplementation of an Olive Extract on the Growth Performance, Hepatic Antioxidant Capacity and Lipid Metabolism of Largemouth Bass (*Micropterus salmoides*). *Antioxidants* **2022**, *11*, 577. [CrossRef] [PubMed]

16. Xu, R.; Wang, T.; Ding, F.-F.; Zhou, N.-N.; Qiao, F.; Chen, L.-Q.; Du, Z.-Y.; Zhang, M.-L. *Lactobacillus plantarum* Ameliorates High-Carbohydrate Diet-Induced Hepatic Lipid Accumulation and Oxidative Stress by Upregulating Uridine Synthesis. *Antioxidants* **2022**, *11*, 1238. [CrossRef] [PubMed]
17. Wang, J.; Mai, K.S.; Ai, Q.H. Conventional Soybean Meal as Fishmeal Alternative in Diets of Japanese Seabass (*Lateolabrax japonicus*): Effects of Functional Additives on Growth, Immunity, Antioxidant Capacity and Disease Resistance. *Antioxidants* **2022**, *11*, 951. [CrossRef] [PubMed]
18. Liu, C.; Liu, H.; Zhu, X.; Han, D.; Jin, J.; Yang, Y.; Xie, S. The Effects of Dietary Arthrospira platensis on Oxidative Stress Response and Pigmentation in Yellow Catfish *Pelteobagrus fulvidraco*. *Antioxidants* **2022**, *11*, 1100. [CrossRef] [PubMed]

antioxidants

MDPI

Article

Effects of High Starch and Supplementation of an Olive Extract on the Growth Performance, Hepatic Antioxidant Capacity and Lipid Metabolism of Largemouth Bass (*Micropterus salmoides*)

Xiaofang Liang [1], Pei Chen [1], Xiaoliang Wu [1], Shujuan Xing [1], Sofia Morais [2,*], Maolong He [3], Xu Gu [1] and Min Xue [1,*]

1 National Aquafeed Safety Assessment Center, Institute of Feed Research, Chinese Academy of Agricultural Sciences, Beijing 100081, China; liangxiaofang01@caas.cn (X.L.); chen_pei1@ctg.com.cn (P.C.); 82101201182@caas.cn (X.W.); shujuan.xing@wur.nl (S.X.); guxu@caas.cn (X.G.)
2 Innovation Division, Animal Science Unit, Lucta S.A., 08193 Bellaterra, Spain
3 Innovation Division, Lucta (Guangzhou) Flavours Co., Ltd., Guangzhou 510530, China; maolong.he@lucta.com
* Correspondence: sofia.morais@lucta.com (S.M.); xuemin@caas.cn (M.X.)

Abstract: An 8-week feeding trial was conducted to investigate the effects of high-starch diets and the supplementation of an olive extract (OE) on the growth performance, liver health and lipid metabolism of largemouth bass (*Micropterus salmoides*). Four isonitrogenous and isolipidic diets were prepared: two basal diets containing low (9.0%) and high (14.4%) levels of starch (named as LS and HS), and 0.125% OE was supplemented to each basal diet (named LSOE and HSOE). The results show that high-starch diets had significant negative effects on growth performance, with lower FR, SGR and higher FCR, whereas OE significantly lowered FCR, determined by two-way ANOVA analysis. High-starch diets induced oxidative stress, inflammatory response and liver function injury, with significant increases in the content of plasmatic AKP, AST, ALT, hepatic SOD and MDA, and up-regulation of hepatic *TNFα*, *IL1β*, and *TGFβ1* gene expression. In addition, a high-starch diet decreased the phosphorylation of AMPK and upregulated the expression of SREBP, together with higher hepatic liver lipid and HSI. The oxidative stress and lipid metabolism disorders indicate metabolic liver disease (MLD) of largemouth bass fed high-starch diets. Feeding on OE-supplemented diets increased the hepatic antioxidant capacity by decreasing the content of MDA and SOD. Fish fed the HSOE diet had an activated phosphorylation of JNK and decreased expression of pro-inflammatory *IL1β* compared with those fed the HS diet, which strongly indicated that the degree of inflammatory responses was reduced after OE supplementation. Interestingly, this study demonstrated that OE regulates hepatic lipid metabolism in fish by inhibiting the expression of hepatic lipogenesis genes (*ACC1* and *FASN*) and promoting lipolysis (*ATGL*) and β-oxidation (*CPT1α*) to prevent TG accumulation. In conclusion, high-starch feed induced oxidative stress and lipid metabolic disorder of largemouth bass, while supplementation with OE improved its antioxidant capacity, anti-inflammatory responses and lipid metabolism. However, hepatic histopathological results suggested that OE supplementation could not completely repair the MLD caused by the high level of starch in largemouth bass.

Keywords: largemouth bass; high starch; antioxidant; lipid metabolism; metabolic liver disease

Citation: Liang, X.; Chen, P.; Wu, X.; Xing, S.; Morais, S.; He, M.; Gu, X.; Xue, M. Effects of High Starch and Supplementation of an Olive Extract on the Growth Performance, Hepatic Antioxidant Capacity and Lipid Metabolism of Largemouth Bass (*Micropterus salmoides*). *Antioxidants* **2022**, *11*, 577. https://doi.org/10.3390/antiox11030577

Academic Editor: Stanley Omaye

Received: 23 February 2022
Accepted: 16 March 2022
Published: 17 March 2022

1. Introduction

World aquaculture production has increased tremendously over the past years. However, due to intensive farming practices, metabolic diseases of fish have posed a great threat to fish production, causing heavy economic losses. Largemouth bass (*Micropterus salmoides*) is a North America-native carnivorous species, which has been one of the most widely cultured and consumed fish species in China because of its rapid growth, good flesh quality, and high market value. Its production in 2020 was up to 600 thousand tons in China. As a typical carnivorous fish, *M. salmoides* was traditionally fed on chilled fish; however, with the requirements for sustainable development of aquaculture, more and more formulated diets are used nowadays. Starch is necessary as a binder and sweller during extrusion processing of formulated, pelleted feeds [1]. However, excess dietary starch in carnivorous fish diets can induce hyperglycemia, glycogen and lipid accumulation, and chronic inflammation response, as well as compromising the immune function and antioxidant capabilities, eventually resulting in metabolic liver disease (MLD) [2–4].

To develop environmentally friendly aquaculture and ensure food safety, various non-antibiotic feed additives have been used to improve animal health [5]. Different plant extracts have been reported to act as immunostimulants and to have antibacterial and anti-parasitic (virus, protozoans, monogeneans) properties in aquaculture due to the presence of active molecules such as alkaloids, terpenoids, saponins and flavonoids [6]. Recently, olive extracts (OEs), either the byproduct of olive oil production or leaf extract (OLE), have been proven as natural products playing very important roles as antimicrobial, anti-inflammatory, antiviral and anti-tumor compounds, in which particular emphasis is placed on the antioxidant activity of the extracts [7]. At present, research on the application of olive extracts in aquaculture is mainly focused on OLE. The main active components of OLE are polyphenols, in which the most abundant is oleuropein [8]. Dietary OLE alters some immune gene expression levels and disease resistance to *Yersinia ruckeri* infection in rainbow trout (*Oncorhynchus mykiss*) [9]. Furthermore, dietary supplementation of OLE can enhance the growth performance of common carp (*Cyprinus carpio*) by activating digestive enzyme activity in the intestine and increasing the expression level of several genes (*GH* and *IGF-I*) related to growth in the brain, liver, head kidney and muscle [10]. On the other hand, the active components of OE derived (as byproduct) from olive oil production also include triterpenes, such as maslinic acid and oleanolic acid, which are bioactive components with anti-oxidative, anti-inflammatory and hepatoprotective activities [11]. Triterpenes are involved in the regulation of oxidative status through the reduction in ROS, raising SOD and CAT activities, the suppression of nuclear factor-κB (NF-κB), and the prevention of lipid peroxidation [7,12–14]. OE has been demonstrated as an immunopotentiator in Black seabream (*Acanthopagrus schlegelii*) [7]. In addition, research performed in mammals has established that triterpene-enriched OE is involved in lipid regulation. For example, OE treatment has been proved to decrease hepatic lipid accumulation by regulating lipid metabolism in vivo and in vitro in mice [15]. Therefore, we hypothesized that OE has the potential to prevent or reduce MLD caused by a high-starch diet in aquatic animals. To our knowledge, information regarding the roles of OE in MLD in largemouth bass have not been reported yet.

The objective of the present study was to investigate effects of high-starch diets and the supplementation of OE on the growth performance, hepatic antioxidant capacity and lipid metabolism of largemouth bass. The results of this study provide insights into the environmentally friendly prevention of MLD in carnivorous fish.

2. Materials and Methods

2.1. Growth Trial and Sample Collection

Two isonitrogenous and isoenergetic experimental diets with 9.0% (low starch, named as LS) and 14.4% (high starch, named as HS) starch were prepared, and 0.125% OE (Lucta S.A., Barcelona, Spain) was added to these two basal diets, respectively (named as LSOE and HSOE). The OE product was prepared from the olive cake remaining from typical two-phase olive oil production, with a standardized composition (determined by HPLC) of minimum 7.5% triterpenes (maslinic acid and oleanolic acid). The product also contains > 2% protein, > 5% fat and > 30% fiber. The feed ingredients were ground into fine powder through a 247 μm mesh. Each diet was processed into 3 mm-diameter floating pellets under the following extrusion condition: feeding section (90 °C/5 s), compression section (130 °C/5 s), and metering section (150 °C/4 s), using a Twin-screwed extruder (EXT50A,

YangGong Machine, Beijing, China) according to our previous studies [4,16]. All diets were air-dried at room temperature and stored at −20 °C until use. The diet formulations and analyzed chemical compositions are shown in Table 1.

Table 1. Formulation and composition of experimental diets (%).

Ingredients	LS	HS	LSOE	HSOE
Fish meal [a]	30.0	30.0	30.0	30.0
Cottonseed protein concentrate [a]	23.5	22.6	23.4	22.5
Microbial protein [a]	4.0	4.0	4.0	4.0
Tapioca starch	5.0	5.0	5.0	5.0
Wheat flour	9.0	18.0	9.0	18.0
Wheat gluten meal	4.0	4.0	4.0	4.0
Soybean meal [a]	2.0	-	2.0	-
Spay-dried blood cell powder	4.0	4.0	4.0	4.0
α-cellulose	4.6	-	4.6	-
$Ca(H_2PO4)_2$	1.7	1.7	1.7	1.7
Lecithin oil	2.0	2.0	2.0	2.0
Fish oil	3.5	3.5	3.5	3.5
Soybean oil	3.5	3.5	3.5	3.5
Vitamin and mineral premix [b]	1.4	1.4	1.4	1.4
Kelp powder	1.5	0	1.5	0
L-Thr	0.1	0.1	0.1	0.1
DL-Met	0.2	0.2	0.2	0.2
Olive extract	0	0	0.125	0.125
Total	100	100	100	100
Analyzed chemical composition (dry matter basis %)				
Moisture	6.10	7.43	7.25	7.34
Crude protein	50.83	51.15	51.17	51.11
Crude lipid	12.36	12.33	12.44	12.41
Crude ash	10.08	10.04	9.84	9.82
Starch [c]	9.00	14.40	9.00	14.40
Gross energy (MJ/Kg)	20.45	20.15	20.49	20.27

[a] Fish meal: crude protein content was 68.8%; cottonseed protein concentrate: crude protein content was 61.5%; microbial protein: crude protein content was 86.7%; soybean meal: crude protein content was 44.7%. [b] Vitamin premix (mg/kg diet): vitamin A 20; vitamin D3 10; vitamin K3 20; vitamin E 400; vitamin B1 10; vitamin B2 15; vitamin B6 15; vitamin B12 (1%) 8; ascorbic acid (35%) 1000; calcium pantothenate 40; niacinamide 100; inositol 200; biotin (2%) 2; folic acid 10; corn gluten meal 150; choline chloride (50%) 4000. Mineral premix (mg/kg diet): $CuSO_4 \cdot 5H_2O$ 10; $FeSO_4 \cdot H_2O$ 300; $ZnSO_4 \cdot H_2O$ 200; $MnSO_4 \cdot H_2O$ 100; KI (10%) 80; $CoCl_2 \cdot 6H_2O$ (10% Co) 5; Na_2SeO_3 (10% Se) 10; $MgSO4 \cdot 5H_2O$ 2000; NaCl 100; zeolite 4995; antioxidant 200. [c] Starch content was estimated based on the starch content of tapioca starch (72% starch) and wheat flour (60% starch).

Largemouth bass were obtained from a commercial aquafarm (Tangshan, Hebei, China). All fish were acclimated and fed the control experimental diet (LS diet) for 4 weeks before the formal feeding trial. After a 24 h fasting period, fish (initial body weight = 35.98 ± 0.21 g) were distributed into 12 cylindrical plastic tanks (capacity: 256 L) with three replicates per treatment and 20 fish per tank, and each diet was randomly assigned to 12 tanks. Fish were fed to apparent satiation twice daily at 08:00 h and 17:00 h. During the experiment, water temperature was maintained at 21–25 °C, pH at 7.2–8.0, dissolved oxygen > 6.0 mg/L, and ammonia-N < 0.3 mg/L.

Growth performance factors (final body weight (FBW), survival rate (SR), specific growth rate (SGR), feed conversion ratio (FCR), feeding rate (FR)) were determined by batch weighing the fish at the end of the 8 weeks after starvation for 24 h. All the sampled fish were anesthetized with chlorobutanol (300 mg/mL). Individual body weight, body length, viscera, liver and visceral adipose tissue weight of four fish in each tank were recorded to calculate the condition factor (CF), viscerosomatic index (VSI), hepatosomatic index (HSI) and visceral adipose index (VAI). Blood was rapidly drawn from the caudal vein and centrifuged (4000× *g*, 10 min, 4 °C) to obtain plasma for the analysis of hematological parameters. Four liver samples from each tank were dissected and immediately frozen in liquid nitrogen, and kept at −80 °C for mRNA isolation and tissue homogenate analysis until used. Four liver samples near the bile duct in each tank were fixed in 4% paraformaldehyde (Solarbio, China) for histology determination. The rest of the livers in each tank were pooled into zip-lock bags and then stored at −20 °C for the assay of crude lipids.

2.2. Chemical Composition Analysis of Diets

The crude protein, crude lipid, crude ash, moisture, starch, and gross energy contents of experimental diets and whole body of fish were analyzed according to standard methods as previously described [4,16].

2.3. Plasma and Hepatic Homogenate Parameters

The content of plasma glucose, triglycerides (TG), total cholesterol (TC), high-density lipoprotein cholesterol (HDL-C), low-density lipoprotein cholesterol (LDL-C), total bile acids (TBA), alkaline phosphatase (AKP), aspartate aminotransferase (AST), alanine aminotransferase (ALT), and hepatic TG, TC, TBA, HDL-C, LDL-C, malondialdehyde (MDA) and activities of total antioxidant capacity (T-AOC) and glutathione peroxidase (GSH-Px), superoxide dismutase (SOD), and catalase (CAT) were determined by commercial assay kits (Nanjing Jiancheng Co., Nanjing, China) following the manufacturer's protocols. Reactive oxygen species (ROS) was determined using a commercial kit (Jiangsu Meimian industrial Co., Ltd., Yancheng, China).

2.4. Hepatic Histopathological and Immunofluorescence Examination

Liver samples were fixed, dehydrated, embedded, and stained for hematoxylin and eosin (H&E) as previously described [4,16]. The results of HE staining were observed using light microscopy (DM2500, Leica, Weztlar, Germany). Immunohistochemistry for NF-κB was conducted as previously described [16]. Briefly, sections were incubated with polyclonal NF-κB (#8242, CST, Boston, USA) overnight at 4 °C. The results were observed using a high-resolution living cell imaging system (DeltaVision, GE, Boston, USA).

2.5. Quantitative Real-Time PCR

Total RNA extraction and cDNA synthesis were carried out as described previously [4,16]. The qPCR analysis was performed using a CFX96TM Real-Time System (Bio-Rad, CA, USA) using iTaqTM Universal SYBR® Green Supermix (Bio-Rad, CA, USA). Each sample was run in triplicate and analyzed using the $2^{-\Delta\Delta Ct}$ method. *EF1α* was used as an endogenous reference gene. Primer sequences are shown in Table 2.

Table 2. Primer sequences for RT-qPCR.

Genes	Forward Primer (5′-3′)	Reverse Primer (5′-3′)	Tm (°C)	E-Values (%)	Accession Number
EF1α	TGCTGCTGGTGTTGGTGAGTT	TTCTGGCTGTAAGGGGGCTC	60.4	102.8	119901934
ACC1	ATCCCTCTTTGCCACTGTTG	GAGGTGATGTTGCTCGCATA	57.5	102.2	119896388
FASN	TGTGGTGCTGAACTCTCTGG	CATGCCTAGTGGGGAGTTGT	57.5	102.1	119915567
ATGL	CCATGATGCTCCCCTACACT	GGCAGATACACTTCGGGAAA	58	99.1	119893301
CPT1α	CATGGAAAGCCAGCCTTTAG	GAGCACCAGACACGCTAACA	60.0	98.8	119893292
TNFα	CTTCGTCTACAGCCAGGCATCG	TTTGGCACACCGACCTCACC	63	105.7	119906688
IL1β	CGTGACTGACAGCAAAAAGAG	GATGCCCAGAGCCACAGTTC	59.4	101.3	119914255
TGFβ1	GCTCAAAGAGAGCGAGGATG	TCCTCTACCATTCGCAATCC	59	95.6	119882881
IL10	CGGCACAGAAATCCCAGAGC	CAGCAGGCTCACAAAATAAACA	62.1	113.6	119885912
SREBP1	AGTCTGAGCTACAGCGACAAGG	TCATCACCAACAGGAGGTCACA	61	98.1	119888831

EF1α, elongation factor-1α; ACC1, acetyl-CoA carboxylase 1; FASN, fatty acid synthase; ATGL, adipose triglyceride lipase; CPT1α, carnitine palmitoyltransferase 1α; TNFα, tumor necrosis factor α; TGFβ1, transforming growth factor β1. SREBP1, Sterol-regulatory element binding protein 1.

2.6. Western Blot

Protein extraction of the liver samples was carried out as described previously [4]. Protein concentration was measured using a BCA Protein Quantification Kit (Bio-Rad, CA, USA). The primary antibodies used were β-tubulin (#2146, CST, Boston, USA), ERK1/2 (#4695, CST, Boston, USA), P-ERK1/2(Thr202/Tyr204) (#4370, CST, Boston, USA), AMPKα (#A11184, Abclonal, Wuhan, China), P-AMPKα (AMPKα1-Thr-183/AMPKα2-Thr172) (#AP0432, Abclonal, Wuhan, China), JNK (#9252, CST, Boston, USA), and P-JNK (Thr-183/Tyr185) (#9252, Genetex, CA, USA). Automated Western blots were performed on a JessTM system (Protein Simple) using pre-filled plates (12–230 kDa) according to the manufacturer's standard instructions.

2.7. Statistical Analysis

All data were presented as the mean value ± standard error of the mean (S.E.M). Statistical analyses were performed using the SPSS software version 22.0 for Windows (IBM Inc., New York, USA). All data means were analyzed after homogeneity of variances were tested. Two-way ANOVA was used to analyze the significant differences among treatment means based on starch levels (9.0 and 14.4%) and OE levels (0 and 0.125%). Meanwhile, one-way ANOVA was also performed for the data analysis of four groups in figures. $p < 0.05$ was considered significantly different. The graphics were drawn using GraphPad Prism 7.0 (GraphPad Software Inc., CA, USA).

3. Results

3.1. Growth Performance and Morphometric Parameters

The results of growth performance and morphometric parameters are presented in Tables 3 and 4. The SR of largemouth bass was the same in all groups ($p > 0.05$). According to the two-way ANOVA analysis, high-starch diets significantly reduced the FBW, SGR, FR and FCR compared with low-starch diets ($p < 0.05$). There were no significant differences in FBW, SGR, and FR after adding 0.125% OE to the diets ($p > 0.05$). However, supplementation of OE significantly decreased the FCR ($p < 0.05$). Both a high starch level and OE supplementation significantly increased the HSI, while OE supplementation significantly decreased the CF and high starch significantly increased the VSI ($p < 0.05$). Both a high starch level and OE supplementation had no effect on the crude protein content of the whole body ($p > 0.05$). OE supplementation significantly increased the crude lipid level of the whole body ($p < 0.05$) (Table 5).

Table 3. Effects of experimental diets on the growth performance in largemouth bass.

	OE (%)	Level of Dietary Starch (%)		OE Level	*P* Values		
		9.0	14.4		OE Level	Starch Level	Interaction
IBW [1]	-	35.98 ± 0.21		-	-	-	-
FBW [2]	0	99.98 ± 1.68	97.66 ± 1.69	98.82 ± 1.18	0.223	0.002	0.008
	0.125	109.59 ± 1.77	93.34 ± 2.68	101.46 ± 3.91			
	starch level	104.78 ± 2.41 ^{B}	95.50 ± 1.72 ^{A}				
SR [3]	0	96.67 ± 1.67	96.67 ± 1.67	96.67 ± 1.05	1.000	1.000	1.000
	0.125	96.67 ± 1.67	96.67 ± 1.67	96.67 ± 1.05			
	starch level	96.67 ± 1.05	96.67 ± 1.05				
SGR [4]	0	1.94 ± 0.03	1.91 ± 0.03	1.93 ± 0.02	0.272	0.002	0.008
	0.125	2.12 ± 0.03	1.82 ± 0.06	1.97 ± 0.07			
	starch level	2.03 ± 0.44 ^{B}	1.86 ± 0.04 ^{A}				

Table 3. *Cont.*

	OE (%)	Level of Dietary Starch (%)		OE Level	*P* Values		
		9.0	14.4		OE Level	Starch Level	Interaction
FCR [5]	0	1.01 ± 0.01	0.96 ± 0.01	0.98 ± 0.01 ^{b}	0.002	0.005	0.327
	0.125	0.95 ± 0.01	0.93 ± 0.01	0.94 ± 0.01 ^{a}			
	starch level	0.98 ± 0.01 ^{B}	0.94 ± 0.01 ^{A}				
FR [6]	0	0.44 ± 0.01	0.42 ± 0.01	0.43 ± 0.01	0.108	0.000	0.010
	0.125	0.46 ± 0.01	0.39 ± 0.01	0.42 ± 0.02			
	starch level	0.45 ± 0.00 ^{B}	0.40 ± 0.01 ^{A}				

Two-way ANOVA was used to analyze the significant differences among treatment means based on starch level (9.0 and 14.4%) and OE level (0 and 0.125%). Different superscript lowercase letters "a" or "b" denote significant differences ($p < 0.05$) among experimental groups fed different OE levels; different capital letters "A" or "B" denote significant differences ($p < 0.05$) between groups with different starch levels. [1] IBW: initial body weight. [2] FBW: final body weight. [3] SR (survival rate, %) = 100 × final fish number/initial fish number. [4] SGR (specific growth rate, %/d) = 100 × [Ln (FBW)−Ln (IBW)]/days. [5] FCR (feed conversion ratio) = feed intake/(W_f + W_d − W_i). W_f is the final total weight, W_d is the total weight of dead fish, and W_i is the initial total weight. The same below. [6] FR (feeding rate, %) = 100 × feed intake/[(W_f + W_i + W_d)/2]/days.

Table 4. Effects of experimental diets on morphometric parameters in largemouth bass.

	OE (%)	Level of Dietary Starch (%)		OE Level	*P* Values		
		9.0	14.4		OE Level	Starch Level	Interaction
CF [1]	0	1.96 ± 0.10	1.75 ± 0.03	1.85 ± 0.06 ^{b}	0.004	0.066	0.054
	0.125	1.68 ± 0.02	1.68 ± 0.03	1.68 ± 0.02 ^{a}			
	starch level	1.82 ± 0.06	1.72 ± 0.02				
VSI [2]	0	7.14 ± 0.10	7.60 ± 0.23	7.37 ± 0.13	0.142	0.001	0.110
	0.125	7.01 ± 0.22	8.30 ± 0.29	7.66 ± 0.22			
	starch level	7.07 ± 0.12 ^{A}	7.95 ± 0.20 ^{B}				
HIS [3]	0	1.72 ± 0.15	2.54 ± 0.14	2.13 ± 0.13 ^{a}	0.007	0.000	0.842
	0.125	2.05 ± 0.12	3.04 ± 0.22	2.57 ± 0.17 ^{b}			
	starch level	1.88 ± 0.10 ^{A}	2.79 ± 0.14 ^{B}				
VAI [4]	0	1.46 ± 0.07	1.74 ± 0.20	1.60 ± 0.11	0.115	0.348	0.489
	0.125	1.85 ± 0.17	1.90 ± 0.21	1.88 ± 0.13			
	starch level	1.66 ± 0.10	1.82 ± 0.14				

Two-way ANOVA was used to analyze the significant differences among treatment means based on starch level (9.0 and 14.4%) and OE level (0 and 0.125%). Different superscript lowercase letters "a" or "b" denote significant differences ($p < 0.05$) among experimental groups fed different OE levels; different capital letters "A" or "B" denote significant differences ($p < 0.05$) between groups with different starch levels. [1] CF (condition factor) = 100 × (body weight, g)/(body length, cm)^3. [2] VSI (viscerosomatic index, %) = 100 × visceral weight/whole body weight. [3] HSI (hepatosomatic index, %) = 100 × liver weight/whole body weight. [4] VAI (visceral adipose index, %) = 100 × visceral adipose weight/whole body weight.

Table 5. Effects of experimental diets on the composition of the whole body in largemouth bass.

	OE (%)	Level of Dietary Starch (%)		OE Level	*P* Values		
		9.0	14.4		OE Level	Starch Level	Interaction
Moisture	0	73.39 ± 1.14	70.60 ± 0.18	72.00 ± 0.81 ^{b}	0.004	0.066	0.054
	0.125	69.87 ± 0.49	69.87 ± 0.49	69.74 ± 0.26 ^{a}			
	starch level	71.63 ± 0.96	70.11 ± 0.27				
Crude protein	0	16.58 ± 0.15	16.94 ± 0.06	16.76 ± 0.11	0.142	0.001	0.110
	0.125	16.70 ± 0.15	16.13 ± 0.05	16.42 ± 0.15			
	starch level	16.64 ± 0.10	16.54 ± 0.19				
Crude lipid	0	5.37 ± 0.84	7.93 ± 0.22	6.65 ± 0.69 ^{a}	0.007	0.000	0.842
	0.125	9.24 ± 0.36	9.16 ± 0.23	9.20 ± 0.19 ^{b}			
	starch level	7.31 ± 0.96	8.54 ± 0.31				

Two-way ANOVA was used to analyze the significant differences among treatment means based on starch level (9.0 and 14.4%) and OE level (0 and 0.125%). Different superscript lowercase letters "a" or "b" denote significant differences ($p < 0.05$) among experimental groups fed different OE levels.

3.2. Hematological and Liver Functions Parameters

The hematological and liver function parameters of largemouth bass are presented in Tables 6 and 7. No significant differences were observed in plasma glucose, TG, TC, HDL-C, LDL-C between largemouth bass fed low-starch and high-starch diets ($p > 0.05$). OE supplementation significantly increased plasma TC, which was mainly induced by the increase in HDL-C ($p < 0.05$), and plasma LDL-C/TC was decreased with 0.125% OE supplementation ($p < 0.05$). Moreover, high levels of dietary starch led to abnormal liver function with significantly higher AST, ALT and AKP, and supplementation with OE significantly decreased ALT in plasma, which indicated that OE may improve liver function.

Table 6. Effects of experimental diets on hematological parameters in largemouth bass.

	OE(%)	Level of Dietary Starch (%)		OE Level	*P* Values		
		9.0	14.4		OE Level	Starch Level	Interaction
Glucose (mM/L)	0	4.82 ± 0.40	4.98 ± 0.70	4.90 ± 0.30	0.097	0.053	0.043
	0.125	5.05 ± 0.37	3.19 ± 0.31	4.12 ± 0.33			
	Starch level	4.93 ± 0.26	4.08 ± 0.36				
TG (mM/L)	0	4.31 ± 0.43	6.59 ± 0.98	5.45 ± 0.60	0.895	0.598	0.038
	0.125	5.82 ± 0.77	4.68 ± 0.54	5.25 ± 0.48			
	starch level	5.07 ± 0.47	5.64 ± 0.59				
TC (mM/L)	0	6.34 ± 0.29	7.13 ± 0.48	6.74 ± 0.29^{a}	0.000	0.587	0.167
	0.125	8.73 ± 0.57	8.33 ± 0.40	8.53 ± 0.34^{b}			
	starch level	7.53 ± 0.44	7.73 ± 0.34				
HDL-C (mM/L)	0	1.47 ± 0.23	1.20 ± 0.21	1.33 ± 0.15^{a}	0.013	0.742	0.511
	0.125	1.86 ± 0.29	2.05 ± 0.25	1.95 ± 0.18^{b}			
	starch level	1.66 ± 0.18	1.59 ± 0.19				
HDL-C/TC	0	0.23 ± 0.03	0.17 ± 0.03	0.20 ± 0.02	0.363	0.522	0.190
	0.125	0.21 ± 0.03	0.22 ± 0.03	0.21 ± 0.02			
	starch level	0.22 ± 0.02	0.19 ± 0.02				
LDL-C (mM/L)	0	1.77 ± 0.22	2.56 ± 0.21	2.16 ± 0.18	0.591	0.059	0.100
	0.125	2.22 ± 0.20	2.26 ± 0.18	2.24 ± 0.13			
	starch level	2.00 ± 0.16	2.41 ± 0.14				
LDL-C/TC	0	0.28 ± 0.03	0.37 ± 0.03	0.32 ± 0.03^{b}	0.021	0.154	0.109
	0.125	0.26 ± 0.02	0.25 ± 0.02	0.25 ± 0.01^{a}			
	starch level	0.27 ± 0.02	0.31 ± 0.02				

Two-way ANOVA was used to analyze the significant differences among treatment means based on starch level (9.0 and 14.4%) and OE level (0 and 0.125%). Different superscript lowercase letters "a" or "b" denote significant differences ($p < 0.05$) among experimental groups fed different OE levels.

Table 7. Effects of experimental diets on hematological liver function in largemouth bass.

	OE (%)	Level of Dietary Starch (%)		OE Level	*P* Values		
		9.0	14.4		OE Level	Starch Level	Interaction
AKP (U/L)	0	41.15 ± 3.51	92.49 ± 8.46	66.82 ± 7.97	0.551	0.000	0.001
	0.125	69.61 ± 9.19	70.34 ± 5.91	69.98 ± 9.86			
	starch level	55.38 ± 6.01 ^{A}	81.42 ± 5.75 ^{B}				
AST (U/L)	0	4.96 ± 0.58	14.58 ± 2.38	9.86 ± 2.30	0.317	0.000	0.087
	0.125	6.22 ± 0.85	10.37 ± 1.11	8.87 ± 1.17			
	starch level	5.33 ± 1.13 ^{A}	12.89 ± 1.77 ^{B}				
ALT (U/L)	0	4.55 ± 0.51	15.86 ± 1.65	9.99 ± 1.85 ^{b}	0.022	0.000	0.000
	0.125	6.98 ± 0.91	8.50 ± 0.74	8.14 ± 1.00 ^{a}			
	starch level	5.73 ± 1.02 ^{A}	12.25 ± 1.45 ^{B}				

Two-way ANOVA was used to analyze the significant differences among treatment means based on starch level (9.0 and 14.4%) and OE level (0 and 0.125%). Different superscript lowercase letters "a" or "b" denote significant differences ($p < 0.05$) among experimental groups fed different OE levels; different capital letters "A" or "B" denote significant differences ($p < 0.05$) between groups with different starch levels.

3.3. Hepatic Antioxidant Responses

The contents of hepatic ROS were significantly higher, and those of SOD and CAT were lower in the group fed the high-starch diet than those fed the low-starch diet. The supplementation with OE significantly decreased the contents of hepatic MDA, SOD and CAT, and SOD/MDA was significantly increased ($p < 0.05$), which indicated that OE may improve hepatic antioxidant capacity (Table 8).

Table 8. Effects of experimental diets on hepatic antioxidant parameters in largemouth bass.

	OE (%)	Level of Dietary Starch (%)		OE Level	*P* Values		
		9.0	14.4		OE Level	Starch Level	Interaction
ROS (U/mg prot)	0	64.42 ± 4.61	86.41 ± 4.42	75.41 ± 4.20	0.105	0.000	0.101
	0.125	64.33 ± 3.73	102.84 ± 6.37	83.59 ± 6.12			
	starch level	64.37 ± 2.86 ^{A}	94.62 ± 4.30 ^{B}				
T-AOC (μM/g prot)	0	69.12 ± 5.80	95.70 ± 8.69	82.41 ± 6.10	0.749	0.185	0.009
	0.125	91.01 ± 7.21	81.60 ± 6.37	86.31 ± 4.80			
	starch level	80.07 ± 5.29	88.65 ± 5.51				
CAT (U/mg prot)	0	49.61 ± 3.02	47.81 ± 4.39	48.71 ± 2.58 ^{b}	0.000	0.000	0.000
	0.125	49.83 ± 3.70	8.17 ± 1.48	29.00 ± 5.71 ^{a}			
	starch level	49.72 ± 2.31 ^{B}	27.99 ± 5.59^{A}				
GSH-Px (U/ug prot)	0	4.14 ± 0.70	4.57 ± 0.74	4.35 ± 0.49	0.137	0.815	0.642
	0.125	4.87 ± 0.54	5.24 ± 0.42	5.06 ± 0.33			
	starch level	4.50 ± 0.44	4.90 ± 0.42				
SOD (U/mg prot)	0	195.90 ± 11.13	192.83 ± 11.57	194.37 ± 7.76 ^{b}	0.004	0.038	0.104
	0.125	185.02 ± 9.71	144.04 ± 7.72	164.53 ± 7.99 ^{a}			
	starch level	190.46 ± 7.27 ^{B}	168.44 ± 9.21^{A}				
MDA (nM/mg prot)	0	4.48 ± 0.49	2.70 ± 0.85	3.53 ± 0.54 ^{b}	0.002	0.452	0.020
	0.125	1.22 ± 0.16	2.17 ± 0.45	1.70 ± 0.26 ^{a}			
	starch level	2.75 ± 0.49	2.43 ± 0.47				
SOD/MDA	0	47.5 ± 6.08	103.8 ± 21.79	77.52 ± 13.74^{a}	0.004	0.819	0.002
	0.125	166.93 ± 19.51	98 ± 23.14	132.47 ± 17.12^{b}			
	starch level	111.2 ± 19.00	100.90 ± 15.37				

Two-way ANOVA was used to analyze the significant differences among treatment means based on starch level (9.0 and 14.4%) and OE level (0 and 0.125%). Different superscript lowercase letters "a" or "b" denote significant differences ($p < 0.05$) among experimental groups fed different OE levels; different capital letters "A" or "B" denote significant differences ($p < 0.05$) between groups with different starch levels.

3.4. Hepatic Pathological Examination Histology

The hepatic histopathological examination results of each group are shown in Figure 1. Four phenotypes of hepatic histopathological examination were defined, with symptoms from light to severe, by H&E staining and immunofluorescence signaling for NF-κB (Figure 1A). Phenotype I showed normal hepatocytes with well-shaped cells and low expression of NF-κB in the nucleus. Phenotype II defined fatty liver tissues with enlarged and vacuolated cells and low expression of NF-κB in the nucleus. Phenotype III defined nuclear dense tissues, which is usually a precursor to liver fibrosis, with unclear liver cord and low expression of NF-κB in the nucleus. Phenotype IV defined liver fibrosis, with severe vacuolation along with hepatic fibrosis symptoms and NF-κB mainly expressed in the nucleus. Twelve samples were observed in each group (except eleven samples in HS group). There were eight and eleven samples generally normal (phenotype I) in the LS and LSOE group. However, only five samples were generally normal in the HS and HSOE group. Fish in the HS and HSOE groups showed a high proportion of fatty liver (five to six samples) and even a fibrosis (one sample) phenotype (Figure 1B).

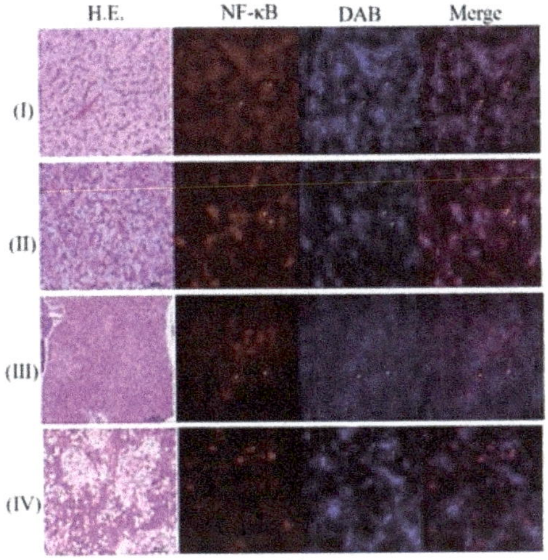

(A)

Liver Phenotype	LS	HS	LSOE	HSOE
(I) No obvious abnormality	8	5	11	5
(II) Fatty liver	1	5	1	6
(III) Nuclear dense	3	0	0	0
(IV) Fibrosis	0	1	0	1
Total	12	11	12	12

(B)

Figure 1. Effects of different diets on hepatic histopathological and inflammatory responses of largemouth bass. (**A**) Four phenotypes of hepatic histopathological examination with symptoms from light to heavy by HE staining for histology examination. Inflammatory response signals of NF-κB were lower (marked by yellow arrows) and mainly (marked by green arrows) expressed in the nucleus (marked with DAPI in blue color) (bar = 15 μm), in which (I) no obvious abnormity, (II) fatty liver, (III) nuclear dense tissue, and (IV) hepatic fibrosis symptoms were observed. (**B**) Statistical results of liver phenotypes (n = 12). Since the samples were damaged during the embedding process, the number of slices was less than 12 of the HS group.

3.5. Hepatic Proliferation and Inflammation Responses

As shown in Figure 2A, the phosphorylated ERK (P-ERK(Thr202/Tyr204)) levels in liver tissues were significantly higher in the high-starch groups than in the low-starch groups ($p < 0.05$). Simultaneously, the P-ERK/ERK ratio showed a tendency towards an increase in the high-starch groups. Supplementation with OE significantly increased ERK levels ($p < 0.05$), but had no significant effect on P-ERK and P-ERK/ERK levels ($p > 0.05$). Phosphorylated JNK (P-JNK(Thr183/Tyr185)) levels, and the P-JNK/JNK ratio in liver tissues, were significantly lower in the high-starch groups than in the low-starch groups ($p < 0.05$). Adding OE significantly decreased phosphorylated JNK (P-JNK) levels ($p < 0.05$), but had no significant effect on P-JNK(Thr183/Tyr185)/JNK levels ($p > 0.05$). Moreover, One-way ANOVA analysis showed that the ratio of P-JNK/JNK was significantly up-regulated in the HSOE group compared to the HS group ($p < 0.05$). We also observed a significant up-regulation of mRNA levels of pro-inflammatory cytokines (TNFα, IL1β) and anti-inflammatory cytokines (TGFβ1) in the high starch groups ($p < 0.05$), while supplementation with OE had no significant effect on the expression of inflammatory cytokines by two-way ANOVA ($p > 0.05$). However, one-way ANOVA analysis showed that the mRNA level of IL1β was significantly decreased in the HSOE group compared with the HS group ($p < 0.05$) (Figure 2B). The results show that high levels of dietary starch induced inflammatory response in largemouth bass, and that adding OE in the high-starch diets could reduce the expression of pro-inflammatory factors to a certain extent, which was achieved by activating the phosphorylation of JNK.

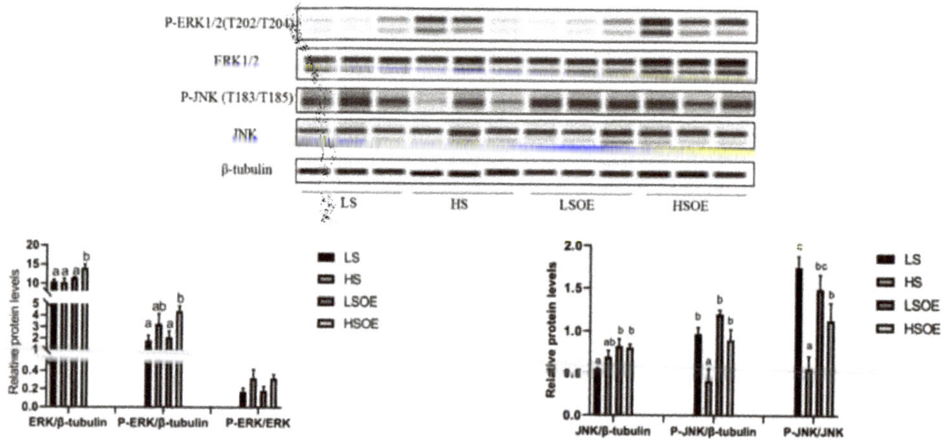

P-value	JNK/β-tubulin	P-JNK/β-tubulin	P-JNK/JNK
Starch	0.332	0.004	0.002
OE	0.014	0.010	0.367
Interaction	0.214	0.271	0.036

P-value	ERK/β-tubulin	P-ERK/β-tubulin	P-ERK/ERK
Starch	0.172	0.015	0.051
OE	0.016	0.280	0.942
Interaction	0.111	0.497	0.910

(A)

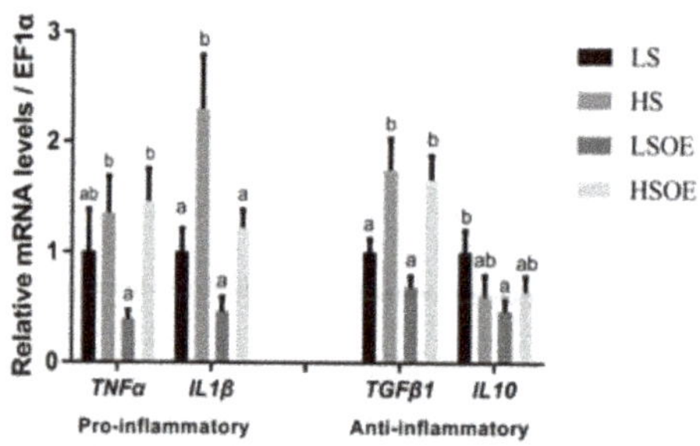

P-value	*TNFα*	*IL1β*	*TGFβ1*	*IL10*
Starch	0.021	0.031	0.000	0.499
OE	0.386	0.107	0.308	0.149
Interaction	0.223	0.187	0.553	0.089

(B)

Figure 2. Effects of different diets on hepatic proliferation and inflammatory responses of largemouth bass, (**A**) Western blot of P-ERK, ERK, P-JNK and JNK in the liver (n = 3). (**B**) Effects of different diets on the transcriptional levels of hepatic pro- and anti-inflammation-related genes (n = 8). Both one-way ANOVA and two-way ANOVA statistics were analyzed. Differences were regarded as significant when $p < 0.05$ (n = 8). Values marked with "a, b and c" are significantly different according to one-way ANOVA.

3.6. Hepatic Lipid Metabolism

As shown in Table 9, no significant differences were observed in hepatic TG, TC, LDL-C and TBA between largemouth bass fed low-starch and high-starch diets ($p > 0.05$). High-starch diets induced a higher content of lipid in the liver ($p < 0.05$). Supplementation of OE significantly decreased hepatic TG, LDL-C, LDL-C/TC and TBA ($p < 0.05$). Although the two-way ANOVA results showed that OE had no significant effect on the level of liver lipids ($p > 0.05$), there was a clear reduction in hepatic lipids after adding OE to the low starch diet.

Table 9. Effects of experimental diets on hepatic lipid metabolism parameters in largemouth bass.

	OE (%)	Level of Dietary Starch (%)		OE Level	*P* Values		
		9.0	14.4		OE Level	Starch Level	Interaction
TG (mM/g prot)	0	0.39 ± 0.06	0.32 ± 0.05	0.36 ± 0.05 ^{b}	0.010	0.838	0.197
	0.125	0.20 ± 0.03	0.25 ± 0.05	0.22 ± 0.03 ^{a}			
	starch level	0.30 ± 0.04	0.29 ± 0.05				
TC (mM/g prot)	0	0.16 ± 0.02	0.15 ± 0.01	0.16 ± 0.01	0.426	0.286	0.041
	0.125	0.12 ± 0.01	0.17 ± 0.01	0.15 ± 0.01			
	starch level	0.14 ± 0.01	0.16 ± 0.01				
LDL-C (μM/g prot)	0	38.03 ± 3.01	42.67 ± 3.04	40.35 ± 2.15 ^{b}	0.010	0.056	0.426
	0.125	26.06 ± 2.29	35.20 ± 3.30	30.63 ± 2.27 ^{a}			
	starch level	32.04 ± 2.39	38.94 ± 2.37				
LDL-C/TC	0	0.24 ± 0.02	0.21 ± 0.02	0.26 ± 0.17	0.050	0.504	0.273
	0.125	0.22 ± 0.02	0.21 ± 0.22	0.22 ± 0.13			
	starch level	0.23 ± 0.02	0.25 ± 0.02				
TBA (μM/g prot)	0	3.14 ± 0.21	3.47 ± 0.91	3.29 ± 0.42 ^{b}	0.001	0.917	0.551
	0.125	1.66 ± 0.14	1.43 ± 0.14	1.54 ± 0.10 ^{a}			
	starch level	2.45 ± 0.24	2.45 ± 0.52				
Liver lipid (%)	0	1.99 ± 0.01	2.01 ± 0.16	2.03 ± 0.06	0.606	0.007	0.006
	0.125	1.66 ± 0.10	2.27 ± 0.04	1.95 ± 0.15			
	starch level	1.77 ± 0.10 ^{A}	2.16 ± 0.07 ^{B}				

Two-way ANOVA was used to analyze the significant differences among treatment means based on starch level (9.0 and 14.4%) and OE level (0 and 0.125%). Different superscript lowercase letters "a" or "b" denote significant differences ($p < 0.05$) among experimental groups fed different OE levels; different capital letters "A" or "B" denote significant differences ($p < 0.05$) between groups with different starch levels.

As shown in Figure 3A, a high starch level decreased the phosphorylated AMPKα (P-AMPKα (AMPKα1 T183/AMPKα2 T172)) and P-AMPKα/AMPKα levels ($p < 0.05$). In terms of mRNA levels, high-starch diets significantly increased *SREBP1* compared with diets containing low starch ($p < 0.05$) (Figure 3B). On the other hand, supplementation with OE had a great influence on lipid metabolism. In particular, mRNA levels of hepatic fatty acid synthesis-related genes (*ACC1* and *FASN*) were significantly down-regulated, and TG hydrolysis (*ATGL*) and fatty acid β-oxidation (*CPT1α*)-related genes were significantly up-regulated by OE supplementation ($p < 0.05$) (Figure 3C).

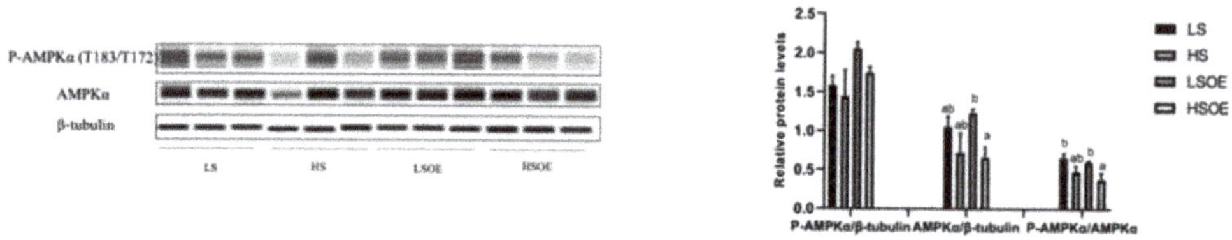

P-value	AMPKα/β-tubulin	P-AMPKα/β-tubulin	P-AMPKα/AMPKα
Starch	0.239	0.013	0.012
OE	0.072	0.729	0.229
Interaction	0.667	0.487	0.753

(A)

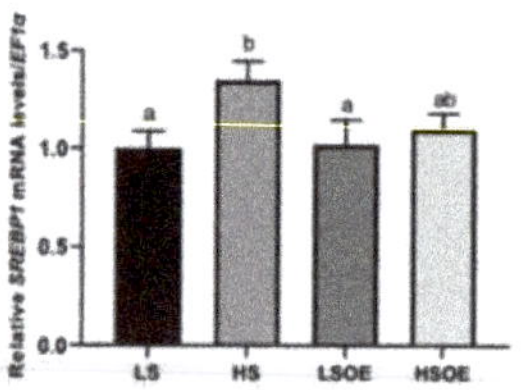

P-value	*SREBP1*
Starch	0.039
OE	0.258
Interaction	0.182

(B)

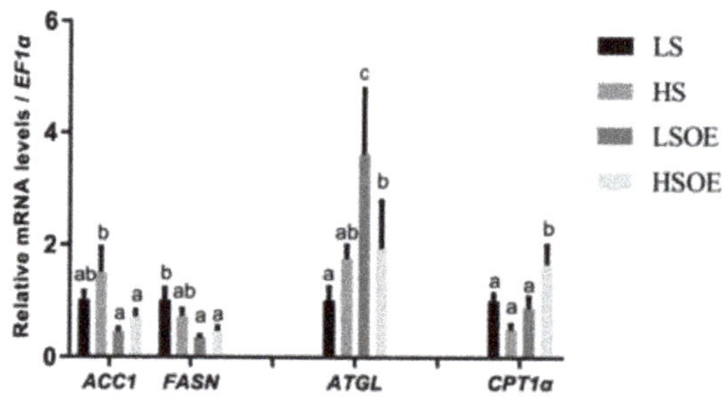

	Lipogenesis		lipolysis	β-oxidation
P-value	*ACC1*	*FASN*	*ATGL*	*CPT1α*
Starch	0.124	0.511	0.135	0.493
OE	0.008	0.001	0.000	0.022
Interaction	0.605	0.097	0.000	0.005

(C)

Figure 3. Effects of different diets on the hepatic lipid metabolism of largemouth bass. (**A**) Western blot of P-AMPK and AMPK in the liver (n = 3). (**B**) Transcriptional levels of *SREBP1* (n = 8). (**C**) Transcriptional levels of hepatic FA synthesis (*ACC1* and *FASN*), TG hydrolysis (*ATGL*), and β-oxidation (*CPT1α*) related genes (n = 8). Both one-way ANOVA and two-way ANOVA statistics were analyzed. Differences were regarded as significant when $p < 0.05$ (n = 8). Values marked with "a, b and c" are significantly different according to one-way ANOVA.

4. Discussion

Starch is typically used as a technical ingredient (stabilizer and swelling agent) in the manufacture of aquatic feeds, and is sometimes used as a cheap energy source in diets for some fish species. However, carnivorous fish having a metabolism adapted to diets low in carbohydrates exhibit symptoms of glucose intolerance after the intake of high-starch diets [17,18]. Previous studies showed that largemouth bass had poor starch utilization capacity, and high dietary inclusion of digestible carbohydrate (over 10%) was recognized as the primary factor inducing MLD [19,20]. Excess dietary starch induced hyperglycemia, glycogen and lipid accumulation, chronic inflammation, and reduced immune functions and antioxidant capabilities, causing MLD in largemouth bass [2,4]. The present study indicated that, compared with low-starch feed (9.0%), high-starch (14.4%) feed decreased the growth performance of largemouth bass and induced liver lipid accumulation, inflammatory response, oxidative stress, liver function injury and higher hepatosomatic index, which is similar to the nonalcoholic steatohepatitis symptom (NASH). Matsumoto et al. developed a NASH model in the ricefish medaka (*Oryzias latipes*), which is based on feeding the fish a high-fat diet (HFD). Medaka that are fed a HFD exhibited macrovesicular fat deposition and liver dysfunction [21]. In aquaculture, various additives are commonly added to the diets to improve nutrient utilization, growth performance and survival of cultured fish, such as probiotics, yeast, amino acids, antioxidants, enzymes, plant extracts and certain organic acids/salts and so on [22–24]. Among these, plants extracts have a broad utilization for growth promotion and appetite stimulation [25]. For instance, garlic and ginger increased SGR and WG, and decreased FCR in rainbow trout [26,27]. Similar effects have been observed in Nile tilapia (*Oreochromis niloticus*) fed diets including extracts of ginseng (Ginsana® G115) [28] or limonene and thymol [29], and in juvenile olive flounder (*Paralichthys olivaceus*) fed diets including extract of green tea [30]. Growth performance and expression levels of growth-related genes in common carp (*Cyprinus carpio*) were enhanced after feeding diets with 0.10–0.25% OLE, but decreased after feeding diets with high levels of OLE (0.50–1.00%) [10]. Some studies also demonstrated that dietary supplementation of 0.00–1.00% of OLE did not affect growth performance and feed utilization in rainbow trout [9]. In this study, 0.125% OE had no significant effect on the weight gain of largemouth bass, but FCR was clearly reduced, leading to decreases in the farming feed cost of largemouth bass.

To avoid metabolic stress, such as that potentially caused by high carbohydrate, organisms rely on an antioxidant protection system to prevent oxidative injury. SOD enzymes control the levels of a variety of ROS and reactive nitrogen, thus limiting the potential toxicity of these molecules and controlling broad aspects of cellular life that are regulated by their signaling functions [31]. MDA is the product from the peroxidation of n-3 and n-6 polyunsaturated fatty acids, and has a strong biotoxicity to cells [32]. Our study suggested that 14.4% starch in the diet reduced the hepatic antioxidant capacity of largemouth bass, with higher ROS and lower CAT and SOD levels, which was consistent with previous studies showing that different starch sources and levels significantly affected the antioxidant status of largemouth bass [19,33]. Significant reductions in SOD, CAT and GSH-Px activities, and an increase in MDA content, were detected in the liver of largemouth bass when dietary corn starch level increased from 0 to 25% [19]. In this study, the increase in ROS with rising starch level, with no differences observed after supplementation with OE, may be the main reason for the decrease in SOD. OE inclusion strongly decreased the hepatic MDA content in both high- and low-starch diets. The lower MDA content may be attributed to the lower oxidative stress and inflammatory degree, which can be demonstrated by the decreased SOD activity observed in this study. Moreover, triterpenes possess anti-inflammatory activity and antioxidant protection properties in vivo [34] and in vitro [35]. In recent years, more and more studies have proved that triterpenes play apoptotic roles against tumor cells, but a principal feature of these compounds is their antioxidant effect [11]. Therefore, triterpene-rich olive extracts are considered a natural source of antioxidant compounds for the prevention of fish diseases related to cell oxidative damage.

At present, the most widely reported application of olive extracts in fish is related to their immunomodulatory effects. Dietary supplementation of 0.1% OLE increased the expression levels of immune-associated genes (*IL-1β* in head kidney tissue and *TNF-α* in spleen tissue) and enhanced the survival rate of common carp juveniles by inhibiting the pathogenicity of *E. tarda* [10]. Adding 0.1% OLE to a rainbow trout's diet activated the expression of immune-related genes in the spleen tissue, including *TNFα*, *IL1-β* and *IL-8*, although the expression levels of these genes decreased with higher doses of OLE, such as 0.25, 0.5, and 1.0% [9]. Triterpene-enriched OE acted as an immunopotentiator in Black Sea bream (*Acanthopagrus schlegelii*), repairing the hampered immune response induced by cadmium exposure [7]. Moreover, OLE could be applied in the control and prevention of white spot virus syndrome in white leg shrimp (*Penaeus vannamei*) [36]. In this study, increased expression levels of *IL1β*, *TNFα* and *TGFβ1* suggested that high starch contents activated the immune response of largemouth bass, while supplementation with OE in the diet containing high starch level reduced transcript levels of pro-inflammatory gene *IL1β* and expression of anti-inflammatory genes remained at a high level.

Oxidative stress induced by oxidized fish oil activated NF-κB signaling and release of inflammatory cytokines in *Megalobrama amblycephala* and *Rhynchocypris lagowskii* Dybowski [37,38]. In mammals, triterpenes have been shown to promote an anti-inflammatory status by suppressing nuclear NF-κB activity [39,40]. In fact, NF-κB pathway is a double-edged sword in the inflammatory response. Once the activity of NF-κB is abnormally increased or continuously activated, it shows an anti-apoptotic effect by inducing anti-inflammatory response and promoting the expression of target genes related to cell proliferation [41]. MAPK family member ERK1/2 plays an important role in cellular proliferation and differentiation. Protein phosphorylation is a process of protein post-translational modification. Usually, the ERK1/2 is located in the cytoplasm and quickly passes through the nuclear membrane once phosphorylated, and then activates transcription factors to further regulate the transcription of their respective target genes, causing changes in the expression or activity of specific proteins and ultimately regulating cell biological function [42]. A high starch level induced liver fibrosis and increased NF-κB in the nucleus, and the phosphorylation level of ERK1/2 showed an increasing trend, which suggests that the fish fed high starch diet was under a "self-repair" status. In the study, OE increased the total ERK1/2 expression, which may be regulated by the post-transcriptional translation of growth factor receptors from upstream signals (or protein tyrosine kinase receptors); however, phosphorylation activation is mainly reflected in the starch level rather than the presence of OE supplementation. In addition, another MAPK family member the JNK signal transduction pathway has been implicated in cellular stress, inflammation and apoptosis. The demethylating substance betaine could repair alcoholic and non-alcoholic fatty liver disease (NAFLD) by inhibiting JNK-mediated signaling [43]. Indeed, JNK plays a dual role in the development of hepatocellular carcinoma. JNK promotes an inflammatory hepatic environment that supports tumor development, but JNK deficiency in hepatocytes increased the tumor burden, and so also functions in hepatocytes to reduce tumor development [44]. Our results show that the high starch diet inhibited JNK phosphorylation, which could protect hepatocytes from apoptosis. On the other hand, the addition of OE to the high-starch diet promoted the activation (phosphorylation) of the JNK pathway in largemouth bass, concomitantly with the decrease in expression level of the proinflammatory factor *IL1β*. However, hepatic histopathological analysis did not show significant effects of OE supplementation on the repair of liver injury induced by the high starch diet, which may be a result of the short feeding time (eight weeks) in the study.

Hepatic lipid content and HSI are important indicators of liver health in fish. In this study, we demonstrated that a high-starch diet led to an over-accumulation of lipid and enlarged liver in largemouth bass, with a higher content of hepatic lipid and HSI. In addition, our results suggested that both AMPK and SREBP were involved in the over-accumulation of lipid. Up-regulation of SREBP increases the uptake and absorption of glycerol by hepatocytes and enhances the synthesis of fatty acids and cholesterol [45]. When the contents of free fatty acids, TG and TC exceed the adaptative capacity of the body, further accumulation in hepatocytes then leads to an overproduction of ROS, cytokines and inflammatory factors, which is the most common pathogenesis pathway of NAFLD [46]. AMPK, a kinase directly targeting SREBP-1, inhibits SREBP-1 cleavage and intranuclear translocation, and suppresses the expression of SREBP-1 target genes in hepatocytes exposed to high levels of glucose, thus decreasing lipogenesis [47]. In the present study, the high-starch diet activated AMPK, as observed by a decrease in phosphorylation of AMPK, which further upregulated the expression of *SREBP-1*. OE showed a tendency to decrease *SREBP-1* expression in the HS diet.

In order to determine the functional consequences of activated SREBP1, we further examined the gene expression levels of several key target enzymes of *SREBP1* involved in fatty acid synthesis, including *ACC1* and *FASN*. Surprisingly, no significant effects of starch level were observed in the expression of these genes. However, one-way ANOVA analysis showed that expression of *ACC1* was the highest in the HS group, which was similar to the expression of *SREBP1*. A significantly lower expression of *ACC1* and *FASN* was observed in the OE-supplemented diets, strongly indicating that OE decreased lipogenesis. In addition, OE supplementation increased the expression levels of the lipolysis gene *ATGL* and β-oxidation gene *CPT1α*. These results are consistent with a previous study showing that a Chinese olive fruit extract improved the metabolic abnormalities of mice associated with fatty liver under high fat challenge by increasing the protein expression of phosphorylated AMPK, ACC1, CPT-1, and PPARα, but downregulating the expression of mature SREBP-1c and FAS [15]. In this study, OE inhibited hepatic lipogenesis and promoted lipolysis and β-oxidation, leading to significant reductions in the content of hepatic TG and LDL-C. However, this pathway was not regulated by AMPK, and the key regulatory factors need to be further identified. Still, our study presents the first report that OE can regulate fat metabolism in aquatic animals.

At the level tested in this study (0.125%), supplementation with OE was associated with several beneficial effects. However, it is important to note that the concentration of OE supplemented into diets should be optimized, as hepatotoxic effects of some triterpenes have been reported in a dose-dependent manner. For instance, the anti-inflammatory pentacyclic triterpene oleanolic acid at about 20% caused body weight loss, inflammation, hepatocellular apoptosis, necrosis, and feathery degeneration (indicative of cholestasis) in mice [48]. Therefore, it is necessary to clarify the optimal addition level of triterpene-rich OE in the feed of aquatic animals in future studies.

5. Conclusions

In summary, the intake of high starch diets induced oxidative stress, inflammatory response, lipid metabolism disorder and even MLD, which significantly reduced the growth performance of largemouth bass. Supplementation with 0.125% of an olive extract obtained as a byproduct of olive oil production significantly reduced FCR, which would result in a lower farming feed cost of largemouth bass. In addition, our results demonstrate that supplementation with OE in the HS diet increased the hepatic antioxidant capacity, promoted the activation of JNK signaling pathway and decreased inflammatory responses. In this study, we present the first evidence that OE regulates hepatic lipid metabolism in fish. The protective mechanisms induced by the supplementation of OE mainly depend on inhibiting hepatic lipogenesis and promoting lipolysis and β-oxidation, leading to the prevention of hepatic TG over-accumulation in fish. A limited effect of OE was observed histologically on the repair of MLD induced by high starch in largemouth bass, but considering the short trial period and single doses tested, it would be worthwhile to further investigate the effects of different OE doses and feeding time on the growth, hepatoprotection and immune status of fish in future studies.

Author Contributions: Conceptualization, M.X. and S.M.; methodology, X.L. and P.C.; validation, M.X., S.M. and X.L.; investigation, X.L.; resources, M.X. and S.M.; data curation, X.L., S.X., X.W. and P.C.; writing—original draft preparation, X.L.; writing—review and editing, M.X., S.M. and M.H.; supervision, M.X.; funding acquisition, M.X. and X.G. All authors have read and agreed to the published version of the manuscript.

Funding: This study was funded by Lucta (Guangzhou) Flavours Co., Ltd., and also supported by the National Key R&D Program of China (2018YFD0900400 and 2019YFD0900200); National Natural Science Foundation of China (32172981 and 31902382); The Agricultural Science and Technology Innovation Program of CAAS, China (CAAS-ASTIP-2017-FRI-08).

Institutional Review Board Statement: All animal care and experimental procedures were approved by the Experimental Animal Ethics Committee of Institute of Feed Research, Chinese Academy of Agricultural Sciences (IFR-CAAS20190501).

Informed Consent Statement: Not applicable.

Data Availability Statement: The data presented in this study are available on request from the corresponding author. The data are not publicly available due to containing information that could compromise the privacy of research participants.

Conflicts of Interest: The author Morais, S. and He, M. are employed by Lucta S.A. and Lucta (Guangzhou) Flavours Co., Ltd., respectively. The other authors declare that the research was conducted in the absence of any commercial or financial relationships that could be construed as a potential conflict of interest.

References

1. Sørensen, M.; Morken, T.; Kosanovic, M.; Øverland, M. Pea and wheat starch possess different processing characteristics and affect physical quality and viscosity of extruded feed for Atlantic salmon. *Aquac. Nutr.* **2011**, *17*, e326–e336. [CrossRef]
2. Lin, S.-M.; Shi, C.-M.; Mu, M.-M.; Chen, Y.-J.; Luo, L. Effect of high dietary starch levels on growth, hepatic glucose metabolism, oxidative status and immune response of *Juvenile largemouth bass, Micropterus salmoides*. *Fish Shellfish Immunol.* **2018**, *78*, 121–126. [CrossRef] [PubMed]
3. Yu, H.; Liang, X.; Chen, P.; Wu, X.; Zheng, Y.; Luo, L.; Qin, Y.; Long, X.; Xue, M. Dietary supplementation of Grobiotic®-A increases short-term inflammatory responses and improves long-term growth performance and liver health in largemouth bass (*Micropterus salmoides*). *Aquaculture* **2018**, *500*, 327–337. [CrossRef]
4. Yu, H.; Zhang, L.; Chen, P.; Liang, X.; Cao, A.; Han, J.; Wu, X.; Zheng, Y.; Qin, Y.; Xue, M. Dietary Bile Acids Enhance Growth, and Alleviate Hepatic Fibrosis Induced by a High Starch Diet via AKT/FOXO1 and cAMP/AMPK/SREBP1 Pathway in Micropterus salmoides. *Front. Physiol.* **2019**, *10*, 1430. [CrossRef]
5. Liu, Y.; Espinosa, C.D.; Abelilla, J.J.; Casas, G.A.; Lagos, L.V.; Lee, S.A.; Kwon, W.B.; Mathai, J.K.; Navarro, D.M.D.L.; Jaworski, N.W.; et al. Non-antibiotic feed additives in diets for pigs: A review. *Anim. Nutr.* **2018**, *4*, 113–125. [CrossRef]
6. Bulfon, C.; Volpatti, D.; Galeotti, M. Current research on the use of plant-derived products in farmed fish. *Aquac. Res.* **2013**, *46*, 513–551. [CrossRef]
7. Rong, J.; Han, Y.; Zha, S.; Tang, Y.; Shi, W.; Guan, X.; Du, X.; He, M.; Liu, G. Triterpene-enriched olive extract as an immunopotentiator in Black Sea Bream (*Acanthopagrus schlegelii*). *J. Ocean U. China* **2020**, *19*, 428–438. [CrossRef]
8. Sudjana, A.N.; D Orazio, C.; Ryan, V.; Rasool, N.; Ng, J.; Islam, N.; Riley, T.V.; Hammer, K.A. Antimicrobial activity of commercial *Olea europaea* (olive) leaf extract. *Int. J. Antimicrob. Ag.* **2009**, *33*, 461–463. [CrossRef]
9. Baba, E.; Acar, Ü.; Yılmaz, S.; Zemheri, F.; Ergün, S. Dietary olive leaf (*Olea europea* L.) extract alters some immune gene expression levels and disease resistance to *Yersinia ruckeri* infection in rainbow trout *Oncorhynchus mykiss*. *Fish Shellfish Immunol.* **2018**, *79*, 28–33. [CrossRef]
10. Zemheri Navruz, F.; Acar, Ü.; Yılmaz, S. Dietary supplementation of olive leaf extract enhances growth performance, digestive enzyme activity and growth-related genes expression in common carp *Cyprinus carpio*. *Gen. Comp. Endocrinol.* **2020**, *296*, 113541. [CrossRef]
11. Sánchez-Quesada, C.; López-Biedma, A.; Warleta, F.; Campos, M.; Beltrán, G.; Gaforio, J.J. Bioactive Properties of the Main Triterpenes Found in Olives, Virgin Olive Oil, and Leaves of *Olea europaea*. *J. Agric. Food Chem.* **2013**, *61*, 12173–12182. [CrossRef] [PubMed]
12. Salminen, A.; Lehtonen, M.; Suuronen, T.; Kaarniranta, K.; Huuskonen, J. Terpenoids: Natural inhibitors of NF-κB signaling with anti-inflammatory and anticancer potential. *Cell Mol. Life Sci.* **2008**, *65*, 2979–2999. [CrossRef] [PubMed]
13. Wang, C.; Liu, X.; Lian, C.; Ke, J.; Liu, J. Triterpenes and aromatic meroterpenoids with antioxidant activity and neuroprotective effects from *Ganoderma lucidum*. *Molecules* **2019**, *24*, 4353. [CrossRef]
14. Xu, J.; Wang, X.; Su, G.; Yue, J.; Sun, Y.; Cao, J.; Zhang, X.; Zhao, Y. The antioxidant and anti-hepatic fibrosis activities of acorns (*Quercus liaotungensis*) and their natural galloyl triterpenes. *J. Funct. Foods* **2018**, *46*, 567–578. [CrossRef]
15. Yeh, Y.-T.; Cho, Y.-Y.; Hsieh, S.-C.; Chiang, A.-N. Chinese olive extract ameliorates hepatic lipid accumulation in vitro and in vivo by regulating lipid metabolism. *Sci. Rep.* **2018**, *8*, 1057. [CrossRef]
16. Zhang, Y.; Chen, P.; Liang, X.F.; Han, J.; Wu, X.F.; Yang, Y.H.; Xue, M. Metabolic disorder induces fatty liver in Japanese seabass, *Lateolabrax japonicas* fed a full plant protein diet and regulated by cAMP-JNK/NF-kB caspase signal pathway. *Fish Shellfish Immun.* **2019**, *90*, 223–234. [CrossRef]
17. Enes, P.; Panserat, S.; Kaushik, S.; Oliva-Teles, A. Nutritional regulation of hepatic glucose metabolism in fish. *Fish Physiol. Biochem.* **2008**, *35*, 519–539. [CrossRef]
18. Hemre, G.-I.; Mommsen, T.P.; Krogdahl, A. Carbohydrates in fish nutrition: Effects on growth, glucose metabolism and hepatic enzymes. *Aquac. Nutr.* **2002**, *8*, 175–194. [CrossRef]

19. Ma, H.-J.; Mou, M.-M.; Pu, D.-C.; Lin, S.-M.; Chen, Y.-J.; Luo, L. Effect of dietary starch level on growth, metabolism enzyme and oxidative status of *Juvenile largemouth bass, Micropterus salmoides*. *Aquaculture* **2018**, *498*, 482–487. [CrossRef]
20. Zhang, Y.; Xie, S.; Wei, H.; Zheng, L.; Liu, Z.; Fang, H.; Xie, J.; Liao, S.; Tian, L.; Liu, Y.; et al. High dietary starch impaired growth performance, liver histology and hepatic glucose metabolism of *Juvenile largemouth bass, Micropterus salmoides*. *Aquac. Nutr.* **2020**, *26*, 1083–1095. [CrossRef]
21. Matsumoto, T.; Terai, S.; Oishi, T.; Kuwashiro, S.; Fujisawa, K.; Yamamoto, N.; Fujita, Y.; Hamamoto, Y.; Furutani-Seiki, M.; Nishina, H.; et al. Medaka as a model for human nonalcoholic steatohepatitis. *Dis. Model. Mech.* **2010**, *3*, 431–440. [CrossRef] [PubMed]
22. Dawood, M.A.; Koshio, S.; Esteban, M.A. Beneficial roles of feed additives as immunostimulants in aquaculture: A review. *Rev. Aquac.* **2017**, *10*, 950–974. [CrossRef]
23. El-Saadony, M.T.; Alagawany, M.; Patra, A.K.; Kar, I.; Tiwari, R.; Dawood, M.A.; Dhama, K.; Abdel-Latif, H.M. The functionality of probiotics in aquaculture: An overview. *Fish Shellfish Immunol.* **2021**, *117*, 36–52. [CrossRef]
24. Lieke, T.; Meinelt, T.; Hoseinifar, S.H.; Pan, B.; Straus, D.L.; Steinberg, C.E.W. Sustainable aquaculture requires environ-mental-friendly treatment strategies for fish diseases. *Rev. Aquacult.* **2020**, *12*, 943–965. [CrossRef]
25. Reverter, M.; Bontemps, N.; Lecchini, D.; Banaigs, B.; Sasal, P. Use of plant extracts in fish aquaculture as an alternative to chemotherapy: Current status and future perspectives. *Aquaculture* **2014**, *433*, 50–61. [CrossRef]
26. Nya, E.J.; Austin, B. Use of dietary ginger, *Zingiber officinale Roscoe*, as an immunostimulant to control *Aeromonas hydrophila* infections in rainbow trout, *Oncorhynchus mykiss* (Walbaum). *J. Fish Dis.* **2009**, *32*, 971–977. [CrossRef]
27. Nya, E.J.; Austin, B. Use of garlic, *Allium sativum*, to control *Aeromonas hydrophila* infection in rainbow trout, Oncorhynchus mykiss (Walbaum). *J. Fish Dis.* **2009**, *32*, 963–970. [CrossRef]
28. Goda, A. Effect of dietary ginseng herb (Ginsana® G115) supplementation on growth, feed utilization, and hemato-logical indices of *Nile tilapia, Oreochromis niloticus* (L.), fingerlings. *J. World Aquacult. Soc.* **2008**, *39*, 205–214. [CrossRef]
29. Aanyu, M.; Betancor, M.B.; Monroig, O. Effects of dietary limonene and thymol on the growth and nutritional physiology of *Nile tilapia* (*Oreochromis niloticus*). *Aquaculture* **2018**, *488*, 217–226. [CrossRef]
30. Cho, S.H.; Lee, S.-M.; Park, B.H.; Ji, S.-C.; Lee, J.; Bae, J.; Oh, S.-Y. Effect of dietary inclusion of various sources of green tea on growth, body composition and blood chemistry of the *juvenile olive flounder, Paralichthys olivaceus*. *Fish Physiol. Biochem.* **2006**, *33*, 49–57. [CrossRef]
31. Wang, Y.; Branicky, R.; Noe, A.; Hekimi, S. Superoxide dismutases: Dual roles in controlling ROS damage and regulating ROS signaling. *J. Cell Biol.* **2018**, *217*, 1915–1928. [CrossRef]
32. Ayala, A.; Muñoz, M.F.; Argüelles, S. Lipid peroxidation: Production, metabolism, and signaling mechanisms of Malondialdehyde and 4-Hydroxy-2-Nonenal. *Oxid. Med. Cell. Longev.* **2014**, *2014*, 360438. [CrossRef]
33. Song, M.; Shi, C.; Lin, S.; Chen, Y.; Shen, H.; Luo, L. Effect of starch sources on growth, hepatic glucose metabolism and an-tioxidant capacity in juvenile largemouth bass, *Micropterus salmoides*. *Aquaculture* **2018**, *490*, 355–361. [CrossRef]
34. Singh, G.B.; Singh, S.; Bani, S.; Gupta, B.D.; Banerjee, S.K. Anti-inflammatory activity of oleanolic acid in rats and mice. *J. Pharm. Pharmacol.* **1992**, *44*, 456–458. [CrossRef] [PubMed]
35. Marquez-Martin, A.; De La Puerta, R.; Fernandez-Arche, A.; Ruiz-Gutierrez, V.; Yaqoob, P. Modulation of cytokine secretion by pentacyclic triterpenes from olive pomace oil in human mononuclear cells. *Cytokine* **2006**, *36*, 211–217. [CrossRef] [PubMed]
36. Gholamhosseini, A.; Kheirandish, M.R.; Shiry, N.; Akhlaghi, M.; Soltanian, S.; Roshanpour, H.; Banaee, M. Use of a methanolic olive leaf extract (*Olea europaea*) against white spot virus syndrome in Penaeus vannamei: Comparing the biochemical, hematological and immunological changes. *Aquaculture* **2020**, *528*. [CrossRef]
37. Song, C.; Liu, B.; Xu, P.; Xie, J.; Ge, X.; Zhou, Q.; Sun, C.; Zhang, H.; Shan, F.; Yang, Z. Oxidized fish oil injury stress in Meg-alobrama amblycephala: Evaluated by growth, intestinal physiology, and transcriptome-based PI3K-Akt/NF-κB/TCR inflam-matory signaling. *Fish Shellfish Immun.* **2018**, *81*, 446–455. [CrossRef] [PubMed]
38. Zhang, D.; Guo, Z.; Zhao, Y.; Wang, Q.; Gao, Y.; Yu, T.; Chen, Y.; Chen, X.; Wang, G. L-carnitine regulated Nrf2/Keap1 activation in vitro and in vivo and protected oxidized fish oil-induced inflammation response by inhibiting the NF-κB signaling pathway in *Rhynchocypris lagowski* Dybowski. *Fish Shellfish Immun.* **2019**, *93*, 1100–1110. [CrossRef] [PubMed]
39. Lee, W.; Yang, E.-J.; Ku, S.-K.; Song, K.-S.; Bae, J.-S. Anti-inflammatory Effects of Oleanolic Acid on LPS-Induced Inflammation In Vitro and In Vivo. *Inflammation* **2012**, *36*, 94–102. [CrossRef]
40. Yang, E.-J.; Lee, W.; Ku, S.-K.; Song, K.-S.; Bae, J.-S. Anti-inflammatory activities of oleanolic acid on HMGB1 activated HUVECs. *Food Chem. Toxicol.* **2012**, *50*, 1288–1294. [CrossRef]
41. Biswas, D.K.; Shi, Q.; Baily, S.; Strickland, I.; Ghosh, S.; Pardee, A.B.; Iglehart, J.D. NF-κB activation in human breast cancer specimens and its role in cell proliferation and apoptosis. *Proc. Natl. Acad. Sci. USA* **2004**, *101*, 10137–10142. [CrossRef]
42. Zhang, W.; Liu, H.T. MAPK signal pathways in the regulation of cell proliferation in mammalian cells. *Cell Res.* **2002**, *12*, 9–18. [CrossRef] [PubMed]
43. Wang, Z.; Yao, T.; Pini, M.; Zhou, Z.; Fantuzzi, G.; Song, Z. Betaine improved adipose tissue function in mice fed a high-fat diet: A mechanism for hepatoprotective effect of betaine in nonalcoholic fatty liver disease. *Am. J. Physiol. Liver Physiol.* **2010**, *298*, G634–G642. [CrossRef]
44. Das, M.; Garlick, D.S.; Greiner, D.L.; Davis, R.J. The role of JNK in the development of hepatocellular carcinoma. *Genes Dev.* **2011**, *25*, 634–645. [CrossRef] [PubMed]

45. Pai, W.Y.; Hsu, C.C.; Lai, C.Y.; Chang, T.Z.; Tsai, Y.L.; Her, G.M. Cannabinoid receptor 1 promotes hepatic lipid accumulation and lipotoxicity through the induction of SREBP-1c expression in zebrafish. *Transgenic Res.* **2013**, *22*, 823–838. [CrossRef]
46. Kohjima, M.; Enjoji, M.; Higuchi, N.; Kato, M.; Kotoh, K.; Yoshimoto, T.; Fujino, T.; Yada, M.; Yada, R.; Harada, N.; et al. Re-evaluation of fatty acid metabolism-related gene expression in nonalcoholic fatty liver disease. *Int. J. Mol. Med.* **2007**, *20*, 351–358. [CrossRef]
47. Li, Y.; Xu, S.; Mihaylova, M.M.; Zheng, B.; Hou, X.; Jiang, B.; Park, O.; Luo, Z.; Lefai, E.; Shyy, J.Y.-J.; et al. AMPK Phosphorylates and inhibits SREBP activity to attenuate hepatic steatosis and atherosclerosis in diet-induced insulin-resistant mice. *Cell Metab.* **2011**, *13*, 376–388. [CrossRef]
48. Liu, J.; Lu, Y.-F.; Wu, Q.; Xu, S.-F.; Shi, F.-G.; Klaassen, C.D. Oleanolic acid reprograms the liver to protect against hepatotoxicants, but is hepatotoxic at high doses. *Liver Int.* **2018**, *39*, 427–439. [CrossRef] [PubMed]

antioxidants

MDPI

Article

Conventional Soybean Meal as Fishmeal Alternative in Diets of Japanese Seabass (*Lateolabrax japonicus*): Effects of Functional Additives on Growth, Immunity, Antioxidant Capacity and Disease Resistance

Jie Wang [1,2,3], Kangsen Mai [1,2] and Qinghui Ai [1,2,*]

1 Key Laboratory of Aquaculture Nutrition and Feed (Ministry of Agriculture and Rural Affairs), Key Laboratory of Mariculture (Ministry of Education), Ocean University of China, 5 Yushan Road, Qingdao 266003, China; wangjie03@caas.cn (J.W.); kmai@ouc.edu.cn (K.M.)

2 Laboratory for Marine Fisheries Science and Food Production Processes, Qingdao National Laboratory for Marine Science and Technology, 1 Wenhai Road, Qingdao 266237, China

3 National Aquafeed Safety Assessment Center, Institute of Feed Research, Chinese Academy of Agricultural Sciences, Beijing 100081, China

* Correspondence: qhai@ouc.edu.cn; Tel.: +86-13969712326

Abstract: Aiming to optimize soymeal-based diets for Japanese seabass (*Lateolabrax japonicas*), a 105-day feeding trial was conducted to evaluate the effects of functional additives, including antioxidants (ethoxyquin, thymol and carvacrol) and chelated trace elements (Cu, Mn and Zn), on the growth, immunity, antioxidant capacity and disease resistance of fish fed diets with conventional soybean meal replacing 50% of fishmeal. Three isonitrogenous (45%) and isolipidic (11%) diets were formulated: (1) standard reference diet (FM, 42% fishmeal); (2) soymeal-based diet (SBM, 21% fishmeal and 30% conventional soybean meal); (3) SBM diet supplemented 0.0665% functional additives (FAS). Each experimental diet was randomly fed to quadruplicate groups of forty feed-trained Japanese seabass (initial average body weight = 125.6 ± 0.6 g) stocked in a saltwater floating cage. Upon the conclusion of the feeding trial, lower feed intake was observed in fish fed SBM compared to those fed FM and FAS. Fish fed FM showed the highest growth performance, estimated as the weight gain rate. Notably, FAS supported faster growth of fish than those fed SBM, indicating the optimal growth performance of dietary functional additives. The feed conversion rate showed the opposite trend among dietary treatments, with the highest value in fish fed SBM. Regarding immunity, fish fed soymeal-based diets suppressed the serum alternative complement pathway activities compared to FM, whereas the respiratory burst activity in macrophages of head kidneys showed a similar picture, but no statistical differences were observed. Further, fish fed soymeal-based diets had lower serum Cu-Zn SOD, CAT and GPx activities as well as liver vitamin E levels and scavenging rates of hydroxyl radical but higher liver MDA contents compared to the FM-fed group. Fish fed FAS had higher serum Cu-Zn SOD and GPx activities and liver vitamin E levels than those fed SBM, suggesting the enhancement of antioxidant capacity of dietary functional additives. For the disease resistance against *Vibrio harveyi* infection, fish fed SBM had the highest cumulative mortality, followed by the FAS and FM groups. Additionally, the biomarkers related to the immune and antioxidant capacities had a positive correlation with the relative abundance of *Paracoccus* and *Pseudomonas*, while liver MDA levels had a negative correlation with the relative abundance of *Pseudomonas* and *Psychrobacter*. Collectively, soymeal replacing 50% of fishmeal suppressed the growth, immunity, antioxidant capacity and disease resistance of Japanese seabass, while dietary supplementation of antioxidants and chelated trace elements could mitigate soymeal-induced adverse effects on growth and disease resistance through the improvement in antioxidant capacity and regulation of gut microbiota.

Keywords: *Lateolabrax japonicas*; functional additives; conventional soybean meal; growth performance; immunity; antioxidant capacity; disease resistance

Citation: Wang, J.; Mai, K.; Ai, Q. Conventional Soybean Meal as Fishmeal Alternative in Diets of Japanese Seabass (*Lateolabrax japonicus*): Effects of Functional Additives on Growth, Immunity, Antioxidant Capacity and Disease Resistance. *Antioxidants* **2022**, *11*, 951. https://doi.org/10.3390/antiox11050951

Academic Editor: Stanley Omaye

Received: 14 April 2022
Accepted: 7 May 2022
Published: 12 May 2022

1. Introduction

Japanese seabass (*Lateolabrax japonicus*), as one euryhaline and carnivorous species, has been widely cultured in Eastern Asia due to its commercial value. In recent decades, the anxieties of unsustainable and costly wild-captured forage fish driven by the expansion of aquaculture production has driven scientists to find economical and nutritious fishmeal alternatives. Studies on fishmeal alternatives have been continually conducted in Japanese seabass, including, but not limited to, defatted black soldier fly larvae meal [1], porcine meal [2], animal protein blend [3], canola meal [4], corn gluten meal [5] and cottonseed meal [6] as well as soymeal [7–13]. Among these fishmeal alternatives, conventional soybean meal and its relevant products have received more focus in Japanese seabass due to its stable production, relatively high digestibility and low market price. However, several obvious disadvantages in conventional soybean meal, such as a deficiency of essential nutrients, poor palatability and the existence of anti-nutritional factors, induced enteritis and metabolic liver disease, have limited its high utilization in aquafeed [6,14,15]. In general, previous studies in Japanese seabass demonstrated that conventional soybean meal replacing more than 50% of fishmeal protein could adversely influence the digestive abilities, immune responses and antioxidant abilities and consequently suppress the growth performance and health of the host [8,9,11,16]. Aiming to optimize soymeal-based diets for Japanese seabass, a range of processing approaches, such as fermentation and gamma-irradiation, have been implemented to partially remove anti-nutritional factors, but replacing more than 50% of fishmeal still induces negative effects on growth performance and health under certain conditions [10,13,16].

Based on this background, to increase the utilization of conventional soybean meal and mitigate its negative influence, expanded evaluations of functional additives on the growth performance and health of fish fed soymeal-based diets are necessary. The administration of functional additives could be achieved alone or in combination. In general, a mixture of functional additives rather than a single substance could provide more effective effects on the growth and health of the host [17–19]. Thus, in the current study, a mixture of antioxidants, i.e., ethoxyquin and a 1:1 standardized combination of thymol and carvacrol and chelated trace elements, i.e., Mintrex® Cu, Mn and Zn, was selected to be tested. The ethoxyquin, as a feed preservative, has been used in aquafeed for many years [20]. The 1:1 standardized combination of thymol and carvacrol are essential oils from thyme and oregano extracts that can improve the growth performance, digestive abilities, antioxidant activity and gut microbiota and boost the general health and immune responses of fish under less desirable conditions, such as lower or non-dietary fishmeal diets, stress or disease [21]. Regarding the chelated trace elements, Mintrex® Cu, Mn and Zn not only keep trace elements balanced but could also improve growth performance and antioxidant ability and balance the requirements of essential amino acids [22]. Further, supplemental chelated trace elements may become more necessary due to the existence of antagonists in soybean meal that would bind with divalent cationic trace elements [23,24].

Therefore, this study aimed to evaluate whether functional additives could mitigate the negative effects on growth, immunity, antioxidant capacity and disease resistance against *Vibrio harveyi* infection in Japanese seabass fed diets with conventional soymeal meal replacing 50% of fishmeal.

2. Materials and Methods

2.1. Fish Husbandry

This experiment was conducted following the Management Rule of Laboratory Animals (Order No. 676 of the Chinese State Council). The Japanese seabass were purchased from a commercial producer located in Qinzhou, Guangxi, China. Before the feeding trial, all fish were stocked and fed a commercial diet for about two weeks to acclimate to the experimental conditions in floating sea cages (4.0 m × 5.0 m × 5.0 m) in Qinzhou, Guangxi, China. Subsequently, fish with an initial average body weight of 125.6 ± 0.6 g were randomly transported into 12 floating sea cages (1.5 m × 3.0 m × 3.0 m) where each treatment was split into quadruplicate cages, each containing 40 fish. Fish were hand-fed continuously to satiation twice daily (06:00 and 18:00) for 105 days. Regarding the water parameters, the water temperature followed a natural fluctuation ranging from 14 to 26 °C, and the water salinity and dissolved oxygen were above 20‰ and 7 mg L^{-1}, respectively, for the entire feeding period.

2.2. Diets

Table 1 shows the formulation and nutrient composition of the experimental diets. Specifically, three isonitrogenous (45% crude protein) and isolipidic (11% crude lipid) experimental diets were formulated. One standard reference diet, named FM, was formulated based on the commercial diet for Japanese seabass containing 42% fishmeal. Two experimental diets with (named FAS) and without functional additives (named SBM) were formulated using conventional soybean meal as the only fishmeal alternative with the substitution level at 50%. The formulation of the functional additives was the mixture of ethoxyquin (0.02% Agrado®, Columbia), a 1:1 standardized combination of thymol and carvacrol (0.006% NE-150®) as well as chelated trace elements (0.0055% Mintrex® Cu, 0.01% Mintrex® Mn and 0.025% Mintrex® Zn) (Novus International Inc., St. Charles, MO, USA). Each diet was produced into proper sizes of floating pellets (4- and 5-mm diameter) using a twin-screwed extruder (EXT50A, Yanggong Machine, Yangzhou, China) and stored at 20 °C until used.

Table 1. Formulation and nutrient composition of the experimental diets.

Ingredients (% Dry Matter)	FM	SBM	FAS
Fishmeal [1]	42	21	21
Conventional soybean meal [1]	12	30	30
Wheat [1]	25.8	22.1	22.1
Peanut meal [1]	10	10	10
Squid meal [1]	2.5	3.5	3.5
Spray blood meal [1]		2.8	2.8
Soy protein concentrate [1]		2	2
Fish oil	4	5.5	5.5
Vitamin and mineral premix	1	1	1
Lecithin	1	1	1
Vitamin C	0.1	0.1	0.1
Calcium propionic acid	0.1	0.1	0.1
Choline	0.5	0.5	0.5
Monocalcium phosphate	1	1	1
MeraMet [2]		0.26	0.26
Threonine [2]		0.1	0.1
Ethoxyquin	0.02	0.02	0.02
Functional additives [3]			0.0665
Proximate analysis			
Crude protein	45.2	45.1	44.9
Crude lipid	11.6	11.4	11.2
Lysine	2.74	2.66	2.68
Methionine	0.96	0.99	1.02

[1] These ingredients were purchased from Great Seven Bio-Tech, Qingdao, China. Fish meal: protein 63.90, lipid 10.15; Squid meal: protein 56.67, lipid 28.85; Peanut meal: protein 55.99, lipid 6.46; Spray blood meal: protein 98.92, lipid 0.99; Soybean meal: protein 54.51, lipid 1.48; Soy protein concentrate: protein 73.32, lipid 1.04. [2] MeraMet is methionine hydroxy calcium. Methionine, threonine and lysine were satisfied for fish in all the treatments. [3] The functional additives were from Novus International Inc., St. Charles, MO, USA.

2.3. Sample Collection

At the end of the feeding trial, fish were anesthetized with eugenol (1:10,000) before tissue sampling. The fork length and body weight of all fish were recorded, and the plasma samples were obtained from the caudal vein of 3 fish per sea cage, i.e., 12 fish per dietary treatment. Plasma was collected after centrifugation at 2000× g for 10 min at 4 °C and frozen in liquid N_2 and then stored at −80 °C prior to analysis. The livers of fish were collected from 3 fish per cage, i.e., 12 fish per dietary treatment, and then frozen at −80 °C until analysis.

2.4. Proximate Composition of Diets and the Whole Body of Fish

Duplicate analyses of the moisture, crude protein and crude lipid as well as the ash of diets and the whole bodies of fish were analyzed according to the standard protocols in our previous descriptions [6].

2.5. Respiratory Burst Activity (RB) Assay

The head kidney macrophages of 3 fish per sea cage, i.e., 12 fish per dietary treatment, were isolated according to the previous description [25]. Briefly, the head kidney was collected and cut into small fragments and then washed with Leibovitz-15 medium (L-15, Sigma, St. Louis, MO, USA) containing 100 IU/mL penicillin, 100 IU/mL streptomycin and 2% fetal calf serum (Gibco, Invitrogen, New York, NY, USA). Subsequently, the cell suspensions were prepared by forcing the head kidney to pass through a 100 μ steel mesh. The resultant cell suspensions were enriched by centrifugation at 400× *g* for half an hour at 4 °C using the 51% (*v/v*) Percoll density gradient (Pharmacia, St. Louis, MO, USA). The bands of cells were collected at the 51% interface and then washed twice using the L-15 medium. Cell viability of the head kidney was determined by the trypan blue exclusion method, and the cell density was measured using the hemocytometer. Finally, the L-15 medium was added to adjust the cell concentration (1×10^7 mL^{-1}) for analysis.

For the respiratory burst activity analysis, the production of intracellular superoxide anion (O^{2-}) was measured using nitroblue tetrazolium (NBT) reduction (Sigma, Louis, MO, USA). A 100 μL cell suspension was stained using 100 μL of 0.3% NBT and 100 μL of 1 mg/mL phorbol 12-myristate 13-acetate (Sigma, Louis, MO, USA) for 40 min, and then the absolute methanol was added to terminate the staining reaction. The 70% methanol was used to wash the cell suspension three times. Subsequently, 120 μL of 2 M KOH and 140 μL of DMSO were added, and then the color was measured at 630 nm using a spectrophotometer (Thermo Fisher Scientific, Waltham, MA, USA) with KOH/DMSO as the blank.

2.6. Serum Alternative Complement Pathway (ACP) Activities

The serum ACP activities were measured as previously described [26]. Briefly, volumes of diluted serum ranging from 0.1 to 0.25 mL were dispensed into 5 mL test tubes. The total reaction volumes were increased up to 0.25 mL using a barbitone buffer with a mixture of thyleneglycolbis (2-aminoethoxy)-tetraacetic acid and Mg^{2+}, and then 0.1 mL of rabbit red blood cells were added to each test tube. After incubation for 1.5 h at 22 °C, 3.15 mL of 0.9% sodium chloride was added to each test tube. Subsequently, the test tubes were centrifuged at 1000× *g* for 5 min at 4 °C to remove unlysed rabbit red blood cells. The optical density of the supernatant was measured at 414 nm using the spectrophotometer (Thermo Fisher Scientific, Waltham, MA, USA). The serum reaction volume producing 50% hemolysis (ACH50) was measured and recorded.

2.7. Serum Antioxidant Enzyme Assay

The serum copper-zinc superoxide dismutase (Cu-Zn SOD) activities were measured spectrophotochemically by the ferricytochrome C method using xanthine/xanthine oxidase as the source of superoxide radicals according to the protocol of the commercial assay kit (No 19160, Sigma, Louis, MO, USA). The results were expressed as Cu-Zn SOD unit mL^{-1} and the unit was defined as the amount of enzyme necessary to create a 50% inhibition of the ferricytochrome C reduction rate measured at 550 nm.

The serum catalase (CAT) activities were determined according to the standard protocol [27]. The initial rate of hydrogen peroxide decomposition was measured, and one unit of serum CAT activity was defined as the amount of enzyme that catalyzes the decomposition of 1.0 μmol of hydrogen peroxide per second.

The serum glutathione peroxidase (GPX) activities were measured according to the previous protocol [28]. Briefly, the serum GPX activities were measured by quantifying the rate of hydrogen-peroxide-induced oxidation of reduced glutathione to oxidized glutathione. The reaction volume of the yellow product had absorbance at 412 nm that could be formed as glutathione reacted with dithiobisnitrobenzoic acid.

2.8. Liver Malondialdehyde (MDA) Levels

The liver MDA levels were measured using the thiobarbituric acid method according to the standard protocol of the commercial kit (MDA detection kits, Jincheng Bioengineering Ltd., Nanjing, China). The MDA could form a red adduct with thiobarbituric acid (TBA) that could be measured at the wavelength of 532 nm by reacting with TBA to form a stable chromophoric production. The MDA was calculated in nmol/mg using the calibration curve of the standard addition method and then expressed as nanomoles per milligram of protein in the liver (nmol/mg protein). The liver protein levels were measured according to the standard protocol of the commercial assay kits (Nanjing Jiancheng Institute, Nanjing, China).

2.9. Liver Vitamin E Concentrations

The liver vitamin E concentrations were measured using reverse-phase high-performance liquid chromatography based on the method in one previous study [29]. Briefly, the mobile phase included a mixture of 95% methanol and 5% water. The solvent was filtered through a 0.45 μm filter and then degassed. The pump flow rate was 1.0 mL/min. The standard solution of DL-α-tocopheryl acetate was prepared in high-performance liquid-chromatography-grade methanol and then stored at 4 °C in the dark. Subsequently, 20 microliters of the prepared sample were injected into the high-performance liquid chromatography system and then the C18 μ bondapack column was used. The absorbance of the reaction volume was measured using the UV detector at 280 nm (Waters, Taunton, UK).

2.10. Liver Scavenging Rate of Hydroxyl Radical (SROH)

The liver scavenging rate of •OH was measured according to the 1, 10-phenanthroline-Fe^{2+} -Fenton reaction method as previously described [30]. Briefly, a 0.5 mL solution of phenanthrene anhydrous ethanol (0.75 mmol/L) was added and mixed with 1 mL of 0.2 mol/L phosphate buffer solution (PBS, pH 7.40) and 0.5 mL of distilled water. Next, the solution was mixed with 0.5 mL of 0.75 mmol/L ferrous sulfate solution ($FeSO_4$) and then added with 0.5 mL of 0.01% hydrogen peroxide (H_2O_2). The reaction mixture was incubated at 37 °C for 60 min in a water bath, and its absorbance was measured at 536 nm. The SROH was expressed as: scavenged rate (%) = $(A_1 - A_0)/(A_s - A_0) \times 100$, where A_1 was the absorbance of the sample, A_s was the absorbance in the presence of a positive control using distilled water, and A_0 was the absorbance in the presence of a negative control using H_2O_2.

2.11. Challenge Test

The *Vibrio harveyi* strain (reference sequence: NZ-JH720471.1, NCBI) was originally isolated from infected Japanese seabass. The 10-day LD50 (*V. harveyi* dose that killed 50% of the test Japanese seabass) was measured by intraperitoneal injection of 60 fish with graded doses of *V. harveyi* (10^6, 10^7, 10^8 and 10^9 CFU/fish) at 20 °C. The results showed that the LD50 was 10^8 CFU/fish [31]. At the termination of the feeding experiment, 15 fish from each cage were collected and transported into 12 tanks with an independent culture system in an indoor facility and then injected intraperitoneally with 0.2 mL of PBS containing 2×10^8 live *V. harveyi* from a 24 h culture at 25 °C. Subsequently, the cumulative mortality was calculated for 10 days.

2.12. Calculations and Statistical Methods

Survival rate (%) = 100 × number of final fish/number of initial fish

Weight gain rate (WG, %) = 100 × ((final body weight − initial body weight)/initial body weight)

Feed intake (FI, %/day) = 100 × feed consumption/((final body weight + initial body weight + weight of dead fish)/2)/days

Feed conversion ratio (FCR) = feed consumed (dry weight)/weight gain

Condition factor (CF, g/cm^3) = 100 × body weight/body length^3

Hepaticsomatic index (HIS, %) = 100 × liver weight/body weight

Viscerasomatic index (VSI, %) = 100 × viscera weight/body weight

Except for the figure balloon plot, all other figures and data analyses were conducted using GraphPad Prism 9 (GraphPad Software, San Diego, CA, USA). Data were evaluated for homogeneity of variances and normality using the Shapiro–Wilk and Barlett's test, respectively. A one-way ANOVA followed Tukey–Kramer HSD multiple comparisons were conducted to compare the mean values among the dietary treatments. All results were presented as means ± S.E.M (standard errors of the mean). The level of significant differences was set at $p < 0.05$.

The Pearson correlation between the gut microbiota and biomarkers in this study from the same sea cage was analyzed using GraphPad Prism 9. The balloon plot was made by TBtools [32] based on the statistical results of the Pearson correlation. Values marked with white symbol "*" are significant correlations ($p < 0.05$), and "**" are very significant correlations ($p < 0.01$).

3. Results

3.1. Growth Performance

In our study, the survival rate of Japanese seabass averaged above 98.75% regardless of the experimental diets ($p > 0.05$, Figure 1A). Fish fed FM showed the highest growth performance, estimated as the final body weight and WG, compared to those fed soymeal-based diets ($p < 0.05$, Figure 1B–D), and fish fed FAS had higher growth performance than those fed SBM ($p < 0.05$, Figure 1B–D).

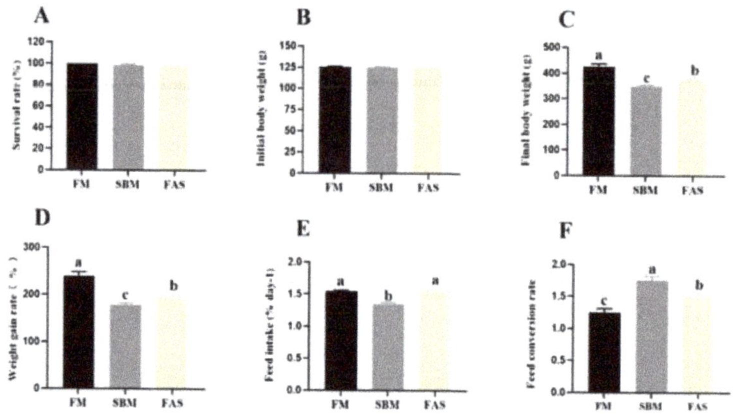

Figure 1. Survival rate (**A**) and growth performance (**B–F**) of Japanese seabass fed diets using soybean meal as fishmeal alternative with or without functional additives. Data are presented as means ± SEM. Bars sharing the same letters are not significantly different ($p > 0.05$, $n = 4$).

A lower FI was observed in fish fed SBM compared to those in the FM and FAS groups ($p < 0.05$, Figure 1E), and no clear difference was observed the between FM and FAS groups ($p > 0.05$, Figure 1E). Regarding FCR, fish fed SBM showed the highest value, followed by the FAS and FM groups ($p < 0.05$, Figure 1F).

3.2. Body Indexes

Similar to the growth performance, the highest CF and VSI values were observed in fish fed FM compared to those fed soymeal-based diets ($p < 0.05$, Figure 2A,C) and these values were not significantly affected by dietary functional additives ($p > 0.05$, Figure 2A,C). No clear difference was found in the HSI of fish among the dietary treatments ($p > 0.05$, Figure 2B).

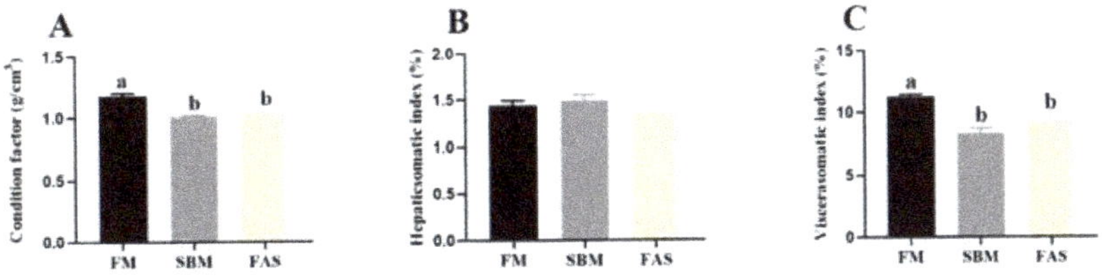

Figure 2. Body indexes of Japanese seabass fed diets using soybean meal as fishmeal alternative with or without functional additives. (**A**) Condition factor. (**B**) Hepaticsomatic index. (**C**) Viscerasomatic index. Data are presented as means ± SEM. Bars sharing the same letters are not significantly different ($p > 0.05$, $n = 12$).

3.3. Proximate Composition of the Whole Body of Fish

Fish fed FM had significantly lower whole-body moisture than those in fish fed SBM and FAS ($p < 0.05$, Table 2), whereas there was not a significant difference in whole-body moisture between fish fed SBM and FAS ($p > 0.05$, Table 2). The whole-body crude lipid showed the opposite trend among dietary treatments, with the highest value in fish fed FM ($p < 0.05$, Table 2). Regarding the whole-body crude protein and ash, there were no significant differences among the dietary treatments ($p > 0.05$, Table 2).

Table 2. The body composition in Japanese sea bass fed functional additives.

(% Wet Weight)	FM	SBM	FAS
Moisture	67.0 ± 0.9^{b}	71.6 ± 0.4^{a}	71.1 ± 0.6^{a}
Crude lipid	11.2 ± 0.7^{a}	7.0 ± 0.4^{b}	8.0 ± 0.3^{b}
Crude protein	16.9 ± 0.7	16.8 ± 0.1	16.5 ± 0.4
Ash	4.4 ± 0.2	5.2 ± 0.1	5.0 ± 0.2

Note: values are represented as means ± S.E.M ($n = 12$); values in the same row with the same superscripts are not significantly different ($p > 0.05$).

3.4. Immune Parameters

Fish fed FM had higher respiratory burst activities in macrophages of head kidneys than those fed soymeal-based diets but there was no statistical difference among the dietary treatments ($p > 0.05$, Figure 3A). The highest serum ACP activities were observed in fish fed FM compared to those fed soybean meal diets, i.e., the SBM and FAS groups ($p < 0.05$, Figure 3B), whereas the ACP activities were not significantly influenced by dietary functional additives ($p > 0.05$, Figure 3B).

3.5. Antioxidant Capacities

The results showed that fish fed FM revealed higher serum Cu-Zn SOD, CAT and GPx activities than those fed soymeal-based diets ($p > 0.05$, Figure 4). Of note, fish fed the soymeal-based diet with functional additives, i.e., the FAS group, had higher serum Cu-Zn SOD and GPx activities than those fed a similar diet without functional additives, i.e., the SBM group ($p < 0.05$, Figure 4A,C).

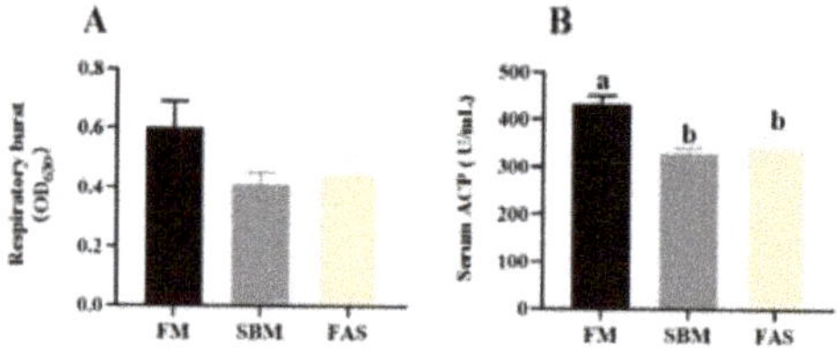

Figure 3. Immune responses of Japanese seabass fed diets using soybean meal as fishmeal alternative with or without functional additives. (**A**) Respiratory burst activity of the head kidney macrophages. (**B**) Serum alternative complement pathway (ACP) activity. Data are presented as means ± SEM. Bars sharing the same letters are not significantly different ($p > 0.05$, $n = 12$).

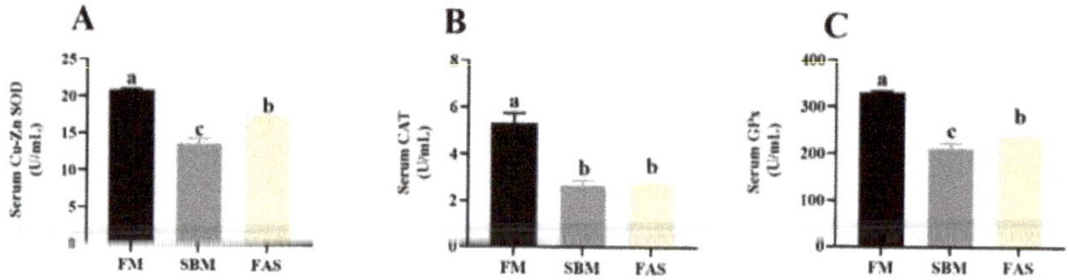

Figure 4. Antioxidant capacity in serum of Japanese seabass fed diets using soybean meal as fishmeal alternative with or without functional additives. (**A**) Serum copper-zinc superoxide dismutase (Cu-Zu SOD) activity. (**B**) Serum catalase (CAT) activity. (**C**) Serum glutathione peroxidase (GPx) activity. Data are presented as means ± SEM. Bars sharing the same letters are not significantly different ($p > 0.05$, $n = 12$).

Although there was an increasing trend in liver MDA levels from FM to FAS, no significant statistical difference was observed among the dietary treatments ($p > 0.05$, Figure 5A). Like the serum antioxidant capacities, the vitamin E levels and scavenging rate of •OH in the liver had higher values in fish fed FM compared to fish fed SBM and FAS ($p < 0.05$, Figure 5B,C). Regarding the effects of dietary functional additives, fish fed FAS significantly increased liver vitamin E levels compared to those fed SBM ($p < 0.05$, Figure 5B).

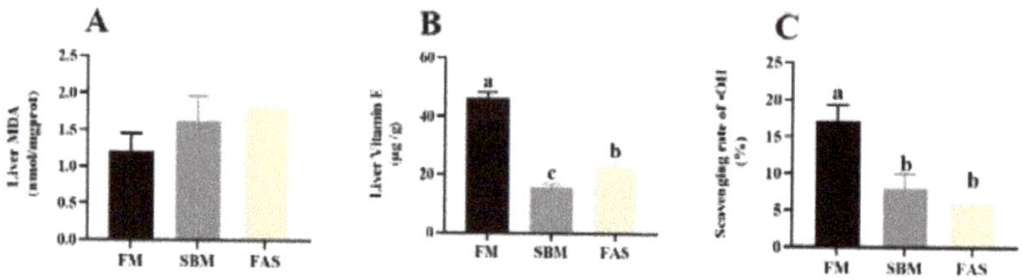

Figure 5. Antioxidant capacity in the liver of Japanese seabass fed diets using soybean meal as fishmeal alternative with or without functional additives. (**A**) Liver malondialdehyde (MDA) level. (**B**) Liver Vitamin E level. (**C**) Liver scavenging rate of hydroxyl radical (•OH). Data are presented as means ± SEM. Bars sharing the same letters are not significantly different ($p > 0.05$, $n = 12$).

3.6. Challenge Test

After the 10-day challenge test, fish fed the SBM diet had the highest cumulative mortality rate of V. harveyi infection (80%), followed by fish fed the FAS (66.7%) and FM (53.3%) diets ($p < 0.05$, Figure 6).

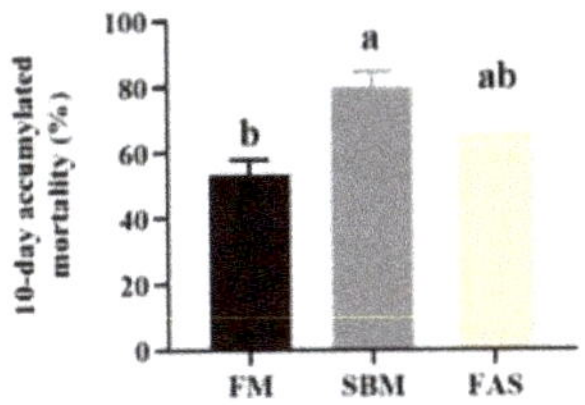

Figure 6. Ten-day accumulated mortality in Vibrio harveyi challenge test of Japanese seabass fed diets using soybean meal as fishmeal alternative with or without functional additives. Data are presented as means ± SEM. Bars sharing the same letters are not significantly different ($p > 0.05$, $n = 4$).

3.7. Pearson Correlation between Gut Microbiota and Host Responses

To help readers to interpret the Pearson correlation results we present here, results of gut microbiota that have been reported elsewhere [7] are briefly summarized as the following: Gut contents from the whole intestine of three fish per sea cage were collected and pooled for microbiota analysis. Ten genera (*Halomonas*, *Shewanella*, *Paracoccus*, *Methylobacterium*, *Ochrobactrum*, *Stappia*, *Corynebacterium*, *Pseudomonas*, *Acinetobacter* and *Psychrobacter)* strongly dominated the gut microbiota of Japanese seabass, accounting for about 70% of total abundance. Further, dietary functional additives could mitigate soymeal-induced enteritis, whereas the composition of gut microbiota was not clearly affected by dietary treatments according to the alpha and beta diversity.

For the results of the Pearson correlation analysis, the respiratory burst activity in macrophages of head kidneys, serum CAT and liver vitamin E had strong positive correlations with the relative abundance of *Paracoccus*. Moreover, the respiratory burst activity in macrophages of head kidneys had a positive correlation with the relative abundance of *Methylobacterium*. Additionally, weight gain, serum ACP, CAT and GPx as well as liver SROH had strong positive correlations with the relative abundance of *Pseudomonas*, while the liver MDA had a negative correlation with the relative abundance of *Pseudomonas* and *Psychrobacter* ($p < 0.05$, Figure 7).

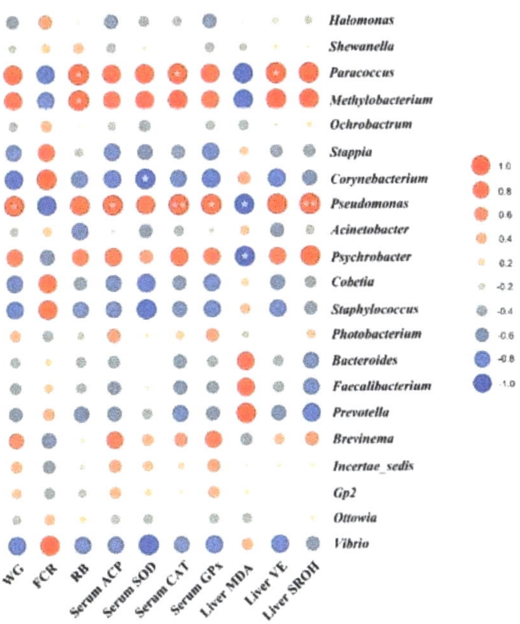

Figure 7. The Pearson correlation between gut microbiota and host responses. Circles in the balloon plot

denoting the direction of the association are colored red (positive) or blue (negative) and are overlaid with the white symbol "*" indicating a significant correlation. Values marked with a white symbol "*" are significant correlations ($p < 0.05$), and "**" are very significant correlations ($p < 0.01$). WG: weight gain rate; FCR: feed conversion ratio; RB: respiratory burst activity; ACP: alternative complement pathway; SOD: copper-zinc superoxide dismutase; CAT: catalase; GPx: glutathione peroxidase; MDA: malondialdehyde; VE: Vitamin E; SROH: scavenging rate of hydroxyl radical.

4. Discussion

4.1. Growth Performance and Body Composition

Fish fed the soymeal based diets with or without functional additives, i.e., the SBM and FAS groups, suppressed the growth performance in the present study, indicating that Japanese seabass cannot tolerate conventional dietary soybean meal as a fish meal substitute at 50%. Overall, previous studies concluded that the optimal replacement level of fishmeal with conventional soybean meal is highly variable for carnivorous fish species. For example, similar to our findings, the optimal replacement level should be less than 50% in some carnivorous fish species, including, but not limited to, Atlantic salmon, *Salmo salar* (30%) [33], Chinese sucker, *Myxocyprinus asiaticus* (30%) [34], European sea bass, *Dicentrarchus labrax* (25%) [35], Japanese flounder, *Paralichthys olivaceus* (21%) [36], spotted rose snapper, *Lutjanus guttatus* (20%) [37], and turbot, *Scophthalmus maximus* L. (40%) [38]. However, certain carnivorous fish species could tolerate conventional dietary soybean meal replacing more than 80% of fish meal without negative effects on growth performance, such as hybrid striped bass, *Morone chrysops* × *M. saxatilis* [39], and largemouth bass, *Micropterus salmoides* [40].

Regarding Japanese seabass, studies related to the use of conventional soybean meal as a fishmeal alternative have been widely reported, but the results varied greatly between studies due to the differences in fish size, feed formulation, feeding strategy and rearing conditions. Some previous studies supported our findings [8,9,13], while other studies found that replacing up to 50% of fishmeal did not influence the growth performance [11,12]. Generally, the three causes of soymeal-based diet inducing inferior growth performance may occur alone or together: (1) low palatability; (2) anti-nutritional factors; and (3) a shortage of essential nutrients.

In our study, a lower FI was observed in fish fed SBM compared to those fed FM, indicating a poor palatability of the SBM diet, as the palatability is an important driver of FI. Further, various anti-nutritional factors in soybean meal, for example, saponin, phytic acid, lectin and protease inhibitor, have been widely reported to decrease palatability, induce enteritis and consequently suppress growth performance [15,33], as also supported by our study reported elsewhere [7]. To eliminate certain anti-nutritional factors of soybean meal, various processing approaches, such as fermentation and gamma-irradiation, have been implemented. These processed soybean meal products could successfully replace more than 50% of fishmeal without an adverse influence on the growth performance and digestibility of Japanese seabass [13]. Furthermore, lacking essential amino acids, e.g., lysine and methionine, and nutrients related to bile acid synthesis, e.g., cholesterol and taurine, has been reported to limit the growth performance of fish [6,34]. In terms of the former, methionine hydroxy calcium, blood meal and Mintrex® chelated trace elements have been supplemented to dietary treatments to ensure the requirement of methionine and lysine in the present study. The dietary methionine (0.96–1.02% of dry diet) and lysine (2.66–2.74% of dry diet) were all above the requirement, i.e., 0.9% and 2.6% of the dry diet, respectively, estimated by Mai et al. [41]. Hence, the deficiency of methionine and lysine might not be the main reasons for the inferior growth performance in our study. Regarding the nutrients related to bile acid synthesis, since the level of these nutrients and bile acid metabolism were not observed in this study, complementary studies warrant further investigation.

Notably, fish fed FAS showed an FI similar to those fed FM and higher than those fed the SBM diet. Meanwhile, a higher growth performance was observed in FAS compared to the SBM group. These findings suggest that dietary functional additives could increase FI and improve growth performance. In addition, fish fed the FAS diet exhibited remarkably higher growth performance and FCR, possibly suggesting the functional additives herein might partially eliminate the negative influence on growth performance caused by antinutritional factors through the improvement in digestibility, immune response and antioxidant capacity [21,22], as discussed below.

Higher values of CF and VSI as well as whole-body crude lipid content were found in fish fed FM compared to soymeal-based diets, which is in line with previous studies in Japanese seabass [11–13,16]. Of note, the increase in VSI and whole-body crude lipid content could be at least partially responsible for the increase in growth performance, possibly due to the increase in feed digestibility and digestive enzyme activity [11,13].

4.2. *Immunity*

Nutritional factors play an important role in immune responses in fish [42]. Previous studies have found that the administration of high levels of soybean meal suppressed immune responses in turbot [43] and red sea bream [44] as well as Japanese seabass [16]. Likewise, fish fed soymeal-based diets suppressed the serum ACP activities in the current study. Further, the respiratory burst activity in macrophages of head kidneys showed a similar picture as serum ACP but there were no statistical differences. Reductions in complement activities in soymeal-based diets were also found in one previous study in Japanese seabass that replaced 80% of fishmeal with soybean meal, causing significant decreases in complement C3 activities [16]. The suppression of immune responses is likely to be the result of the presence of anti-nutritional factors and antigens in soybean meal that damage the immune system in fish [15,16]. On the other hand, fish fed SBM had the highest cumulative mortality of *Vibrio harveyi* infection compared to soymeal-based diets. This is in line with the previous studies in yellowtail, *Seriola quinqueradiata*, where a high dietary level of soymeal suppressed immune functions and increased host–pathogen susceptibility [45].

Additionally, in our study, supplementing these functional additives in the SBM group might boost immune responses of Japanese seabass in the FAS group, as concluded in one previous review [21]. Results of decreased cumulative mortality in the current study and induced enteritis [7] in the FAS group could further confirm that these selected functional dietary additives could partially mitigate the suppression of immunity.

4.3. *Antioxidant Capacity*

The antioxidant capacity is an important indicator of the health status of fish. Oxidative stress occurs in fish when the reactive oxygen species (ROS) generation rate exceeds that of their removal abilities [46]. On one hand, a wide range of antioxidant substances, such as Vitamin E, play the first line of defense against the adverse effects of oxidative stress and other free radicals [47]. The FM-fed group was found to increase the level of liver Vitamin E compared to those fed soymeal-based diets in our study, suggesting dietary soybean meal reduced the number of antioxidant substances and suppressed the antioxidant capacity, as further supported by the results of the scavenging rate of •OH in the present study. This is also consistent with the previous claim that soybean meal could affect antioxidative defense due to the existence of anti-nutritional components in a plant-based diet, such as tocopherols [48]. On the other hand, various radical scavenging antioxidant enzymes, for example, Cu-Zn SOD, CAT and GPx, play the second line of defense to prevent the cascade of oxidant reactions and serve as significant biomarkers for the immune and antioxidant status of the host [46]. In our study, soymeal-based groups exhibited lower levels of these antioxidant enzyme activities in serum compared to the FM group. These findings are paralleled by previous studies in turbot [43] and Japanese seabass [16], further suggesting dietary soybean meal suppressed antioxidant capacity. Moreover, the MDA is the final product of lipid peroxidation, which could be a valuable biomarker for endogenous oxidative damage. In this study, high liver MDA levels were observed in the FM group, but there was no significant statistical difference. Collectively, these results in our study further support the theory that soybean meal replacing 50% of fishmeal led to a significant reduction in antioxidant capacities.

One of the key purposes of dietary functional additives, especially natural products, is to improve the antioxidant capacities of fish. The application of thymol and carvacrol as natural antioxidant feed additives in aquafeed plays an important role in improving antioxidant activities and health [21]. Moreover, dietary chelated trace minerals (Cu, Mn and Zn) rather than inorganic minerals could also improve antioxidant ability, such as Cu-Zn SOD activities, which have been widely reported in fish and shrimp [22–24]. Likewise, in our study fish fed FAS had higher serum Cu-Zn SOD and GPx activities as well as liver vitamin E levels than those fed SBM, indicating that dietary functional additives could significantly enhance antioxidant capacities via the increase in antioxidant substance levels and antioxidant enzyme activities. The increase in antioxidant activity could be at least partially responsible for the improvement in growth performance and disease resistance against *Vibrio harveyi* infection in this study.

4.4. Associations between Gut Microbiota and Host Responses

Most fish microbiota studies published so far are descriptive studies on the taxonomic composition and its changes under different experimental conditions. To identify the major microbial clades that benefit fish health and welfare, characterizing the associations between gut microbiota and host responses is a necessary step [49–51]. In the current study, we found that the relative abundance of the genus *Pseudomonas* showed a strong positive correlation with the biomarkers related to the immune and antioxidant capacities, while there was a negative correlation with liver MDA levels, suggesting dietary treatments could shape key intestinal microbial clades and thereby regulate host immune and antioxidant functions. However, the literature reporting gut microbiota and its correlation with immunity and antioxidant capacities in Japanese seabass is still limited so far, and the mechanisms behind these correlations remain unknown. Of note, the *Pseudomonas* has been widely found in the gut microbiota of Atlantic salmon using the culturing and molecular approaches [51,52], and it was found that its relative abundance was negatively correlated with the distal intestine somatic index [51]. These findings call for more investigation to increase the information on the associations between *Pseudomonas*, immune responses and antioxidant capacities in Japanese seabass.

5. Conclusions

The findings in the present study showed that Japanese seabass fed the diet with conventional soybean meal replacing 50% of fishmeal exhibited suppressed growth performance, immunities, antioxidant capacities and disease resistance against *Vibrio harveyi* infection. Moreover, supplementation of a mixture of antioxidants (ethoxyquin, thymol and carvacrol) and chelated trace elements (Cu, Mn and Zn) in the diet with conventional soybean meal replacing 50% of fishmeal could mitigate soymeal-induced adverse effects on growth and disease resistance via improvements in the antioxidant capacities and gut microbiota.

Author Contributions: Conceptualization, Q.A. and K.M.; methodology, J.W.; validation, Q.A. and J.W.; investigation, J.W.; resources, Q.A. and J.W.; data curation, J.W.; writing—original draft preparation, J.W.; writing—review and editing, Q.A. and J.W.; supervision, Q.A. and K.M.; funding acquisition, Q.A. and K.M. All authors have read and agreed to the published version of the manuscript.

Funding: This work was supported by the National Department Public Benefit Research Foundation (Ministry of Agriculture of the People's Republic of China, Grant no: 201003020).

Institutional Review Board Statement: This experiment was conducted following the Management Rule of Laboratory Animals (Order No. 676 of the Chinese State Council).

Informed Consent Statement: Not applicable.

Data Availability Statement: The data presented in this study are available on request from the corresponding author. The data are not publicly available due to containing information that could compromise the privacy of research participants.

Acknowledgments: Special thanks should be given to Zhang Qin from the Key Laboratory of Marine Biotechnology of Guangxi for her help in the challenge test.

Conflicts of Interest: The authors declare no conflict of interest.

References

1. Wang, G.; Peng, K.; Hu, J.; Yi, C.; Chen, X.; Wu, H.; Huang, Y. Evaluation of defatted black soldier fly (*Hermetia illucens* L.) larvae meal as an alternative protein ingredient for juvenile Japanese seabass (*Lateolabrax japonicus*) diets. *Aquaculture* **2019**, *507*, 144–154. [CrossRef]
2. Wang, J.; Yun, B.; Xue, M.; Wu, X.; Zheng, Y.; Li, P. Apparent digestibility coefficients of several protein sources, and replacement of fishmeal by porcine meal in diets of Japanese seabass, *Lateolabrax japonicus*, are affected by dietary protein levels. *Aquac. Res.* **2012**, *43*, 117–127. [CrossRef]
3. Hu, L.; Yun, B.; Xue, M.; Wang, J.; Wu, X.; Zheng, Y.; Han, F. Effects of fish meal quality and fish meal substitution by animal protein blend on growth performance, flesh quality and liver histology of Japanese seabass (*Lateolabrax japonicus*). *Aquaculture* **2013**, *372*, 52–61. [CrossRef]
4. Cheng, Z.; Ai, Q.; Mai, K.; Xu, W.; Ma, H.; Li, Y.; Zhang, J. Effects of dietary canola meal on growth performance, digestion and metabolism of Japanese seabass, *Lateolabrax japonicus*. *Aquaculture* **2010**, *305*, 102–108. [CrossRef]
5. Men, K.; Ai, Q.; Mai, K.; Xu, W.; Zhang, Y.; Zhou, H. Effects of dietary corn gluten meal on growth, digestion and protein metabolism in relation to IGF-I gene expression of Japanese seabass, *Lateolabrax japonicus*. *Aquaculture* **2014**, *428*, 303–309. [CrossRef]
6. Zhang, Y.; Chen, P.; Liang, X.F.; Han, J.; Wu, X.F.; Yang, Y.H.; Xue, M. Metabolic disorder induces fatty liver in Japanese seabass, *Lateolabrax japonicas* fed a full plant protein diet and regulated by cAMP-JNK/NF-kB-caspase signal pathway. *Fish Shellfish Immunol.* **2019**, *90*, 223–234. [CrossRef]
7. Wang, J.; Tao, Q.; Wang, Z.; Mai, K.; Xu, W.; Zhang, Y.; Ai, Q. Effects of fish meal replacement by soybean meal with supplementation of functional compound additives on intestinal morphology and microbiome of Japanese seabass (*Lateolabrax japonicus*). *Aquac. Res.* **2016**, *48*, 2186–2197. [CrossRef]
8. Li, Y.; Ai, Q.; Mai, K.; Xu, W.; Cheng, Z. Effects of the partial substitution of dietary fish meal by two types of soybean meals on the growth performance of juvenile Japanese seabass, *Lateolabrax japonicus* (Cuvier 1828). *Aquac. Res.* **2012**, *43*, 458–466. [CrossRef]
9. Li, Y.; Ai, Q.; Mai, K.; Xu, W.; Deng, J.; Cheng, Z. Comparison of high-protein soybean meal and commercial soybean meal partly replacing fish meal on the activities of digestive enzymes and aminotransferases in juvenile Japanese seabass, *Lateolabrax japonicus* (Cuvier, 1828). *Aquac. Res.* **2014**, *45*, 1051–1060. [CrossRef]
10. Liang, X.F.; Hu, L.; Dong, Y.C.; Wu, X.F.; Qin, Y.C.; Zheng, Y.H.; Shi, D.D.; Xue, M.; Liang, X.F. Substitution of fish meal by fermented soybean meal affects the growth performance and flesh quality of Japanese seabass (*Lateolabrax japonicus*). *Anim. Feed Sci. Technol.* **2017**, *229*, 1–12. [CrossRef]
11. Zhang, C.; Rahimnejad, S.; Wang, Y.-r.; Lu, K.; Song, K.; Wang, L.; Mai, K. Substituting fish meal with soybean meal in diets for Japanese seabass (*Lateolabrax japonicus*): Effects on growth, digestive enzymes activity, gut histology, and expression of gut inflammatory and transporter genes. *Aquaculture* **2018**, *483*, 173–182. [CrossRef]
12. Zhang, Y.; Ji, W.; Wu, Y.; Han, H.; Qin, J.; Wang, Y. Replacement of dietary fish meal by soybean meal supplemented with crystalline methionine for Japanese seabass (*Lateolabrax japonicus*). *Aquac. Res.* **2016**, *47*, 243–252. [CrossRef]
13. Zhang, Y.Q.; Wu, Y.B.; Jiang, D.L.; Qin, J.G.; Wang, Y. Gamma-irradiated soybean meal replaced more fish meal in the diets of Japanese seabass (*Lateolabrax japonicus*). *Anim. Feed Sci. Technol.* **2014**, *197*, 155–163. [CrossRef]
14. Refstie, S.; Sahlstrom, S.; Brathen, E.; Baeverfjord, G.; Krogedal, P. Lactic acid fermentation eliminates indigestible carbohydrates and antinutritional factors in soybean meal for Atlantic salmon (*Salmo salar*). *Aquaculture* **2005**, *246*, 331–345. [CrossRef]
15. Krogdahl, A.; Gajardo, K.; Kortner, T.M.; Penn, M.; Gu, M.; Berge, G.M.; Bakke, A.M. Soya Saponins Induce Enteritis in Atlantic Salmon (*Salmo salar* L.). *J. Agric. Food Chem.* **2015**, *63*, 3887–3902. [CrossRef] [PubMed]
16. Rahimnejad, S.; Lu, K.; Wang, L.; Song, K.; Mai, K.; Davis, D.A.; Zhang, C. Replacement of fish meal with *Bacillus pumillus* SE5 and *Pseudozyma aphidis* ZR1 fermented soybean meal in diets for Japanese seabass (*Lateolabrax japonicus*). *Fish Shellfish Immunol.* **2019**, *84*, 987–997. [CrossRef] [PubMed]
17. Papatryphon, E.; Soares, J.H. The effect of dietary feeding stimulants on growth performance of striped bass, Morone saxatilis, fed-a-plant feedstuff-based diet. *Aquaculture* **2000**, *185*, 329–338. [CrossRef]
18. Hai, N.V. The use of medicinal plants as immunostimulants in aquaculture: A review. *Aquaculture* **2015**, *446*, 88–96.
19. Wang, J.; Kortner, T.M.; Chikwati, E.M.; Li, Y.; Jaramillo-Torres, A.; Jakobsen, J.V.; Ravndal, J.; Brevik, O.J.; Einen, O.; Krogdahl, A. Gut immune functions and health in Atlantic salmon (*Salmo salar*) from late freshwater stage until one year in seawater and effects of functional ingredients: A case study from a commercial sized research site in the Arctic region. *Fish Shellfish Immunol.* **2020**, *106*, 1106–1119. [CrossRef]
20. Beth, K.; Hellwig, D.; Gill, D.; Owens, F.; Murapa, P.; Gandhapudi, S.; Skaggs, H.S.; Sarge, K.D. Effect of Agrado® on performance and health of calves new to the feedlot environment. *Res. Ser.-Ark. Agric. Exp. Stn.* **2000**, 84–87.

21. Alagawany, M.; Farag, M.R.; Abdelnour, S.A.; Elnesr, S.S. A review on the beneficial effect of thymol on health and production of fish. *Rev. Aquac.* **2020**, *13*, 632–641. [CrossRef]
22. Apines-Amar, M.J.S.; Satoh, S.; Caipang, C.M.A.; Kiron, V.; Watanabe, T.; Aoki, T. Amino acid-chelate: A better source of Zn, Mn and Cu for rainbow trout, *Oncorhynchus mykiss*. *Aquaculture* **2004**, *240*, 345–358. [CrossRef]
23. Katya, K.; Lee, S.; Bharadwaj, A.S.; Browdy, C.L.; Vazquez-Anon, M.; Bai, S.C. Effects of inorganic and chelated trace mineral (Cu, Zn, Mn and Fe) premixes in marine rockfish, *Sebastes schlegeli* (Hilgendorf), fed diets containing phytic acid. *Aquac. Res.* **2017**, *48*, 4165–4173. [CrossRef]
24. Katya, K.; Lee, S.; Yun, H.; Dagoberto, S.; Browdy, C.L.; Vazquez-Anon, M.; Bai, S.C. Efficacy of inorganic and chelated trace minerals (Cu, Zn and Mn) premix sources in Pacific white shrimp, *Litopenaeus vannamei* (Boone) fed plant protein based diets. *Aquaculture* **2016**, *459*, 117–123. [CrossRef]
25. Secombes, C.J. Isolation of salmonid macrophages and analysis of their killing activity. *Tech. Fish Immunol.* **1990**, *1*, 137–154.
26. Nilsson, U.R.; Nilsson, B. Simplified assays of hemolytic activity of the classical and alternative complement pathways. *J. Immunol. Methods* **1984**, *72*, 49–59. [CrossRef]
27. Abei, H. Catalase in vitro. *Methods Enzymol.* **1984**, *105*, 121–126.
28. Zhang, X.D.; Zhu, Y.F.; Cai, L.S.; Wu, T.X. Effects of fasting on the meat quality and antioxidant defenses of market-size farmed large yellow croaker (*Pseudosciaena crocea*). *Aquaculture* **2008**, *280*, 136–139. [CrossRef]
29. Tangney, C.C.; Driskell, J.A. Effects of vitamin E deficiency on the relative incorporation of 14C-arachidonate into platelet lipids of rabbits. *J. Nutr.* **1981**, *111*, 1839–1847. [CrossRef]
30. Wang, F.; Le, G.; Shi, Y.; Chen, Z. Effect of mannose-oligosaccharides on scavenging free radicals and the activities of antioxidant enzymes in liver of rats. *Acta Nutr.* **2007**, *5*, 15.
31. Ai, Q.H.; Xu, H.G.; Mai, K.S.; Xu, W.; Wang, J.; Zhang, W.B. Effects of dietary supplementation of Bacillus subtilis and fructooligosaccharide on growth performance, survival, non-specific immune response and disease resistance of juvenile large yellow croaker, *Larimichthys crocea*. *Aquaculture* **2011**, *317*, 155–161. [CrossRef]
32. Chen, C.; Chen, H.; Zhang, Y.; Thomas, H.R.; Frank, M.H.; He, Y.; Xia, R. TBtools: An Integrative Toolkit Developed for Interactive Analyses of Big Biological Data. *Mol. Plant* **2020**, *13*, 1194–1202. [CrossRef] [PubMed]
33. Van den Ingh, T.S.; Krogdahl, Å.; Olli, J.J.; Hendriks, H.G.; Koninkx, J.G.J.F. Effects of soybean-containing diets on the proximal and distal intestine in Atlantic salmon (*Salmo salar*): A morphological study. *Aquaculture* **1991**, *94*, 297–305. [CrossRef]
34. Yu, D.H.; Gong, S.Y.; Yuan, Y.C.; Luo, Z.; Lin, Y.C.; Li, Q. Effect of partial replacement of fish meal with soybean meal and feeding frequency on growth, feed utilization and body composition of juvenile Chinese sucker, *Myxocyprinus asiaticus* (Bleeker). *Aquac. Res.* **2013**, *44*, 388–394. [CrossRef]
35. Tibaldi, E.; Hakim, Y.; Uni, Z.; Tulli, F.; de Francesco, M.; Luzzana, U.; Harpaz, S. Effects of the partial substitution of dietary fish meal by differently processed soybean meals on growth performance, nutrient digestibility and activity of intestinal brush border enzymes in the European sea bass (*Dicentrarchus labrax*). *Aquaculture* **2006**, *261*, 182–193. [CrossRef]
36. Ye, J.D.; Liu, X.H.; Wang, Z.J.; Wang, K. Effect of partial fish meal replacement by soybean meal on the growth performance and biochemical indices of juvenile Japanese flounder *Paralichthys olivaceus*. *Aquac. Int.* **2011**, *19*, 143–153. [CrossRef]
37. Silva-Carrillo, Y.; Hernandez, C.; Hardy, R.W.; Gonzalez-Rodriguez, B.; Castillo-Vargasmachuca, S. The effect of substituting fish meal with soybean meal on growth, feed efficiency, body composition and blood chemistry in juvenile spotted rose snapper *Lutjanus guttatus* (Steindachner, 1869). *Aquaculture* **2012**, *364*, 180–185. [CrossRef]
38. Peng, M.; Xu, W.; Ai, Q.H.; Mai, K.S.; Liufu, Z.G.; Zhang, K.K. Effects of nucleotide supplementation on growth, immune responses and intestinal morphology in juvenile turbot fed diets with graded levels of soybean meal (*Scophthalmus maximus* L.). *Aquaculture* **2013**, *392*, 51–58. [CrossRef]
39. Rossi, W.; Tomasso, J.R.; Gatlin, D.M. Performance of Cage-Raised, Overwintered Hybrid Striped Bass Fed Fish Meal- or Soybean-Based Diets. *N. Am. J. Aquac.* **2015**, *77*, 178–185. [CrossRef]
40. Rossi, W.; Allen, K.M.; Habte-Tsion, H.-M.; Meesala, K.-M. Supplementation of glycine, prebiotic, and nucleotides in soybean meal-based diets for largemouth bass (*Micropterus salmoides*): Effects on production performance, whole-body nutrient composition and retention, and intestinal histopathology. *Aquaculture* **2021**, *532*, 736031. [CrossRef]
41. Mai, K.S.; Zhang, L.; Ai, Q.H.; Duan, Q.Y.; Zhang, C.X.; Li, H.T.; Wan, J.L.; Liufu, Z.G. Dietary lysine requirement of juvenile Japanese seabass, *Lateolabrax japonicus*. *Aquaculture* **2006**, *258*, 535–542. [CrossRef]
42. Kiron, V. Fish immune system and its nutritional modulation for preventive health care. *Anim. Feed Sci. Technol.* **2012**, *173*, 111–133. [CrossRef]
43. Wang, L.; Zhou, H.H.; He, R.J.; Xu, W.; Mai, K.S.; He, G. Effects of soybean meal fermentation by Lactobacillus plantarum P8 on growth, immune responses, and intestinal morphology in juvenile turbot (*Scophthalmus maximus* L.). *Aquaculture* **2016**, *464*, 87–94. [CrossRef]
44. Khosravi, S.; Rahimnejad, S.; Herault, M.; Fournier, V.; Lee, C.R.; Bui, H.T.D.; Jeong, J.B.; Lee, K.J. Effects of protein hydrolysates supplementation in low fish meal diets on growth performance, innate immunity and disease resistance of red sea bream *Pagrus major*. *Fish Shellfish Immun.* **2015**, *45*, 858–868. [CrossRef] [PubMed]
45. Maita, M.; Maekawa, J.; Satoh, K.; Futami, K.; Satoh, S. Disease resistance and hypocholesterolemia in yellowtail Seriola quinqueradiata fed a non-fishmeal diet. *Fish. Sci.* **2006**, *72*, 513–519. [CrossRef]

46. Martinez-Alvarez, R.M.; Morales, A.E.; Sanz, A. Antioxidant defenses in fish: Biotic and abiotic factors. *Rev. Fish Biol. Fish.* **2005**, *15*, 75–88. [CrossRef]
47. Limon-Pacheco, J.; Gonsebatt, M.E. The role of antioxidants and antioxidant-related enzymes in protective responses to environmentally induced oxidative stress. *Mutat. Res./Genet. Toxicol. Environ. Mutagen.* **2009**, *674*, 137–147. [CrossRef]
48. Olsvik, P.A.; Torstensen, B.E.; Hemre, G.I.; Sanden, M.; Waagbo, R. Hepatic oxidative stress in Atlantic salmon (*Salmo salar* L.) transferred from a diet based on marine feed ingredients to a diet based on plant ingredients. *Aquac. Nutr.* **2011**, *17*, E424–E436. [CrossRef]
49. Wang, J.; Jaramillo-Torres, A.; Li, Y.; Brevik, Ø.J.; Jakobsen, J.V.; Kortner, T.M.; Krogdahl, Å. Gut Health and Microbiota in Out-of-Season Atlantic Salmon (*Salmo salar* L.) Smolts Before and After Seawater Transfer Under Commercial Arctic Conditions: Modulation by Functional Feed Ingredients. *Front. Mar. Sci.* **2022**, *9*. [CrossRef]
50. Wang, J.; Jaramillo-Torres, A.; Li, Y.; Kortner, T.M.; Gajardo, K.; Brevik, O.J.; Jakobsen, J.V.; Krogdahl, A. Microbiota in intestinal digesta of Atlantic salmon (*Salmo salar*), observed from late freshwater stage until one year in seawater, and effects of functional ingredients: A case study from a commercial sized research site in the Arctic region. *Anim. Microbiome* **2021**, *3*, 14. [CrossRef]
51. Li, Y.; Bruni, L.; Jaramillo-Torres, A.; Gajardo, K.; Kortner, T.M.; Krogdahl, A. Differential response of digesta- and mucosa-associated intestinal microbiota to dietary insect meal during the seawater phase of Atlantic salmon. *Anim. Microbiome* **2021**, *3*, 8. [CrossRef] [PubMed]
52. Hatje, E.; Neuman, C.; Stevenson, H.; Bowman, J.P.; Katouli, M. Population dynamics of Vibrio and Pseudomonas species isolated from farmed Tasmanian Atlantic salmon (*Salmo salar* L.): A seasonal study. *Microb. Ecol.* **2014**, *68*, 679–687. [CrossRef] [PubMed]

antioxidants

MDPI

Article

The Effects of Dietary *Arthrospira platensis* on Oxidative Stress Response and Pigmentation in Yellow Catfish *Pelteobagrus fulvidraco*

Cui Liu [1,2], Haokun Liu [1,*], Xiaoming Zhu [1,2,3], Dong Han [1,2,3], Junyan Jin [1], Yunxia Yang [1] and Shouqi Xie [1,2,3,4]

1 State Key Laboratory of Freshwater Ecology and Biotechnology, Institute of Hydrobiology, Chinese Academy of Sciences, Wuhan 430072, China; liucui@ihb.ac.cn (C.L.); xmzhu@ihb.ac.cn (X.Z.); hand21cn@ihb.ac.cn (D.H.); jinjunyan@ihb.ac.cn (J.J.); yxyang@ihb.ac.cn (Y.Y.); sqxie@ihb.ac.cn (S.X.)
2 College of Advanced Agricultural Sciences, University of Chinese Academy of Sciences, Beijing 100049, China
3 The Innovative Academy of Seed Design, Chinese Academy of Sciences, Beijing 100101, China
4 Hubei Engineering Research Center for Aquatic Animal Nutrition and Feed, Wuhan 430072, China
* Correspondence: liuhaokun@ihb.ac.cn; Tel.: +86-276-878-0060

Abstract: In aquaculture, fish are often exposed to several stress conditions, which will cause oxidative disorder and bring about health and quality problems. *Arthrospira platensis* contains abundant bioactive ingredients, which are beneficial for animal health. This study was conducted to investigate the effects of *A. platensis* on pigmentation, antioxidant capacity, and stress response after air exposure of fish. A total of 120 yellow catfish *Pelteobagrus fulvidraco* (initial weight 70.19 ± 0.13 g) were divided into three tanks per treatment and fed diets supplemented with 0 g kg^{-1} *A. platensis* (CON) and 20 g kg $^{-1}$ *A. platensis* (AP) for 65 days. The results indicated that dietary *A. platensis* had no effects on the growth of yellow catfish. The AP diet significantly reduced lactic acid (LD) and cortisol levels stimulated by air exposure stress ($p < 0.05$). Dietary *A. platensis* significantly increased plasma superoxide dismutase (SOD) and glutathione peroxidase (GPX) activities and glutathione (GSH) contents, and the relative expression levels of *sod* and *cat*, to protect against oxidative stress caused by air exposure ($p < 0.05$). The AP diet significantly improved the relative expression level of *nrf2* (nuclear factor erythroid-2 related factor 2), while the relative expression level of *keap1* (kelch-like ECH associated protein 1) was downregulated, and the protein levels of liver Nrf2 were significantly increased after air exposure stimuli ($p < 0.05$). Dietary *A. platensis* significantly increased skin lutein contents, increased skin redness, yellowness and chroma ($p < 0.05$), and improved body color abnormalities after oxidative stress caused by air exposure stimuli. Skin yellowness was associated with lutein contents and the expression levels of some antioxidant genes to varying degrees. Overall, dietary *A. platensis* could be utilized as a feed additive to activate the antioxidant response, as well as alleviate oxidative stress and pigmentation disorder induced by air exposure.

Keywords: aquaculture; Spirulina; antioxidant; redox enzyme

Citation: Liu, C.; Liu, H.; Zhu, X.; Han, D.; Jin, J.; Yang, Y.; Xie, S. The Effects of Dietary *Arthrospira platensis* on Oxidative Stress Response and Pigmentation in Yellow Catfish *Pelteobagrus fulvidraco*. *Antioxidants* **2022**, *11*, 1100. https://doi.org/10.3390/antiox11061100

Academic Editors: Alessandra Napolitano, Stanley Omaye and Isabel Seiquer

Received: 15 April 2022
Accepted: 30 May 2022
Published: 31 May 2022

1. Introduction

Stress is a general term that applies to a situation in which a person or an animal is subjected to a challenge that may result in real or symbolic danger to its integrity [1]. In aquaculture, fish are often exposed to several stress conditions due to farming practices (high densities, transport or handling, or changes in abiotic factors) and environmental factors (temperature, salinity, biochemical water quality, etc.) [2]. The response to stress in fish is often characterized as primary, which is the activation of the hypothalamic–pituitary–interrenal axis that eventually results in a hormonal response to the release of cortisol and catecholamine; secondary, which includes the changes in metabolic, hematological, hydromineral, and structural due to the action of cortisol and catecholamine; and tertiary, which are associated with the performance of the organism, exemplified by inhibited growth, hampered reproduction, and immunosuppression [3,4]. Extreme conditions also produce reactive oxygen species (ROS), which lead to DNA, protein, and lipid damage [5].

Air exposure is a common acute stressor with a subsequent physiological disturbance in aquaculture since handling procedures during animal husbandry and the production cycle require exposing fish to air environments [2]. Air exposure always brings a series of problems to aquatic animals, like exacerbating blood chemical conditions, causing oxidative and antioxidant responses, and adversely affecting survival [6–9]. In recent years, there has been increasing interest in using natural compounds as functional nutrients to reduce stress responses and improve the immunity of aquatic animals [10–12].

Arthrospira platensis, a blue-green alga that is gaining worldwide popularity as a food supplement, contains high contents of protein, polyunsaturated fatty acids, vitamins, and minerals [13]. In addition, *A. platensis* is plenitudinous in antioxidant compounds such as carotenoids, phycocyanin, and tocopherols [14]. In previous studies, it has been demonstrated that *A. platensis* can enhance antioxidant and immunity abilities as well as ensure skin pigmentation in fish [15–17].

Yellow catfish *Pelteobagrus fulvidraco* is a commercially important fish species in aquaculture in East and South Asia because of its excellent meat quality and perfect flavor [18,19]. However, farmed fish are always influenced by handling, weighing, crowding, grading, and transporting, which cause fish exposure to air and bring about a stress response. Hence, the main objectives of the study were to evaluate the antioxidant ability and stress responses of yellow catfish fed an *A. platensis*-supplemented diet and exposed to air as an experimental stressor causing oxidative stress. In addition, we also investigated the effect of *A. platensis* on the pigmentation of fish and detected the correlation between skin color and skin lutein content and oxidative stress. We are expecting that dietary *A. platensis* can reduce the stress responses and bring welfare to fish during fish farming.

2. Materials and Methods

2.1. Experimental Diets

Two isonitrogenous (420 g crude protein kg^{-1} diet) and isolipidic (80 g crude lipid kg^{-1} diet) diets were designed as practical diets, using white fishmeal, soybean meal, and rapeseed meal as blended protein sources (Table 1). This study consisted of a control group (CON-without supplementing *A. platensis*) and an *A. platensis* group (AP, 20 g kg^{-1} *A. platensis*). All ingredients were completely mixed with appropriate water and then made into pellets using an SLP-45 laboratory granulator (Fishery Mechanical Facility Research Institute, Shanghai, China). The diets were oven-dried at 60 °C for 12 h and stored at 4 °C until used for experimentation. Moisture, crude protein, crude lipid, and lutein contents of diets were analyzed following the same procedures as the previous study [20]. Moisture content was determined by oven drying at 105 °C to a constant weight. A 2300 Kjeltec Analyzer Unit machine (FOSS Tecator, Haganas, Sweden) was used to measure crude protein content. Crude lipid content was determined by ether extraction in a Soxtec System HT6 (Tecator Ltd., Haganas, Sweden). Lutein was extracted by mixed extractant (n-Hexane: acetone: ethanol: toluene = 50:35:30:35, $v/v/v/v$) and analyzed by HPLC.

Table 1. Diet formulation and chemical compositions of the experimental diets.

Ingredient (g kg^{-1} Dry Matter)	CON	AP
White fishmeal [1]	300	300
A. platensis [2]	0	20
Soybean meal [3]	187.8	175.8
Rapeseed meal [3]	200	185.5
Wheat flour	150	150
Fish oil	25	25
Soybean oil	25	25
Cellulose	27.2	33.7
Vitamin premix [4]	3.9	3.9

Table 1. *Cont.*

Ingredient (g kg^{-1} Dry Matter)	CON	AP
Mineral premix [5]	50	50
CMC [6]	30	30
Choline chloride	1.1	1.1
Chemical composition (g kg^{-1})		
Moisture	108.55	108.15
In dry matter (g kg^{-1})		
Crude protein	418.30	416.56
Crude lipid	80.99	75.22
Lutein (μg/g)	4.96	8.08

[1] White fishmeal: Seafood white fishmeal imported from the United States. [2] *A. platensis*: Center for Microalgal Biotechnology and Biofuels, Wuhan, China. [3] Soybean meal and rapeseed meal were purchased from Wuhan Gaolong Feed Co., Ltd., Wuhan, Hubei, China. [4] Vitamin premix (mg kg^{-1} diet): Vitamin B_1, 20; Vitamin B_2, 20; Vitamin B_6, 20; Vitamin B_{12}, 0.02; folic acid, 5; calcium pantothenate, 50; inositol, 100; niacin, 100; biotin, 0.1; cellulose, 3522; Vitamin A, 11; Vitamin D, 2; Vitamin E, 100; Vitamin K, 10. [5] Mineral premixes (mg kg^{-1} diet): NaCl, 500.0; $MgSO_4 \cdot 7H_2O$, 8155.6; $NaH_2PO_4 \cdot 2H_2O$, 12,500.0; KH_2PO_4, 16,000.0; $Ca(H_2PO_4)\ 2H_2O$, 7650.6; $FeSO_4 \cdot 7H_2O$, 2286.2; $C_6H_{10}CaO_6 \cdot 5H_2O$, 1750.0; $ZnSO_4 \cdot 7H_2O$, 178.0; $MnSO_4 \cdot H_2O$, 61.4; $CuSO_4 \cdot 5H_2O$, 15.5; $CoSO_4 \cdot 7H_2O$, 0.91; KI, 1.5; Na_2SeO_3, 0.60; Corn starch, 899.7. [6] CMC: Carboxymethyl cellulose.

2.2. Fish, Experimental Conditions, and Feeding Procedures

All experimental animal care protocols were approved by the ethics committee of the Institute of Hydrobiology, Chinese Academy of Sciences. Yellow catfish were obtained from the Dengjia State fish farm (Dengjiazhou, Jiangxia, Wuhan, China). Two weeks before the feeding trials, all fish were temporarily domesticated in a fiberglass cylinder (1500 L) and fed twice a day at 08:30 and 16:30. Feeding trials were conducted in an indoor recirculating system. At the beginning of each trial, the fish fasted for 24 h. Healthy and similarly sized fish (n = 120, 70.19 ± 0.13 g) were randomly selected, batch weighed, and placed into 6 fiberglass tanks (3 tanks per treatment, 20 individuals per tank, water volume, 400 L, diameter, 70 cm). The experimental fish were hand-fed to apparent satiation twice daily (8:30 and 16:30) for 65 days. Each tank was provided with continuous aeration. During the experiment, the water temperature was maintained at 30.5 ± 1.0 °C. Total ammonia-nitrogen was maintained at <0.1 mg L^{-1}, dissolved oxygen at >6 mg L^{-1}, and residual chloride at <0.01 mg L^{-1}. The light period was from 8:00 to 20:00, and the light intensity was approximately 2.20–3.70 μ$mols^{-1}m^{-2}$ (at water surface).

2.3. Sample Collection and Air Exposure Stress Challenge

At the end of the feeding trial, the experimental fish were batch-weighed. Three fish from each tank were randomly selected and anesthetized with 80 mg L^{-1} MS-222 (Sigma-Aldrich, St. Louis, MO, USA) and then measured coloration and sampled for blood, liver, kidney, dorsal, and abdominal skin tissues. Blood samples were collected from the caudal vein with heparinized syringes. After centrifugation (3500× g, 15 min, 4 °C), plasma was collected and stored at −80 °C for further analysis. After blood sampling, the liver, kidney, dorsal, and abdominal skin were dissected on ice and stored at −80 °C.

At the end of the trial, three fish were randomly selected from each tank for a stress challenge. Fish were exposed to air for 5 min and then sampled for blood, liver, and kidney. The methods of plasma collection and sample storage were the same as those for normal sampling (without air exposure).

2.4. Fish Skin Color Determination

The dorsal and abdominal skin color parameters of the fish were measured by a Konica Minolta CR-400 tristimulus colorimeter (Minolta, Osaka, Japan). The L* value represents lightness (0 for black, 100 for white), a* represents the red/green dimension, and b* represents the yellow/blue dimension, the value of chroma C is the distance from the lightness axis (L*) and starts at 0 in the center, in accordance with the recommendations of the International Commission on Illumination [21]. Chroma is an expression of the saturation or intensity of the color and is calculated by the equation C = $(a^{*2} + b^{*2})^{1/2}$ [22]. Fish from different experimental groups were photographed using a Nikon D5100 camera (Japan).

2.5. Lutein Extraction and Quantification

The extraction of lutein from the skin was determined by following the method described previously [23]. Skin samples (200–300 mg) were separately mixed into a 0.7 mL solution consisting of 5% sodium chloride; 1 mL ethanol was added to homogenize the samples. During homogenization, 2 mL hexane was added. Then, the samples were centrifuged (4 °C, 4600 rpm, 10 min), and the hexane phase was collected. Extraction with hexane was performed twice, and the combined phase was evaporated under nitrogen to obtain pigment samples. The analytical conditions of lutein were based on those reported previously [24,25] with some modifications. The pigment samples were dissolved in an isocratic solvent system, methanol/methyl-tert-butyl ether = 86/14 (*v*/*v*), the lutein standard (07168, sigma) was diluted into 100, 50, 20, 10, 5, 2, 1, 0.5, 0.1 μg/mL with the same solvent system. The obtained solution was used for HPLC (Waters e2695, Milford, Delaware, USA) analyses immediately after passing through a 0.22 μm membrane filter. The HPLC was equipped with a Waters YMC Carotenoid C30 column (5 μm, 4.6 × 250 mm); the mobile phase consisted of solvent A (methanol: methyl-tert-butyl ether: H_2O = 81:15:4) and solvent B (methanol: methyl-tert-butyl ether = 600:90). The gradient procedure was performed at a flow rate of 1 mL/min. The total run time was 40 min, and the injection volume was 10 μL.

2.6. Plasma LD, Cortisol, Glucose, and Antioxidant Assays

The contents of lactic acid (LD), cortisol, glucose (GLU), malondialdehyde (MDA), and reduced glutathione (GSH) and the activities of superoxide dismutase (SOD) and glutathione peroxidase (GPX) were tested using commercial kits (Nanjing Jiancheng Bioengineering Institute, Nanjing, Jiangsu, China).

2.7. Quantitative Real-Time PCR Analysis

Total RNA was extracted from the liver and kidney using TRIzol reagent according to the manufacturer's instructions (Invitrogen, Carlsbad, CA, USA). The quality of total RNA was evaluated by 1% agarose gel electrophoresis. The purity and concentration were assessed by a NanoDrop® ND-2000 UV–Vis Spectrophotometer (NanoDrop Technologies, Wilmington, DE, USA). The total RNA was then reverse-transcribed with an M-MLV FirstStrand Synthesis Kit (Invitrogen, Shanghai, China). The polymerase chain reaction (PCR) primer sequences for *sod, gpx, gr, cat, nrf2, keap1,* and the reference gene β-actin, which were designed based on the cDNA sequences of yellow catfish, are shown in Table 2. A LightCycler 480 System (Roche, Germany) with SYBR® Green I Master Mix (Roche, Germany) was used to perform quantitative RT–PCR. Each sample was run in duplicate, and the relative expression was calculated [26].

Table 2. Sequences of the primers used for qRT-PCR analysis in yellow catfish.

Gene	Forward Primer (5′–3′)	Reverse Primer (5′–3′)	Amplicon Size (bp)	Tm (°C)	PCR Efficiency [a/b]	Accession No
sod	TTGGAGACAATACAAATGGGTG	CATCGGAATCGGCAGTCA	129	57	2.007/1.968	XM_027171881
gpx	ATCTACATTGGCTTGGAAAC	GAAAGTAGGGACTGAGGTGA	257	58	1.966/1.950	XM_027163146
gr	CAGTCGCTTTGTTTGTTCTA	TCCTCCGATACACTTCTCAC	280	57	1.992/2.050	XM_027152663
cat	TCTGTTCCCGTCCTTCATCC	ATATCCGTCAGGCAATCCAC	151	58	1.964/2.001	XM_027163801
nrf2	TCTCGCCCAGTTACAGCTTG	GTTCCGTGAACGCCACATTC	128	60	1.992/1.952	XM_027164284
keap1	CGCAGCCGGGCTTTTATTTT	AGGCAGAAACGGGTTCAAGT	286	59	1.989/1.959	XM_027133478.1
β-actin	TTCGCTGGAGATGATGCT	CGTGCTCAATGGGGTACT	136	58	2.044/2.003	EU161066

[a] Liver, [b] Kidney. *sod*, superoxide dismutase; *gpx*, glutathione peroxidase; *gr*, glutathione reductase; *cat*, catalase; *nrf2*, nuclear factor erythroid-2 related factor 2; *keap1*, kelch-like ECH associated protein 1.

2.8. Western Blot Analysis

Liver tissues were lysed with RIPA lysis buffer (Beyotime Biotechnology, China) containing protease inhibitor cocktail and phosphatase inhibitor cocktail (Roche, Basel, Switzerland). Equal amounts of protein were separated on sodium dodecyl sulfate–polyacrylamide gel electrophoresis (SDS–PAGE) gels and transferred to polyvinylidene fluoride (PVDF) membranes. The PVDF membranes were blocked for 1 h with 5% milk in TBST buffer and then incubated overnight at 4 °C with Nrf2 (1:1000, ab62352, Abcam) and β-actin (1:1000, #8457, CST) antibodies. β-actin was used as an internal reference protein. Horseradish peroxidase-labelled secondary antibodies were used to generate a chemiluminescent signal that was detected by ImageQuant LAS 4000 mini (GE Healthcare Life Sciences) and quantified using ImageJ software (National Institutes of Health).

2.9. Statistical Analyses

All data were statistically analyzed with SPSS 19.0 and subjected to one-way ANOVA. Before any statistical analysis, normality and homoscedasticity assumptions were confirmed. Duncan's multiple range test was used to detect the significance of differences in mean values among different treatments. The results are presented as the mean ± standard error (SEM). The Pearson's correlation coefficient was calculated to analyze the significance of linear relationships between skin yellowness (b*) and the studied variables. The significance difference level was set at $p < 0.05$.

3. Results

3.1. Growth and Feed Utilization

The survival rate of fish was 100% in both groups during the feeding trial; there were no significant differences in final body weight (FBW), specific growth rate (SGR), feeding rate (FR), or feed efficiency (FE) in either the CON or AP groups (Table 3). At the end of the experimental period, no significant differences were found in the condition factor, hepatosomatic index, or viscerosomatic index between the CON and AP groups (Table 3).

Table 3. Growth, feed utilization, and morphological indices of yellow catfish fed different experimental diets.

Indices/Diets	CON	AP
IBW (g)	69.8 ± 0.10	70.23 ± 0.15
FBW (g)	103.23 ± 0.35	105.32 ± 1.07
Survival rate (%)	100	100
FR (%BW d^{-1})	2.52 ± 0.08	2.63 ± 0.02
SGR (% d^{-1})	0.68 ± 0.05	0.62 ± 0.02
FE (%)	30.51 ± 3.82	30.69 ± 0.49
Condition indices		
Condition factor (g cm^{-3})	1.89 ± 0.08	1.81 ± 0.04
Hepatosomatic index (%)	1.40 ± 0.17	1.68 ± 0.11
Viscerosomatic index (%)	11.14 ± 0.85	10.8 ± 0.56

IBW: initial body weight. FBW: final body weight. FR: feeding rate (% body weight day^{-1}) = 100 × (feed intake in dry matter)/[days × (initial body weight + final body weight)/2]. SGR: specific growth rate (% d^{-1}) = 100 × [ln (final body weight) − ln (initial body weight)]/days. FE: feed efficiency (%) = (final body weight-initial body weight)/feed intake in dry matter. Condition factor (g cm^{-3}) = whole body weight/(body length)^3 × 100. Hepatosomatic index (%) = liver weight/whole body weight × 100. Viscerosomatic index (%) = visceral weight/whole body weight × 100. Data are presented as the mean ± SEM (n = 3). No significant differences were found between two feeding experiments ($p > 0.05$).

3.2. Stress Response Markers in Plasma

Stress response markers such as LD, cortisol, and glucose were significantly increased after air exposure stress in fish fed the CON diet ($p < 0.05$, Figure 1). However, the AP diet significantly reduced LD and cortisol levels stimulated by air exposure stress ($p < 0.05$, Figure 1).

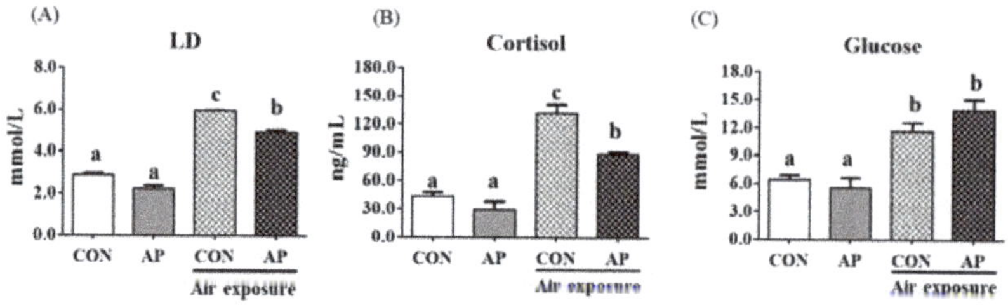

Figure 1. Plasma stress biomarkers in the two groups before and after air exposure stress. LD (**A**), cortisol (**B**), and glucose (**C**). Values are represented as the mean ± SEM ($n = 6$). Bars with different lowercase letters mean significant differences among groups ($p < 0.05$).

3.3. Antioxidant-Related Parameters in Plasma

To test whether dietary AP affected fish antioxidant capacity, we analyzed antioxidant-related parameters in plasma. As shown in Figure 2, the AP diet significantly increased plasma SOD and GPX activities after air exposure to protect against oxidative stress ($p < 0.05$). The contents of plasma GSH were significantly decreased in the CON group after air exposure stimuli ($p < 0.05$), while the AP diet significantly improved GSH contents to protect against oxidative stress caused by air exposure ($p < 0.05$). The MDA content showed no obvious changes.

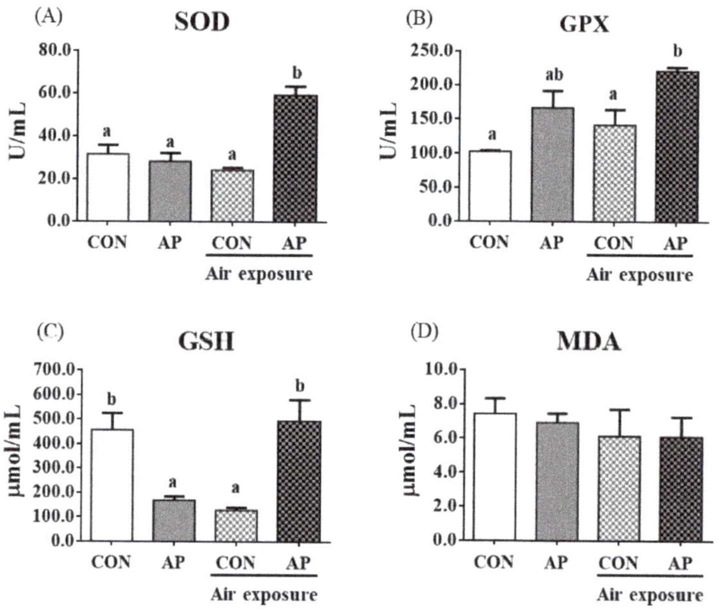

Figure 2. Plasma antioxidant related parameters in the two groups before and after air exposure stress. SOD (**A**), GPX (**B**), GSH (**C**), and MDA (**D**). Values are represented as the mean ± SEM ($n = 6$). Bars with different lowercase letters mean significant differences among groups ($p < 0.05$).

3.4. Antioxidant Related Gene Expression and the Nrf2 Signaling Pathway in the Liver

As presented in Figure 3, the relative expression levels of liver *sod* and *gpx* were significantly downregulated in the CON group after air exposure stimulation ($p < 0.05$), and there were no obvious differences in the relative expression levels of liver *gr* and *cat* between CON and AP groups after air exposure stimulation.

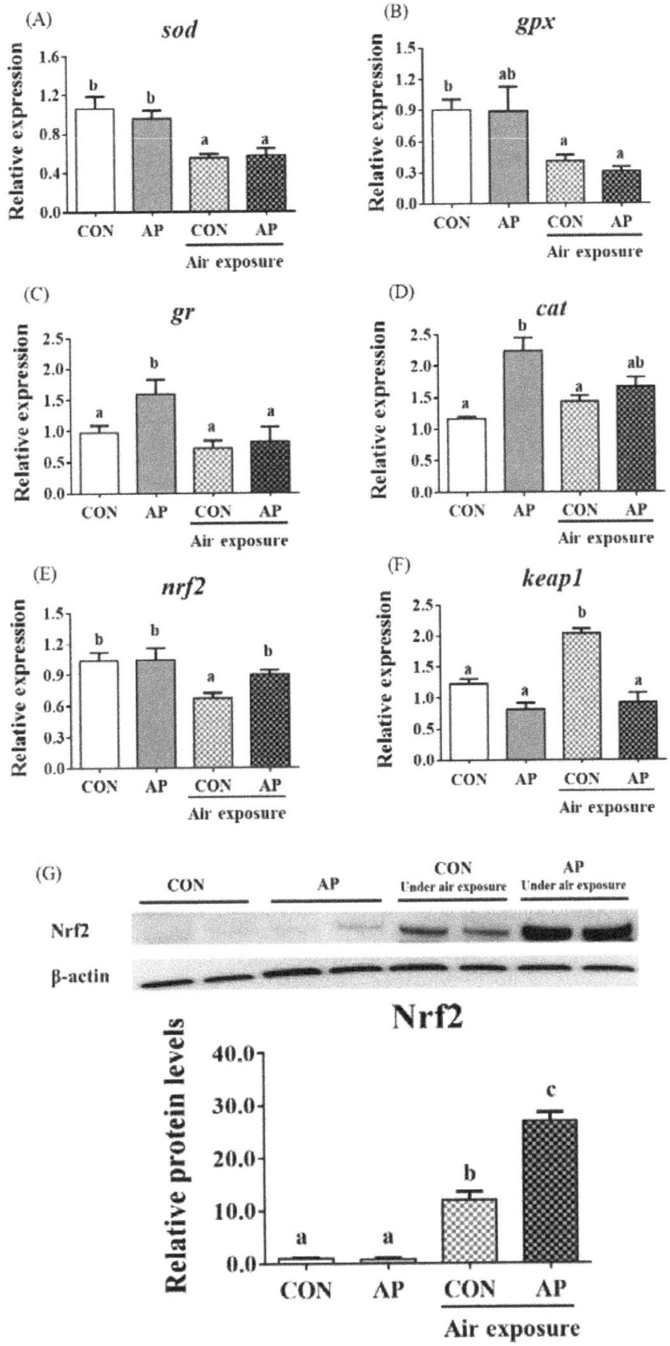

Figure 3. *sod* (**A**), *gpx* (**B**), *gr* (**C**), *cat* (**D**), *nrf2* (**E**), and *keap1* (**F**) gene relative expressions levels and Nrf2 (**G**) protein relative expressions level of liver in the two groups before and after air exposure stress. Data are indicated as mean ± SEM (n = 6). Bars with different lowercase letters mean significant differences among groups ($p < 0.05$).

The relative expression level of *nrf2* in the liver was significantly downregulated in the CON group after air exposure stimulation ($p < 0.01$), while the AP diet significantly improved the relative expression level of *nrf2* to protect against oxidative stress caused by air exposure stimuli ($p < 0.05$, Figure 3). The relative expression level of *keap1* in the liver was significantly upregulated in the CON group after air exposure stimulation ($p < 0.05$), while the AP diet significantly reduced the relative expression level of *keap1* to protect against oxidative stress caused by air exposure ($p < 0.05$, Figure 3). The protein levels of liver Nrf2 were significantly higher in the AP group than that in the CON group after air exposure stimulation ($p < 0.01$, Figure 3).

3.5. *Antioxidant Related Gene Expression in Kidney*

The relative expression levels of *sod*, *gpx*, and *cat* were significantly downregulated in the CON group after air exposure stimulation ($p < 0.05$, Figure 4); however, the AP diet significantly improved the relative expression levels of *sod* and *cat* to protect against oxidative stress caused by air exposure ($p < 0.05$) but did not significantly affect the relative expression levels of *gr* in both groups after air exposure stimulation (Figure 4).

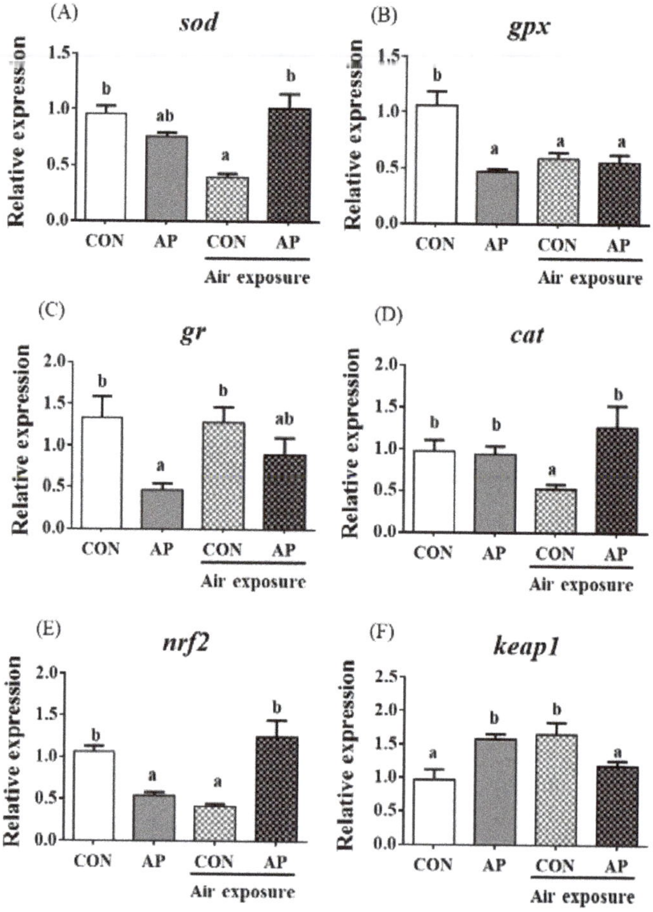

Figure 4. *sod* (**A**), *gpx* (**B**), *gr* (**C**), *cat* (**D**), *nrf2* (**E**), and *keap1* (**F**) gene relative expressions levels of kidney in the two groups before and after air exposure stress. Data are indicated as mean ± SEM ($n = 6$). Bars with different lowercase letters mean significant differences among groups ($p < 0.05$).

The relative expression level of *nrf2* in the kidney was significantly downregulated in the CON group after air exposure stimuli ($p < 0.01$), while the AP diet significantly improved the relative expression level of *nrf2* to protect against oxidative stress caused by air exposure stimuli ($p < 0.05$, Figure 4). The relative expression level of *keap1* in the kidney was significantly upregulated in the CON group after air exposure stimuli ($p < 0.05$), while the AP diet significantly reduced the relative expression level of *keap1* to protect against oxidative stress caused by air exposure ($p < 0.05$, Figure 4).

3.6. Skin Lutein Content and Body Color of Fish

The lutein contents of the Con and SP diets were 4.96 and 8.08 μg g^{-1}, respectively. At the end of the feeding trial, the abdominal and dorsal skin lutein contents increased significantly in fish fed the AP diet ($p < 0.05$, Figure 5A).

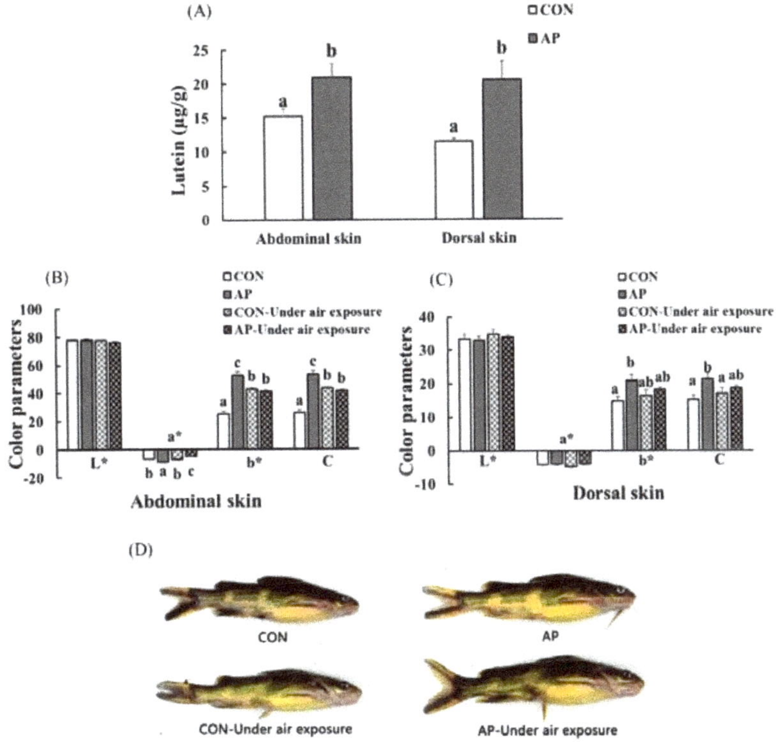

Figure 5. The lutein contents in the skin of yellow catfish fed on different experimental diets (**A**), color parameters of abdominal (**B**) and dorsal skin (**C**), and body color of fish (**D**) in the two groups before and after air exposure stress. Data are indicated as mean ± SEM ($n = 6$). Bars with different lowercase letters mean significant differences among groups ($p < 0.05$).

As shown in Figure 5, there was no significant difference in skin lightness (L*) among all groups. At the end of the experimental period, dietary *A. platensis* supplementation significantly increased the abdominal skin redness value ($p < 0.05$), and air exposure led to a significantly decreased abdominal skin redness value ($p < 0.05$). Skin yellowness and chroma were significantly higher in catfish fed the AP diet at the end of the feeding trial ($p < 0.05$). In the CON group, the abdominal skin yellowness and chroma values of fish were significantly increased after air exposure, while the abdominal skin yellowness and chroma values of fish were significantly decreased in the AP group after air exposure

($p < 0.05$), and fish fed the AP diet had clearer black spots and saturated yellow coloration both before and after air exposure.

3.7. Skin Yellowness Correlation Analysis

A positive and strong correlation between skin yellowness and skin lutein contents was detected ($p < 0.05$, Table 4). We found a negative but not very strong correlation between skin yellowness and plasma LD content ($p < 0.05$, Table 4). Abdominal skin yellowness presented a positive but not significant correlation with liver *sod* and *gr* relative mRNA levels, while dorsal skin yellowness presented a very positive and significant correlation with liver *sod* and *gr* relative mRNA levels ($p < 0.05$, Table 4). Positive and significant correlations were observed between abdominal skin yellowness and kidney *gpx* relative mRNA levels ($p < 0.05$), while positive but nonsignificant correlations were observed between dorsal skin yellowness and kidney *gpx* relative mRNA levels (Table 4). Abdominal skin yellowness presented a positive but not significant correlation with kidney *gr* relative mRNA levels, while dorsal skin yellowness presented a strong positive and significant correlation with kidney *gr* relative mRNA levels ($p < 0.05$, Table 4).

Table 4. Correlation coefficients (r) between yellow catfish skin yellowness (b*-values) and skin lutein and oxidative stress response.

	Yellowness (b*)-Values	
	Abdominal Skin	**Dorsal Skin**
Dorsal skin lutein content	0.82 *	0.87 *
Abdominal skin lutein content	0.90 **	0.67 *
Plasma LD content	−0.59 *	−0.66 *
Liver *sod* relative mRNA level	0.76	0.95 *
Liver *gr* relative mRNA level	0.91	0.99 **
Kidney *gpx* relative mRNA level	0.96 *	0.99
Kidney *gr* relative mRNA level	0.80	0.98 *

* $p < 0.05$; ** $p < 0.01$.

4. Discussion

The present study supported that air exposure induced oxidative stress in fish. Oxidative stress usually causes health and quality problems in fish. Our previous research showed that *A. platensis* can enhance the antioxidant and immune response capacities of fish [17,20]. In the present study, we focused on investigating whether dietary *A. platensis* could alleviate oxidative stress and disordered pigmentation caused by air exposure stimulation in fish.

4.1. *A. platensis* Did Not Affect the Growth Performance and Feed Utilization of Fish

The yellow catfish is an omnivorous freshwater fish [20], which, when fed on diets supplemented with *A. platensis,* did not exhibit differences in growth and feed efficiency. These results are in agreement with a previous study [17]. The previous study also indicated that the dietary inclusion of 25 g kg^{-1} *A. platensis* did not influence the growth of the great sturgeon *Huso huso*; however, higher content of diet *A. platensis* could effectively improve growth performance [27]. We demonstrated that *A. platensis* in the diets of yellow catfish resulted in an apparent digestibility coefficient (ADC) of dry matter and protein up to 70% and 90%, respectively [20]. The above results further validated that it is acceptable to supplement *A. platensis* in yellow catfish diets.

4.2. *A. platensis* Enhanced Stress Response

Cortisol is regarded as a primary physiological stress response in fish [28], and glucose and lactic acid will increase during stress responses [29]. In this study, the plasma cortisol levels increased significantly in yellow catfish due to exposure to the air for 5 min, with similar increases in plasma glucose and lactic acid levels. Air exposure is a stress factor for fish, which is in agreement with the previous studies [30,31]. Notably, the present study showed that *A. platensis* supplementation was effective against air exposure-induced stress responses by reducing plasma cortisol and lactic acid levels. In addition, the reduction of stress markers was dose-dependent on the *A. platensis* supplement [14].

4.3. *A. platensis* Enhanced the Antioxidant Capacity of Fish

The antioxidant enzymes SOD and GPX and the nonenzymatic antioxidant GSH play important roles in free radical scavenging to protect against oxidative damages [32]. In our study, dietary *A. platensis* increased the activities of SOD and GPX and GSH contents to protect against oxidative stress caused by air exposure stimuli. *A. platensis* enhanced antioxidant enzyme activities to protect fish from stress [15,33]. β-carotene, phycocyanin, algal polysaccharides, and polyphenol from *A. platensis* might work to activate antioxidant response [34–37].

Nrf2 plays a critical role in antioxidative defense responses [38]. Usually, *nrf2* transcription is inhibited by binding to *keap1*; when the *keap1*/*nrf2* complex is activated, *nrf2* is translocated into antioxidative factors in the nucleus, where it can regulate antioxidant enzymes, such as *sod*, *cat*, and *gpx* [39–41]. In the present study, the gene and protein expression levels of *nrf2* after air exposure were efficiently upregulated by *A. platensis* supplementation. Concurrently, the expression levels of *sod* and *cat* were obviously activated to improve the antioxidant status of yellow catfish, which is in accordance with a previous study [17]. *A. platensis* involves various proteins, which could have biological activities to induce *nrf2* stabilization and antioxidative enzymes [42]. Additionally, another component of *A. platensis*, polyphenol, has been found to modulate Nrf2-mediated antioxidant events [43,44]. Hence, *A. platensis* can activate the Nrf2 signaling pathway and may be advantageous for its use as an antioxidant nutritional supplement in aquatic animals.

4.4. *A. platensis* Improved Body Color Disorder Caused by Oxidative Stress

A. platensis supplementation enhanced the body color of fish at the end of the experimental trial in accord with previous studies [45,46]. It is known that fish need carotenoids to maintain normal body color from feed [47]. *A. platensis* is considered a carotenoid-producing organism, and lutein and zeaxanthin are mainly found in *A. platensis* [48,49]. The carotenoid pigments of *A. platensis* can also enhance the natural mucus layer, which is responsible for maintaining the glowing appearance of skin in fish [45]. The present results are in accordance with previous studies, suggesting that *A. platensis* is a proper color enhancer.

In this study, *A. platensis* addition promoted increased lutein contents in the skin, and we demonstrated that lutein is mainly deposited in the skin of yellow catfish earlier [50]. This study also confirmed the results. The correlational approach used in this study showed that skin yellowness (b*) was significantly associated with skin lutein contents. Lutein belongs to the xanthophylls, which are a type of carotenoid with antitumor and anti-inflammatory activities. Due to their chemical structure being rich in double bonds that provide them with antioxidant properties, lutein can protect other molecules from oxidative stress by turning off singlet oxygen damage through various mechanisms [51].

Skin color change has the potential to be a useful real-time indicator of stress [52]. Our research also found that skin yellowness was associated with plasma lactate. Notably, stress-induced responses of abnormal body color fish were stronger than those of normal fish, and the albino European catfish *Silurus glanis* showed more obvious behavioral and physiological responses to short-term stress induced by a combination of air exposure and novel environmental stressors than pigmented fish [53]. In this study, dietary *A. platensis* improved body color abnormalities after oxidative stress caused by air exposure stimuli, and skin yellowness was associated with the expression levels of some antioxidant genes to varying degrees. The results may suggest that the antioxidant components (such as lutein) in *A. platensis* successfully help fish defend against body color disorders caused by oxidative stress.

5. Conclusions

Our overall results demonstrated that 20 g kg^{-1} dietary *A. platensis* had no negative effects on the growth of fish but could reduce LD and cortisol levels, increase the activities of antioxidant enzymes, upregulate the expression levels of certain antioxidant-related genes, and regulate the Nrf2 signaling pathway, which aimed to protect against oxidative stress caused by air exposure stimuli. In addition, *A. platensis* could alleviate body color disorder caused by air exposure stress by increasing skin lutein contents. In terms of cost-benefit, the inclusion of 20 g kg^{-1} dietary *A. platensis* can enhance the ability of antioxidants and improve the body color of yellow catfish, as a result, the risk of suffering disease and farming losses will be reduced and market value will be enhanced. Moreover, in our previous study, the *A. platensis* can replace fishmeal in the diets of yellow catfish [20]. With global fisheries approaching unsustainable limits, current fishmeal production will inadequately support the cost-effective demands of aquafeeds; *A. platensis* could be a potential protein source to replace fishmeal in the future.

Author Contributions: C.L.: Investigation, methodology, data curation, validation, writing—original draft. H.L.: Conceptualization, supervision, writing—review & editing. X.Z.: Conceptualization, supervision, funding acquisition. D.H.: Investigation. S.X.: Investigation, supervision. J.J.: Investigation. Y.Y.: Methodology. All authors have read and agreed to the published version of the manuscript.

Funding: This study was funded by the National Key R & D Program of China (2018YFD0900400, 2019YFD0900703), the China Agriculture Research System of MOF and MARA (CARS-46), and the National Natural Science Foundation of China (32061133009, U21A20266).

Institutional Review Board Statement: Yellow catfish were obtained from the Dengjia State fish farm, Hubei Province, China. All of the experiments in the study were conducted according to the regulations of the Guide for Care and Use of Laboratory Animals approved by the Animal Care and Use Committee of the Institute of Hydrobiology, Chinese Academy of Sciences (IHB, CAS, Protocol No. 2016-018; Approval code is YP78021; Approval date is 13 March 2017).

Informed Consent Statement: Not applicable.

Data Availability Statement: Data is contained within the article.

Conflicts of Interest: The authors declare no conflict of interest.

References

1. Tort, L. Stress and immune modulation in fish. *Dev. Comp. Immunol.* **2011**, *35*, 1366–1375. [CrossRef] [PubMed]
2. Skrzynska, A.K.; Maiorano, E.; Bastaroli, M.; Naderi, F.; Míguez, J.M.; Martínez-Rodríguez, G.; Mancera, J.M.; Martos-Sitcha, J.A. Impact of air exposure on vasotocinergic and isotocinergic systems in gilthead sea bream (*Sparus aurata*), new insights on fish stress response. *Front. Physiol.* **2018**, *9*, 96. [CrossRef] [PubMed]
3. Barton, B.A.; Iwama, G.K. Physiological changes in fish from stress in aquaculture with emphasis on the response and effects of corticosteroids. *Annu. Rev. Fish Dis.* **1991**, *1*, 3–26. [CrossRef]
4. Arends, R.; Mancera, J.M.; Muñoz, J.L.; Bonga, S.E.W.; Flik, G. The stress response of the gilthead sea bream (*Sparus aurata* L.) to air exposure and confinement. *J. Endocrinol.* **1999**, *163*, 149. [CrossRef] [PubMed]
5. Xu, Z.H.; Regenstein, J.M.; Xie, D.D.; Lu, W.J.; Ren, X.C.; Yuan, J.J.; Mao, L.C. The oxidative stress and antioxidant responses of *Litopenaeus vannamei* to low temperature and air exposure. *Fish Shellfish Immunol.* **2018**, *72*, 564–571. [CrossRef] [PubMed]
6. Jiang, D.L.; Wu, Y.B.; Huang, D.; Ren, X.; Wang, Y. Effect of blood glucose level on acute stress response of grass carp *Ctenopharyngodon idella*. *Fish Physiol. Biochem.* **2017**, *43*, 1433–1442. [CrossRef]
7. Arlinghaus, R.; Hallermann, J. Effects of air exposure on mortality and growth of undersized pikeperch, *Sander lucioperca*, at low water temperatures with implications for catch-and-release fishing. *Fish Manag. Ecol.* **2007**, *14*, 155–160. [CrossRef]
8. Duan, Y.F.; Zhang, J.S.; Dong, H.B.; Wang, Y.; Liu, Q.S.; Li, H. Effect of desiccation and resubmersion on the oxidative stress response of the kuruma shrimp *Marsupenaeus japonicus*. *Fish Shellfish Immunol.* **2016**, *49*, 91–99. [CrossRef]
9. Paital, B. Modulation of redox regulatory molecules and electron transport chain activity in muscle of air breathing fish *Heteropneustes fossilis* under air exposure stress. *J. Comp. Physiol. B* **2014**, *184*, 65–76. [CrossRef]
10. de Oliveira, S.T.L.; Soares, R.A.N.; de Negreiros Sousa, S.M.; Fernandes, W.A.C.; Gouveia, G.V.; da Costa, M.M. Natural products as functional food ingredients for Nile tilapia challenged with *Aeromonas hydrophila*. *Aquacult. Int.* **2020**, *28*, 913–926. [CrossRef]
11. Kumar, V.; Bossier, P. Importance of plant-derived compounds and/or natural products in aquaculture. *Aquafeed* **2018**, *10*, 28–31.
12. Rodriguez, M.G.M.; Pohlenz, C.; Gatlin, D.M., III. Supplementation of organic acids and algae extracts in the diet of red drum *Sciaenops ocellatus*, immunological impacts. *Aquac. Rep.* **2017**, *48*, 1778–1786. [CrossRef]

13. Şimşek, N.; Karadeniz, A.; Karaca, T. Effects of the *Spirulina platensis* and *Panax ginseng* oral supplementation on peripheral blood cells in rats. *Rev. Méd. Vét.* **2007**, *158*, 483–488.
14. Yeganeh, S.; Teimouri, M.; Amirkolaie, A.K. Dietary effects of *Spirulina platensis* on hematological and serum biochemical parameters of rainbow trout (*Oncorhynchus mykiss*). *Res. Vet. Sci.* **2015**, *101*, 84–88. [CrossRef]
15. Abdelkhalek, N.K.; Ghazy, E.W.; Abdel-Daim, M.M. Pharmacodynamic interaction of *Spirulina platensis* and deltamethrin in freshwater fish Nile tilapia, *Oreochromis niloticus*, impact on lipid peroxidation and oxidative stress. *Environ. Sci. Pollut. Res.* **2015**, *22*, 3023–3031. [CrossRef]
16. Gogoi, S.; Mandal, S.C.; Patel, A.B. Effect of dietary *Wolffia arrhiza* and *Spirulina platensis* on growth performance and pigmentation of Queen loach *Botia dario* (Hamilton, 1822). *Aquacult. Nutr.* **2018**, *24*, 285–291. [CrossRef]
17. Liu, C.; Liu, H.K.; Han, D.; Xie, S.Q.; Jin, J.Y.; Yang, Y.X.; Zhu, X.M. Effects of dietary *Arthrospira platensis* supplementation on the growth performance, antioxidation and immune related-gene expression in yellow catfish (*Pelteobagrus fulvidraco*). *Aquac. Rep.* **2020**, *17*, 100297. [CrossRef]
18. Zhang, X.T.; Zhang, X.T.; Zhang, G.R.; Shi, Z.C.; Yuan, Y.J.; Zheng, H.; Lin, L.; Wei, K.J.; Ji, W. Expression analysis of nine Toll-like receptors in yellow catfish (*Pelteobagrus fulvidraco*) responding to *Aeromonas hydrophila* challenge. *Fish Shellfish Immunol.* **2017**, *63*, 384–393. [CrossRef] [PubMed]
19. Liu, F.; Shi, H.Z.; Guo, Q.S.; Yu, Y.B.; Wang, A.M.; Lv, F.; Shen, W.B. Effects of astaxanthin and emodin on the growth, stress resistance and disease resistance of yellow catfish (*Pelteobagrus fulvidraco*). *Fish Shellfish Immunol.* **2016**, *51*, 125–135. [CrossRef] [PubMed]
20. Liu, C.; Liu, H.K.; Xu, W.J.; Han, D.; Xie, S.Q.; Jin, J.Y.; Yang, Y.X.; Zhu, X.M. Effects of dietary *Arthrospira platensis* supplementation on the growth, pigmentation, and antioxidation in yellow catfish (*Pelteobagrus fulvidraco*). *Aquaculture* **2019**, *510*, 267–275. [CrossRef]
21. CIE. *Recommendations on Uniform Color Spaces, Color Difference Equations, Psychometric Color Terms*; Supplement No. 2 to CIE Publication No. 15; Colorimetry; Bureau Central de la CIE; Paris International Commission on Illumination: Paris, France, 1976.
22. Hunt, R.W. The specification of colour appearance. I. Concepts and terms. *Color Res. Appl.* **1977**, *2*, 55–68. [CrossRef]
23. Karadas, F.; Grammenidis, E.; Surai, P.; Acamovic, T.; Sparks, N.H.C. Effects of carotenoids from lucerne, marigold and tomato on egg yolk pigmentation and carotenoid composition. *Br. Poult. Sci.* **2006**, *47*, 561–566. [CrossRef] [PubMed]
24. Moros, E.; Darnoko, D.; Cheryan, M.; Perkins, E.G.; Jerrell, J. Analysis of xanthophylls in corn by HPLC. *J. Agric. Food Chem.* **2002**, *50*, 5787–5790. [CrossRef] [PubMed]
25. Hsu, Y.W.; Tsai, C.F.; Chen, W.K.; Ho, Y.C.; Lu, F.J. Determination of lutein and zeaxanthin and antioxidant capacity of supercritical carbon dioxide extract from daylily (*Hemerocallis disticha*). *Food Chem.* **2011**, *129*, 1813–1818. [CrossRef]
26. Vandesompele, J.; Preter, K.D.; Pattyn, F.; Poppe, B.; Roy, N.V.; Paepe, A.D.; Speleman, F. Accurate normalization of real-time quantitative RT-PCR data by geometric averaging of multiple internal control genes. *Genome Biol.* **2002**, *3*, 0034.1–0034.11. [CrossRef]
27. Adel, M.; Yeganeh, S.; Dadar, M.; Sakai, M.; Dawood, M.A.O. Effects of dietary *Spirulina platensis* on growth performance, humoral and mucosal immune responses and disease resistance in juvenile great sturgeon (*Huso huso* Linnaeus, 1754). *Fish Shellfish Immunol* **2016**, *56*, 436–444. [CrossRef]
28. Barton, B.A. Salmonid fishes differ in their cortisol and glucose responses to handling and transport stress. *N. Am. J. Aquacult.* **2000**, *62*, 12–18. [CrossRef]
29. Tandon, R.; Joshi, B. Blood glucose and lactic acid levels in the fresh water fish, *Heteropneustes Fossilis*, following Stress 1. *Z. Für Tierphysiol. Tierernährung Futterm.* **1973**, *31*, 210–216. [CrossRef]
30. Lim, H.K.; Hur, J.W. Effects of acute and chronic air exposure on growth and stress response of juvenile olive flounder, *Paralichthys olivaceus*. *Turk. J. Fish. Aquat. Sci.* **2018**, *18*, 143–151. [CrossRef]
31. Hur, J.W.; Kang, K.H.; Kang, Y.J. Effects of acute air exposure on the hematological characteristics and physiological stress response of olive flounder (*Paralichthys olivaceus*) and Japanese croaker (*Nibea japonica*). *Aquaculture* **2019**, *502*, 142–147. [CrossRef]
32. Wu, P.; Zhang, L.; Jiang, W.; Liu, Y.; Jiang, J.; Kuang, S.; Li, S.; Tang, L.; Tang, W.; Zhou, X.; et al. Dietary vitamin A improved the flesh quality of grass carp (*Ctenopharyngodon idella*) in relation to the enhanced antioxidant capacity through Nrf2/Keap 1a signaling pathway. *Antioxidants* **2022**, *11*, 148. [CrossRef] [PubMed]
33. Sheikhzadeh, N.; Mousavi, S.; Oushani, A.K.; Firouzamandi, M.; Mardani, K. *Spirulina platensis* in rainbow trout (*Oncorhynchus mykiss*) feed, effects on growth, fillet composition, and tissue antioxidant mechanisms. *Aquacult. Int.* **2019**, *27*, 1613–1623. [CrossRef]
34. Cao, S.P.; Zhang, P.Y.; Zou, T.; Fei, S.Z.; Han, D.; Jin, J.Y.; Liu, H.K.; Yang, Y.X.; Zhu, X.M.; Xie, S.Q. Replacement of fishmeal by spirulina *Arthrospira platensis* affects growth, immune related-gene expression in gibel carp (*Carassius auratus gibelio* var. CAS III), and its challenge against *Aeromonas hydrophila* infection. *Fish Shellfish Immunol.* **2018**, *79*, 265–273. [CrossRef] [PubMed]
35. Casazza, A.A.; Ferrari, P.F.; Aliakbarian, B.; Converti, A.; Perego, P. Effect of UV radiation or titanium dioxide on polyphenol and lipid contents of *Arthrospira* (*Spirulina*) platensis. *Algal Res.* **2015**, *12*, 308–315. [CrossRef]
36. Li, X.L.; Xu, G.; Chen, T.F.; Wong, Y.S.; Zhao, H.L.; Fan, R.R.; Gu, X.M.; Tong, P.C.; Chan, J.C. Phycocyanin protects INS-1E pancreatic beta cells against human islet amyloid polypeptide-induced apoptosis through attenuating oxidative stress and modulating JNK and p38 mitogen-activated protein kinase pathways. *Int. J. Biochem. Cell Biol.* **2009**, *41*, 1526–1535. [CrossRef]

37. Hassaan, M.S.; Hassaan, M.S.; Mohammady, E.Y.; Soaudy, M.R.; Sabae, S.A.; Mahmoud, A.M.A.; El-Haroun, E.R. Comparative study on the effect of dietary β-carotene and phycocyanin extracted from *Spirulina platensis* on immune-oxidative stress biomarkers, genes expression and intestinal enzymes, serum biochemical in Nile tilapia, *Oreochromis niloticus*. *Fish Shellfish Immunol.* **2021**, *108*, 63–72. [CrossRef]
38. Zou, B.; Xiao, G.S.; Xu, Y.J.; Wu, J.J.; Yu, Y.S.; Fu, M.Q. Persimmon vinegar polyphenols protect against hydrogen peroxide-induced cellular oxidative stress via Nrf2 signalling pathway. *Food Chem.* **2018**, *255*, 23–30. [CrossRef]
39. Esam, F.; Esam, F.; Khalafalla, M.M.; Gewaily, M.S.; Abdo, S.; Hassan, A.M.; Dawood, M.A.O. Acute ammonia exposure combined with heat stress impaired the histological features of gills and liver tissues and the expression responses of immune and antioxidative related genes in Nile tilapia. *Ecotoxicol. Environ. Saf.* **2022**, *231*, 113187. [CrossRef]
40. Chen, X.L.; Kunsch, C. Induction of cytoprotective genes through Nrf2/antioxidant response element pathway, a new therapeutic approach for the treatment of inflammatory diseases. *Curr. Pharm. Des.* **2004**, *10*, 879–891. [CrossRef]
41. Nguyen, T.; Nioi, P.; Pickett, C.B. The Nrf2-antioxidant response element signaling pathway and its activation by oxidative stress. *J. Biol. Chem.* **2009**, *284*, 13291–13295. [CrossRef]
42. Okamoto, T.; Kawashima, H.; Osada, H.; Toda, E.; Homma, K.; Nagai, N.; Imai, Y.Y.; Tsubota, K.; Ozawa, Y. Dietary Spirulina supplementation protects visual function from photostress by suppressing retinal neurodegeneration in mice. *Transl. Vis. Sci. Technol.* **2019**, *8*, 20. [CrossRef] [PubMed]
43. Na, H.K.; Surh, Y.J. Modulation of Nrf2-mediated antioxidant and detoxifying enzyme induction by the green tea polyphenol EGCG. *Food Chem. Toxicol.* **2008**, *46*, 1271–1278. [CrossRef]
44. Martín, M.A.; Ramos, S.; Granado-Serrano, A.B.; Rodríguez-Ramiro, I.; Trujillo, M.; Bravo, L.; Goya, L. Hydroxytyrosol induces antioxidant/detoxificant enzymes and Nrf2 translocation via extracellular regulated kinases and phosphatidylinositol-3-kinase/protein kinase B pathways in HepG2 cells. *Mol. Nutr. Food Res.* **2010**, *54*, 956–966. [CrossRef] [PubMed]
45. Roohani, A.M.; Kenari, A.A.; Kapoorchali, M.F.; Borani, M.S.; Zoriezahra, S.J.; Smiley, A.H.; Esmaeili, M.; Rombenso, A.N. Effect of spirulina *Spirulina platensis* as a complementary ingredient to reduce dietary fish meal on the growth performance, whole-body composition, fatty acid and amino acid profiles, and pigmentation of Caspian brown trout (*Salmo trutta caspius*) juveniles. *Aquacult. Nutr.* **2019**, *25*, 633–645.
46. Teimouri, M.; Amirkolaie, A.K.; Yeganeh, S. The effects of *Spirulina platensis* meal as a feed supplement on growth performance and pigmentation of rainbow trout (*Oncorhynchus mykiss*). *Aquaculture* **2013**, *396–399*, 14–19. [CrossRef]
47. Habib, B.; Parvin, M.; Huntington, T.C.; Hasan, M.R. *A Review on Culture, Production and Use of Spirulina as Food for Humans and Feeds for Domestic Animals and Fish*; Food & Agriculture Organization of the United Nations: Rome, Italy, 2008; p. 41.
48. Leema, J.T.M.; Kirubagaran, R.; Vinithkumar, N.V.; Dheenan, P.S.; Karthikayulu, S. High value pigment production from *Arthrospira* (*Spirulina*) platensis cultured in seawater. *Bioresour. Technol.* **2010**, *101*, 9221–9227. [CrossRef] [PubMed]
49. Tudor, C.; Gherasim, E.C.; Dulf, F.V.; Pintea, A. In vitro bioaccessibility of macular xanthophylls from commercial microalgal powders of *Arthrospira platensis* and *Chlorella pyrenoidosa*. *Food Sci. Nutr.* **2021**, *9*, 1896–1906. [CrossRef]
50. Liu, C.; Li, Y.H.; Chen, Z.; Yuan, L.; Liu, H.K.; Han, D.; Jin, J.Y.; Yang, Y.X.; Hu, Q.; Zhu, X.M.; et al. Effects of dietary whole and defatted *Arthrospira platensis* (Cyanobacterium) on growth, body composition and pigmentation of the yellow catfish *Pelteobagrus fulvidraco*. *J. Appl. Phycol.* **2021**, *33*, 2251–2259. [CrossRef]
51. Pereira, A.G.; Otero, P.; Echave, J.; Carreira-Casais, A.; Chamorro, F.; Collazo, N.; Jaboui, A.; Lourenço-Lopes, C.; Simal-Gandara, J.; Prieto, M.A. Xanthophylls from the sea, algae as source of bioactive carotenoids. *Mar. Drugs* **2021**, *19*, 188. [CrossRef]
52. Tveit, G.M.; Anders, N.; Bondø, M.S.; Mathiassen, J.R.; Breen, M. Atlantic mackerel (*Scomber scombrus*) change skin colour in response to crowding stress. *J. Fish Biol.* **2022**, *100*, 738–747. [CrossRef]
53. Slavík, O.; Horký, P.; Valchářová, T.; Pfauserová, N.; Velíšek, J. Comparative study of stress responses, laterality and familiarity recognition between albino and pigmented fish. *Zoology* **2022**, *150*, 125982. [CrossRef] [PubMed]

antioxidants

MDPI

Article

Vitamin C Attenuates Oxidative Stress, Inflammation, and Apoptosis Induced by Acute Hypoxia through the Nrf2/Keap1 Signaling Pathway in Gibel Carp (*Carassius gibelio*)

Liyun Wu [1,2], Wenjie Xu [1], Hongyan Li [1], Bo Dong [1,2], Hancheng Geng [1,2], Junyan Jin [1,*], Dong Han [1,2], Haokun Liu [1], Xiaoming Zhu [1], Yunxia Yang [1] and Shouqi Xie [1,2,3]

1 State Key Laboratory of Freshwater Ecology and Biotechnology, Institute of Hydrobiology, Chinese Academy of Sciences, Wuhan 430072, China; wuliyun@ihb.ac.cn (L.W.); wenjie@ihb.ac.cn (W.X.); lihongyan@prfri.ac.cn (H.L.); dongbo@ihb.ac.cn (B.D.); genghancheng@ihb.ac.cn (H.G.); hand21cn@ihb.ac.cn (D.H.); liuhaokun@ihb.ac.cn (H.L.); xmzhu@ihb.ac.cn (X.Z.); yxyang@ihb.ac.cn (Y.Y.); sqxie@ihb.ac.cn (S.X.)

2 University of Chinese Academy of Sciences, Beijing 100049, China

3 The Innovative Academy of Seed Design, Chinese Academy of Sciences, Beijing 100101, China

* Correspondence: jinjunyan@ihb.ac.cn

Abstract: Previous studies have found that vitamin C (VC) has protective effects in fish. However, the efficacy of VC on hypoxia-induced liver injury in fish remains unknown. Therefore, to investigate the protective mechanism of VC on liver injury after acute hypoxic stimulation in fish, gibel carp were fed a diet containing VC for eight weeks, then were subjected to acute hypoxia stimulation. The specific growth rate of fish was increased by the supplementation of VC. Plasma stress markers (glucose, lactic acid, and cortisol) were decreased by the VC supplementation. Moreover, the levels of the inflammatory cytokines (*tnf-α*, *il-2*, *il-6*, and *il-12*) were increased by enhancing the Nrf2/Keap1 signaling pathway. Upregulation of the antioxidant enzymes activity (CAT, SOD, and GPx); T-AOC; and anti-inflammatory factors (*il-4* and *tgf-β*) highlighted the antioxidant and anti-inflammatory activities of VC. The results showed that VC reduced the apoptotic index of the fish hypothalamus. The expression of GRP78 protein in the liver and endoplasmic reticulum stress and apoptosis induced by hypoxia were inhibited by VC. Taken together, the results indicate that VC can attenuate oxidative damage, inflammation, and acute hypoxia induced apoptosis in gibel carp via the Nrf2/Keap1 signaling pathway. The results identify a new defense strategy of gibel carp in response to hypoxic conditions.

Keywords: hypoxia; vitamin C; Nrf2/Keap1 signaling pathway; oxidative stress; inflammation; apoptosis

Citation: Wu, L.; Xu, W.; Li, H.; Dong, B.; Geng, H.; Jin, J.; Han, D.; Liu, H.; Zhu, X.; Yang, Y.; et al. Vitamin C Attenuates Oxidative Stress, Inflammation, and Apoptosis Induced by Acute Hypoxia through the Nrf2/Keap1 Signaling Pathway in Gibel Carp (*Carassius gibelio*). *Antioxidants* **2022**, *11*, 935. https://doi.org/10.3390/antiox11050935

Academic Editors: Min Xue, Junmin Zhang, Zhenyu Du, Jie Wang and Wei Si

Received: 15 April 2022
Accepted: 6 May 2022
Published: 9 May 2022

1. Introduction

In mammals, oxygen is critical for survival, and it is an indispensable substrate for cell metabolism, energy balance, and signal transmission. Hypoxia-inducible factor-1α (HIF-1α), a major regulator of oxygen homeostasis, is highly expressed under hypoxic conditions [1]. Hypoxic conditions trigger the sharp increase of reactive oxygen species and oxidative stress, resulting in protein, DNA, and lipid damage, mitochondrial dysfunction, and apoptosis [2]. HIF can mediate the expression of inflammatory cytokines in brain and renal ischemic diseases, an important factor in resisting inflammatory infection [3,4].

Reactive oxygen species (ROS) formation can be affected by changes in oxygen concentration, causing oxidative stress when the levels exceed the removal ability of defense mechanisms [5]. ROS have been shown to regulate the active expression of HIF [6]. The liver is the main site of ROS production and the primary target of injury. ROS are critical in signal transduction, but excess ROS can trigger oxidative stress and disrupt intracellular homeostasis. Increased or decreased activity of antioxidant defense enzymes, such as superoxide dismutase (SOD), catalase (CAT), and glutathione peroxidase (GPx), can be used to maintain organism balance and reduce cell damage [7,8]. As one of the most important cellular defense mechanisms, nuclear factor erythroid 2-related factor (Nrf2) is a key transcription factor regulating the expression of genes involved in the antioxidant and anti-inflammatory responses. The key pathway of cellular resistance to oxidative stress was found to be Nrf2 (nuclear factor erythroid 2-related factor 2)/Keap1 (kelch-like ECH-associated protein 1 [9]. The Nrf2/Keap1 pathway has detoxification and neutralization effects, and many antioxidant enzymes regulated by the Nrf2/Keap1 pathway can remove reactive oxygen species; this not only resists oxidative damage from the external environment but also enhances the antioxidant capacity of the body [10]. It has been reported that resveratrol alleviates endoplasmic reticulum (ER) stress and apoptosis in brain tissue induced by explosion injury in mice through the Nrf2/Keap1 pathway [11]. In lipopolysaccharide-induced acute lung injury in mice, hispolon inhibited ER stress-mediated apoptosis and autophagy by regulating the Nrf2/Keap1 pathway [12]. In addition, studies on Wuchang bream (*Megalobrama amblycephala*) and gibel carp (*Carassius gibelio*) showed that emodin can mediate the Nrf2/Keap1 signaling pathway to improve the body's antioxidant capacity and protect it from oxidative stress [13,14]. In recent years, a large amount of evidence has demonstrated that the Nrf2/Keap1 pathway is activated in various cellular and animal models of oxidative stress injury and is involved in tissue homeostasis and inflammatory responses [15].

An anoxic environment is closely related to the ER stress response. The ER has the structural and functional characteristics of organelles that can trigger stress responses and activate the unfolded protein (UPR) signaling pathway under hypoxia. It has been reported that the UPR primarily protects against oxidative stress damage caused by hypoxia through triggering autophagy. This mechanism attempts to rescue cells from damage from ER stress, and when this pathway is absent or disabled, cells are more sensitive to hypoxia-induced death [16]. In human cancer cells, severe hypoxia and ER stress trigger the transcription of autophagy-related genes by activating transcription factor 4 (ATF4) expression, and this upregulation is crucial for hypoxia and metabolic stress in tumors [17].

As an essential micronutrient and powerful antioxidant, vitamin C (VC) has been shown to have multiple biological properties that regulate antioxidant, antiaging, and anti-inflammatory activities and mediate cell death pathways [18]. Previous studies have shown that VC protects against chemically induced liver damage in mice [19]. In addition, VC deficiency can increase the levels of proinflammatory factors in grass carp and reduce the levels of anti-inflammatory factors, such as interleukin (IL) 10 and transforming growth factor (TGF) 1β. In addition, VC also promoted the expression of caspase-3, caspase-7, and caspase-9 and aggravated the apoptosis of grass carp (*Ctenopharyngodon idella*) cells [20]. It has been reported that VC can improve the antioxidant, antiapoptotic, and immune abilities of the abalone *Haliotis discus hannai* Ino [21].

When aquatic animals are starved of oxygen, they exhibit metabolic disorders and oxidative damage, ultimately leading to death [22]. Therefore, reducing tissue oxidative damage caused by hypoxia is the key to improving the survival of aquatic animals. Gibel carp (*Carassius gibelio*) is an economically important freshwater fish in China. Gibel carp may experience hypoxia in intensive culture; this not only harms the ecosystem and affects the diversity of aquatic resources, but it also causes large-scale asphyxia death of cultured fish. To date, there are no reports concerning whether VC treatment can alleviate acute hypoxic stimulation in gibel carp. In addition, the relationship between VC and the Nrf2/Keap1 pathway in response to acute hypoxia in gibel carp remains unknown. The purpose of this study was to explore whether dietary VC could alleviate acute hypoxia stress suffered by gibel carp. To this end, we performed a systematic study of the response of gibel carp to hypoxia from the aspects of immune response, endoplasmic reticulum stress, autophagy, and apoptosis, and the regulation of the Nrf2/Keap1 pathway in order to provide effective strategies for alleviating the fish response to hypoxic stress.

2. Materials and Methods

2.1. Ethical Statement

Procedures related to animal treatment in this study were approved by the ethics committee of the Institute of Hydrobiology, Chinese Academy of Sciences. All of the experiments were strictly in accordance with the Guidelines for the Management and Use of Laboratory Animals.

2.2. Experimental Diets

White fishmeal, rapeseed meal, and soybean meal were used as the main protein sources, and fish oil and soybean oil (1:1) were used as lipid sources to prepare two isonitrogenous and isolipic experimental diets; 1200 mg/kg VC was added to the basal diet (CON). VC (purity: 35%) was purchased from Guangzhou Liankun Biotechnology Co., Ltd. (Guangdong, China) and was added to the formula in the form of L-ascorbyl-2 monophosphate. The formulation and basic chemical composition of the diets are listed in Table 1. All of the ingredients were crushed through a 40-mesh sieve and thoroughly mixed according to the relevant formula ratio. Finally, oil was added and mixed. After mixing with water, a granulating machine (SLR-45, Fishery Machinery and Instrument Research Institute, Chinese Academy of Fishery Sciences, Shanghai, China) was used for granulating the feed, and the feed was then stored in a refrigerator at 4 °C.

Table 1. Ingredients and proximate composition of the experimental diets (% dry matter).

Ingredients	CON	VC
White fish meal [1]	15	15
Rapeseed meal [2]	20	20
Soybean meal [2]	25	25
Wheat flour	25.6	25.6
Oil mixture [3]	5.5	5.5
Vitamin C	0	0.12
Vitamin premix [4]	0.39	0.39
Choline chloride	0.11	0.11
Mineral premix [5]	5	5
Carboxy methyl cellulose sodium	3	3
Cellulose	0.40	0.28
Proximate analysis (dry matter)		
Crude protein (%)	37.21	37.42
Crude lipid (%)	6.77	6.63
Moisture (%)	9.08	9.17
Ash (%)	10.02	10.34
Gross energy (kJ g^{-1})	19.27	19.22

[1] White fish meal: Purchased from American Seafood Company, Seattle, WA, USA. [2] Soybean and rapeseed meal: Purchased from Coland Feed Co. Ltd., Wuhan, Hubei, China. [3] Oil mixture: soybean oil: fish oil = 1:1. [4] Vitamin premix (mg kg^{-1} diet): Vitamin B1, 20; Vitamin B2, 20; Vitamin B6, 20; Vitamin B12, 0.02; folic acid, 5; calcium pantothenate, 50; inositol, 100; niacin, 100; biotin, 0.1; cellulose, 3522; Vitamin C, 100; Vitamin A, 110; Vitamin D, 20; Vitamin E, 50; Vitamin K, 10. [5] Mineral salt premix (mg kg^{-1} diet): NaCl, 500.0; $MgSO_4 \cdot 7H_2O$, 8155.6; $NaH_2PO_4 \cdot 2H_2O$, 12500.0; KH_2PO_4, 16000; $Ca(H_2PO_4) \cdot 2H_2O$, 7650.6; $FeSO_4 \cdot 7H_2O$, 2286.2; $C_6H_{10}CaO_6 \cdot 5H_2O$, 1750.0; $ZnSO_4 \cdot 7H_2O$, 178.0; $MnSO_4 \cdot H_2O$, 61.4; $CuSO_4 \cdot 5H_2O$, 15.5; $CoSO_4 \cdot 7H_2O$, 0.91; KI, 1.5; Na_2SeO_3, 0.60; Corn starch, 899.7.

2.3. Experimental Fish and Feeding Management

All of the gibel carp used in this experiment were provided by the Institute of Hydrobiology, Chinese Academy of Sciences (Wuhan, Hubei). Before the formal experiment, the experimental fish were temporarily raised in a tank and fed a temporary diet (36% crude protein and 7% crude lipid) for 14 days to adapt to the experimental conditions. Before the start of the experiment, the experimental fish were removed after being starved for 24 h, and healthy fish with similar specifications (6.66 ± 0.01 g) were randomly assigned to indoor fiberglass tanks for the experiment, with three tanks for each treatment and 30 fish in each tank. During the experiment, the experimental fish were fed to apparent satiety at 8:30, 13:30 and 18:30. The water temperature was measured every day; the fluctuation range was 29–31 °C. Dissolved oxygen, ammonia nitrogen, and pH were monitored weekly, with dissolved oxygen >6.0 mg/L, total ammonia nitrogen <0.1 mg/L, and pH 7.0–7.4. The aquaculture experiment lasted for 56 days.

2.4. Hypoxia Stress Test

After the eight-week feeding period, fish were randomly assigned to a CON group (control group), VC group (vitamin C group), HYPO group (hypoxia group), or HYPO + VC group (hypoxia + vitamin C group). During the hypoxia stress experiment, six healthy and uniform gibel carp were selected from each treatment and placed in a 1 L transparent plastic tank. The oxygen concentration of the hypoxia workstation (Ruskinn INVIVO$_2$ 400, Beijing Longfujia Biotechnology Co., Ltd, Beijing, China) was adjusted to 2% to create a closed hypoxic environment, and the fish were moved into the hypoxia workstation for close monitoring. Before the stress test began, the oxygen concentration (6.14 ± 0.09 mg/L) in the fish tank was measured with an LDO101 probe (HQ30D, HACH). After being stressed in the hypoxic incubator for 3 h, the oxygen concentration in the cylinder was measured again; the oxygen concentration at this point was 0.08 ± 0.02 mg/L.

2.5. Sampling Procedures

After the feeding period, the fish were starved for 24 h, and the contents of the digestive tract were emptied. Then, all of the fish in each tank were weighed. Two fish with similar measurements were selected from each tank to be wiped dry, weighed, and stored at −20 °C for body composition analysis. Then, two fish were randomly chosen from each cylinder and put into a hypoxia incubator for three hours. Blood was taken from the caudal vein using a syringe soaked with heparin sodium (concentration: 0.2%), put into an anticoagulation centrifuge tube, and centrifuged (3000× *g* at 4 °C) for 15 min. The supernatant plasma was taken and stored in a refrigerator at −80 °C for the determination of plasma glucose, lactic acid, and cortisol. After the blood was drawn, the fish was dissected, and the liver tissues samples were taken, wrapped in sterilized tin foil paper, and immediately put into liquid nitrogen for subsequent analyses. In addition, brain tissue (hypothalamus) was collected and promptly placed in 4% paraformaldehyde fixative for histological analysis.

2.6. Biochemical Parameters

In this experiment, the determination of basic composition of all of the feed and fish body end samples was strictly in accordance with the AOAC (2003) standard methods [23]. The specific operation process was carried out according to Li et al. (2019) [24]. The moisture content was determined by the weight loss method. The ash content was determined after calcination in a muffle furnace (Hubei, China) at 550 °C for 3 h. Crude protein was determined by the Kjeldahl method (FOSS Tecator, Haganas, Sweden). Crude lipid was extracted and determined by a Soxtec system (Soxtec System HT6, Tecator, Haganas, Sweden). Gross energy was measured using an oxygen bomb calorimeter (Calorimeter, Parr instrument Company, Moline, IL, USA).

The liver enzyme activity needed to be measured before the experimental treatment. Physiological saline was added to liver samples according to the ratio of weight (g): volume (mL) = 1:9. After centrifugation, the supernatant was removed for later enzyme activity analysis. Protein concentration in a liver homogenate was determined by the Coomassie bright blue method (Bradford, P0006, Beyotime, Shanghai, China). Plasma lactic acid (LD) and total anti-oxidant capacity (T-AOC), superoxide dismutase (SOD), catalase (CAT), reduced glutathione (GSH), glutathione peroxidase (GPX), and the concentration of malondialdehyde (MDA) were determined using commercial kits (Nanjing Jiancheng Bioengineering Institute, Catalog: A019-2-1, A015-2, A001-3-2, A007-1-1, A006-2-1, and A005-1-2 and A003-1-2). The activities of Caspase 3 (C1116) and Caspase 9 (C1158) in the liver were detected using commercial kits (Beytime Biotechnology, Shanghai, China). The content of plasma cortisol was detected by enzyme linked immunosorbent assay (ELISA) (H094) provided by the Nanjing Jiancheng Bioengineering Research Institute. The principle of the method is to detect the content of cortisol in the sample by competition, i.e., adding the sample to the enzyme-labeled wells containing antibodies, then adding biotin-labeled recognition antigens that competed with the solid antibodies to form an immune complex, then adding avidin-HRP after washing, thus producing a yellow color under the action of acid with an absorption peak at a wavelength of 450 nm. Plasma glucose was determined using commercial kits (Fujifilm, Wako Pure Chemical Corporation, Osaka, Japan). The plasma samples were placed on ice, and a 96-well microenzyme assay was used. The plasma samples of 2 μL and different standard liquid concentrations were added into the 300-μL reaction mixture and mixed evenly. The plasma samples were placed in an incubator at 37 °C for 5 min away from light, and then the absorbance was read at 505 nm with a microenzyme reader.

2.7. Tissue Total RNA Extraction and Real-Time Fluorescence Quantitative PCR

In this experiment, the mRNA expression level of the liver tissue was analyzed, and the total RNA was extracted using the Trizol reagent (Invitrogen, Carlsbad, CA, USA). The detection of total RNA integrity primarily relied on 1.0% agarose electrophoresis, and then, 1 μL was taken to detect the purity and concentration of the sample by NanoDrop®ND-2000 ultra-micro spectrophotometry (NanoDrop Technologies, Wilmington, DE, USA). The reverse transcription was performed using an M-MLV First-Strand Synthesis System according to the manufacturer's instructions. Real-time fluorescence quantification was performed on a LightCycle 480 II (Roche Diagnostics, Basel, Switzerland) instrument to determine the amplification efficiency in a pre-experiment and to make standard curves using internal reference and target genes. SYBR Green I Master Mix (Roche Diagnostics, Indianapolis, IN, USA) fluorescent staining was used to determine the expression levels of all of the target genes. The calculation method of the relative expression refers to Pfaffl [25]. In this study, three housekeeping genes were employed initially: *β-actin*, *tubulin*, and *ef-1α*. After the analysis, it was determined that *ef-1α* had the highest stability, and thus, it was selected as a housekeeping gene. The target gene primers used in the experiment refer to a previous study [14], and the other primers are listed in Table 2.

Table 2. Sequences of primers applied for quantitative real-time PCR analysis in gibel carp.

Gene Name	Sense and Antisense Primer (5′–3′)	Gene Bank Accession No.	Product Length (bp)
Tumor necrosis factor-α (*tnf-α*)	TTGAGCAGGAGATGGGAACCG AGAGCCTCAGGGCAACGGAAA	XM_026282152.1	115
Interleukin-2 (*il-2*)	GACCACAAAGGTAGACCCATCC GAGGTTTGTGCGGAATGGAC	MN338056	212
Interleukin-6 (*il-6*)	TGTTCTCAGGGCATTCGCTT GGAGTTGTAGTGCCCTTGGT	XM_026289280.1	161
Interleukin-12 (*il-12*)	CTTCAGAAGCAGCTTTGTTGTTG CAGTTTTTGAGAGCTCACCAATATC	LN592213.1	77
Interleukin-1β (*il-1β*)	TTTGTGAAGATGCGCTGCTC CCAATCTCGACCTTCCTGGTG	AB757758.1	133
Interleukin-4 (*il-4*)	CGATTGTAGCCGTTACTGGGT TGGCAAATGTGTTCCTCCG	KX574595	166
Transforming growth factorβ (*tgf-β*)	ATGAGGGTGGAGAGTTTAT AGTCGTAGTTTGCTGAGAA	EU086521.1	155
Nuclear factor of kappa light polypeptide gene enhancer in B-cells inhibitor, alpha (*iκbα*)	TTGCGAATCCAAAGGGGACA TCTGTGATGACGGCGAGATG	XM_026291433.1	196

2.8. TUNEL Staining

Fresh brain tissue (hypothalamus) was immobilized in 4% paraformaldehyde solution and then made into paraffin sections. The sections were repaired and permeated with protease K (2 mg/mL) and Trition-X-100/PBS (0.2%) solutions, and the TUNEL reagent was mixed with appropriate reagents 1 (TdT) and 2 (dUTP) at a ratio of 2:29 according to the tissue size, then incubated at 37 °C for 2 h. Slices were washed with PBS three times, each time for 5 min, and then DAPI dye solution (4′,6-diamidino-2-phenylindole, 0.3 mmol/L) was added and incubated at room temperature in the dark for 10 min. The sections were examined under a fluorescence microscope, and images were collected. Three fields were randomly selected for each section to be photographed. Image-Pro Plus 4.1 software was used to count and analyze the apoptosis rate calculated as follows:

Apoptosis rate (%) = The number of apoptotic nuclei/The number of observed nuclei × 100%.

2.9. Western Blot Analysis

The protein levels of HIF-1α, Nrf2, Keap1, and BiP/GRP78 in the liver tissue after hypoxic stress were detected by Western blotting. The liver samples were weighed and put into 1.5-mL centrifuge tubes, and protein lysate was added in proportion. The lysate consisted of a protease inhibitor, phosphatase inhibitor (Roche, Basel, Switzerland), and RIPA cleavage buffer (Beyotime Biotechnology, Shanghai, China). Then, ultrasonic crushing homogenization was carried out on ice, followed by centrifugation at 4 °C for 10 min at 13,000× *g*, and the supernatant was taken for later use. The sample protein concentration was measured according to the instructions of a BCA protein quantitative kit (Beyotime Biotechnology, Shanghai, China), and 50% RIPA lysate was used to adjust the protein concentration of all of the liver tissue samples to be consistent. The protein samples were then separated and transferred to polyvinylidene fluoride (PVDF) membranes after sodium dodecyl sulfate-polyacrylamide gel electrophoresis (SDS-PAGE). TBST was prepared with 5% skim milk powder at room temperature, and the PVDF membranes were sealed for 1 h. The specific primary antibody was then incubated at 4 °C for 12 h, followed by shaking cleaning with a TBST room temperature shaker three times for 20 min each time. Specific antibodies HIF-1α, Nrf2, BiP/GRP78, and GAPDH refer to a previous study [14]. Keap1 antibody (ab119403, Abcam) was incubated with a secondary antibody at room temperature for 2 h. The signal intensity was measured using an imager, and protein bands were obtained. The optical density of the protein bands was quantified using ImageJ software (National Institutes of Health, Bethesda, MD, USA). All of the values were normalized to the internal control (GAPDH) values.

2.10. Statistical Analysis

All of the data were verified for normality and homogeneity of variance before analysis using SPSS (SPSS Inc., Chicago, IL, USA) software. Data were expressed as the mean ± SE (standard error), and p-values < 0.05 or <0.01 were considered to be significant and extremely significant differences, respectively. The one-way ANOVA used in this experiment targeted the CON and the HYPO groups, as well as the HYPO + VC group.

3. Results

3.1. Growth Results and Body Composition

The results of the growth performance of gibel carp are shown in Table 3. The final body weight and specific growth rate of gibel carp fed the VC diet were significantly higher than those of the control group after eight weeks. Compared with the control group, no significant differences in feed utilization rate or feed efficiency were found in the VC group. Similarly, no significant differences were found in the crude protein, crude lipid, ash, or moisture content between the control group and the VC group.

Table 3. Effects of dietary VC on growth performance and body composition of gibel carp.

Category	Parameter	Group	
		CON	VC
Growth performance	Initial body weight (g)	6.60 ± 0.00	6.67 ± 0.03
	Final body weight (g)	26.40 ± 0.21 [a]	31.40 ± 1.36 [b]
	WGR [1] (%)	219.80 ± 12.98	261.38 ± 17.86
	FE [2] (%)	44.19 ± 0.42	50.79 ± 2.81
	SGR [3] (% d^{-1})	4.93 ± 0.03 [a]	5.54 ± 0.16 [b]
Body composition	Crude protein (%)	15.18 ± 0.16	14.91 ± 0.1
	Crude lipid (%)	7.35 ± 0.19	7.22 ± 0.04
	Ash (%)	2.68 ± 0.12	2.60 ± 0.13
	Moisture (%)	71.95 ± 0.54	71.81 ± 0.27

The data listed in the table are all expressed as the means ± SEM, and the superscripts (a or b) of different letters in the same row indicate significant differences ($p < 0.05$). [1] WGR: Feeding rate (%) = [(final weight (g) − initial weight (g))/initial weight (g)] × 100]. [2] FE: Feeding efficiency (%) = (100 × fresh body weight gain)/dry feed intake. [3] SGR: Specific growth rate (% d^{-1}) = 100 × [ln (final weight) - ln (initial weight)]/day.

3.2. Plasma Metabolites

The plasma metabolite levels of gibel carp after acute hypoxia stimulation are shown in Figure 1. The plasma glucose, lactic acid, and cortisol levels were increased significantly by acute hypoxia. However, all of the levels were significantly decreased in the VC-supplemented group.

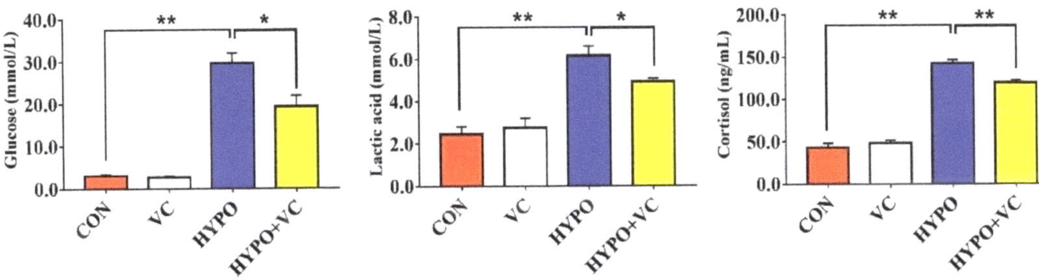

Figure 1. Plasma metabolites of gibel carp in the control and normal groups and hypoxia groups after feeding with VC for 56 days. The * or ** at the top of the bar chart indicates significant differences between treatments; * indicates a significant difference ($p < 0.05$), and ** indicates an extremely significant difference ($p < 0.01$). All of the data are presented as the mean ± standard error (n = 6).

3.3. Changes of HIF-1α Protein in the Liver

Expression of the HIF-1α protein in the liver tissue is illustrated in Figure 2. The expression of HIF-1α protein in the liver tissue of gibel carp increased significantly after the hypoxia treatment. However, the HIF-1α levels were significantly decreased after VC supplementation.

3.4. Expression of Nrf2/Keap1 Signaling Pathway and Antioxidant-Related Genes in the Liver

As shown in Figure 3, the protein levels of Nrf2 and Keap1 increased significantly in the liver after acute hypoxia stimulation. Moreover, acute hypoxia induced a significant increase in the expression level of *hsp70* in the liver and significantly decreased the expression of antioxidant-related genes, such as *sod* and *gpx*, while there was no significant effect on the expression of *cat*. Meanwhile, in the HYPO + VC group, Nrf2 protein expression was further activated, and Keap1 protein expression was decreased. In terms of genes, VC supplementation significantly reduced the expression of *hsp70* and increased the expression of *sod*, but it had no significant effect on the expression of *gpx* or *cat*.

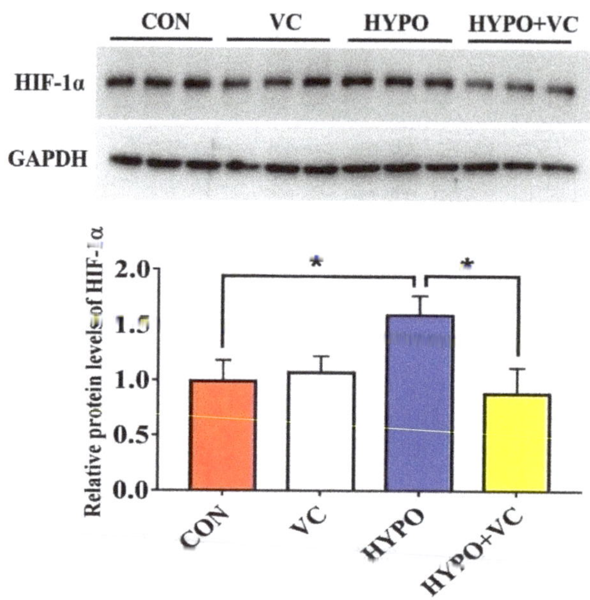

Figure 2. Changes in HIF-1α protein expression in the liver of gibel carp in normal and hypoxia groups after 56 days of feeding with VC. The * at the top of the bar chart indicates significant differences between treatments, * indicates a significant difference ($p < 0.05$). All of the data are presented as the mean ± standard error (n = 6).

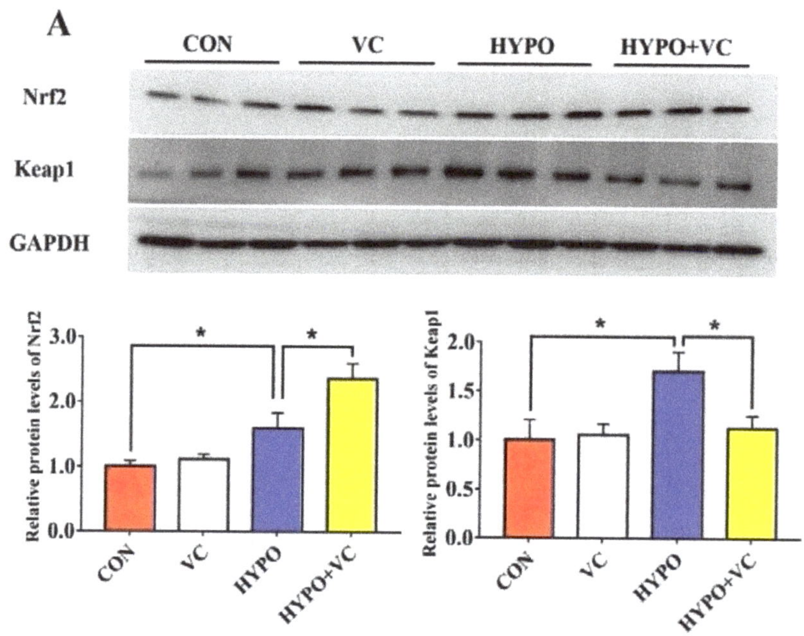

Figure 3. *Cont.*

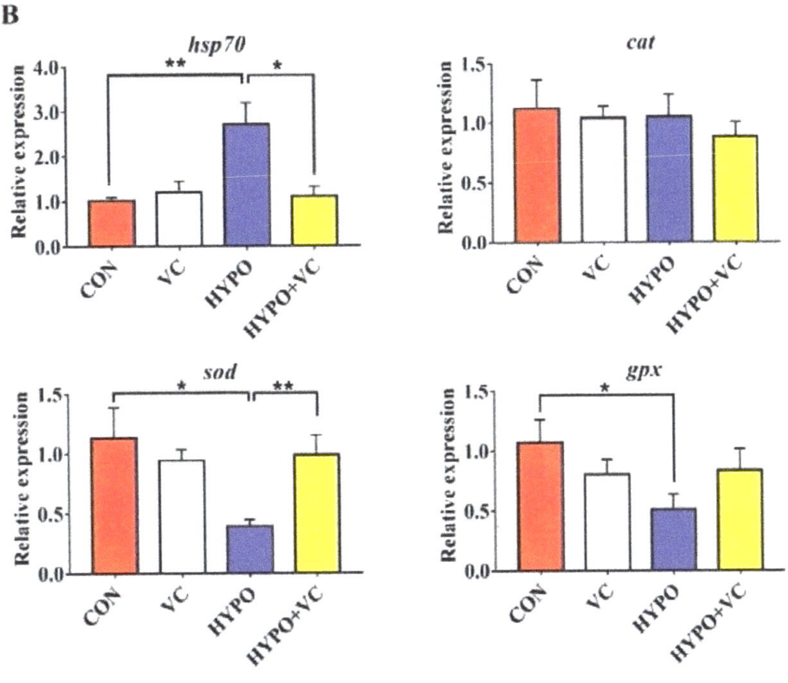

Figure 3. After 56 days of diet supplemented with VC, the expression levels of Nrf2 and Keap1 protein (**A**) and the expression of related genes (**B**) involved in antioxidant the Nrf2/Keap1 pathway in the liver of gibel carp were altered between the normal and hypoxia groups. The * or ** at the top of the bar chart indicate significant differences between treatments; * indicates a significant difference ($p < 0.05$), and ** indicates an extremely significant difference ($p < 0.01$). All of the data are presented as the mean ± standard error (n = 6). *hsp70*—heat shock protein 70; *cat*—catalase; *sod*—superoxide dismutase; *gpx*—glutathione peroxidase.

3.5. Antioxidant Enzyme Parameters in the Liver

The results of the antioxidant enzyme parameters in the liver are shown in Figure 4. The activities of the antioxidant enzymes CAT, SOD, and GPx in fish liver were significantly decreased after acute hypoxia treatment. T-AOC was also decreased, while the content of MDA was significantly increased. However, the antioxidant parameters were significantly enhanced, while the MDA content was significantly reduced in the HYPO + VC group. However, no change in the GSH content was found in this group.

3.6. Expression of Inflammation-Related Genes in the Liver

Changes of the inflammation-related genes in the liver are shown in Figure 5. Gene expression levels of proinflammatory cytokines *tnf-α*, *il-2*, and *il-6* in gibel carp liver were significantly increased in the acute hypoxia group compared with the control group. The VC treatment significantly decreased the gene expression levels of *tnf-α*, *il-2*, *il-6*, and *il-12*. The transcript levels of anti-inflammatory cytokines *il-4* and *tgf-β* were inhibited by acute hypoxia. The gene expression levels of *il-4* and *tgf-β* were significantly increased in the VC-fed group after acute hypoxic stress. The mRNA levels of *il-1β* and *iκbα* were unaltered after the hypoxia treatment in all of the groups.

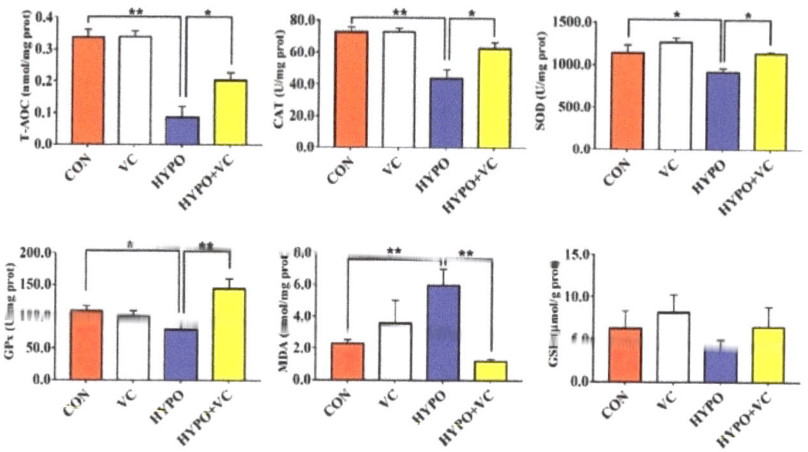

Figure 4. Changes in antioxidant enzyme parameters in the liver of gibel carp in the normal and hypoxia groups after 56 days of feeding with VC. The * or ** at the top of the bar chart indicates a significant difference between treatments; * indicates a significant difference ($p < 0.05$), and ** indicates an extremely significant difference ($p < 0.01$). All of the data are presented as mean ± standard error (n = 6). T-AOC—Total antioxidant capacity; CAT—Catalase; SOD—Superoxide dismutase; GP—Glutathione peroxidase; MDA—Malondialdehyde; GSH—Reduced glutathione.

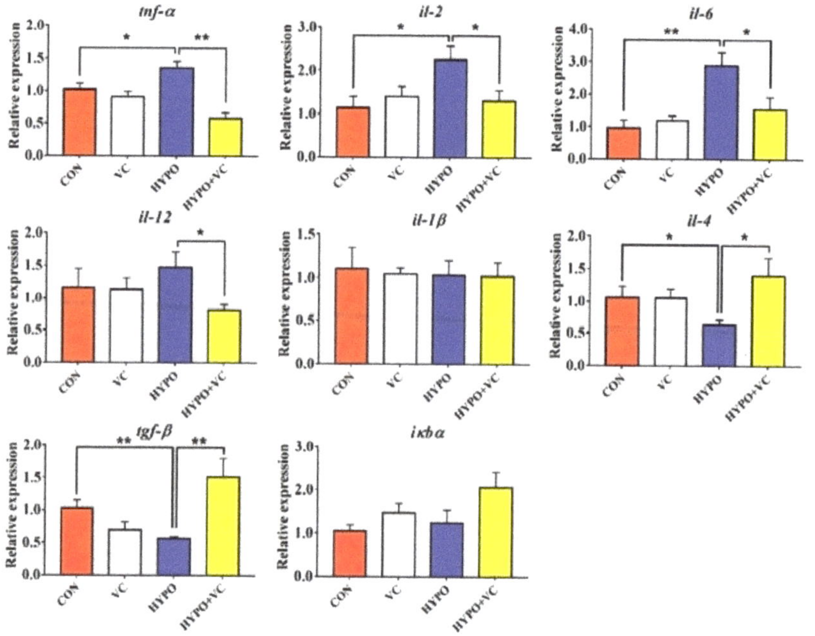

Figure 5. Changes of inflammation-related genes in the liver of gibel carp in the normal and hypoxia groups after 56 days of feeding with VC. The * or ** at the top of the bar chart indicates significant differences between treatments; * indicates a significant difference ($p < 0.05$), and ** indicates the extremely significant difference ($p < 0.01$). All of the data are presented as the mean ± standard error (n = 6). *tnf-α*—tumor necrosis factor-α; *il-2*—interleukin 2; *il-6*—interleukin 6; *il-12*—interleukin 12; *il-1β*—interleukin 1β; *tgf-β*—transforming growth factor-β; *il-4*—interleukin 4; *iκbα*—nf-κb inhibitorα.

3.7. Expression of ER Stress Key Proteins and Related Genes in the Liver

As shown in Figure 6A, the expression levels of GRP78 protein in gibel carp liver were significantly increased after acute hypoxia stimulation. However, the GRP78 protein levels were significantly decreased in fish fed the VC diet. The transcript levels of *ire1*, *perk*, *atf6*, *bip*, atf6, and *chop* were significantly higher in the hypoxia group compared with the control group (Figure 6B). However, *xbp1* and *eif2a* mRNA levels were not affected by hypoxic stress. In the HYPO + VC group, the expression of the ER stress-related GRP78 protein and *ire1*, *perk*, and *bip* genes was significantly decreased. There were no significant differences in *atf4*, *atf6*, or *chop* mRNA levels. In addition, the gene expression levels of *xbp1* and *eif2a* were unaltered among all of the treatments.

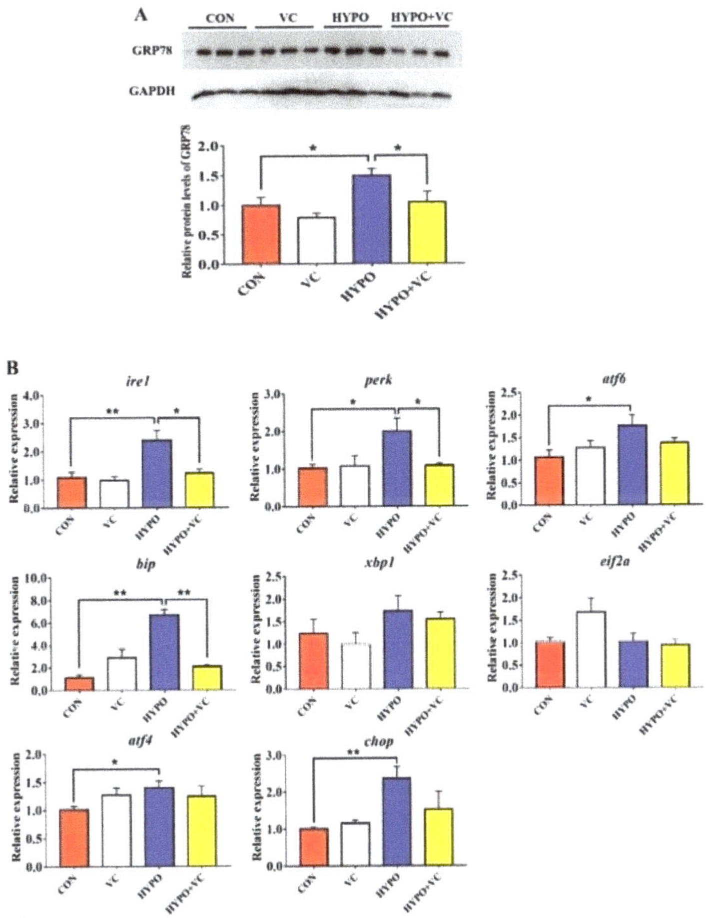

Figure 6. Changes in the expression levels of endoplasmic reticulum stress key protein, (**A**) and related genes (**B**) in the liver of gibel carp in normal and hypoxia groups after 56 days of feeding with VC. The * or ** at the top of the bar chart indicates significant differences between treatments; * indicates a significant difference ($p < 0.05$), and ** indicates an extremely significant difference ($p < 0.01$). All of the data are presented as the mean ± standard error (n = 6). GRP78—glucose-regulated protein 78; *ire1*—inositol-requiring protein-1a; *perk*—eukaryotic translation initiation factor 2-alpha kinase 3; *atf6*—activating transcription factor 6; *bip*—binding immunoglobulin protein; *xbp1*—X-box-binding protein 1; *ei2fa*—eukaryotic translation initiation factor 2A; *atf4*—activating transcription factor 4; *chop*—DNA damage-inducible transcript 3 protein.

3.8. Expression of Autophagy and Apoptosis-Related Genes in the Liver

The changes of autophagy-related genes, including *atg12*, *p62*, *beclin1*, *lc3b*, and *atg5*, in gibel carp liver after acute hypoxia stimulation are presented in Figure 7A. In the liver, the expression levels of *beclin1*, lc3b, and *atg5* were significantly increased after the acute hypoxia stimulation compared with the control group. The opposite change was found in the *p62* mRNA levels. Dietary supplementation with VC significantly inhibited the gene expression levels of *lc3b* and *atg5*, while there were no significant differences in the gene expression levels of *p62* or *beclin1*. In addition, the expression levels of *atg12* were unchanged, irrespective of treatment.

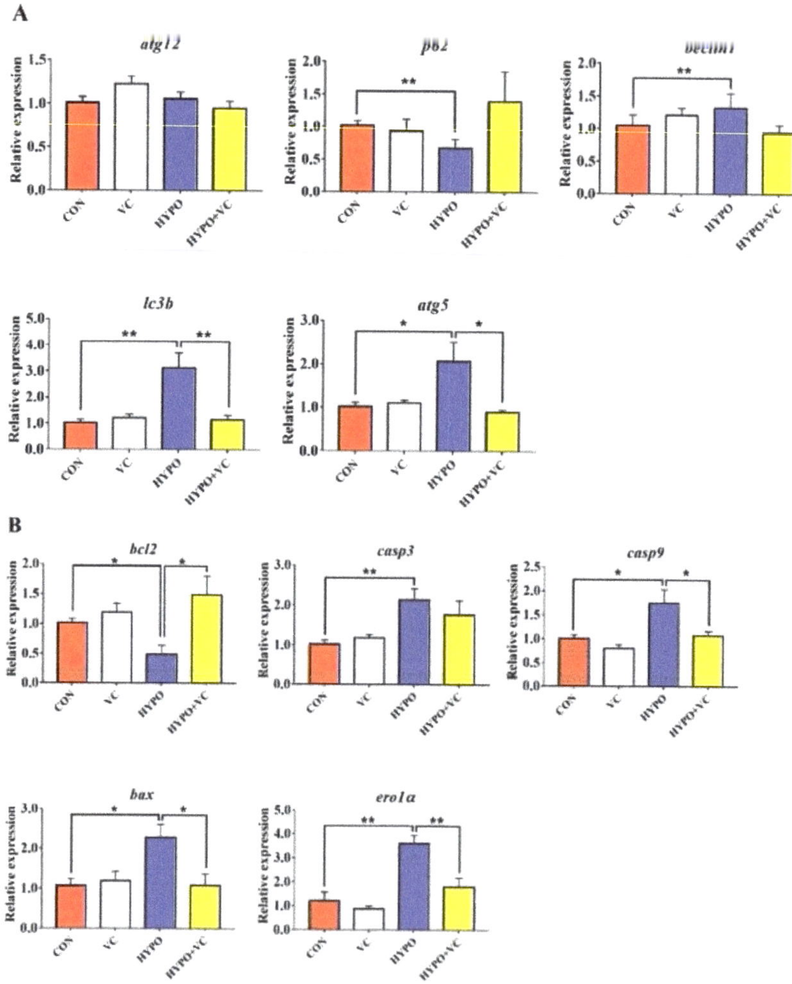

Figure 7. Changes of the expression levels of autophagy (**A**) and apoptosis (**B**)-related genes in the liver of gibel carp in the normal group and hypoxia group after 56 days of feeding with VC. The * or ** at the top of the bar chart indicates significant difference between treatments; * indicates a significant difference ($p < 0.05$), and ** indicates an extremely significant difference ($p < 0.01$). All of the data are presented as the mean ± standard error (n = 6).

The relative expression of apoptosis-related genes in the liver tissue is shown in Figure 7B. The gene expression of *bcl2* was decreased significantly after acute hypoxia. The expression levels of proapoptotic genes *bax* and *casp3*, *casp9* and *ero1α* were significantly induced by hypoxia. However, the expression levels of *bcl2* and *casp9* were significantly increased after VC supplementation, while the *bax* and *ero1α* transcript levels were significantly decreased.

3.9. Determination of Casp 3 and 9 Activities in the Liver

Activities of Casp 3 and Casp 9 in the liver are shown in Figure 8. Acute hypoxia induced the activities of Casp 3 and Casp 9 compared with the control group. However, VC treatment significantly decreased the activities of Casp 3 and Casp 9 in the liver of gibel carp.

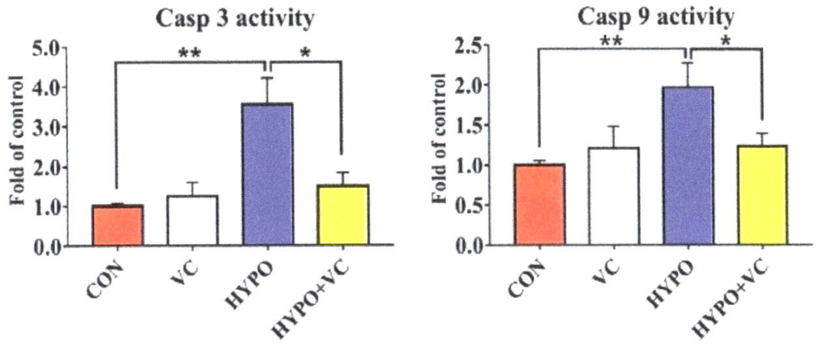

Figure 8. Changes of Casp 3 and Casp 9 activities in the livers of gibel carp in normal group and hypoxia group after 56 days of feeding with VC. The * or ** at the top of the bar chart indicates significant differences between treatments; * indicates a significant difference ($p < 0.05$), and ** indicates an extremely significant difference ($p < 0.01$). All of the data are presented as the mean ± standard error (n = 6).

3.10. TUNEL Observations

The results of TUNEL and DAPI staining in the hypothalamus are presented in Figure 9. Apoptosis signals were increased significantly after acute hypoxia stimulation, and these were decreased in fish fed with VC diet after hypoxia.

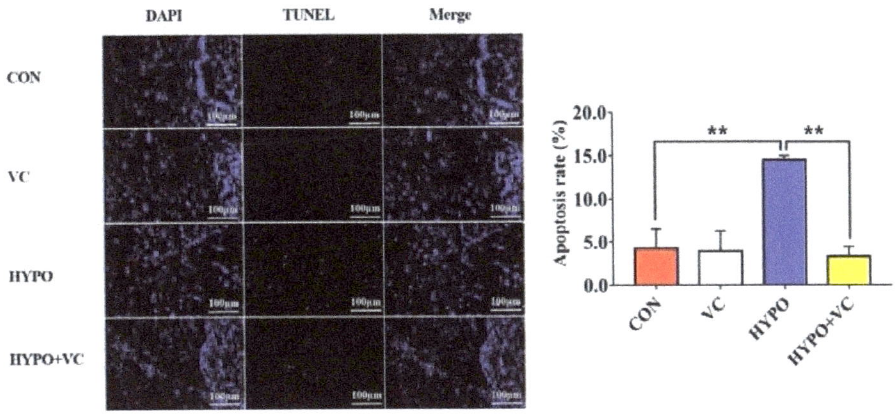

Figure 9. Changes of DAPI and TUNEL staining from representative sections in the hypothalamus of gibel carp in the normal group and hypoxia group after 56 days of feeding with VC. Positive apoptotic nuclei in green and normal nuclei are shown in blue. The magnification and scale of the microscope were 200× and 100 μm, respectively. The ** at the top of the bar chart indicates significant differences between treatments; ** indicates an extremely significant difference ($p < 0.01$). All of the data are presented as the mean ± standard error (n = 6).

4. Discussion

Vitamin C is extremely critical to maintaining normal growth and the antistress ability of fish. VC participates in a variety of physiological processes in the body, such as growth, development, and stress response, and it can protect the important functions of the immune system from oxidative damage. High doses of VC significantly improved the growth performance of Asian catfish (*Clarias batrachus*) and Wuchang bream (*Megalobrama amblycephala*) [26,27]. In line with this, a higher specific growth rate was found in fish fed the VC diet, indicating that VC improves the growth of gibel carp. However, no effect of supplementation with VC was found on the growth of large yellow croaker (*Pseudosciaena crocea*) [28]. The differences in the effects of VC may be related to the fish species and supplemental VC concentration.

As an energy substance metabolized by various tissues, the concentration of glucose in the blood is essential to maintaining life activities of fish. Normally, glycemia can be induced by stress in fish [29]. Consistent with this, glycemia was induced by hypoxia in gibel carp. As a metabolite, lactate content increases rapidly under hypoxic conditions [30]. In biochemical tests, the cortisol level is used as a characteristic index of fish stress intensity. It is one of the hormones secreted by the adrenal gland and can enter the blood directly [31]. The levels of cortisol in the hypoxia group were significantly increased, suggesting that hypoxia could induce a severe stress response in gibel carp. However, in the treatment group fed the VC supplemented diet, the levels of glucose, lactate, and cortisol in the plasma were significantly decreased, indicating that VC alleviated the hypoxic stimulation of gibel carp.

Hypoxia inducible factors (HIFs) are the most important transcription factor family that can regulate a variety of genes in response to decreases of intracellular oxygen concentration. HIF-1α overexpression can inhibit the apoptosis of K562 cells induced by reactive oxygen species [32]. In the present study, HIF-1α protein expression in the liver was significantly upregulated after acute hypoxic stimulation. Interestingly, VC supplementation alleviated the induction of HIF-1α protein expression by hypoxia, indicating a decreased response of fish to hypoxic stress. Additionally, VC has antioxidant and anti-inflammatory activities [18]. CAT and SOD activities were significantly downregulated after 12 h of acute hypoxia in the liver of largemouth bass (*Micropterus salmoides*) [33]. In the present study, compared with the control group, the activities of CAT, SOD, and GPx in the liver of gibel carp in the hypoxia group were downregulated, together with decreased T-AOC, implying that hypoxia leads to an excessive accumulation of ROS and the body possible severe oxidative damage. In hypoxia/reoxygenated rats, SOD and GSH activities in the brain were significantly downregulated, and the MDA content was significantly upregulated after stress [34]. CAT and GPx activities in the brain were induced by hypoxia in common carp (*Cyprinus carpio*); however, no significant difference in CAT activity in the liver or muscle was found, and GPx activity was significantly downregulated after hypoxia treatment [35]. Thus, the response of antioxidant levels may differ among tissues. In addition, excessive oxidative stress can cause hepatocyte apoptosis or necrosis, thereby indirectly initiating liver injury [36]. In the present study, oxidative stress was effectively inhibited by VC, consistent with previous reports [37,38]. The increases of ROS and oxidative stress are positively correlated with hypoxia, both of which are considered to be the main etiological factors leading to body dysfunction and damage and are important parts of the series of events leading to proinflammatory mediators and cell apoptosis. Hypoxia causes a massive accumulation of ROS that directly or indirectly oxidizes or damages DNA, proteins, and lipids; induces the formation of lipid peroxidation; and damages cell membranes [39]. Lipid peroxidation was the most sensitive indicator respond to hypoxia in goldfish [35]. MDA was significantly upregulated by hypoxia in gibel carp. MDA concentration was increased in largemouth bass after 24 h of acute hypoxia exposure [33]. The MDA content in the liver tissue was downregulated in VC-fed fish, indicating that VC plays a potential protective role in alleviating acute hypoxia-induced oxidative stress.

Nrf2 is a common regulator of the antioxidant system and immune inflammatory responses. The protein dissociates from Keap1 into the nucleus and plays a key role in antioxidant capacity, anti-inflammation, and anti-apoptosis when oxidative stress occurs [40–42]. The antioxidant capacity was improved through Nrf2/Keap1 signaling by a VC-supplemented diet in abalone [21]. In the present study, the Nrf2/Keap1 pathway was activated by acute hypoxia, as indicated by increased protein levels of Nrf2. Since increased protein expression of Nrf2 and decreased protein expression of Keap1 were found in VC-supplemented group, this indicated that VC was involved in the Nrf2/Keap1 pathway regulation of hypoxia. Inflammation is an important immune defense mechanism of the body, a state of self-protection or damage repair initiated when tissues are subjected to environmental stress or bacterial infection. In macrophages, Nrf2 activation prevents the onset of inflammatory responses [43]. In the present study, the expression levels of pro-inflammatory genes *tnf-α*, *il-2*, and *il-6* were upregulated in the hypoxia group, while the expression levels of anti-inflammatory genes *il-4* and *tgf-β* were downregulated, suggesting that acute hypoxia stimulated an inflammatory response in gibel carp. Upregulated expression of *tnf-α*, *il-2*, *il-6*, and *il-12* and downregulated expression of anti-inflammatory genes *il-4* and *tgf-β* induced by VC suggested that VC attenuated the expression of proinflammatory factors [44]. Thus, oxidative stress, inflammation, and apoptosis were attenuated by VC via the Nrf2/Keap1 pathway in gibel carp.

Various endogenous and exogenous stresses can destroy protein homeostasis in organelles, leading to ER damage and activating the UPR response [45]. UPR activation is closely related to pathological processes, such as inflammation and autophagy [46–48]. Protein kinase RNA-activated-like ER kinase (PERK), inositol-requiring enzyme 1α (IRE1α), and activated transcription factor 6 (ATF6) act as ER-located protein stress markers that can coordinate the response to harmful accumulation of unfolded or misfolded proteins. In the present study, VC treatment reduced the expression of GRP78, a marker protein of ER stress induced by acute hypoxia. Similarly, chrysophanol treatment also reduced hypoxia/reoxygenation-induced GRP78 protein expression [49]. Molecular chaperone binding immunoglobulin (BiP) is recognized as one of the most sensitive markers of ER stress, as it facilitates tumor growth through a variety of mechanisms such as promoting the maturation and secretion of growth factors, inhibiting apoptosis, and stimulating angiogenesis. In mice, VC treatment reduced ER stress induced by perfluorooctane sulfonate and downregulated the expression of ATF6 and GRP78 proteins [44]. Herein, the expression levels of *ire1*, *perk*, *atf6*, *bip*, *atf4*, and *chop* in the liver were significantly upregulated after acute hypoxia, indicating that acute hypoxia stimulation induced the activation of hepatic ER stress in gibel carp. However, *ire1*, *perk*, and *bip* genes in the liver tissue were significantly downregulated by VC supplementation. Therefore, VC may exert its pharmacological protective effect in the liver by inhibiting ER stress in fish.

Autophagy is a process by which eukaryotic cells utilize lysosomes to degrade cytosolic proteins and damaged organelles under the regulation of autophagy-related genes. Hypoxia can induce autophagy, and autophagy has dual functions in tissue damage caused by hypoxia. Under hypoxia, autophagy or autophagy-related proteins induce apoptosis by activating Caspase or by reducing endogenous apoptosis inhibitors, ultimately aggravating cell injury [50]. Under severe hypoxia, the PERK pathway upregulates the expression of *lc3b* by activating the expression of *atf4* and *chop* [51]. In myocytes, hypoxia causes upregulation of *beclin1* expression that simultaneously prompts the conversion of LC3B-I to LC3B-II and is recruited to the autophagosome outer membrane [52]. In the process of autophagy, LC3B-II extends throughout the autophagic process and is a currently recognized marker of autophagy. In the present study, *beclin1*, *lc3b*, and *atg5* mRNA levels were significantly upregulated in the hypoxia group, indicating that the autophagy process was initiated. However, the gene expression levels of *lc3b* and *atg5* were significantly downregulated in the VC-diet-fed group, implying that VC attenuated autophagic process induced by hypoxia in gibel carp. Severe autophagy induces apoptosis, and caspase is an essential part of the apoptosis mechanism that can reflect whether apoptosis occurs in the body [53]. Bax and Bcl2 play regulatory roles in apoptosis; both belong to the Bcl family, and they ultimately mediate apoptosis by controlling the permeability of mitochondria and sequentially activating caspase 9 and caspase 3 [54,55]. In the present study, apoptosis was activated by hypoxia, as indicated by increased expression levels of *bax* and *ero1α*. Increased expression levels of *bcl2*, concurrent with inhibited activities of Casp3 and Casp9 together with downregulated *casp9* expression induced by VC indicated that VC inhibited apoptosis induced by hypoxia in gibel carp, consistent with results in abalone [21]. Additionally, TUNEL-positive cells were significantly reduced in gibel carp fed a VC diet under hypoxia compared with fish in the control group, verifying the anti-apoptosis effects of VC. Taken together, the results suggest that VC enhanced the anti-apoptotic ability of gibel carp through inhibiting ER stress and caspase-dependent pathways.

5. Conclusions

Acute hypoxia stimulates oxidative damage, inflammation, and ER stress, resulting in apoptosis in gibel carp. VC protected fish from acute hypoxia injury by enhancing Nrf2/Keap1 signaling and downregulating the level of HIF-1α protein, thereby leading to inhibition of oxidative stress, inflammation, and apoptosis. This study provides new evidence for the hypothesis that VC supplementation can effectively alleviate oxidative stress and inflammation induced by acute hypoxia in fish.

Author Contributions: Investigation, Methodology, Writing—Original Draft Preparation, L.W.; Data curation, Formal Analysis, H.L. (Hongyan Li) and W.X.; Validation, B.D. and H.G.; Conceptualization, Funding Acquisition, Supervision, Writing—Review & Editing, J.J.; Supervision, D.H., H.L. (Haokun Liu) and X.Z.; Methodology, investigation, Y.Y.; Supervision, Funding Acquisition, S.X. All authors have read and agreed to the published version of the manuscript.

Funding: This work was financially supported by the National Natural Science Foundation of China (32122089; U19A2041; 31972805), the National Key R&D Program of China (2018YFD0900605; 2019YFD0900200), the Strategic Priority Research Program of Chinese Academy of Sciences (XDA24010206) and the China Agriculture Research System of MOF and MARA (CARS-45-09).

Institutional Review Board Statement: All animal care and experimental procedures were approved by the Experimental Animal Ethics Committee of Institute of Hydrobiology, Chinese Academy of Sciences (approval ID: IHB20140724).

Informed Consent Statement: Not applicable.

Data Availability Statement: Data are contained within the article.

Acknowledgments: The authors thank Guanghan Nie for his technical support with the research system and the Analysis and Testing Center of Institute of Hydrobiology, Chinese Academy of Sciences for assistance in the experimental treatment of acute hypoxia stress.

Conflicts of Interest: The authors declare no conflict of interest.

References

1. Park, J.; Shim, M.K.; Jin, M.; Rhyu, M.R.; Lee, Y. Methyl syringate, a TRPA1 agonist represses hypoxia-induced cyclooxygenase-2 in lung cancer cells. *Phytomedicine* **2016**, *23*, 324–329. [CrossRef] [PubMed]
2. Farias, J.G.; Molina, V.M.; Carrasco, R.A.; Zepeda, A.B.; Figueroa, E.; Letelier, P.; Castillo, R.L. Antioxidant therapeutic strategies for cardiovascular conditions associated with oxidative stress. *Nutrients* **2017**, *9*, 966. [CrossRef] [PubMed]
3. Koh, H.S.; Chang, C.Y.; Jeon, S.B.; Yoon, H.J.; Ahn, Y.H.; Kim, H.S.; Kim, I.H.; Jeon, S.H.; Johnson, R.S.; Park, E.J. The HIF-1/glial TIM-3 axis controls inflammation-associated brain damage under hypoxia. *Nat. Commun.* **2015**, *6*, 6340. [CrossRef] [PubMed]
4. Zhang, J.; Han, C.; Dai, H.; Hou, J.; Dong, Y.; Cui, X.; Xu, L.; Zhang, M.; Xia, Q. Hypoxia-inducible factor-2α limits natural killer T cell cytotoxicity in renal ischemia/reperfusion injury. *J. Am. Soc. Nephrol.* **2016**, *27*, 92–106. [CrossRef]
5. Chen, R.L.; Lai, U.H.; Zhu, L.L.; Singh, A.; Ahmed, M.; Forsyth, N.R. Reactive oxygen species formation in the brain at different oxygen levels: The role of hypoxia inducible factors. *Front. Cell Dev. Biol.* **2018**, *6*, 132. [CrossRef]
6. Masson, N.; Singleton, R.S.; Sekirnik, R.; Trudgian, D.C.; Ambrose, L.J.; Miranda, M.X.; Tian, Y.M.; Kessler, B.M.; Schofield, C.J.; Ratcliffe, P.J. The FIH hydroxylase is a cellular peroxide sensor that modulates HIF transcriptional activity. *EMBO Rep.* **2012**, *13*, 251–257. [CrossRef]

7. Cecerska-Heryc, E.; Surowska, O.; Heryc, R.L.; Serwin, N.; Napiontek-Balinska, S.; Dolegowska, B. Are antioxidant enzymes essential markers in the diagnosis and monitoring of cancer patients—A review. *Clin. Biochem.* **2021**, *93*, 1–8. [CrossRef]
8. Peng, Y.B.; Ren, Y.; Zhu, H.; An, Y.; Chang, B.S.; Sun, T.L. Ultrasmall copper nanoclusters with multi-enzyme activities. *Rsc. Adv.* **2021**, *11*, 14517–14526. [CrossRef]
9. Sajadimajd, S.; Khazaei, M. Oxidative stress and cancer: The role of nrf2. *Curr. Cancer Drug Targets* **2018**, *18*, 538–557. [CrossRef]
10. Bartolini, D.; Dallaglio, K.; Torquato, P.; Piroddi, M.; Galli, F. Nrf2-p62 autophagy pathway and its response to oxidative stress in hepatocellular carcinoma. *Transl. Res.* **2018**, *193*, 54–71. [CrossRef]
11. Cong, P.; Wang, T.; Tong, C.; Liu, Y.; Shi, L.; Mao, S.; Shi, X.; Jin, H.; Liu, Y.; Hou, M. Resveratrol ameliorates thoracic blast exposure-induced inflammation, endoplasmic reticulum stress and apoptosis in the brain through the Nrf2/Keap1 and NF-κB signaling pathway. *Injury* **2021**, *52*, 2795–2802. [CrossRef] [PubMed]
12. Huang, C.Y.; Deng, J.S.; Huang, W.C.; Jiang, W.P.; Huang, G.J. Attenuation of lipopolysaccharide-induced acute lung injury by hispolon in mice, through regulating the TLR4/PI3K/Akt/mTOR and Keap1/Nrf2/HO-1 pathways, and suppressing oxidative stress-mediated ER stress-induced apoptosis and autophagy. *Nutrients* **2020**, *12*, 1742. [CrossRef] [PubMed]
13. Song, C.Y.; Liu, B.; Xu, P.; Ge, X.P.; Zhang, H.M. Emodin ameliorates metabolic and antioxidant capacity inhibited by dietary oxidized fish oil through PPARs and Nrf2-Keap1 signaling in Wuchang bream (*Megalobrama amblycephala*). *Fish Shellfish Immunol.* **2019**, *94*, 842–851. [CrossRef] [PubMed]
14. Wu, L.Y.; Li, L.Y.; Xu, W.J.; Dong, B.; Geng, H.C.; Jing, J.Y.; Han, D.; Liu, H.K.; Yang, Y.X.; Xie, S.Q. Emodin alleviates acute hypoxia-induced apoptosis in gibel carp (*Carassius gibelio*) by upregulating autophagy through modulation of the AMPK/mTOR pathway. *Aquaculture* **2022**, *548*, 737689. [CrossRef]
15. Itoh, K.; Tong, K.I.; Yamamoto, M. Molecular mechanism activating Nrf2-Keap1 pathway in regulation of adaptive response to electrophiles. *Free Radic. Biol. Med.* **2004**, *36*, 1208–1213. [CrossRef] [PubMed]
16. Diaz-Bulnes, P.; Saiz, M.L.; Lopez-Larrea, C.; Rodriguez, R.M. Crosstalk between hypoxia and ER stress response: A key regulator of macrophage polarization. *Front. Immunol.* **2020**, *10*, 2951. [CrossRef] [PubMed]
17. Rouschop, K.M.A.; van den Beucken, T.; Dubois, L.; Niessen, H.; Bussink, J.; Savelkouls, K.; Keulers, T.; Mujcic, H.; Landuyt, W.; Voncken, J.W.; et al. The unfolded protein response protects human tumor cells during hypoxia through regulation of the autophagy genes MAP1LC3B and ATG5. *J. Clin. Investig.* **2010**, *120*, 127–141. [CrossRef]
18. Travica, N.; Ried, K.; Sali, A.; Scholey, A.; Hudson, I.; Pipingas, A. Vitamin C status and cognitive function: A systematic review. *Nutrients* **2017**, *9*, 960. [CrossRef]
19. Su, M.; Chao, G.; Liang, M.Q.; Song, J.H.; Wu, K. Anticytoproliferative effect of vitamin C on rat hepatic stellate cell. *Am. J. Transl. Res.* **2016**, *8*, 2820–2825.
20. Xu, H.J.; Jiang, W.D.; Feng, L.; Liu, Y.; Wu, P.; Jiang, J.; Kuang, S.Y.; Tang, L.; Tang, W.N.; Zhang, Y.A.; et al. Dietary vitamin C deficiency depresses the growth, head kidney and spleen immunity and structural integrity by regulating NF-κB, TOR, Nrf2, apoptosis and MLCK signaling in young grass carp (*Ctenopharyngodon idella*). *Fish Shellfish Immunol.* **2016**, *52*, 111–138. [CrossRef]
21. Luo, K.; Li, X.; Wang, L.; Rao, W.; Wu, Y.; Liu, Y.; Pan, M.; Huang, D.; Zhang, W.; Mai, K. Ascorbic acid regulates the immunity, anti-oxidation and apoptosis in abalone *Haliotis discus hannai* Ino. *Antioxidants* **2021**, *10*, 1449. [CrossRef] [PubMed]
22. Majmundar, A.J.; Wong, W.J.; Simon, M.C. Hypoxia-inducible factors and the response to hypoxic stress. *Mol. Cell* **2010**, *40*, 294–309. [CrossRef] [PubMed]
23. AOAC. *Official Methods of Analysis of the Association of Official Analytical Chemists*, 17th ed.; Association of Official Analytical Chemists: Arlington, VA, USA, 2003.
24. Li, H.Y.; Xu, W.J.; Jin, J.Y.; Zhu, X.M.; Yang, Y.X.; Han, D.; Liu, H.K.; Xie, S.Q. Effects of dietary carbohydrate and lipid concentrations on growth performance, feed utilization, glucose, and lipid metabolism in two strains of gibel carp. *Front. Vet. Sci.* **2019**, *6*, 165. [CrossRef] [PubMed]
25. Pfaffl, M.W. A new mathematical model for relative quantification in real-time RT-PCR. *Nucleic Acids Res.* **2001**, *29*, e45. [CrossRef] [PubMed]
26. Kumari, J.; Sahoo, P.K. High dietary vitamin C affects growth, non-specific immune responses and disease resistance in asian catfish, *Clarias batrachus*. *Mol. Cell. Biochem.* **2005**, *280*, 25–33. [CrossRef] [PubMed]
27. Ming, J.; Xie, J.; Xu, P.; Ge, X.; Liu, W.; Ye, J. Effects of emodin and vitamin C on growth performance, biochemical parameters and two HSP70s mRNA expression of Wuchang bream (*Megalobrama amblycephala* Yih) under high temperature stress. *Fish Shellfish Immunol.* **2012**, *32*, 651–661. [CrossRef] [PubMed]
28. Ai, Q.H.; Mai, K.S.; Tan, B.P.; Xu, W.; Zhang, W.B.; Ma, H.M.; Liufu, Z.G. Effects of dietary vitamin C on survival, growth, and immunity of large yellow croaker, *Pseudosciaena crocea*. *Aquaculture* **2006**, *261*, 327–336. [CrossRef]
29. Yang, S.; Wu, H.; He, K.; Yan, T.; Zhou, J.; Zhao, L.L.; Sun, J.L.; Lian, W.Q.; Zhang, D.M.; Du, Z.J.; et al. Response of AMP-activated protein kinase and lactate metabolism of largemouth bass (*Micropterus salmoides*) under acute hypoxic stress. *Sci. Total Environ.* **2019**, *666*, 1071–1079. [CrossRef]
30. Lee, D.C.; Sohn, H.A.; Park, Z.Y.; Oh, S.H.; Kang, Y.K.; Lee, K.M.; Kang, M.H.; Jang, Y.J.; Yang, S.J.; Hong, Y.K.; et al. A lactate-induced response to hypoxia. *Cell* **2015**, *161*, 595–609. [CrossRef]
31. Hsieh, S.L.; Chen, Y.N.; Kuo, C.M. Physiological responses, desaturase activity, and fatty acid composition in milkfish (*Chanos chanos*) under cold acclimation. *Aquaculture* **2003**, *220*, 903–918. [CrossRef]

32. Sun, R.L.; Meng, X.; Pu, Y.Q.; Sun, F.X.; Man, Z.D.; Zhang, J.; Yin, L.H.; Pu, Y.P. Overexpression of HIF-1a could partially protect K562 cells from 1,4-benzoquinone induced toxicity by inhibiting ROS, apoptosis and enhancing glycolysis. *Toxicol. Vitr.* **2019**, *55*, 18–23. [CrossRef] [PubMed]
33. Zhao, L.L.; Cui, C.; Liu, Q.; Sun, J.L.; He, K.; Adam, A.A.; Luo, J.; Li, Z.Q.; Wang, Y.; Yang, S. Combined exposure to hypoxia and ammonia aggravated biological effects on glucose metabolism, oxidative stress, inflammation and apoptosis in largemouth bass (*Micropterus salmoides*). *Aquat. Toxicol.* **2020**, *224*, 105514. [CrossRef] [PubMed]
34. Coimbra-Costa, D.; Alva, N.; Duran, M.; Carbonell, T.; Rama, R. Oxidative stress and apoptosis after acute respiratory hypoxia and reoxygenation in rat brain. *Redox Biol.* **2017**, *12*, 216–225. [CrossRef]
35. Lushchak, V.I.; Bagnyukova, T.V.; Husak, V.V.; Luzhna, L.I.; Lushchak, O.V.; Storey, K.B. Hyperoxia results in transient oxidative stress and an adaptive response by antioxidant enzymes in goldfish tissues. *Int. J. Biochem. Cell Biol.* **2005**, *37*, 1670–1680. [CrossRef] [PubMed]
36. Luangmonkong, T.; Suriguga, S.; Mutsaers, H.A.M.; Groothuis, G.M.M.; Olinga, P.; Boersema, M. Targeting oxidative stress for the treatment of liver fibrosis. *Rev. Physiol. Biochem. Pharmacol.* **2018**, *175*, 71–102. [PubMed]
37. Chen, Y.Y.; Luo, G.Y.; Yuan, J.; Wang, Y.Y.; Yang, X.Q.; Wang, X.Y.; Li, G.P.; Liu, Z.G.; Zhong, N.C. Vitamin C mitigates oxidative stress and tumor necrosis factor-alpha in severe community-acquired pneumonia and LPS-induced macrophages. *Mediat. Inflamm.* **2014**, *2014*, 426740. [CrossRef]
38. Liang, X.P.; Li, Y.; Hou, Y.M.; Qiu, H.; Zhou, Q.C. Effect of dietary vitamin C on the growth performance, antioxidant ability and innate immunity of juvenile yellow catfish (*Pelteobagrus fulvidracoRichardson*). *Aquacult. Res.* **2015**, *48*, 149–160. [CrossRef]
39. Quinonez-Flores, C.M.; Gonzalez-Chavez, S.A.; Pacheco-Tena, C. Hypoxia and its implications in rheumatoid arthritis. *J. Biomed. Sci.* **2016**, *23*, 62. [CrossRef]
40. Hou, D.X.; Korenori, Y.; Tanigawa, S.; Yamada-Kato, T.; Nagai, M.; He, X.; He, J.H. Dynamics of Nrf2 and Keap1 in ARE-Mediated NQO1 expression by wasabi 6-(methylsulfinyl)hexyl isothiocyanate. *J. Agric. Food Chem.* **2011**, *59*, 11975–11982. [CrossRef]
41. Strom, J.; Xu, B.B.; Tian, X.Q.; Chen, Q.M. Nrf2 protects mitochondrial decay by oxidative stress. *FASEB J.* **2016**, *30*, 66–80. [CrossRef]
42. Ren, X.Y.; Yu, J.G.; Guo, L.L.; Ma, H. TRIM16 protects from OGD/R-induced oxidative stress in cultured hippocampal neurons by enhancing Nrf2/ARE antioxidant signaling via downregulation of Keap1. *Exp. Cell Res.* **2020**, *391*, 111998. [CrossRef]
43. Kobayashi, E.H.; Suzuki, T.; Funayama, R.; Nagashima, T.; Hayashi, M.; Sekine, H.; Tanaka, N.; Moriguchi, T.; Motohashi, H.; Nakayama, K.; et al. Nrf2 suppresses macrophage inflammatory response by blocking proinflammatory cytokine transcription. *Nat. Commun.* **2016**, *7*, 11624. [CrossRef] [PubMed]
44. Su, M.; Liang, X.L.; Xu, X.X.; Wu, X.M.; Yang, B. Hepatoprotective benefits of vitamin C against perfluorooctane sulfonate-induced liver damage in mice through suppressing inflammatory reaction and ER stress. *Environ. Toxicol. Pharmacol.* **2019**, *65*, 60–65. [CrossRef] [PubMed]
45. Flores-Santibanez, F.; Medel, B.; Bernales, J.I.; Osorio, F. Understanding the role of the unfolded protein response sensor ire1 in the biology of antigen presenting cells. *Cells* **2019**, *8*, 1563. [CrossRef] [PubMed]
46. Ghosh, K.; De, S.; Mukherjee, S.; Das, S.; Ghosh, A.N.; Sengupta, S. Withaferin A induced impaired autophagy and unfolded protein response in human breast cancer cell-lines MCF-7 and MDA-MB-231. *Toxicol. Vitr.* **2017**, *44*, 330–338. [CrossRef] [PubMed]
47. Amen, O.M.; Sarker, S.D.; Ghildyal, R.; Arya, A. Endoplasmic reticulum stress activates unfolded protein response signaling and mediates inflammation, obesity, and cardiac dysfunction: Therapeutic and molecular approach. *Front. Pharmacol.* **2019**, *10*, 977. [CrossRef] [PubMed]
48. Kubra, K.T.; Akhter, M.S.; Uddin, M.A.; Barabutis, N. Unfolded protein response in cardiovascular disease. *Cell. Signal.* **2020**, *73*, 109699. [CrossRef]
49. Lin, C.H.; Tseng, H.F.; Hsieh, P.C.; Chiu, V.; Lin, T.Y.; Lan, C.C.; Tzeng, I.S.; Chao, H.N.; Hsu, C.C.; Kuo, C.Y. Nephroprotective role of chrysophanol in hypoxia/reoxygenation-induced renal cell damage via apoptosis, ER stress, and ferroptosis. *Biomedicines* **2021**, *9*, 1283. [CrossRef]
50. Schaaf, M.B.E.; Cojocari, D.; Keulers, T.G.; Jutten, B.; Starmans, M.H.; de Jong, M.C.; Begg, A.C.; Savelkouls, K.G.M.; Bussink, J.; Vooijs, M.; et al. The autophagy associated gene, ULK1, promotes tolerance to chronic and acute hypoxia. *Radiother. Oncol.* **2013**, *108*, 529–534. [CrossRef]
51. Rzymski, T.; Milani, M.; Pike, L.; Buffa, F.; Mellor, H.R.; Winchester, L.; Pires, I.; Hammond, E.; Ragoussis, I.; Harris, A.L. Regulation of autophagy by ATF4 in response to severe hypoxia. *Oncogene* **2010**, *29*, 4424–4435. [CrossRef]
52. Baskaran, R.; Poornima, P.; Priya, L.B.; Huang, C.Y.; Padma, V.V. Neferine prevents autophagy induced by hypoxia through activation of Akt/mTOR pathway and Nrf2 in muscle cells. *Biomed. Pharmacother.* **2016**, *83*, 1407–1413. [CrossRef] [PubMed]
53. Redza-Dutordoir, M.; Averill-Bates, D.A. Activation of apoptosis signalling pathways by reactive oxygen species. *Biochim. Biophys. Acta.* **2016**, *1863*, 2977–2992. [CrossRef] [PubMed]
54. Imam, F.; Al-Harbi, N.O.; Al-Harbi, M.M.; Ansari, M.A.; Al-Asmari, A.F.; Ansari, M.N.; Al-Anazi, W.A.; Bahashwan, S.; Almutairi, M.M.; Alshammari, M.; et al. Apremilast prevent doxorubicin-induced apoptosis and inflammation in heart through inhibition of oxidative stress mediated activation of NF-κB signaling pathways. *Pharmacol. Rep.* **2018**, *70*, 993–1000. [CrossRef] [PubMed]
55. Zhang, L.M.; Xia, Q.Q.; Zhou, Y.Y.; Li, J. Endoplasmic reticulum stress and autophagy contribute to cadmium-induced cytotoxicity in retinal pigment epithelial cells. *Toxicol. Lett.* **2019**, *311*, 105–113. [CrossRef] [PubMed]

antioxidants

MDPI

Article

Docosahexaenoic Acid Alleviates Palmitic Acid-Induced Inflammation of Macrophages via TLR22-MAPK-PPARγ/Nrf2 Pathway in Large Yellow Croaker (*Larimichthys crocea*)

Dan Xu [1], Kun Cui [1], Qingfei Li [1], Si Zhu [1], Junzhi Zhang [1], Shengnan Gao [1], Tingting Hao [1], Kangsen Mai [1,2] and Qinghui Ai [1,2,*]

1 Key Laboratory of Aquaculture Nutrition and Feed (Ministry of Agriculture and Rural Affairs), Key Laboratory of Mariculture (Ministry of Education), Ocean University of China, 5 Yushan Road, Qingdao 266003, China; danxu_ouc@163.com (D.X.); 15764210558@163.com (K.C.); qfli0310@163.com (Q.L.); dscdove@163.com (S.Z.); zhjun123@hunau.edu.cn (J.Z.); gaoshengnan21@gmail.com (S.G.); rachelhaotingting@163.com (T.H.); kmai@ouc.edu.cn (K.M.)

2 Laboratory for Marine Fisheries Science and Food Production Processes, Qingdao National Laboratory for Marine Science and Technology, 1 Wenhai Road, Qingdao 266237, China

* Correspondence: qhai@ouc.edu.cn; Tel.: +86-13969712326

Abstract: Palmitic acid (PA) is a saturated fatty acid (SFA) that can cause an inflammatory response, while docosahexaenoic acid (DHA) is always used as a nutritional modulator due to its anti-inflammatory properties. However, the potential molecular mechanism is still not completely elucidated in fish. Herein, the PA treatment induced an inflammatory response in macrophages of large yellow croaker (*Larimichthys crocea*). Meanwhile, the mRNA expression of Toll-like receptor (TLR)-related genes, especially *tlr22*, and the phosphorylation of the mitogen-activated protein kinase (MAPK) pathway were significantly upregulated by PA. Further investigation found that the PA-induced inflammatory response was suppressed by *tlr22* knockdown and MAPK inhibitors. Moreover, the results of the peroxisome proliferator-activated receptor γ (PPARγ) agonist and inhibitor treatment proved that PPARγ was involved in the PA-induced inflammation. PA treatment decreased the protein expression of PPARγ, while *tlr22* knockdown and MAPK inhibitors recovered the decreased expression. Besides, the PA-induced activation of Nrf2 was regulated by p38 MAPK. Furthermore, DHA-executed anti-inflammatory effects by regulating the phosphorylation of the MAPK pathway and expressions of PPARγ and Nrf2. Overall, the present study revealed that DHA alleviated PA-induced inflammation in macrophages via the TLR22-MAPK-PPARγ/Nrf2 pathway. These results could advance the understanding of the molecular mechanism of the SFA-induced inflammatory response and provide nutritional mitigative strategies.

Keywords: palmitic acid; inflammatory response; Toll-like receptor; docosahexaenoic acid; *Larimichthys crocea*

Citation: Xu, D.; Cui, K.; Li, Q.; Zhu, S.; Zhang, J.; Gao, S.; Hao, T.; Mai, K.; Ai, Q. Docosahexaenoic Acid Alleviates Palmitic Acid-Induced Inflammation of Macrophages via TLR22-MAPK-PPARγ/Nrf2 Pathway in Large Yellow Croaker (*Larimichthys crocea*). *Antioxidants* **2022**, *11*, 682. https://doi.org/10.3390/antiox11040682

Academic Editors: Min Xue, Junmin Zhang, Zhenyu Du, Jie Wang and Wei Si

Received: 17 December 2021
Accepted: 28 March 2022
Published: 31 March 2022

1. Introduction

Palmitic acid (PA), one of the most common saturated fatty acids (SFAs), can be transported in cells and converted into phospholipids, diacylglycerol and ceramides [1–3]. PA has been reported to induce an inflammatory response of macrophages in mammals, which triggers the activation of various signaling pathways and induces the production of cytokines [4]. Palm oil (PO), enriched with PA, is increasingly used as an alternative to fish oil in aquaculture [5]. Although partially replacing fish oil with PO in diets has no significant effects on the growth performance of cultured fish, the overuse of PO often induces an inflammatory response and suppresses the antioxidant capacity [6,7]. Moreover, PA has been proven to induce an inflammatory response in fish [8]. However, the molecular mechanism of the inflammatory response induced by PA is still not completely understood.

As a family of pattern recognition receptors (PRRs), Toll-like receptors (TLRs) play an essential role in initiating and regulating innate immunity both in mammals and fish [9,10]. The activation of TLRs recruits its adaptor proteins, such as the myeloid differentiation factor 88 (MyD88) and TIR domain containing adaptor-inducing interferon-β (TRIF). Then, downstream signaling cascades are triggered and activated, which could induce the expression of inflammatory genes [11,12]. Mitogen-activated protein kinase (MAPK) and nuclear transcription factor kappa-B (NF-κβ) are the widely studied downstream pathways of TLR. In mammals, PA could act as a strong agonist of TLR2 and TLR4 and activate downstream inflammatory signaling pathways [13,14]. However, the composition of the TLR family in teleost is different from that in mammals, which has some members considered to be fish-specific TLRs. Among all the fish-specific TLRs, TLR22 is widely explored as the typical member in many fish species [15–18]. Previous studies on large yellow croaker (*Larimichthys crocea*) have revealed that TLR22 could respond to fatty acids [19,20]. However, the role and downstream regulation mechanism of TLR22 in the PA-induced inflammatory response remain unclear in fish.

N-3 polyunsaturated fatty acids (PUFAs), such as docosahexaenoic acid (DHA) and eicosapentaenoic acid (EPA), can inhibit inflammation by regulating the activity of inflammatory signaling pathways and influencing the production of lipid mediators [21,22]. PUFAs could inhibit SFA-induced cyclooxygenase-2 expression by mediating a common signaling pathway derived from TLR [23]. Moreover, PUFA supplementation in diets results in a lower incidence of metabolic disease [24]. Previous studies on large yellow croaker have revealed that DHA supplementation in diets significantly improved fish health, and the anti-inflammatory effect of DHA was much stronger than EPA [25,26]. However, the anti-inflammatory molecular mechanism of DHA in fish needs further investigation.

Large yellow croaker is an important mariculture fish in China. In large yellow croaker feed, PO is widely used as a promising alternative to fish oil. However, it is still limited in the mechanism of inflammatory response induced by PA. Nutritional regulation of the immune system provides a potentially powerful strategy for improving fish health and quality. Studies on nutritional regulation in large yellow croaker have been intensively explored [27,28]. Large yellow croaker can be used as a good model animal in nutrition research. Therefore, this study aimed to investigate the molecular mechanism of PA-induced inflammation and its mitigative strategy mediated by DHA in macrophages of large yellow croaker. These results could advance the understanding of the molecular mechanism of the inflammatory response induced by SFAs and provide nutritional strategies against inflammation.

2. Materials and Methods

2.1. Macrophages Culture and Treatment

Large yellow croaker (weight 500 ± 50.26 g) was purchased from a commercial fish farm in Ningbo, China. All experimental procedures performed on fish were in strict accordance with the Management Rule of Laboratory Animals (Chinese Order No. 676 of the State Council, revised 1 March 2017). Macrophages were isolated from fish head kidneys and maintained according to the previous procedure with some modifications [29]. The primary macrophage was cultured in DMEM/F12 medium supplemented with 10% fetal bovine serum (FBS; Gibco, Carlsbad, CA, USA), 100-U/mL penicillin and 100-mg/mL streptomycin at 28 °C. Before stimulation, macrophages were seeded in six-well plates (Corning, Corning, NY, USA) at 2.0×10^6 cells per well. PA (Sigma-Aldrich, St. Louis, MO, USA) was combined with 1% fatty acid-free bovine serum albumin (BSA; Equitech-Bio, Kerrville, TX, USA) to reach a final concentration at 1 mm. Before the fatty acids treatment, cells were starved with DMEM/F12 alone for 1 h. Macrophages were incubated with different concentration of PA at 250 or 500 μM for 12 h. Cells treated with 1% fatty acid-free BSA were considered as the control group.

To confirm the role of the MAPK pathway in PA-induced inflammation, SB203580 (p38 inhibitor; MCE, Jersey City, NJ, USA) and SP600125 (JNK inhibitor; MCE) were used at a final concentration of 10 μm for 2 h before PA treatment. Furthermore, to confirm the role of PPARγ in PA-induced inflammation, troglitazone (PPARγ agonist; MCE) and GW9662 (PPARγ inhibitor; MCE) were incubated at a final concentration of 10 μm for 12 h before PA treatment. The control group was incubated with the same concentration of dimethyl sulfoxide (DMSO; Solarbio, Beijing, China).

To investigate the effect of DHA (Sigma-Aldrich, St. Louis, MO, USA) on PA-induced inflammation, macrophages were incubated with 200-μm DHA for 12 h before PA stimulation. After culturing, the macrophages were harvested for further analysis.

2.2. RNA Isolation and Quantitative Real-Time PCR (RT-qPCR)

Total RNA was extracted from macrophages by RNAiso Plus (Takara, Tokyo, Japan) according to the manufacturer's instructions. The 1.5% denaturing agarose gel was used to measure the integrity of the RNA. The concentration and quality of extracted total RNA were confirmed by a NanoDrop®2000 spectrophotometer (Thermo Scientific, Waltham, MA, USA). Then, complementary DNA (cDNA) was reverse-transcribed from the extracted RNA with the PrimeScript™ RT reagent kit (Takara, Tokyo, Japan).

The RT-qPCR primers used in this study are shown in Table 1 according to the nucleotide sequences of target genes in large yellow croaker. *β-actin* was used as the housekeeping gene. RT-qPCR was performed on a CFX96 Touch real-time PCR detection system (Bio-Rad, Hercules, CA, USA) using the SYBR Premix Ex Taq kit (Takara, Tokyo, Japan). The amplification was performed in a total volume of 20 μL, containing 10 μL of SYBR qPCR Master Mix, 6 μL of DEPC water, 2 μL of cDNA and 1 μL of F/R primer. The PCR temperature profile was performed at 95 °C for 2 min and, afterward, 39 cycles of 95 °C for 10 s, 58 °C for 15 s and 72 °C for 10 s. The gene expression levels were calculated via the $2^{-\Delta\Delta CT}$ method [30].

Table 1. Primers used for RT-qPCR and the gene accession number.

Gene	Forward (5′-3′)	Reverse (5′-3′)	Accession Number
β-actin	GACCTGACAGACTACCTCATG	AGTTGAAGGTGGTCTCGTGGA	GU584189
tnfα	ACACCTCTCAGCCACAGGAT	CCGTGTCCCACTCCATAGTT	NM_001303385
il1β	CATAGGGATGGGGACAACGA	AGGGGACGGACACAAGGGTA	XM_010736551
il6	CGACACACCCACTATTTACAAC	TCCCATTTTCTGAACTGCCTC	XM_010734753
cox2	CTGGAAAGGCAACACAAGC	CGGTGAGAGTCAGGGACAT	XM_010734489
arg1	AACCACCCGCAGGATTACG	AAACTCACTGGCATCACCTCA	XM_19269015
il10	AGTCGGTTACTTTCTGTGGTG	TGTATGACGCAATATGGTCTG	XM_010738826
cd68	GCAGGGCTTCAATCTGACCAA	AGGATGAGCACCAGCAATGTC	NM_001319937
cd86	TGTGCGTCTTAGTCTACCTTCT	AAACTCTTCCGTCATCTTGC	XM_010756962
cd209	GATGGGTGTATTTCAGCGGTAG	TGTTGATAATCACCAGGTCTGC	XM_027278935
tlr1	TGTGCCACCGTTTGGATA	TTCAGGGCGAACTTGTCG	KF318376
tlr2	TCTGCTGGTGTCAGAGGTCA	GGTGAATCCGCCATAGGA	XM_027287556
tlr3	ACTTAGCCCGTTTGTGGAAG	CCAGGCTTAGTTCACGGAGG	XM_019274877
tlr7	ATGCAATGAGCCAAAGTCT	CATGTGAGTCAATCCCTCC	XM_010743042
tlr13	CCTCCTGTTTATGGTAGTGTCC	GCTCGTCATGGGTGTTGTAG	XM_010743101
tlr21	CTTTGCCTACATCACAGGGACT	GAAACACGAGCAGGAGAACATC	KY025428
tlr22	TATGCGAGCAGGAAGACC	CAGAAACACCAGGATCAGC	GU324977
myd88	TACGAAGCGACCAATAACCC	ATCAATCAAAGGCCGAAGAT	EU978950
trif	TACAATACTGTTATCCCTCTGCTGC	TCTCTTCTGTTTTCTAATCCTCGCG	MK863372
pparγ	TGTCCGAGCTGGAAGACAAC	TGGGGTCATAGGGCATACCA	XM_010731330
nrf2	GATGGAAATGGAGGTGATGC	CATGTTCTTTCTGTCGGTGG	XM_010737768
si*tlr22*	GCAAGUUUGGUGGUGCUUUTT	AAAGCACCACCAAACUUGCTT	GU324977

2.3. Flow Cytometry Analysis

Macrophages treated with PA and BSA (the control group) were harvested and incubated with CD68 or CD209 antibodies at 37 °C for 1 h. CD68 and CD209 antibodies were produced by immunizing rabbits with synthetic recombinant proteins according to sequences from large yellow croaker, and the specificity was verified in previous studies [31]. Alexa Flour 488 goat anti-rabbit IgG (Beyotime Biotechnology, Shanghai, China) was incubated for 45 min at 37 °C to combine with primary antibodies. The percentage of positive cells was detected, and the data was analyzed by a flow cytometer (BD Accuri™ C6, Franklin Lakes, NJ, USA). Gated represent macrophages (R1) were selected. M1 and M2 represented CD68^{+} and CD209^{+} populations compared with the control group, respectively.

2.4. Western Blotting

RIPA reagent (Solarbio, Beijing, China) with a supplementation of protease and phosphatase inhibitors (Thermo Fisher Scientific, USA) was used to obtain the proteins of macrophages. Protein concentrations were measured by the BCA Protein Assay Kit (Beyotime Biotechnology, Shanghai, China). All protein concentrations were adjusted to the same level before heating. Western blot experiments were performed as follows: 20 µg of proteins were loaded on 10% SDS-PAGE and then transferred to polyvinylidene difluoride (PVDF) membranes (Merck Millipore, Berlin, Germany). The PVDF membranes were incubated with 5% skim milk for 2 h at room temperature and then incubated with the targeting antibody overnight at 4 °C. The primary antibodies against ERK1/2 (Cat. No. 4695), phospho-ERK1/2 (Cat. No. 4370), JNK1/2 (Cat. No. 9252), phospho-JNK1/2 (Cat. No. 4668), p38 (Cat. No. 8690), phospho-p38 (Cat. No. 9215), IKKβ (Cat. No. 2678), phospho-IKKα/β (Cat. No. 2697) and PPARγ (Cat. No. 2443) were obtained from Cell Signaling Technology (Boston, MA, USA). Anti-GAPDH antibody (Cat. No. TA-08; Golden Bridge Biotechnology, Beijing, China) was used as the reference. Then, the membrane was incubated with the secondary antibody (HRP-labeled Goat Anti-Rabbit IgG (H + L)) for 2 h at room temperature and then was visualized by an electrochemiluminescence kit (Beyotime Biotechnology, Shanghai, China). The target proteins were quantified using ImageJ software (National Institutes of Health, Bethesda, MD, USA).

2.5. RNA Interference

Large yellow croaker TLR22-specific small interfering RNA (siRNA; sense and antisense sequences of siRNA are shown in Table 1; Gene Pharma, Shanghai, China) was transfected into cells using the Xfect™ RNA Transfection Reagent (Takara, Tokyo, Japan) according to the manufacture's protocol for 36 h to knock down the expression of *tlr22* in macrophages. The RNAi negative control (NC) was used as the control group. After transfection for 36 h, cells were treated with PA for another 12 h.

2.6. Statistical Analysis

SPSS 22.0 (IBM, Armonk, NY, USA) was used to perform the statistical analysis, and the results were presented as means ± standard error of the mean (S.E.M.). All data were subjected to independent sample *t*-tests or one-way analysis of variance (ANOVA) followed by Tukey's multiple range test. A value of $p < 0.05$ was considered statistically significant.

3. Results

3.1. PA-Induced Inflammatory Response in Macrophages of Large Yellow Croaker

The PA treatment significantly upregulated the mRNA expression levels of proinflammatory genes, including *tnfα*, *il1β*, *il6* and *cox2* ($p < 0.05$) (Figure 1A–D). The mRNA expression level of anti-inflammatory gene *arg1* was significantly decreased in the PA treatment compared with that in the control group ($p < 0.05$), while the mRNA expression of *il10* showed no significant differences ($p > 0.05$) (Figure 1E,F). Moreover, mRNA expression levels of *cd68* and *cd86*, markers of proinflammatory phenotype macrophages, were significantly increased after PA treatment, whereas the mRNA expression of *cd209*, a marker of the anti-inflammatory phenotype macrophage, was significantly decreased ($p < 0.05$) (Figure 1G–I). Gated macrophages (R1) in forward and side scatter (FS-SS) dot plots and combined fluorescence histograms were shown (Figure 1J). The flow cytometry analysis showed that the PA treatment significantly increased the CD68^{+} population

(Figure 1K,L). These results demonstrated that PA treatment induced an inflammatory response in macrophages of large yellow croaker.

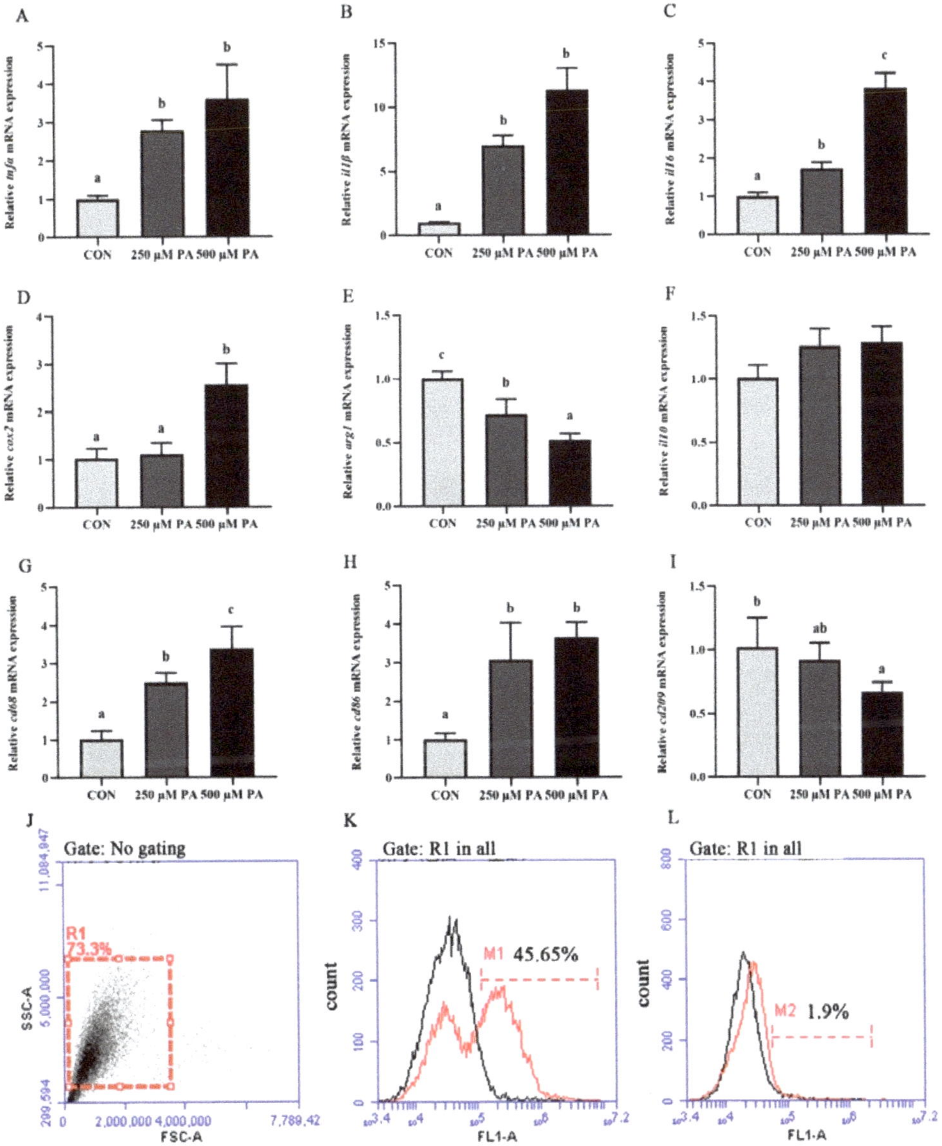

Figure 1. PA induced inflammatory response in macrophages of large yellow croaker. (**A–I**) mRNA expression levels of inflammatory genes (*tnfα*, *il1β*, *il6*, *cox2*, *arg1* and *il10*) and macrophage markers (*cd68*, *cd86* and *cd209*) after PA treatment (n = 6). (**J–L**) Fluorescence histograms of $CD68^+$ (**K**) and $CD209^+$ (**L**) populations after PA treatment (scale of M1). Represent macrophages gated (R1) on a forward scatter (FSC) versus side scatter (SSC) dot plot. Data are presented as the means ± SEM and are analyzed using one-way ANOVA, followed by Tukey's test. Bars labeled with the same letters are not significantly different ($p > 0.05$).

3.2. PA Activated the TLR-Related Genes Expression and MAPK Signaling Pathway in Macrophages

To investigate whether the TLR signaling pathway was activated by PA, TLR-related genes expression and downstream signaling pathways were detected in macrophages after PA treatment. PA treatment significantly upregulated the mRNA expression levels of *tlr3*, *tlr7*, *tlr13*, *tlr22*, *myd88* and *trif* ($p < 0.05$), while the mRNA expression levels of *tlr1*, *tlr2* and *tlr21* had no significant changes ($p > 0.05$) (Figure 2A). As a typical specific TLR, the increased expression level of *tlr22* was more conspicuous than the others. Moreover, the phosphorylation level of MAPK, including JNK and p38, was significantly increased after the PA treatment ($p < 0.05$), while the phosphorylation levels of ERK and IKK showed no significant differences ($p > 0.05$) (Figure 2B). These results revealed that the PA treatment induced TLRs gene expressions, especially *tlr22*, and activated the MAPK pathway in macrophages.

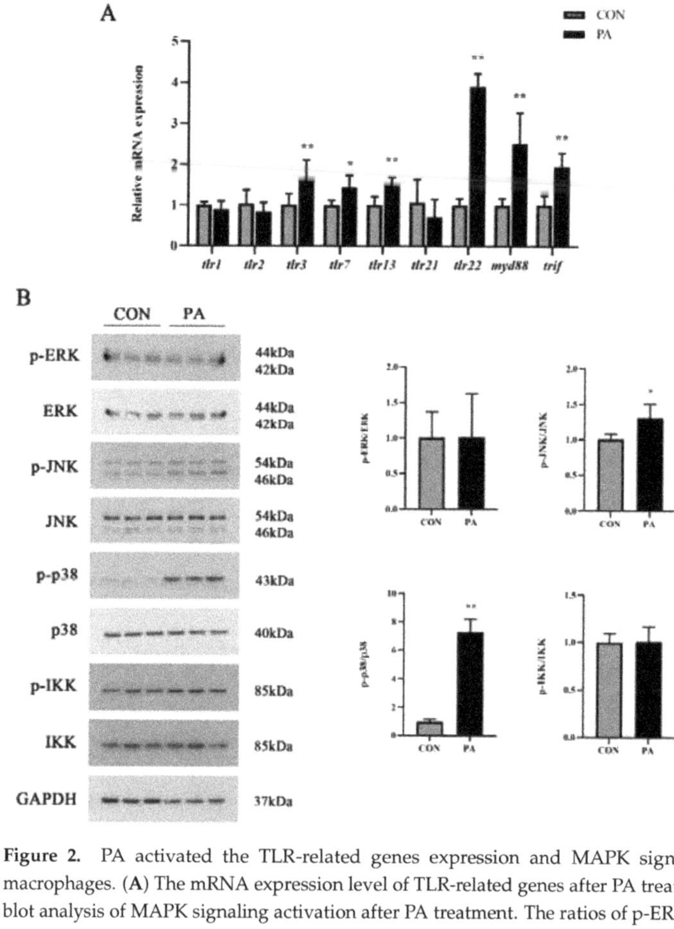

Figure 2. PA activated the TLR-related genes expression and MAPK signaling pathway in macrophages. (**A**) The mRNA expression level of TLR-related genes after PA treatment. (**B**) Western blot analysis of MAPK signaling activation after PA treatment. The ratios of p-ERK to ERK, p-p38 to p38, p-JNK to JNK and p-IKK to IKK were determined. The data are presented as the means ± SEM ($n = 6$) and are analyzed using independent t-tests. * $p < 0.05$ and ** $p < 0.01$ indicate significant differences compared with the control group.

3.3. Effects of TLR22-MAPK Signaling Pathway on PA-Induced Inflammation in Macrophages

To confirm whether the TLR22-MAPK pathway participated in PA-induced inflammation, TLR22 siRNA was transfected, and MAPK inhibitors were incubated in macrophages.

The transfection of siRNA into macrophages significantly inhibited the *tlr22* mRNA expression ($p < 0.05$) (Figure 3A). Meanwhile, *tlr22* knockdown significantly suppressed the PA-induced mRNA expression levels of proinflammatory genes, including *il1β*, *il6* and *cox2* ($p < 0.05$), while the mRNA expression level of *arg1* showed no significant differences ($p > 0.05$) (Figure 3B). As the downstream pathway of TLR, the change of MAPK was detected after *tlr22* knockdown under PA treatment. The results showed that the PA-induced phosphorylation levels of JNK and p38 were significantly inhibited after *tlr22* knockdown ($p < 0.05$) (Figure 3C). Furthermore, JNK and p38 inhibitor incubation significantly suppressed the PA-induced mRNA expression levels of proinflammatory genes, including *il1β*, *il6* and *cox2* ($p < 0.05$) (Figure 4). These results demonstrated that PA induced an inflammatory response of macrophages via the TLR22-MAPK signaling pathway.

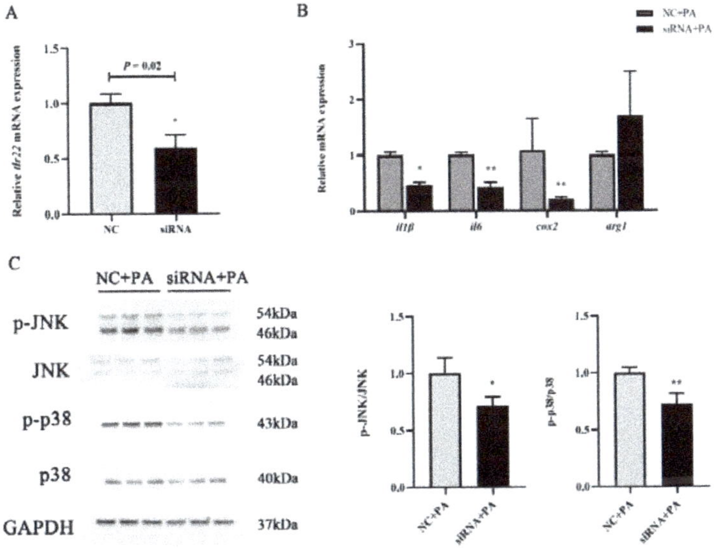

Figure 3. Effects of *tlr22* knockdown on PA-induced inflammation in macrophages. (**A**) The mRNA expression of *tlr22* after *tlr22* knockdown. (**B**) Effects of *tlr22* knockdown on the mRNA expression levels of inflammatory genes induced by PA (**C**) Effects of *tlr22* knockdown on the phosphorylation of the MAPK pathway induced by PA. The ratios of p-p38 to p38 and p-JNK to JNK were determined. The data are presented as the means ± SEM ($n = 3$) and are analyzed using independent *t*-tests. * $p < 0.05$ and ** $p < 0.01$ indicate significant differences compared with the control group.

3.4. *PPARγ Participated in PA-Induced Inflammation via TLR22-MAPK Pathway*

As an integral part of inflammatory responses, we further investigated whether PPARγ was involved in regulating PA-induced inflammation. The PA treatment significantly decreased the protein expression level of PPARγ ($p < 0.05$), while the mRNA expression level of *pparγ* showed no significant differences ($p > 0.05$) (Figure 5A,B). The PPARγ agonist and inhibitor were used to explore the role of PPARγ in PA-induced inflammation. The activation of PPARγ significantly suppressed the PA-induced mRNA expression levels of proinflammatory genes, including *tnfα*, *il1β*, *il6* and *cox2* ($p < 0.05$) (Figure 5C). Meanwhile, the inhibition of PPARγ significantly enhanced the mRNA expression levels of *il1β* and *il6* ($p < 0.05$) (Figure 5C). Moreover, *tlr22* knockdown significantly enhanced the mRNA expression level of *pparγ* induced by PA ($p < 0.05$) but was not significantly different in the protein expression ($p > 0.05$) (Figure 6A,B). However, p38 and JNK inhibitor incubation significantly enhanced the protein expression level of PPARγ induced by PA ($p < 0.05$) (Figure 6C). These results revealed that PPARγ participated in PA-induced inflammation via the TLR22-MAPK pathway.

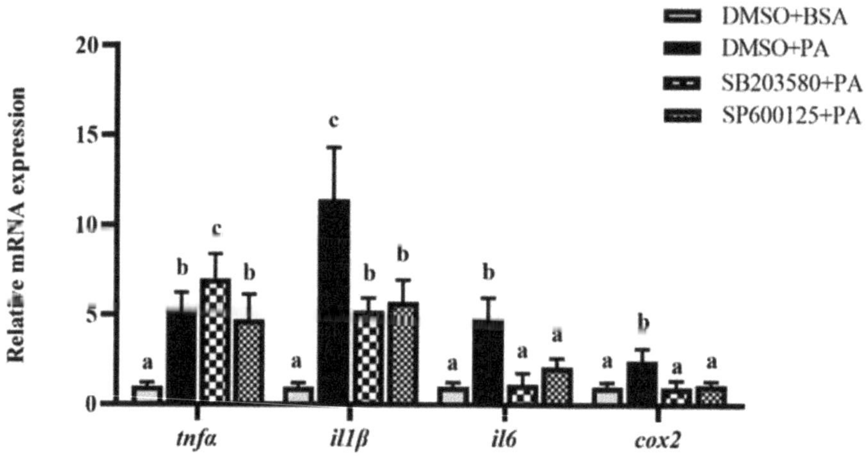

Figure 4. Effects of p38 (SB203580) and JNK (SP600125) inhibitors on the mRNA expression levels of inflammatory genes induced by PA. Data are presented as the means ± SEM (n = 6) and are analyzed using Tukey's test. Bars labeled with the same letters are not significantly different (p > 0.05).

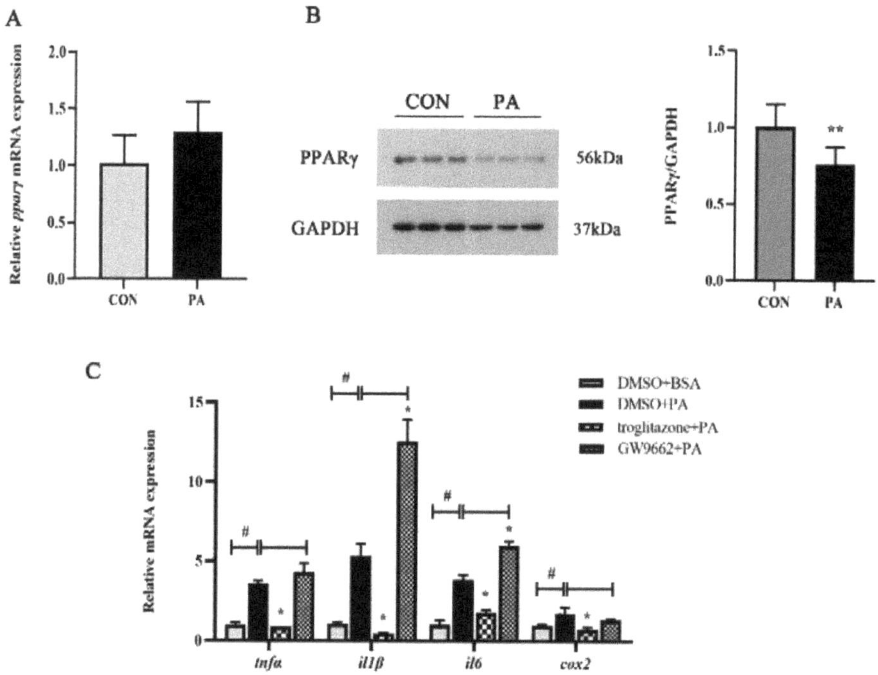

Figure 5. PPARγ was involved in PA-induced inflammation. (**A**,**B**) mRNA and protein expression levels of PPARγ after PA treatment (n = 6). The ratio of PPARγ to GAPDH was determined. (**C**) Effects of the PPARγ activator (troglitazone) and inhibitor (GW9662) on the mRNA expression levels of the inflammatory genes induced by PA (n = 3). Data are presented as the means ± SEM and analyzed using independent t-tests. # $p < 0.05$ indicates significant differences compared with the BSA group as the negative control. * $p < 0.05$ and ** $p < 0.01$ indicate significant differences compared with the control group.

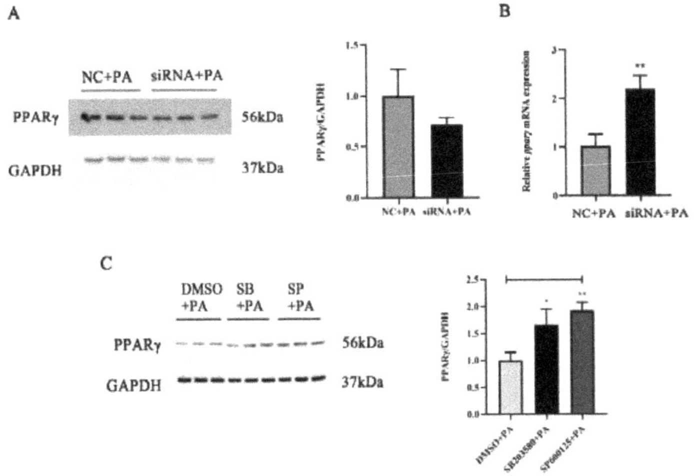

Figure 6. PPARγ participated in PA-induced inflammation via the TLR22-MAPK pathway. (**A**,**B**) Effects of *tlr22* knockdown on the protein and mRNA expression levels of PPARγ induced by PA ($n = 3$). (**C**) Effects of the p38 (SB203580) and JNK (SP600125) inhibitors on the protein expression level of PPARγ induced by PA ($n = 3$). Data are presented as the means ± SEM and analyzed using independent *t*-tests. * $p < 0.05$ and ** $p < 0.01$ indicate significant differences compared with the control group.

3.5. p38 MAPK Regulated the PA-Induced Activation of Nrf2

Nrf2, which was widely known to be the major regulation of antioxidant processes, also played a role in regulating inflammation. The PA treatment significantly upregulated the mRNA expression level of *nrf2* ($p < 0.05$) (Figure 7A). p38 inhibitor incubation significantly suppressed the PA-induced mRNA expression level of *nrf2* ($p < 0.05$), while the JNK inhibitor had no effects (Figure 7B). These results showed that p38 MAPK regulated the PA-induced activation of Nrf2.

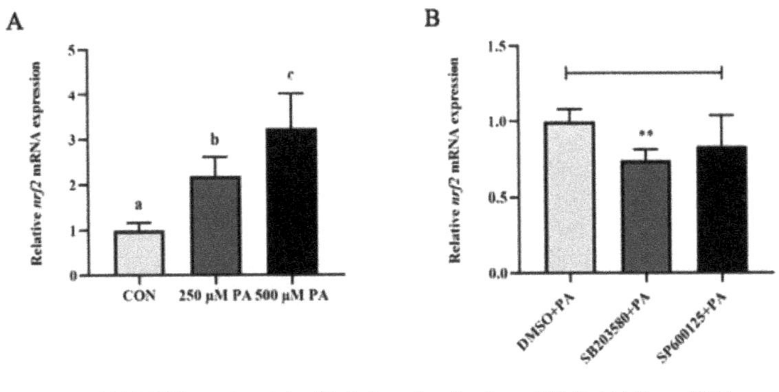

Figure 7. p38 MAPK regulated the PA-induced activation of Nrf2. (**A**) The mRNA expression level of *nrf2* after PA treatment. (**B**) Effects of the p38 (SB203580) and JNK (SP600125) inhibitors on the mRNA expression level of *nrf2* induced by PA. Data are presented as the means ± SEM ($n = 6$) and analyzed using one-way ANOVA, followed by Tukey's test and independent *t*-tests. Bars labeled with the same letters are not significantly different ($p > 0.05$ and ** $p < 0.01$).

3.6. The Protective Effect of DHA against PA-Induced Inflammation

N-3 PUFAs are widely known to exert anti-inflammatory effects. Hence, the protective effect of DHA against PA-induced inflammation in macrophages was investigated.

DHA treatment significantly suppressed the PA-induced mRNA expression levels of proinflammatory genes, including *tnfα*, *il1β*, *il6* and *cox2* ($p < 0.05$), while the mRNA expression of anti-inflammatory gene *arg1* was significantly upregulated ($p < 0.05$) (Figure 8A). However, the mRNA expression level of *tlr22* had no significant differences after DHA treatment ($p > 0.05$) (Figure 8B). Furthermore, the DHA treatment significantly reduced the PA-induced phosphorylation levels of JNK and p38 ($p < 0.05$) (Figure 8C). Moreover, DHA significantly recovered the decrease of PPARγ protein expression caused by PA ($p < 0.05$) (Figure 8C). DHA also significantly inhibited the increase of *nrf2* mRNA expression induced by PA ($p < 0.05$) (Figure 8D). These results demonstrated that the protective effect of DHA against the PA-induced inflammatory response could perform via the TLR22-MAPK-PPARγ/Nrf2 signaling pathway.

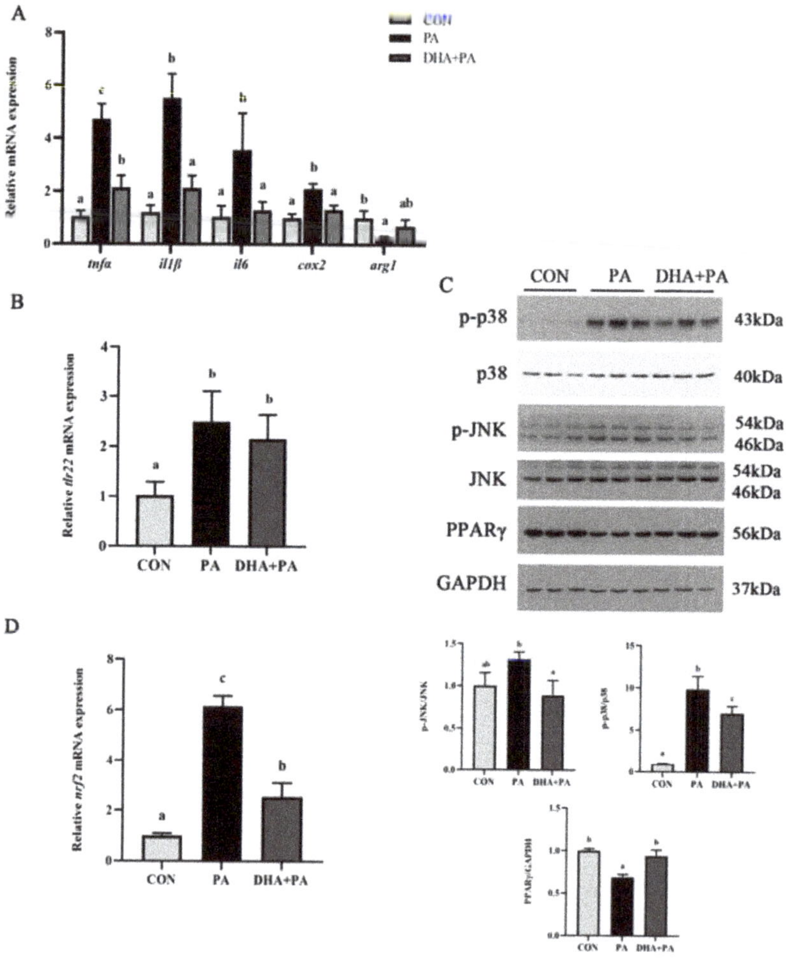

Figure 8. The protective effect of DHA against PA-induced inflammation. (**A**) mRNA expression levels of inflammatory genes after DHA treatment ($n = 4$). (**B**) The mRNA expression level of *tlr22* after DHA treatment ($n = 4$). (**C**) Western blot analysis of MAPK signaling activation and the PPARγ protein level after DHA treatment ($n = 3$). The ratios of p-p38 to p38, p-JNK to JNK and PPARγ to GAPDH were determined. (**D**) The mRNA expression level of *nrf2* after DHA treatment ($n = 4$). Data are presented as the means ± SEM and analyzed using one-way ANOVA, followed by Tukey's test. Bars labeled with the same letters are not significantly different ($p > 0.05$).

4. Discussion

This study utilized the primary head kidney macrophages, an important immune cell in fish, to investigate the potential molecular mechanism of the PA-induced inflammatory response. In the present study, we found that PA induced proinflammatory gene expressions and promoted proinflammatory macrophage polarization, which was consistent with the observations in mammals [32,33].

Fish rely more on their innate immune system to resist the invasion of bacteria because of their evolutionarily less well-developed adaptive immune system [34]. One of mechanisms by which the innate immune system senses stimulation is through TLR signaling. As the TLR family in teleost was different from that in mammals, the change of TLR-related genes in large yellow croaker was detected according to a previous study [19]. The results showed that PA significantly activated the expressions of TLRs and adaptor proteins. Among them, TLR22 may act as the main response receptor to PA. However, TLR22 orthologs in humans serve as nonfunctional pseudogenes [35]. Furthermore, TLR22 knockdown suppressed PA-induced inflammation in macrophages. The current study, for the first time, revealed the role of TLR22 in PA-induced inflammation in fish. Further investigation found that PA induced the activation of JNK and p38 pathways. Although, a previous study in large yellow croaker showed that PO and PA could activate the NF-κβ pathway in the liver and hepatocytes, the present study found that PA activated MAPK but not NF-κβ in macrophages [36,37]. Meanwhile, the transcriptome analysis in the spleen of large yellow croaker revealed multiple signaling pathways were involved in the antiviral response, including the MAPK, NF-β and JAK pathways [38,39]. Hence, previous studies do not have uniform results on the changes of downstream signaling in response to different stimulations [40–42]. These inconsistent results may be due to the fact that different tissues had different regulation mechanisms of TLRs in response to different stimulations. Therefore, the signaling pathways regulated by TLR22 in different stimulations remain to be further explored in fish.

Since PPARγ could be modulated by TLRs and perform an anti-inflammatory effect, we further confirmed whether PPARγ was involved in PA-induced inflammation. The results showed that PA decreased the PPARγ expression at the translation level. The effects of the agonist and inhibitor of PPARγ on inflammatory gene expressions indicated that it presented an anti-inflammatory effect in regulating PA-induced inflammation, which was consistent with previous studies in mammals [43]. Previous studies in large yellow croaker revealed that PPARγ could negatively regulate the expression of inflammatory genes by affecting the promoter activity of genes [44]. Furthermore, the *tlr22* knockdown and MAPK inhibitors recovered the decrease of PPARγ expression caused by PA. These results demonstrated that PPARγ participated in PA-induced inflammation via the TLR22-MAPK pathway in macrophages of large yellow croaker.

There is a complex interaction between the inflammatory response and antioxidant systems. Nrf2 is an important transcriptional regulator regulating the expression of varieties of anti-inflammatory and antioxidant genes and plays a central role in the response to stimulation [45]. In the current study, Nrf2 could be activated by PA, and p38 regulated the PA-induced activation of Nrf2. The results in mammals have reported that Nrf2 could participate in the alleviation of chronic inflammation by the MAPK pathway [46,47]. Therefore, the PA-induced activation of Nrf2 in the present study may be to compete and suppress the inflammatory response induced by PA.

N-3 PUFAs perform anti-inflammatory and antioxidant properties through various mechanisms and are always used as nutritional regulation strategies in mammals [48,49]. After understanding the molecular mechanism of PA in regulating inflammation, we further investigated whether DHA could alleviate PA-induced inflammation. The results showed that DHA inhibited the effect of PA on the mRNA expression of proinflammatory genes and the phosphorylation of JNK and p38, which was consistent with the results in mammals [50,51]. However, the mRNA expression of *tlr22* showed no differences. Previous studies in mammals suggested that the effect of PUFAs on impacting TLR activation might perform by regulating the lipid and protein compositions of the raft membrane, not directly affecting the TLR mRNA expression [52,53]. The results in our lab also found that DHA might regulate the activation of TLR22 by changing the biophysical properties of the cell membrane (unpublished data). Furthermore, DHA recovered the decrease of PPARγ protein expression induced by PA. DHA may affect PPARγ expression by regulating the signaling pathways or directly activating PPARγ as a ligand [54]. Together, these results revealed that the protective effect of DHA against PA-induced inflammation was executed through the TLR22-MAPK-PPARγ/Nrf2 signaling pathway.

5. Conclusions

In conclusion, the present study indicated the PA-induced inflammatory response of macrophages via the TLR22-MAPK-PPARγ/Nrf2 signaling pathway in large yellow croaker. Meanwhile, DHA was found to alleviate the PA-induced inflammation, thereby performing anti-inflammatory and antioxidation properties (Figure 9). These results could advance the understanding of the molecular mechanism of the inflammatory response induced by SFAs and provide nutritional strategies against inflammation, thereby improving the utilization rate of PO in aquafeed.

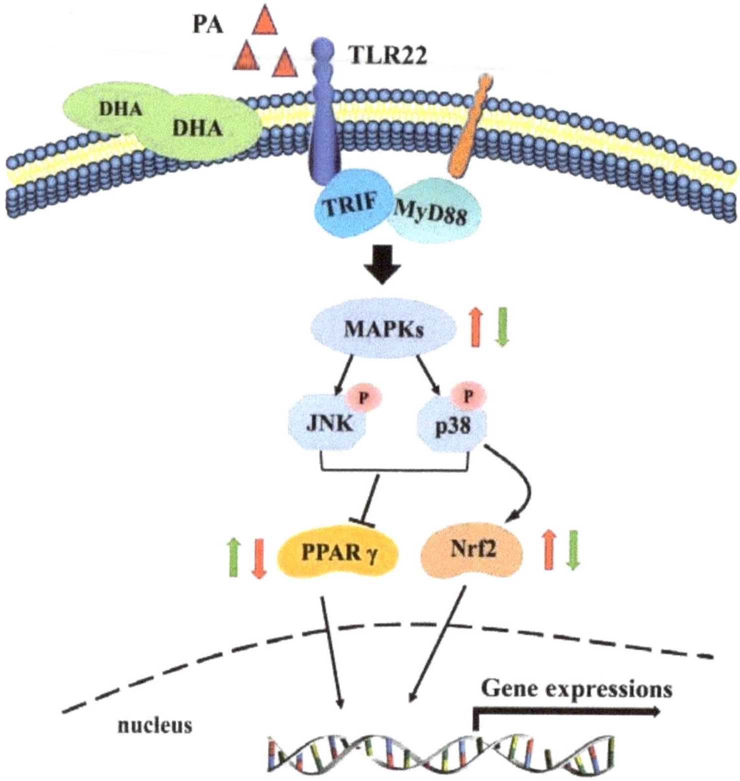

Figure 9. A working model showed the molecular mechanism of the PA-induced inflammatory response and the protective effect of DHA against the inflammation in macrophages of large yellow croaker.

Author Contributions: D.X.: conceptualization, data curation and writing—review and editing; K.C. and Q.L.: methodology and visualization; S.Z. and J.Z.: software and analyze; S.G. and T.H.:

methodology and editing and K.M. and Q.A.: project administration and funding acquisition. All authors have read and agreed to the published version of the manuscript.

Funding: This work was supported by the Key Program of National Natural Science Foundation of China (Grant No. 31830103), the National Science Fund for Distinguished Young Scholars of China (Grant No. 31525024), the Ten-thousand Talents Program (Grant No. 2018-29), the Scientific and Technological Innovation of Blue Granary (Grant No. 2018YFD0900402) and the Agriculture Research System of China (Grant No. CARS-47-11).

Institutional Review Board Statement: All animal care and handling procedures performed in the present study were approved by the Animal Care Committee of the Ocean University of China (Approval number SPXY2020012).

Informed Consent Statement: Not applicable.

Data Availability Statement: Data is contained within the article.

Conflicts of Interest: The authors declare no competing financial interest.

References

1. Chavez, J.A.; Summers, S.A. Characterizing the effects of saturated fatty acids on insulin signaling and ceramide and diacylglycerol accumulation in 3T3-L1 adipocytes and C2C12 myotubes. *Arch. Biochem. Biophys.* **2003**, *419*, 101–109. [CrossRef] [PubMed]
2. Peng, G.; Li, L.; Liu, Y.; Pu, J.; Zhang, S.; Yu, J.; Zhao, J.; Liu, P. Oleate blocks palmitate-induced abnormal lipid distribution, endoplasmic reticulum expansion and stress, and insulin resistance in skeletal muscle. *Endocrinology* **2011**, *152*, 2206–2218. [CrossRef] [PubMed]
3. Leamy, A.K.; Egnatchik, R.A.; Shiota, M.; Ivanova, P.T.; Myers, D.S.; Brown, H.A.; Young, J.D. Enhanced synthesis of saturated phospholipids is associated with ER stress and lipotoxicity in palmitate treated hepatic cells. *J. Lipid Res.* **2014**, *55*, 1478–1488. [CrossRef] [PubMed]
4. Korbecki, J.; Bajdak-Rusinek, K. The effect of palmitic acid on inflammatory response in macrophages: An overview of molecular mechanisms. *Inflamm. Res.* **2019**, *68*, 915–932. [CrossRef]
5. Bell, J.G.; Henderson, R.J.; Tocher, D.R.; McGhee, F.; Dick, J.R.; Porter, A.; Smullen, R.P.; Sargent, J.R. Substituting fish oil with crude palm oil in the diet of Atlantic salmon (*Salmo salar*) affects muscle fatty acid composition and hepatic fatty acid metabolism. *J. Nutr.* **2002**, *132*, 222–230. [CrossRef]
6. Bahurmiz, O.M.; Ng, W.-K. Effects of dietary palm oil source on growth, tissue fatty acid composition and nutrient digestibility of red hybrid tilapia, *Oreochromis* sp., raised from stocking to marketable size. *Aquaculture* **2007**, *262*, 382–392. [CrossRef]
7. Han, Y.Z.; Jiang, Z.Q.; Ren, T.J.; Koshio, S.; Gao, J.; Komilus, C.F.; Jiang, B.Q. Effect of dietary fish oil replacement with palm oil on growth performance, hematology and liver anti-oxidative enzymes of juvenile Japanese flounder *Paralichthys olivaceus* (Temminck & Schlegel, 1846). *J. Appl. Ichthyol.* **2015**, *31*, 518–524. [CrossRef]
8. Zhang, J.; Liu, Q.; Pang, Y.; Xu, X.; Cui, K.; Zhang, Y.; Mai, K.; Ai, Q. Molecular cloning and the involvement of IRE1alpha-XBP1s signaling pathway in palmitic acid induced-Inflammation in primary hepatocytes from large yellow croaker (*Larimichthys croceu*). *Fish Shellfish. Immunol.* **2020**, *98*, 112–121. [CrossRef]
9. Shi, H.; Kokoeva, M.V.; Inouye, K.; Tzameli, I.; Yin, H.; Flier, J.S. TLR4 links innate immunity and fatty acid-induced insulin resistance. *J. Clin. Investig.* **2006**, *116*, 3015–3025. [CrossRef]
10. Rebl, A.; Goldammer, T.; Seyfert, H.M. Toll-like receptor signaling in bony fish. *Vet. Immunol. Immunopathol.* **2010**, *134*, 139–150. [CrossRef]
11. Takeda, K.; Akira, S. TLR signaling pathways. *Semin. Immunol.* **2004**, *16*, 3–9. [CrossRef] [PubMed]
12. Akira, S.; Takeda, K. Toll-like receptor signalling. *Nat. Rev. Immunol.* **2004**, *4*, 499–511. [CrossRef] [PubMed]
13. Rocha, D.M.; Caldas, A.P.; Oliveira, L.L.; Bressan, J.; Hermsdorff, H.H. Saturated fatty acids trigger TLR4-mediated inflammatory response. *Atherosclerosis* **2016**, *244*, 211–215. [CrossRef] [PubMed]
14. Huang, S.; Rutkowsky, J.M.; Snodgrass, R.G.; Ono-Moore, K.D.; Schneider, D.A.; Newman, J.W.; Adams, S.H.; Hwang, D.H. Saturated fatty acids activate TLR-mediated proinflammatory signaling pathways. *J. Lipid Res.* **2012**, *53*, 2002–2013. [CrossRef] [PubMed]
15. Panda, R.P.; Chakrapani, V.; Patra, S.K.; Saha, J.N.; Jayasankar, P.; Kar, B.; Sahoo, P.K.; Barman, H.K. First evidence of comparative responses of Toll-like receptor 22 (TLR22) to relatively resistant and susceptible Indian farmed carps to *Argulus siamensis* infection. *Dev. Comp. Immunol.* **2014**, *47*, 25–35. [CrossRef] [PubMed]
16. Ding, X.; Lu, D.Q.; Hou, Q.H.; Li, S.S.; Liu, X.C.; Zhang, Y.; Lin, H.R. Orange-spotted grouper (*Epinephelus coioides*) toll-like receptor 22: Molecular characterization, expression pattern and pertinent signaling pathways. *Fish Shellfish Immunol.* **2012**, *33*, 494–503. [CrossRef]
17. Li, H.; Yang, G.; Ma, F.; Li, T.; Yang, H.; Rombout, J.H.; An, L. Molecular characterization of a fish-specific toll-like receptor 22 (TLR22) gene from common carp (*Cyprinus carpio* L.): Evolutionary relationship and induced expression upon immune stimulants. *Fish Shellfish Immunol.* **2017**, *63*, 74–86. [CrossRef]

18. Ji, J.; Ramos-Vicente, D.; Navas-Perez, E.; Herrera-Ubeda, C.; Lizcano, J.M.; Garcia-Fernandez, J.; Escriva, H.; Bayes, A.; Roher, N. Characterization of the TLR family in *Branchiostoma lanceolatum* and discovery of a novel TLR22-like involved in dsRNA recognition in Amphioxus. *Front. Immunol.* **2018**, *9*, 2525. [CrossRef]
19. Tan, P.; Dong, X.; Mai, K.; Xu, W.; Ai, Q. Vegetable oil induced inflammatory response by altering TLR-NF-kappaB signalling, macrophages infiltration and polarization in adipose tissue of large yellow croaker (*Larimichthys crocea*). *Fish Shellfish Immunol.* **2016**, *59*, 398–405. [CrossRef]
20. Xiao, X.; Qin, Q.; Chen, X. Molecular characterization of a Toll-like receptor 22 homologue in large yellow croaker (*Pseudosciaena crocea*) and promoter activity analysis of its 5′-flanking sequence. *Fish Shellfish Immunol.* **2011**, *30*, 224–233. [CrossRef]
21. Calder, P.C. Omega-3 polyunsaturated fatty acids and inflammatory processes: Nutrition or pharmacology? *Br. J. Clin. Pharmacol.* **2013**, *75*, 645–662. [CrossRef] [PubMed]
22. Calder, P.C. n−3 Polyunsaturated fatty acids, inflammation, and inflammatory diseases. *Am. J. Clin. Nutr.* **2006**, *83*, 1505S–1519S. [CrossRef] [PubMed]
23. Lee, J.Y.; Plakidas, A.; Lee, W.H.; Heikkinen, A.; Chanmugam, P.; Bray, G.; Hwang, D.H. Differential modulation of Toll-like receptors by fatty acids: Preferential inhibition by n-3 polyunsaturated fatty acids. *J. Lipid Res.* **2003**, *44*, 479–486. [CrossRef] [PubMed]
24. Norris, P.C.; Dennis, E.A. Omega-3 fatty acids cause dramatic changes in TLR4 and purinergic eicosanoid signaling. *Proc. Natl. Acad. Sci. USA* **2012**, *109*, 8517–8522. [CrossRef]
25. Li, Q.; Cui, K.; Wu, M.; Xu, D.; Mai, K.; Ai, Q. Polyunsaturated fatty acids influence LPS-induced inflammation of fish macrophages through differential modulation of pathogen recognition and p38 MAPK/NF-kappaB signaling. *Front. Immunol.* **2020**, *11*, 559332. [CrossRef]
26. Zuo, R.; Ai, Q.; Mai, K.; Xu, W.; Wang, J.; Xu, H.; Liufu, Z.; Zhang, Y. Effects of dietary n-3 highly unsaturated fatty acids on growth, nonspecific immunity, expression of some immune related genes and disease resistance of large yellow croaker (*Larmichthys crocea*) following natural infestation of parasites (*Cryptocaryon irritans*). *Fish Shellfish Immunol.* **2012**, *32*, 249–258. [CrossRef]
27. Xu, H.; Turchini, G.M.; Francis, D.S.; Liang, M.; Mock, T.S.; Rombenso, A.; Ai, Q. Are fish what they eat? A fatty acid's perspective. *Prog. Lipid Res.* **2020**, *80*, 101064. [CrossRef]
28. Du, J.; Xiang, X.; Li, Y.; Ji, R.; Xu, H.; Mai, K.; Ai, Q. Molecular cloning and characterization of farnesoid X receptor from large yellow croaker (*Larimichthys crocea*) and the effect of dietary CDCA on the expression of inflammatory genes in intestine and spleen. *Comp. Biochem. Physiol. B-Biochem. Mol. Biol.* **2018**, *216*, 10–17. [CrossRef]
29. Li, Q.; Ai, Q.; Mai, K.; Xu, W.; Zheng, Y. A comparative study: In vitro effects of EPA and DHA on immune functions of head-kidney macrophages isolated from large yellow croaker (*Larmichthys crocea*). *Fish Shellfish Immunol.* **2013**, *35*, 933–940. [CrossRef]
30. Livak, K.J.; Schmittgen, T.D. Analysis of relative gene expression data using real-time quantitative PCR and the 2(-Delta Delta C(T)) Method. *Methods* **2001**, *25*, 402–408. [CrossRef]
31. Xu, D.; Li, Q.; Zhou, Y.; Shen, Y.; Lai, W.; Hao, T.; Ding, Y.; Mai, K.; Ai, Q. Functional analysis and regulation mechanism of interferon gamma in macrophages of large yellow croaker (*Larimichthys crocea*). *Int. J. Biol. Macromol.* **2022**, *194*, 153–162. [CrossRef] [PubMed]
32. Liu, S.P.; Li, X.Y.; Li, Z.; He, L.N.; Xiao, Y.; Yan, K.; Zhou, Z.G. Octanoylated Ghrelin Inhibits the Activation of the Palmitic Acid-Induced TLR4/NF-kappaB Signaling Pathway in THP-1 Macrophages. *ISRN Endocrinol.* **2012**, *2012*, 237613. [CrossRef] [PubMed]
33. Suganami, T.; Tanimoto-Koyama, K.; Nishida, J.; Itoh, M.; Yuan, X.; Mizuarai, S.; Kotani, H.; Yamaoka, S.; Miyake, K.; Aoe, S.; et al. Role of the Toll-like receptor 4/NF-kappaB pathway in saturated fatty acid-induced inflammatory changes in the interaction between adipocytes and macrophages. *Arterioscler. Thromb. Vasc. Biol.* **2007**, *27*, 84–91. [CrossRef] [PubMed]
34. Uribe, C.; Folch, H.; Enríquez, R.; Moran, G. Innate and adaptive immunity in teleost fish: A review. *Vet. Med.* **2011**, *56*, 486. [CrossRef]
35. Roach, J.C.; Glusman, G.; Rowen, L.; Kaur, A.; Purcell, M.K.; Smith, K.D.; Hood, L.E.; Aderem, A. The evolution of vertebrate Toll-like receptors. *Proc. Natl. Acad. Sci. USA* **2005**, *102*, 9577–9582. [CrossRef]
36. Li, X.; Ji, R.; Cui, K.; Chen, Q.; Chen, Q.; Fang, W.; Mai, K.; Zhang, Y.; Xu, W.; Ai, Q. High percentage of dietary palm oil suppressed growth and antioxidant capacity and induced the inflammation by activation of TLR-NF-kappaB signaling pathway in large yellow croaker (*Larimichthys crocea*). *Fish Shellfish Immunol.* **2019**, *87*, 600–608. [CrossRef]
37. Wang, T.; Yang, B.; Ji, R.; Xu, W.; Mai, K.; Ai, Q. Omega-3 polyunsaturated fatty acids alleviate hepatic steatosis-induced inflammation through Sirt1-mediated nuclear translocation of NF-kappaB p65 subunit in hepatocytes of large yellow croaker (*Larmichthys crocea*). *Fish Shellfish Immunol.* **2017**, *71*, 76–82. [CrossRef]
38. Mu, Y.; Li, M.; Ding, F.; Ding, Y.; Ao, J.; Hu, S.; Chen, X. De novo characterization of the spleen transcriptome of the large yellow croaker (*Pseudosciaena crocea*) and analysis of the immune relevant genes and pathways involved in the antiviral response. *PLoS ONE* **2014**, *9*, e97471. [CrossRef]
39. Mu, Y.; Ding, F.; Cui, P.; Ao, J.; Hu, S.; Chen, X. Transcriptome and expression profiling analysis revealed changes of multiple signaling pathways involved in immunity in the large yellow croaker during *Aeromonas hydrophila* infection. *BMC Genom.* **2010**, *11*, 506. [CrossRef]

40. Ding, X.; Liang, Y.; Peng, W.; Li, R.; Lin, H.; Zhang, Y.; Lu, D. Intracellular TLR22 acts as an inflammation equalizer *via* suppression of NF-κB and selective activation of MAPK pathway in fish. *Fish Shellfish Immunol.* **2018**, *72*, 646–657. [CrossRef]
41. Matsuo, A.; Oshiumi, H.; Tsujita, T.; Mitani, H.; Kasai, H.; Yoshimizu, M.; Matsumoto, M.; Seya, T. Teleost TLR22 recognizes RNA duplex to induce IFN and protect cells from birnaviruses. *J. Immunol.* **2008**, *181*, 3474–3485. [CrossRef] [PubMed]
42. Ji, J.; Liao, Z.; Rao, Y.; Li, W.; Yang, C.; Yuan, G.; Feng, H.; Xu, Z.; Shao, J.; Su, J. Thoroughly remold the localization and signaling pathway of TLR22. *Front. Immunol.* **2020**, *10*, 3003. [CrossRef] [PubMed]
43. Li, H.; Ruan, X.Z.; Powis, S.H.; Fernando, R.; Mon, W.Y.; Wheeler, D.C.; Moorhead, J.F.; Varghese, Z. EPA and DHA reduce LPS-induced inflammation responses in HK-2 cells: Evidence for a PPAR-γ–dependent mechanism. *Kidney Int.* **2005**, *67*, 867–874. [CrossRef] [PubMed]
44. Wu, M.; Li, Q.; Mai, K.; Ai, Q. Regulation of free fatty acid receptor 4 on inflammatory gene induced by LPS in large yellow croaker (*Larimichthys crocea*). *Front. Immunol.* **2021**, *12*, 703914. [CrossRef] [PubMed]
45. Kaspar, J.W.; Niture, S.K.; Jaiswal, A.K. Nrf2:INrf2 (Keap1) signaling in oxidative stress. *Free Radic. Biol. Med.* **2009**, *47*, 1304–1309. [CrossRef]
46. Lee, M.S.; Lee, B.; Park, K.E.; Utsuki, T.; Shin, T.; Oh, C.W.; Kim, H.R. Dieckol enhances the expression of antioxidant and detoxifying enzymes by the activation of Nrf2-MAPK signalling pathway in HepG2 cells. *Food Chem.* **2015**, *174*, 538–546. [CrossRef]
47. Yao, P.; Nussler, A.; Liu, L.; Hao, L.; Song, F.; Schirmeier, A.; Nussler, N. Quercetin protects human hepatocytes from ethanol-derived oxidative stress by inducing heme oxygenase-1 *via* the MAPK/Nrf2 pathways. *J. Hepatol.* **2007**, *47*, 253–261. [CrossRef]
48. De Caterina, R. n–3 fatty acids in cardiovascular disease. *N. Engl. J. Med.* **2011**, *364*, 2439–2450. [CrossRef]
49. Breslow, J.L. n−3 Fatty acids and cardiovascular disease. *Am. J. Clin. Nutr.* **2006**, *83*, 1477S–1482S. [CrossRef]
50. Kennedy, A.; Martinez, K.; Chuang, C.C.; LaPoint, K.; McIntosh, M. Saturated fatty acid-mediated inflammation and insulin resistance in adipose tissue: Mechanisms of action and implications. *J. Nutr.* **2009**, *139*, 1–4. [CrossRef]
51. White, P.J.; Mitchell, P.L.; Schwab, M.; Trottier, J.; Kang, J.X.; Barbier, O.; Marette, A. Transgenic omega-3 PUFA enrichment alters morphology and gene expression profile in adipose tissue of obese mice: Potential role for protectins. *Metabolism* **2015**, *64*, 666–676. [CrossRef] [PubMed]
52. Wong, S.W.; Kwon, M.J.; Choi, A.M.; Kim, H.P.; Nakahira, K.; Hwang, D.H. Fatty acids modulate Toll-like receptor 4 activation through regulation of receptor dimerization and recruitment into lipid rafts in a reactive oxygen species-dependent manner. *J. Biol. Chem.* **2009**, *284*, 27384–27392. [CrossRef] [PubMed]
53. Fessler, M.B.; Rudel, L.L.; Brown, J.M. Toll-like receptor signaling links dietary fatty acids to the metabolic syndrome. *Curr. Opin. Lipidol.* **2009**, *20*, 379–385. [CrossRef] [PubMed]
54. Wang, L.; Waltenberger, B.; Pferschy-Wenzig, E.M.; Blunder, M.; Liu, X.; Malainer, C.; Blazevic, T.; Schwaiger, S.; Rollinger, J.M.; Heiss, E.H.; et al. Natural product agonists of peroxisome proliferator-activated receptor gamma (PPARgamma): A review. *Biochem. Pharmacol.* **2014**, *92*, 73–89. [CrossRef] [PubMed]

antioxidants

MDPI

Article

Dietary Vitamin A Improved the Flesh Quality of Grass Carp (*Ctenopharyngodon idella*) in Relation to the Enhanced Antioxidant Capacity through Nrf2/Keap 1a Signaling Pathway

Pei Wu [1,2,3,†], Li Zhang [1,†], Weidan Jiang [1,2,3], Yang Liu [1,2,3], Jun Jiang [4], Shengyao Kuang [5], Shuwei Li [5], Ling Tang [5], Wuneng Tang [5], Xiaoqiu Zhou [1,2,3,*,‡] and Lin Feng [1,2,3,*]

[1] Animal Nutrition Institute, Sichuan Agricultural University, Chengdu 611130, China; wupei0911@sicau.edu.cn (P.W.); lilizhang09@gmail.com (L.Z.); WDJiang@sicau.edu.cn (W.J.); 11081@sicau.edu.cn (Y.L.)
[2] Fish Nutrition and Safety Production University Key Laboratory of Sichuan Province, Sichuan Agricultural University, Chengdu 611130, China
[3] Key Laboratory for Animal Disease-Resistance Nutrition of China Ministry of Education, Sichuan Agricultural University, Chengdu 611130, China
[4] College of Animal Science and Technology, Sichuan Agricultural University, Chengdu 611130, China; jjun@sicau.edu.cn
[5] Animal Nutrition Institute, Sichuan Academy of Animal Science, Chengdu 610066, China; ksy_cd@163.com (S.K.); lishuwei84511614@126.com (S.L.); tingling@vip.163.com (L.T.); WNTANG@163.com (W.T.)
* Correspondence: zhouxq@sicau.edu.cn (X.Z.); fenglin@sicau.edu.cn (L.F.)
† Both authors contributed equally to this work.
‡ Lead contact.

Citation: Wu, P.; Zhang, L.; Jiang, W.; Liu, Y.; Jiang, J.; Kuang, S.; Li, S.; Tang, L.; Tang, W.; Zhou, X.; et al. Dietary Vitamin A Improved the Flesh Quality of Grass Carp (*Ctenopharyngodon idella*) in Relation to the Enhanced Antioxidant Capacity through Nrf2/Keap 1a Signaling Pathway. *Antioxidants* **2022**, *11*, 148. https://doi.org/10.3390/antiox11010148

Academic Editors: Min Xue, Junmin Zhang, Zhenyu Du, Jie Wang and Wei Si

Received: 23 November 2021
Accepted: 4 January 2022
Published: 12 January 2022

Publisher's Note: MDPI stays neutral with regard to jurisdictional claims in published maps and institutional affiliations.

Abstract: Fish is an important animal-source food for humans. However, the oxidative stress-induced by intensive aquaculture usually causes deterioration of fish meat quality. The nutritional way has been considered to be a useful method for improving fish flesh quality. This study using the same growth experiment as our previous study was conducted to investigate whether vitamin A could improve flesh quality by enhancing antioxidative ability via Nrf2/Keap1 signaling in fish muscle. Six diets with different levels of vitamin A were fed to grass carp (*Ctenopharyngodon idella*) (262.02 ± 0.45 g) for 10 weeks. Dietary vitamin A significantly improved flesh sensory appeal and nutritional value, as evident by higher pH_{24h} value, water-holding capacity, shear force, contents of protein, lipid, four indispensable amino acids (lysine, methionine, threonine, and arginine) and total polyunsaturated fatty acid in the muscle. Furthermore, dietary vitamin A reduced oxidative damage, as evident by decreased levels of muscle reactive oxygen species, malondialdehyde, and protein carbonyl, enhanced activities of antioxidative enzyme (catalase, copper/zinc superoxide dismutase (CuZnSOD), MnSOD, glutathione peroxidase, and glutathione reductase), as well as increased content of glutathione, which was probably in relation to the activation of nuclear factor erythroid 2-related factor 2 (Nrf2) signaling. These findings demonstrated that dietary vitamin A improved flesh quality probably by enhancing antioxidant ability through Nrf2/Keap 1a signaling in fish.

Keywords: vitamin A; flesh quality; grass carp; antioxidant; Nrf2 signaling

1. Introduction

Vitamin A, which is an unsaturated monohydric alcohol with β-ionone ring, is an essential nutrient for fish and capable of scavenging peroxyl radicals thus inhibiting lipid peroxidation in vitro [1]. Our previous study showed that vitamin A reduced the oxidative damage of lipid and protein in grass carp intestine [2]. To our knowledge, fish is an important animal-source food providing essential nutrients with high bioavailability, such as balanced amino acid and omega 3 long-chain polyunsaturated fatty acids (n-3 LC PUFA) for humans [3]. However, fish meat quality could be deteriorated by oxidative stress that is usually caused by intensive aquaculture. The nutritional way has been considered to be a useful method for enhancing fish flesh quality. Whether vitamin A could enhance fish flesh quality via its antioxidative benefits is not known so far.

It is well known that sensory property is one of the main sets of characteristics that make up fish flesh quality as perceived by the consumer [4]. Sensory acceptability of meat is primarily determined by water-holding capacity (WHC) and tenderness, which can be affected by pH value [5]. However, muscle pH value and WHC were reduced by oxidative stress in common carp (*Cyprinus carpio*) [6]. To date, no report has shown the effect of vitamin A on sensory appeal of fish flesh in relation to antioxidant capacity. In terrestrial animals, dietary vitamin A improved the juiciness and tenderness of longissimus thoracis from Holstein bulls and steers [7], the tenderness of longissimus lumborum from Angus crossbred steers [8], and increased the shear force, and decreased the pH_{24h} and drip loss of breast muscle in broiler [9]. Accordingly, dietary vitamin A might change the sensory appeal of fish flesh via antioxidation, which warrants further investigate.

In addition, nutritional value is another main characteristic of fish flesh quality [4]. A safe food supply provides nutritional benefit while posing minimal risks to consumers' health [10]. However, protein oxidative change induces loss of essential amino acids and decrease in digestibility, ultimately reducing the nutritional quality of muscle [11], while lipid oxidation results in loss of nutrient value, off-flavor development, and accumulation of toxic compounds, which may be detrimental to the health of consumers [12]. Yet, little is known about whether vitamin A regulated nutritional value of fish flesh via antioxidant capacity. Studies have found that certain levels of dietary vitamin A enhanced the body crude protein content of juvenile hybrid tilapia (*Oreochromis niloticus* × *O. aureus*) [13], and enhanced the perirenal fat in finishing pigs [14]. Furthermore, certain contents of dietary vitamin A up-regulated fatty acid synthase mRNA level and enzyme activity in the liver of orange spotted grouper (*Epinephelus coioides*) [15]. Thus, vitamin A might affect the nutritional value of fish flesh, which awaits further characterization.

The present study used the same animal trial as our previous study, which reported that vitamin A deficiency depressed the growth and intestinal immunity of on-growing grass carp [16]. The present study aimed to investigate the effects of vitamin A on the flesh sensory appeal, nutritional quality, antioxidative ability, and the possible mechanisms in grass carp, which may be useful for elucidating the mechanisms whereby dietary vitamin A influenced muscle quality in fish. Additionally, the dietary vitamin A requirements for on-growing grass carp based on the muscle antioxidative parameters were also evaluated.

2. Materials and Methods

2.1. Animals, Diets, and Experimental Design

Healthy grass carp were procured from a commercial farm (Sichuan, China) and acclimatized to experimental conditions for four weeks in cages (1.4 m × 1.4 m × 1.4 m). Following another two weeks of vitamin A depletion period, 540 similarly-sized fish (262.02 ± 0.45 g) were randomly allocated into eighteen cages (three cages per treatment) and fed with six experimental diets differing in vitamin A content (18.69 (un-supplemented control), 606.8, 1209, 1798, 2805, and 3796 IU/kg), respectively. The experimental diets were formulated by supplementing retinyl acetate (500,000 IU/g) at concentrations of 0 (un-supplemented control), 600, 1200, 1800, 2800, and 3800 IU/kg into the basal diet (Table 1), which contained 30% crude protein according to the study of Khan et al. [17]. The dietary vitamin A contents were assayed by high-performance liquid chromatographic (HPLC) as described by Moren et al. [18]. During the 70 days experimental period, the fish were kept in natural light and dark cycle, and fed four times daily to apparent satiation. Uneaten feed was removed by using a disc equipped in the bottom of cage. Water temperature was 28 ± 2 °C, pH 7.0 ± 0.2, and dissolved oxygen ≥ 6.0 mg/L.

Table 1. Formulation and nutrient content of the basal diet.

Ingredients	%	Nutrients Content	%
Fish meal	15.55	Crude protein [4]	29.71
Soybean protein concentrate	26.25	Crude lipid [4]	3.58
Gelatin	3.13	n-3 [5]	0.50
α-starch	24.00	n-6 [5]	1.00
Maize starch	16.32	Available phosphorus [5]	0.84
Soybean oil	1.93		
Cellulose	5.00		
L-Met (98%)	0.40		
$Ca(H_2PO_4)_2$	2.87		
Vitamin premix [1]	1.00		
Mineral premix [2]	2.00		
Vitamin A premix [3]	1.00		
Choline chloride (60%)	0.50		
Ethoxyquin (30%)	0.05		

[1] Per kilogram of vitamin premix (g/kg): cholecalciferol (172 mg/g), 0.40; DL-α-tocopherol acetate (50%), 12.58; menadione (22.9%), 0.83; cyanocobalamin (1%), 0.94; D-biotin (2%), 0.75; folic acid (95%), 0.42; thiamine nitrate (98%), 0.11; ascorbic acetate (95%), 4.31; niacin (99%), 2.58; meso-inositol (98%), 19.39; calcium-D-pantothenate (98%), 2.56; riboflavin (80%), 0.63; pyridoxine hydrochloride (98%), 0.62. All ingredients were diluted with maize starch to 1 kg. [2] Per kilogram of mineral premix (g/kg): $MnSO_4{\cdot}H_2O$ (31.8% Mn), 1.8900; $MgSO_4{\cdot}H_2O$ (15.0% Mg), 200.0000; $FeSO_4{\cdot}H_2O$ (30.0% Fe), 24.5700; $ZnSO_4{\cdot}H_2O$ (34.5% Zn), 8.2500; $CuSO_4{\cdot}5H_2O$ (25.0% Cu), 0.9600; KI (76.9% I), 0.0668; Na_2SeO_3 (44.7% Se), 0.0168. All ingredients were diluted with maize starch to 1 kg. [3] Vitamin A premix: premix was added to obtain graded level of vitamin A and the amount of maize starch was reduced to compensate. [4] Crude protein and crude lipid contents were measured value. [5] Available phosphorus, n-3, and n-6 contents were calculated according to NRC [19].

2.2. Sample Collection and Biochemical Analysis

Fishes were anaesthetized with benzocaine before sampling, following procedures from Chen et al. [20]. After sacrifice, the left-side muscle of two fish in each cage were quickly manually filleted, frozen, and preserved at −80 °C until needed for analysis, while the right-side muscle was used for analysis of sensory appeal parameters. Parts of the muscle were fixed in 10% neutral formalin for morphological observation, similar to the previous study from our laboratory [2]. pH value, cooking loss, and shear force of fish muscle were assayed according to Brinker and Reiter [21]. Briefly, muscle pH value was detected using a calibrated pH probe (Testo AG Company, Lenzkirch, Germany) after slaughter and then at 24 h post-mortem (pH_{24h}). Cooking loss was determined by weight changes before and after cooking (sealed in PE-bag and heated at 70 °C for 20 min). Flesh shear force was assayed using an Instron 4411 material testing instrument (Instron Corporation, Canton, MI, USA). Proximate composition, free amino acids, and fatty acids contents of muscle were analyzed according to the method of AOAC [22], using a L-8800 amino acid analyzer (Hitachi Ltd., Tokyo, Japan) and gas chromatography method similar to Carbonera et al. [23], respectively.

For antioxidant-related parameters assay, muscle tissue homogenates were prepared according to the kit instructions. Reactive oxygen species (ROS), malondialdehyde (MDA), protein carbonyl (PC), anti-superoxide anion (ASA) and anti-hydroxyl radical (AHR), superoxide dismutase (SOD), catalase (CAT), glutathione peroxidase (GPx), glutathione-S-transferase (GST), and glutathione reductase (GR), as well as reduced glutathione (GSH) were determined using the commercial detection kits (Nanjing Jiancheng Bioengineering Institute, Nanjing, China). Activities of cathepsin B and L were assayed by the method described by Bahuaud et al. [24]. Measurement of hydroxyproline, lactic acid, and carnosine contents in muscle followed Periago et al. [25], Hultmann et al. [26], and Elbarbary et al. [27], respectively. Muscle vitamin A contents were measured by HPLC method.

2.3. Histology Observation

Histological observation of muscle was performed as described in our previous study [28]. In brief, the fixed muscle tissues were dehydrated in alcohol and embedded in paraffin wax. Afterwards, the tissues were serially sectioned to 4 mm, and stained with hematoxylin and eosin (H & E). The muscle morphological was observed using a Nikon TS100 light microscope (Nikon, Tokyo, Japan).

2.4. Quantitative Real-Time PCR Analysis

Procedures of total RNA isolation, reverse transcription, and quantitative real-time PCR (qPCR) were similar to our previous study [16]. In brief, an RNAiso Plus kit (Takara, Dalian, China) was used to isolate total RNA from muscle. One percent agarose gel electrophoresis and spectrophotometric analysis (A260: 280 nm) were used to assay RNA purity and concentration, respectively. After this, cDNA synthesis was performed using a PrimeScript™ RT reagent Kit (Takara, Dalian, China). Finally, qPCR was performed using a CFX96™ Real-Time PCR System (Bio-Rad, Hercules, CA, USA) with SYBR Green (Takara, Dalian, China). According to the preliminary experiment, β-actin was chosen as the reference gene (data not shown). Specific primers for qPCR were listed in Table 2. The amplification efficiencies of all primers were verified to be approximately 100%. The relative gene expressions were analyzed using the $2^{-\Delta\Delta CT}$ method as described by Livak and Schmittgen [29]

Table 2. Real-time PCR primer sequences.

Genes	Forward (5′→3′)	Reverse (5′→3′)	Temperature (°C)	Accession Number
CuZnSOD	CGCACTTCAACCCTTACA	ACTTTCCTCATTGCCTCC	61.5	GU901214
MnSOD	ACGACCCAAGTCTCCCTA	ACCCTGTGGTTCTCCTCC	60.4	GU218534
CAT	GAAGTTCTACACCGATGAGG	CCAGAAATCCCAAACCAT	58.7	FJ560431
GPx1a	GGGCTGGTTATTCTGGGC	AGGCGATGTCATTCCTGTTC	61.5	EU828796
GPx1b	TTTTGTCCTTGAAGTATGTCCGTC	GGGTCGTTCATAAAGGGCATT	60.3	KT757315
GPx4a	TACGCTGAGAGAGGTTTACACAT	CTTTTCCATTGGGTTGTTCC	60.4	KU255598
GPx4b	CTGGAGAAATACAGGGGTTACG	CTCCTGCTTTCCGAACTGGT	60.3	KU255599
GSTr	TCTCAAGGAACCCGTCTG	CCAAGTATCCGTCCCACA	58.4	EU107283
GSTp1	ACAGTTGCCCAAGTTCCAG	CCTCACAGTCGTTTTTTCCA	59.3	KM112099
GSTp2	TGCCTTGAAGATTATGCTGG	GCTGGCTTTTATTTCACCCT	59.3	KP125490
GR	GTGTCCAACTTCTCCTGTG	ACTCTGGGGTCCAAAACG	59.4	JX854448
Nrf2	CTGGACGAGGAGACTGGA	ATCTGTGGTAGGTGGAAC	62.5	KF733814
keap1a	TTCCACGCCCTCCTCAA	TGTACCCTCCCGCTATG	63.0	KF811013
keap1b	TCTGCTGTATGCGGTGGGC	CTCCTCCATTCATCTTTCTCG	57.9	KJ729125
TOR	TCCCACTTTCCACCAACT	ACACCTCCACCTTCTCCA	61.4	JX854449
S6K1	TGGAGGAGGTAATGGACG	ACATAAAGCAGCCTGACG	54.0	EF373673
β-actin	GGCTGTGCTGTCCCTGTA	GGGCATAACCCTCGTAGAT	61.4	M25013

CuZnSOD = copper/zinc superoxide dismutase; MnSOD = manganese superoxide dismutase; CAT = catalase; GPx1a = glutathione peroxidase 1a; GPx1b = glutathione peroxidase 1b; GPx4a = glutathione peroxidase 4a; GPx4b = glutathione peroxidase 4b; GSTr = glutathione-S-transferase r; GSTp1 = glutathione-S-transferase p1; GSTp2 = glutathione-S-transferase p2; GR = glutathione reductase; Nrf2 = nuclear factor erythroid 2-related factor 2; Keap1a = Kelch-like ECH-associated protein 1a; Keap1b = Kelch-like ECH-associated protein 1b; TOR = target of rapamycin; S6K1 = ribosomal protein s6 kinase polypeptide 1.

2.5. Western Blotting Measurement

The procedure of Western blotting was the same as previous study from our laboratory [30]. Shortly thereafter, extracted total and nuclear protein concentrations from muscle were measured using a protein quantification kit (Bio-Rad, Hercules, CA, USA). After this, protein samples with equal amounts were separated by SDS-PAGE and transferred to a PVDF membrane. Membranes were blocked at room temperature for 1 h, incubated with primary antibody at 4 °C overnight, and then with HRP-conjugated secondary antibodies for 2 h. Anti-Nrf2, total TOR (T-TOR), phospho-TOR Ser2448 (p-TOR), β-actin, and lamin B1 antibodies were the same as previous studies from our laboratory [31,32]. Finally, the bands were visualized and quantified by using an ECL kit (Millipore, Billerica, MA, USA) and Image J software (NIH, Bethesda, MD, USA), respectively. Protein levels in vitamin A supplemented groups were expressed relative to those in the vitamin A-deficient group. The analysis was repeated three times, and similar results were obtained each time.

2.6. Statistical Analysis

Data were treated by using Excel 2019 (Microsoft Inc., Redmond, WA, USA). The data from the individual fish in the same replicate were averaged, and then this mean for the replicate was used in the analysis. Prior to any statistical analysis, normality and homoscedasticity assumptions were confirmed. One-way ANOVA followed by Tukey's HSD test was used for vitamin A effects statistical analyses with SAS 9.4 (SAS Institute Inc., Cary, NC, USA). p-value < 0.05 was considered as statistically significant. The linear and quadratic effect of vitamin A were assayed by orthogonal polynomial contrasts in SAS 9.4. The results are presented as mean and SEM. Data visualization was performed by using the GraphPad Prim 8.0 (GraphPad Inc., La Jolla, CA, USA) and Excel 2019 (Microsoft Inc., Redmond, WA, USA).

3. Results

3.1. Proximate Compositions and Physicochemical Characteristics of Muscle

As presented in Table 3, contents of crude protein and crude lipid in muscle were linearly ($p < 0.05$) and quadratically ($p < 0.05$) enhanced by increase in dietary VA, and the highest in the group with 1798 IU/kg VA, while muscle moisture content was linearly ($p < 0.05$) and quadratically ($p < 0.05$) reduced with increase in dietary VA contents, and the lowest in the group with 1798 IU/kg VA. Compared to the VA deficiency group, increased levels of dietary VA linearly ($p < 0.05$) and quadratically ($p < 0.05$) enhanced shear force, pH_{24h}, as well as carnosine content in grass carp muscle, while linearly ($p < 0.05$) and quadratically ($p < 0.05$) reduced cooking loss, lactic acid content, and cathepsin B and L activities in muscle. Meanwhile, muscle hydroxyproline content showed a quadratically ($p < 0.05$) increase as dietary VA levels increased, and were the highest in the groups with 1209 and 1798 IU/kg VA.

3.2. Free Amino Acid Contents and Fatty Acid Profile in Muscle

In order to determine the effects of VA on flesh flavor, we focused on the free amino acid contents in grass carp muscle. The results showed that the free lysine, methionine, glutamic acid, threonine, and arginine contents in muscle were linearly ($p < 0.05$) and quadratically ($p < 0.05$) increased as dietary VA levels increased. As for lysine and arginine contents, they were significantly increased with increase in dietary VA levels up to 1209 IU/kg ($p < 0.05$), and then plateaued. Methionine content in group with 1798 IU/kg VA was significantly higher than that in the VA deficiency group ($p < 0.05$). Threonine content was significantly improved with the increasing VA levels up to 606.8 IU/kg ($p < 0.05$), and plateaued thereafter. Glutamic acid content was significantly enhanced by 606.8–3796 IU/kg VA ($p < 0.05$), and the highest in the group with 1798 IU/kg VA. However, the other amino acids contents were not significantly affected by dietary VA (Table 4).

Fatty acids are important precursors of flesh flavor. Accordingly, we also evaluated the muscle fatty acids profile. We observed that dietary VA linearly ($p < 0.05$) enhanced the C18: 3n − 3, C20: 3n − 3, C22: 6 (docosahexaenoic acid, DHA) and total polyunsaturated fatty acid (PUFA) contents in muscle, linearly ($p < 0.05$) and quadratically ($p < 0.05$) enhanced the total unsaturated fatty acid content in muscle, whereas linearly ($p < 0.05$) reduced muscle C160 content, and linearly ($p < 0.05$) and quadratically ($p < 0.05$) decreased the C16:1 and total saturated fatty acid contents in muscle (Table 5). C16:1 content significantly decreased with increase in dietary VA levels up to 1209 IU/kg, and plateaued thereafter. Compared to the VA deficiency group, C18: 3n − 3 was significantly higher in groups with 2805 and 3796 IU/kg VA ($p < 0.05$), and C20: 3n − 3, C22: 6, ΣUFA and ΣPUFA contents were significantly higher in group with 2805 and 1798 IU/kg VA ($p < 0.05$), respectively, while ΣSFA content was significantly lower in group with 1798 IU/kg VA ($p < 0.05$).

Table 3. Effects of dietary vitamin A on muscle proximate composition and physicochemical characteristics of on-growing grass carp [1].

	Dietary VA Levels, IU/kg Diet						SEM	*p*-Values	
	18.69	606.8	1209	1798	2805	3796		Linear	Quadratic
Moisture, %	79.67 [a]	78.16 [b]	77.75 [bc]	77.15 [c]	77.93 [bc]	78.04 [bc]	0.20	0.0001	<0.0001
Protein, %	15.20 [d]	16.51 [c]	17.11 [b]	17.78 [a]	17.00 [b]	17.02 [b]	0.06	<0.0001	<0.0001
Lipids, %	10.06 [b]	10.92 [ab]	12.07 [a]	12.17 [a]	11.94 [a]	11.15 [ab]	0.31	0.01	<0.01
Cooking loss, %	16.00 [a]	12.56 [b]	10.12 [c]	10.25 [c]	12.25 [b]	13.85 [b]	0.39	<0.01	<0.0001
Shear force, N	1.18 [d]	1.30 [c]	1.53 [a]	1.51 [a]	1.42 [b]	1.31 [c]	0.02	<0.0001	<0.0001
pH_{24h}	6.43 [b]	6.54 [b]	6.72 [a]	6.79 [a]	6.77 [a]	6.53 [b]	0.04	<0.01	<0.0001
Hydroxyproline, mg/g tissue	0.38 [d]	0.46 [b]	0.58 [a]	0.57 [a]	0.43 [bc]	0.42 [c]	0.01	0.09	<0.0001
Carnosine, ng/g tissue	348.51 [c]	409.53 [bc]	485.00 [a]	485.22 [a]	489.94 [a]	431.10 [ab]	14.96	0.0002	<0.0001
Lactic acid, mmol/g protein	2.58 [a]	2.16 [b]	2.09 [b]	1.63 [d]	1.86 [c]	1.85 [c]	0.04	<0.0001	<0.0001
Cathepsin B, U/g protein	3.98 [a]	3.57 [b]	2.90 [c]	2.89 [c]	3.23 [bc]	3.29 [b]	0.08	<0.0001	<0.0001
Cathepsin L, U/g protein	1.85 [a]	1.72 [b]	1.53 [cd]	1.44 [d]	1.54 [cd]	1.58 [c]	0.03	<0.0001	<0.0001

[1] Data are means of three replicate groups, two fish for each replicate ($n = 3$), SEM = standard error of the mean. [a,b,c,d] within a row, means without a common lowercase superscript differ ($p < 0.05$).

Table 4. Effects of dietary vitamin A on muscle amino acid composition (mg/100 g dry) of on-growing grass carp [1].

	Dietary VA Levels, IU/kg Diet						SEM	*p*-Values	
	18.69	606.8	1209	1798	2805	3796		Linear	Quadratic
Glu	3.89 [d]	4.13 [c]	4.34 [ab]	4.52 [a]	4.20 [bc]	4.20 [bc]	0.04	0.0001	<0.0001
Asp	2.10	2.05	2.11	2.07	2.08	2.08	0.04	0.96	0.89
Gly	25.27	24.80	24.88	24.93	25.36	25.46	0.63	0.62	0.48
Ser	3.66	3.74	3.79	3.70	3.68	3.71	0.09	0.98	0.53
Ala	11.74	11.37	11.54	11.43	11.68	11.59	0.27	0.98	0.50
Met	5.13 [b]	5.49 [b]	5.82 [ab]	6.26 [a]	5.73 [ab]	5.76 [ab]	0.15	<0.01	<0.01
Thr	9.46 [b]	10.84 [a]	11.51 [a]	11.76 [a]	11.40 [a]	11.42 [a]	0.25	<0.01	<0.01
Lys	31.19 [c]	34.91 [bc]	38.82 [ab]	40.19 [a]	38.32 [ab]	37.83 [ab]	0.95	0.0001	0.0003
Arg	19.56 [c]	21.15 [bc]	23.76 [a]	23.61 [a]	22.13 [ab]	22.11 [ab]	0.49	<0.01	0.0002
His	169.71	173.39	177.13	174.86	173.25	172.57	2.62	0.61	0.10
Val	6.17	6.29	6.22	6.28	6.31	6.32	0.15	0.52	0.92
Ile	2.14	2.20	2.26	2.22	2.16	2.19	0.05	0.77	0.24
Leu	3.09	3.06	3.13	3.10	3.14	3.12	0.07	0.55	0.89
Phe	3.34	3.25	3.30	3.24	3.23	3.28	0.07	0.50	0.53
Tyr	5.21	5.21	5.24	5.23	5.24	5.20	0.11	0.95	0.80

[1] Data are means of three replicate groups, two fish for each replicate ($n = 3$), SEM = standard error of the mean. [a,b,c,d] within a row, means without a common lowercase superscript differ ($p < 0.05$).

Table 5. Effect of dietary vitamin A on the fillet fatty acid (FA) profile (% of total FA methyl esters) of on-growing grass carp [1].

	Dietary VA Levels, IU/kg Diet						SEM	*p*-Values	
	18.69	606.8	1209	1798	2805	3796		Linear	Quadratic
C14: 0	4.56	4.56	4.64	4.54	4.74	4.91	0.12	0.05	0.25
C15: 0	0.26	0.26	0.27	0.26	0.26	0.27	0.01	0.60	0.49
C16: 0	23.00 ab	23.22 a	22.61 ab	21.23 b	21.22 b	22.05 ab	0.39	<0.01	0.15
C17: 0	0.24	0.25	0.24	0.24	0.25	0.25	0.02	0.85	0.85
C18: 0	5.59	5.63	5.42	5.07	5.39	5.61	0.13	0.36	<0.05
C20: 0	0.23	0.23	0.23	0.22	0.23	0.22	0.01	0.49	0.91
C23: 0	0.28	0.29	0.29	0.28	0.29	0.31	0.02	0.28	0.72
C24: 0	0.77	0.78	0.75	0.74	0.82	0.82	0.04	0.29	0.24
C14: 1	0.21	0.22	0.21	0.22	0.26	0.25	0.02	0.10	0.86
C16: 1	14.02 a	12.94 ab	12.27 bc	11.69 bc	11.70 bc	11.55 c	0.28	<0.0001	<0.05
C17: 1	0.38	0.35	0.37	0.34	0.33	0.38	0.02	0.73	0.11
C18: 1c + t	22.51	23.20	23.70	24.37	24.57	23.52	0.54	0.05	0.07
C20: 1	1.88	1.82	1.80	1.84	1.76	1.70	0.05	<0.05	0.59
C22: 1	0.20	0.19	0.20	0.21	0.21	0.19	0.01	0.98	0.32
C18: 2c + t	7.83	7.78	7.99	8.50	8.52	8.03	0.18	<0.05	0.09
C20: 2	0.43	0.43	0.40	0.40	0.41	0.38	0.02	0.06	0.97
C18: 3n − 6	0.67	0.65	0.67	0.66	0.66	0.64	0.02	0.53	0.63
C18: 3n − 3	4.60 b	5.00 ab	5.03 ab	5.40 ab	5.56 a	5.54 a	0.19	<0.01	0.37
C20: 3n − 6 + C21: 0	0.37	0.39	0.37	0.38	0.38	0.35	0.02	0.38	0.41
C20: 3n-3	1.06 b	1.10 ab	1.19 ab	1.20 ab	1.25 a	1.21 ab	0.04	<0.01	0.13
C20: 4	0.41	0.42	0.40	0.40	0.40	0.39	0.02	0.25	0.93
C20: 5 + C22: 0	1.11	1.08	1.04	1.03	1.06	1.00	0.04	0.13	0.79
C22: 6	9.39 b	9.20 b	9.92 ab	10.78 a	10.14 ab	10.42 ab	0.27	<0.01	0.24
ΣSFA	34.94 a	35.22 a	34.46 ab	32.58 b	33.19 ab	34.45 ab	0.48	<0.05	<0.05
ΣUFA	64.65 b	64.36 b	65.15 ab	67.02 a	66.41 ab	65.17 ab	0.48	<0.05	<0.05
ΣMUFA	39.19	38.73	38.53	38.66	38.83	37.59	0.49	0.09	0.59
ΣPUFA	25.87 b	26.04 b	27.01 ab	28.76 a	27.98 ab	27.97 ab	0.45	<0.01	0.08

[1] Data are means of three replicate groups, two fish for each replicate (n = 3), SEM = standard error of the mean. a,b,c within a row, means without a common lowercase superscript differ ($p < 0.05$). ΣSFA = Total saturated fatty acid; ΣUFA = Total unsaturated fatty acid; ΣMUFA = Total monounsaturated fatty acid; ΣPUFA = Total polyunsaturated fatty acid.

3.3. Antioxidant Related Parameters in Muscle

To test whether dietary VA affected muscle antioxidant capacity, we analyzed antioxidant related parameters in muscle. The histological results indicated that an obvious rupture in the muscle fiber occurred in the dietary VA deficiency group, but was not observed in other groups (Figure 1). Furthermore, muscle MDA, PC, and ROS contents were linearly ($p < 0.05$) and quadratically ($p < 0.05$) reduced by the enhanced levels of dietary VA (Figure 2). Muscle CAT, CuZnSOD, GPx and GR activities, as well as GSH and VA content were linearly ($p < 0.05$) and quadratically ($p < 0.05$) improved by the increase in dietary VA levels, while ASA, AHR capacities, and MnSOD activity showed a quadratically ($p < 0.05$) enhancement as dietary VA increased (Table 6). However, GST activity in muscle was not significantly affected by dietary VA. Muscle ROS, MDA, and PC contents were significantly reduced by dietary VA supplementation in comparison with the VA deficiency group ($p < 0.05$), and was the lowest in the group with 1798 IU/kg VA. Compared to the VA deficiency group, supplementation of 1209 and 1798 IU/kg VA significantly enhanced the ASA and AHR capacities in muscle ($p < 0.05$). MnSOD and GR activities in group with 1798 IU/kg VA ($p < 0.05$) was significantly higher than those in the VA deficiency group ($p < 0.05$). CuZnSOD, CAT, and GPx activities, and GSH content in muscle, significantly increased with the increase in dietary VA levels up to 1209 IU/kg ($p < 0.05$), and plateaued thereafter. Muscle VA content was the highest in the group with 3796 IU/kg VA ($p < 0.05$).

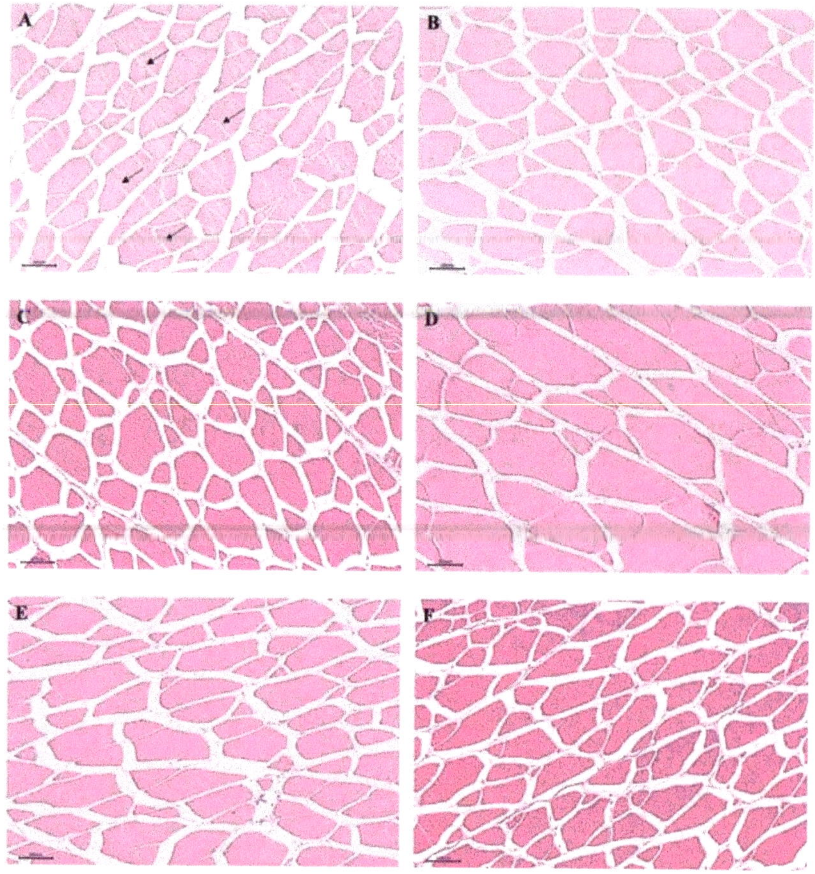

Figure 1. The histology of on-growing grass carp muscle (H&E 100×): (**A**) The vitamin A-deficient group. Arrowhead showed the rupture in muscle fiber. (**B**) The group with vitamin A at 606.8 IU/kg. (**C**) The group with vitamin A at 1209 IU/kg. (**D**) The group with vitamin A at 1798 IU/kg. (**E**) The group with vitamin A at 2805 IU/kg. (**F**) The group with vitamin A at 3796 IU/kg.

To fully characterize the effects of dietary VA on antioxidative enzymes, we tested the relative expressions of antioxidant enzymes genes in muscle. As presented in Figure 3, dietary VA linearly ($p < 0.05$) and quadratically ($p < 0.05$) up-regulated the relative gene expressions of *CAT* and *GPx1b* in muscle, linearly ($p < 0.05$) improved the relative mRNA levels of *CuZnSOD, GPx4a, GPx4b,* and *GSTr* in muscle, and quadratically ($p < 0.05$) increased the relative gene expressions of *MnSOD* and *GPx1a*, but did not significantly affect the relative mRNA levels of *GSTp1, GSTp2* and *GR*. Compared to the VA deficiency group, the relative mRNA levels of *CuZnSOD, GPx4a,* and *GSTr* were significantly higher in groups with 2805 and 3796 IU/kg VA ($p < 0.05$), the relative mRNA levels of *MnSOD* and *GPx1a* were significantly higher in groups with 1209 and/or 1798 IU/kg VA ($p < 0.05$). The relative mRNA levels of *CAT* in muscle significantly increased with increase in dietary VA levels up to 1209 IU/kg ($p < 0.05$), and then plateaued.

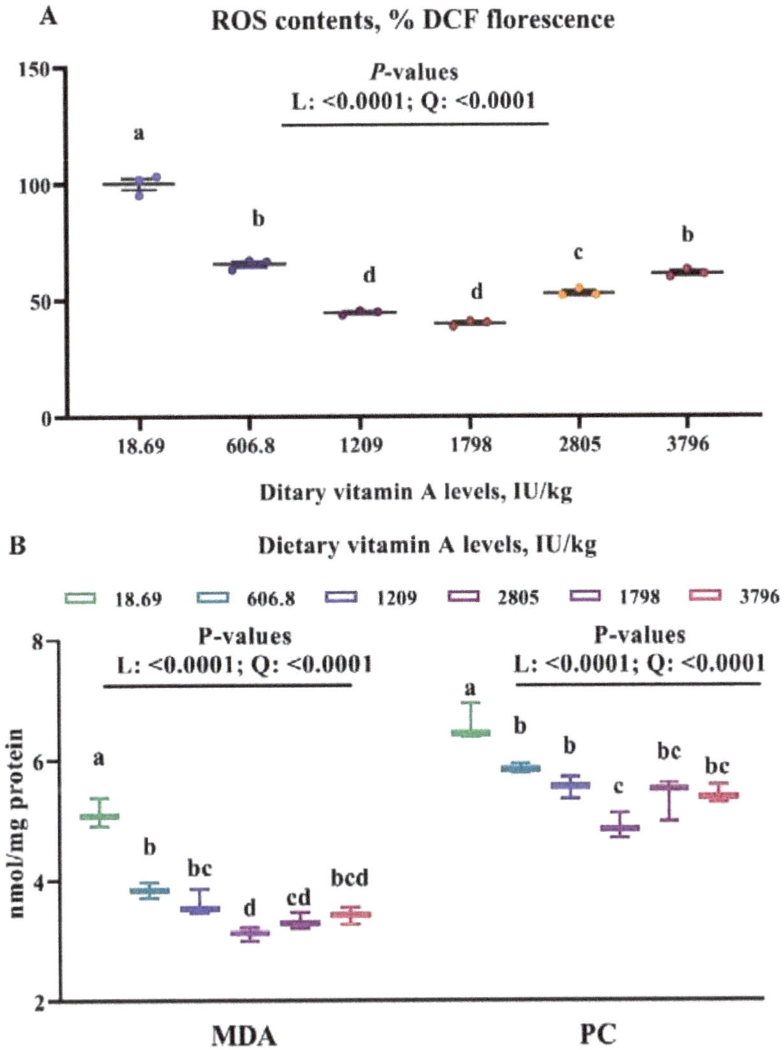

Figure 2. Effects of dietary vitamin A on contents of ROS (**A**), MDA and PC (**B**) in muscle of on-growing grass carp. Data are means ± SEM of three replicate groups, two fish for each replicate (n = 3). [a,b,c,d] within a column, means without a common lowercase superscript differ ($p < 0.05$). p-values underlined with a solid line indicate a linear and quadratic response to dietary vitamin A levels. SEM = standard error of the mean; L = linear; Q = quadratic; ROS = reactive oxygen species, %DCF florescence; MDA = malondialdehyde, nmol/mg prot; PC = protein carbonyl, nmol/mg prot.

Table 6. Effect of dietary vitamin A on antioxidant parameters in muscle of on-growing grass carp [1].

	Dietary VA Levels, IU/kg Diet						SEM	*p*-Values	
	18.69	606.8	1209	1798	2805	3796		Linear	Quadratic
ASA, U/g protein	75.76 ^{c}	83.48 ^{bc}	96.85 ^{a}	92.20 ^{ab}	82.36 ^{bc}	82.68 ^{bc}	2.33	0.20	<0.0001
AHR, U/mg protein	89.54 ^{c}	94.78 ^{abc}	99.75 ^{ab}	101.54 ^{a}	91.49 ^{bc}	90.89 ^{c}	1.83	0.93	0.0002
CuZnSOD, U/mg protein	2.83 ^{c}	2.96 ^{bc}	3.36 ^{ab}	3.65 ^{a}	3.46 ^{a}	3.30 ^{ab}	0.09	0.0001	0.0005
MnSOD, U/mg protein	3.82 ^{b}	4.30 ^{ab}	4.35 ^{ab}	4.48 ^{a}	4.20 ^{ab}	4.17 ^{ab}	0.12	0.14	<0.01
CAT, U/mg protein	1.06 ^{c}	1.28 ^{b}	1.48 ^{a}	1.64 ^{a}	1.61 ^{a}	1.62 ^{a}	0.04	<0.0001	<0.0001
GPx, U/mg protein	84.96 ^{b}	93.14 ^{ab}	103.73 ^{a}	104.63 ^{a}	104.43 ^{a}	103.25 ^{a}	2.92	0.0002	<0.01
GST, U/mg protein	49.17	50.11	51.64	51.14	50.11	50.19	1.38	0.70	0.27
GR, U/g protein	17.44 ^{b}	19.64 ^{ab}	19.70 ^{ab}	22.06 ^{a}	21.35 ^{ab}	20.54 ^{ab}	0.83	<0.01	<0.05
GSH, mg/g protein	1.33 ^{b}	1.50 ^{ab}	1.63 ^{a}	1.62 ^{a}	1.62 ^{a}	1.60 ^{a}	0.04	0.0003	<0.01
Vitamin A, μg/kg tissue	1.63 ^{c}	5.43 ^{b}	6.74 ^{ab}	5.84 ^{ab}	6.56 ^{ab}	8.29 ^{a}	0.52	<0.0001	<0.05

[1] Data are means of three replicate groups, two fish for each replicate ($n = 3$), SEM = standard error of the mean. ^{a,b,c} within a row, means without a common lowercase superscript differ ($p < 0.05$). AHR = anti-hydroxyl radical; ASA = anti-superoxide anion; CuZnSOD = copper/zinc superoxide dismutase; MnSOD = manganese superoxide dismutase; CAT = catalase; GPx = glutathione peroxidase; GST = glutathione-S-transferase; GR = glutathione reductase; GSH = glutathione.

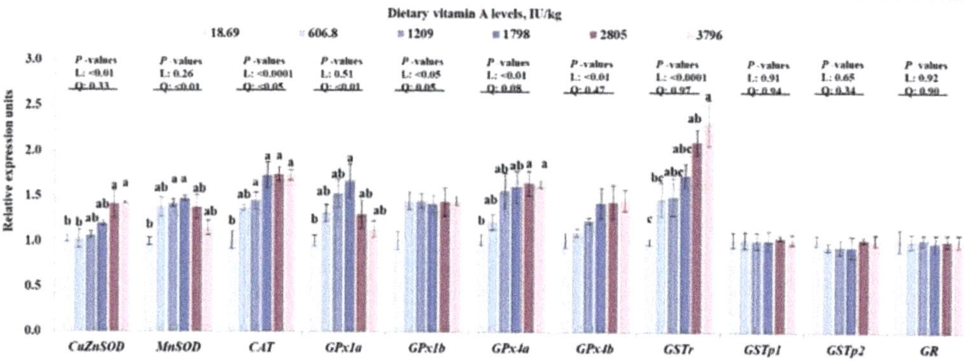

Figure 3. Effects of dietary vitamin A on relative mRNA levels of antioxidant enzymes genes in muscle of on-growing grass carp. Data are means ± SEM of three replicate groups, two fish for each replicate ($n = 3$). $^{a, b, c}$ within a column, means without a common lowercase superscript differ ($p < 0.05$). *p*-values underlined with a solid line indicate a linear and quadratic response to dietary vitamin A levels. SEM = standard error of the mean; L = linear; Q = quadratic; CuZnSOD = copper/zinc superoxide dismutase; MnSOD = manganese superoxide dismutase; CAT = catalase; GPx1a = glutathione peroxidase 1a; GPx1b = glutathione peroxidase 1b; GPx4a = glutathione peroxidase 4a; GPx4b = glutathione peroxidase 4b; GSTr = glutathione-S-transferase r; GSTp1 = glutathione-S-transferase p1; GSTp2 = glutathione-S-transferase p2; GR = glutathione reductase.

3.4. Nrf2 and TOR Signaling in Muscle

To clarify the signaling involved in VA-regulated antioxidant capacity, we studied the TOR and Nrf2 signaling in muscle. For Nrf2 signaling pathway, as dietary VA increased, *Nrf2* gene expressions in the muscle were quadratically ($p < 0.05$) increased, while Kelch-like ECH-associating protein a (*Keap1a*) gene expression was quadratically ($p < 0.05$) down-regulated (Figure 4A). However, the relative mRNA levels of *Keap1b* in muscle were not significantly changed by dietary VA (Figure 4A). Furthermore, the total Nrf2 protein level in muscle was linearly ($p < 0.05$) and quadratically ($p < 0.05$) increased by increasing levels of dietary VA, and the nuclear Nrf2 level in muscle was quadratically ($p < 0.05$) enhanced by increase in dietary VA (Figure 4B). Compared to the VA deficiency group, the relative mRNA levels of *Nrf2*, protein levels of nuclear Nrf2 and total Nrf2 were significantly higher in groups with 606.8, 1209, and/or 1798 IU/kg VA ($p < 0.05$), while the relative mRNA levels of *Keap1a* was significantly lower in group with 1209 IU/kg VA ($p < 0.05$).

As presented in Figure 5A, dietary vitamin A quadratically ($p < 0.05$) and linearly ($p < 0.05$) up-regulated the relative gene expressions of *TOR* and *S6K1* in muscle, respectively. Meanwhile, the protein levels of T-TOR and p-TOR^{ser2448} in muscle were linearly ($p < 0.05$) and quadratically ($p < 0.05$) enhanced by dietary VA, respectively. Muscle p-TOR/T-TOR was linearly ($p < 0.05$) and quadratically ($p < 0.05$) changed by dietary VA (Figure 5B). Compared to the VA deficiency group, the relative gene expressions of *TOR* and *S6K1* in muscle were significantly up-regulated by 606.8-3796 and 2805-3796 IU/kg VA ($p < 0.05$), respectively, the protein levels of T-TOR and p-TOR^{ser2448} were significantly higher in groups with 1798-3796 and 1209-1798 IU/kg VA ($p < 0.05$), respectively.

3.5. Dietary Vitamin A Requirements for On-Growing Grass Carp

As presented in Figure 6, based on muscle shear force and ROS contents, the dietary VA requirements for on-growing grass carp (262.02–996.67 g) were determined to be 2080 and 2244 IU/kg diet, respectively.

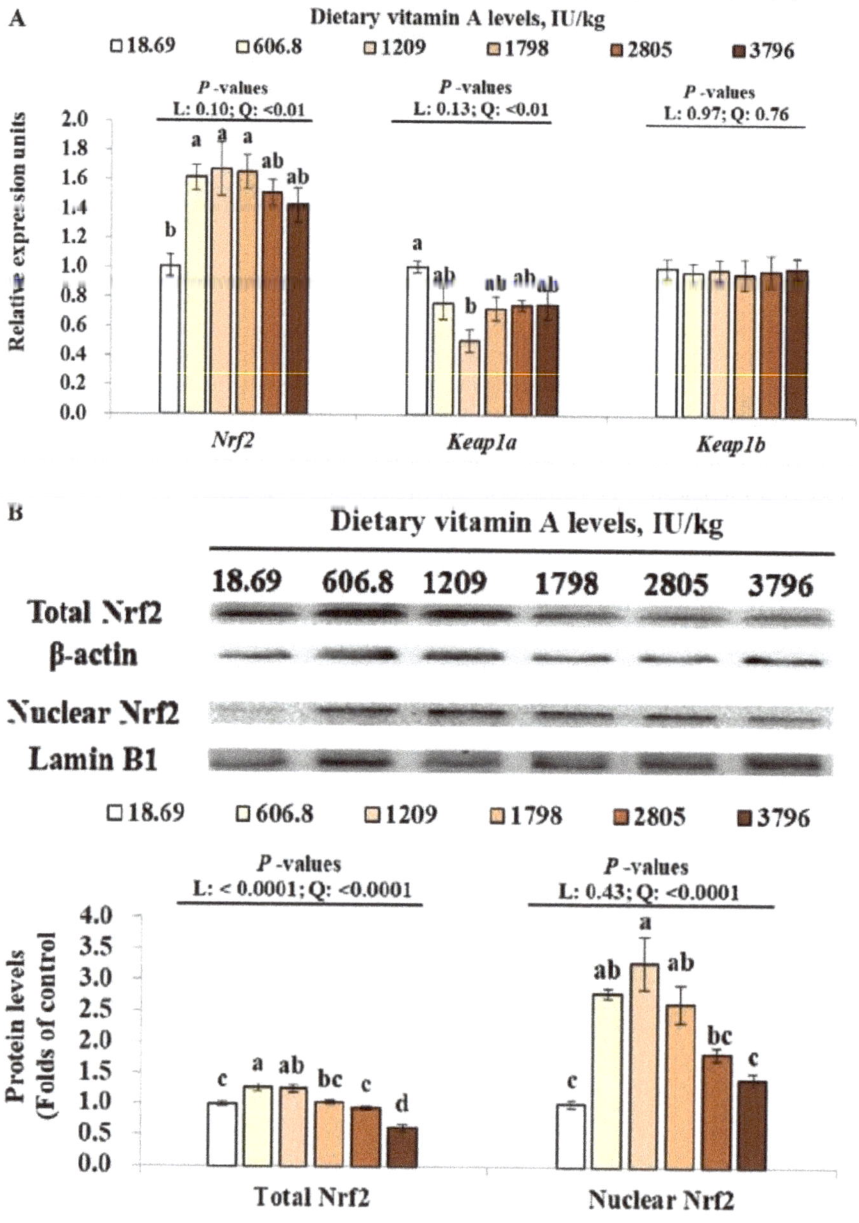

Figure 4. Effects of dietary vitamin A on relative mRNA levels of *Nrf2*, *keap1a*, and *keap1b* (**A**); total and nuclear levels of Nrf2 protein (**B**) in muscle of on-growing grass carp. Data are means ± SEM of three replicate groups, two fish for each replicate ($n = 3$). [a,b,c,d] within a column, means without a common lowercase superscript differ ($p < 0.05$). p-values underlined with a solid line indicate a linear and quadratic response to dietary vitamin A levels. SEM = standard error of the mean; L = linear; Q = quadratic; Nrf2 = nuclear factor erythroid 2-related factor 2; Keap1a = Kelch-like ECH-associated protein 1a; Keap1b = Kelch-like ECH-associated protein 1b.

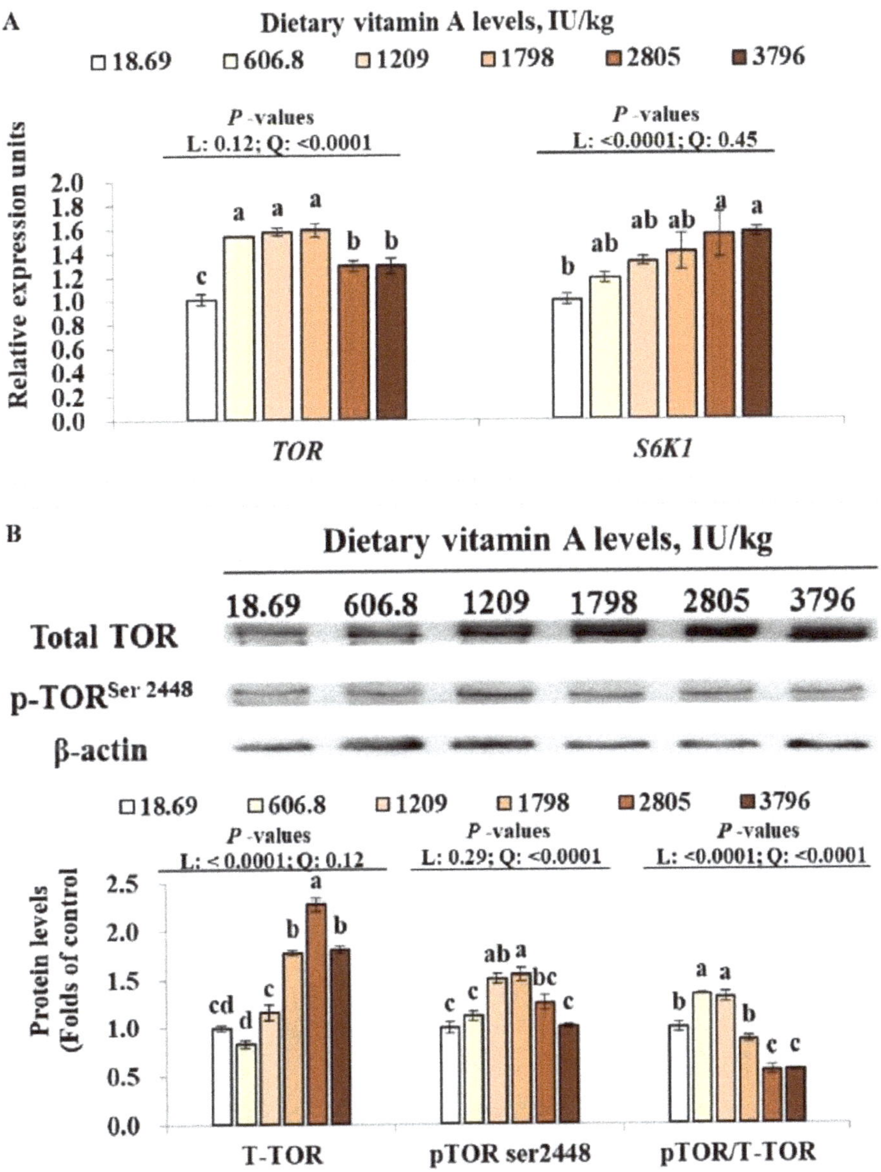

Figure 5. Effects of dietary vitamin A on relative mRNA levels of *TOR* and *S6K1* (**A**), total and phosphorylation levels of TOR protein (**B**) in muscle of on-growing grass carp. Data are means ± SEM of three replicate groups, two fish for each replicate (n = 3). [a,b,c,d] within a column, means without a common lowercase superscript differ ($p < 0.05$). p-values underlined with a solid line indicate a linear and quadratic response to dietary vitamin A levels. SEM = standard error of the mean; L = linear; Q = quadratic; TOR = target of rapamycin; S6K1 = ribosomal protein s6 kinase polypeptide 1.

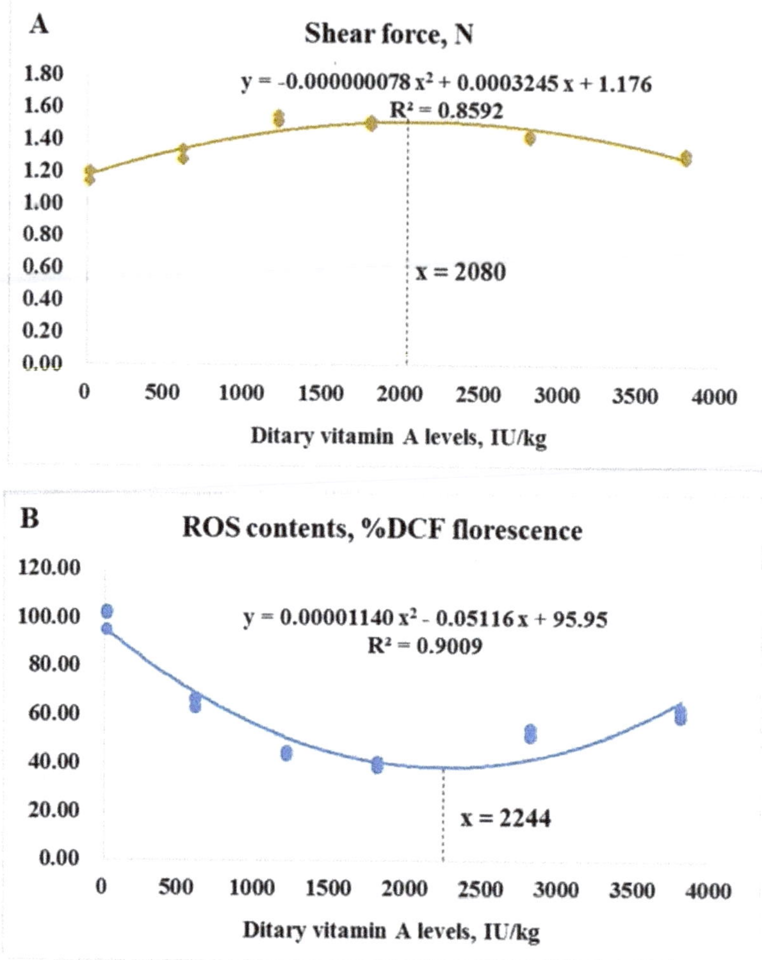

Figure 6. The dietary vitamin A requirements for on-growing grass carp based on muscle shear force (**A**) and contents of reactive oxygen species (ROS, (**B**)).

4. Discussion

This present study used the same growth experiment as our previous research [2,16], which indicated that VA deficiency induced poor growth, impaired intestinal antioxidant capacity, induced apoptosis, and depressed intestinal immunity of on-growing grass carp. Oxidant damage usually can cause deterioration of fish flesh quality. Therefore, the present study focused on clarifying whether dietary VA could improve fish flesh quality through increasing antioxidative capacity.

4.1. Vitamin A Improved Fish Flesh Quality

One of the main sets of flesh quality characteristics is sensory appeal, which can be reflected by WHC, tenderness, and pH value [5]. In the present study, dietary VA deficiency induced a drop of shear force and pH_{24h} value, as well as an increase in cooking loss in grass carp muscle; however, optimal levels of dietary VA reversed these undesirable changes, indicating that VA is helpful for the improvement of fish flesh quality. Although no more information about the effects of VA on sensory appeal was found in fish, our results were similar to studies in terrestrial animals, which indicated that dietary VA improved juiciness and tenderness of longissimus thoracis from Holstein bulls and steers [7], tenderness of longissimus lumborum from Angus crossbred steers [8], and the sensory appeal of breast muscle in broiler [9]. In fish, the decrease in muscle pH value is closely related to the lactic acid production [33]. The present study found that lactic acid content in grass carp muscle was reduced by dietary VA. Meanwhile, correlation analysis showed that muscle pH_{24h} value was negative in relation to lactic acid content (Table 7), indicating that VA-increased muscle pH_{24h} value was partly ascribed to the lactic acid reduction in muscle. Muscle firmness is positively associated with collagen content (reflected by hydroxyproline content) [34], and negatively associated with activities of cathepsins B and L, which play important roles in post-mortem degradation of tissue proteins [35] and are positively correlated to detachments in the muscle structure in fish [26]. In this study, dietary VA enhanced the hydroxyproline content and decreased the cathepsin B and L activities in grass carp muscle. Further analysis indicated that muscle shear force was positively related to hydroxyproline content and negatively related to cathepsin B and L activities (Table 7), suggesting that VA increased fish muscle firmness partly through enhancing collagen content and decreasing cathepsin B and L activities.

Table 7. Correlations of different indices in the muscle of on-growing grass carp.

Dependent Parameters	Independent Parameters	r	*p*
pH_{24h}	Lactic acid content	−0.758	=0.08
Shear force	Hydroxyproline content	+0.894	<0.05
	Cathepsin B activity	−0.964	<0.01
	Cathepsin L activity	−0.906	<0.05
MnSOD activity	*MnSOD* mRNA level	+0.928	<0.01
CAT activity	*CAT* mRNA level	+0.980	<0.01
GPx activity	*GPx4a* mRNA level	+0.999	<0.01
	GPx4b mRNA level	+0.936	<0.01
MnSOD mRNA	*Nrf2* mRNA level	+0.938	<0.01
GPx1a mRNA		+0.842	<0.05

MnSOD = manganese superoxide dismutase; CAT = catalase; GPx = glutathione peroxidase; Nrf2 = nuclear factor erythroid 2-related factor 2.

The nutritional composition of muscle is another major quality aspect in fresh fish. Meanwhile, muscle free amino acids and fatty acids are major flavor contributors and important flavor precursors in fish, respectively [4]. Data in this study showed that contents of muscle protein, lipid, some free amino acids (Lys, Met, Thr, Arg, and Glu), DHA, and total PUFA were decreased by the VA deficiency and increased by optimal levels of dietary VA, suggesting that VA improved muscle nutritional composition of fish. These results were similar to studies that indicated that dietary VA increased body crude protein content of juvenile hybrid tilapia [13], and perirenal fat in finishing pigs [14]. However, information available on the effects of VA on amino acids and fatty acids contents in fish is scarce. Elongases of very long-chain fatty acid (Elovl) are involved in LC-PUFA biosynthesis by catalyzing the rate-limiting condensation step in elongation process [36]. Study found that VA upregulated fatty acid elongase-4 (Elovl4) mRNA levels in WNIN/Ob obese rat retina [37]. This might partially explain the VA-increased muscle DHA and PUFA contents in the present study.

4.2. *Vitamin A Enhanced the Muscle Antioxidant Capacity of Fish*

Lipid and protein oxidation, not only is one of the major causes of meat quality deterioration [12], but it also leads to damage of the structural integrity of cells [38]. High unsaturated lipid levels in fish flesh make it more susceptible to oxidative deterioration, which is mainly caused by an imbalance between ROS production and antioxidative defense [12]. MDA and PC are good biomarkers of protein oxidation and lipid peroxidation in animal tissues [39]. In the present study, dietary VA deficiency-induced rupture in the muscle fiber, while VA supplementation groups did not show this change. Furthermore, compared with the VA-deficient group, optimal levels of VA reduced ROS, MDA, and PC contents in grass carp muscle, showing that VA depressed the oxidative damage of fish muscle. Superoxide anion ($O_2^{\bullet-}$) and hydroxyl radical (•OH−) are two important toxic ROS involving in oxidative damage [40]. Antioxidant enzymes (like SOD and GPx) and non-enzymatic antioxidant (like GSH) play important roles in free radicals scavenging. SOD is the main element of the first level of antioxidant defense against superoxide radical, CAT, and GPx play important roles in detoxifying the hydrogen peroxide, GST, and GR are also important glutathione-dependent enzymes and able to counteract the peroxidative damage [39]. Meanwhile, GSH is a low-molecular-mass thiol that involves in scavenging peroxyl radicals in cells [39]. In this study, VA deficiency decreased superoxide anion and hydroxyl radical scavenging capacities (ASA and AHR, respectively), CAT, CuZnSOD, MnSOD, GPx, and GR activities, as well as GSH and VA contents in grass carp muscle, while optimal levels of vitamin A reversed these changes, indicating that vitamin A decreased ROS content possibly via improving the antioxidant defense in fish muscle. However, GST activity in muscle was not significantly affected by dietary VA in this study. This result was different from the previous study from our laboratory, which found that VA deficiency reduced intestinal GST activity in grass carp [2]. The different results between muscle and intestine might be partly related to the tissue distribution of GST. In river pufferfish (*Takifugu obscurus*), the *GST* genes expressions in intestine were higher than those in muscle, which is associated with the fact that intestine is more susceptible to oxidant damage than muscle [41]. However, further investigation is necessary to clarify the exact mechanisms behind these findings.

As we know, activities of antioxidative enzymes were partly relied on their gene expression. In the current study, dietary VA deficiency resulted in a decrease in *CAT, CuZnSOD, MnSOD, GPx1a, GPx4a* and *GSTr* mRNA levels in muscle, which might partly explain the decrease in CAT, CuZnSOD, MnSOD, and GPx activities in the VA-deficient group. However, *GSTp1, GSTp2,* and *GR* mRNA levels in muscle showed no difference among groups. The difference change pattern of *GST* isoforms in muscle might be partly related to their tissue-specifically expression. In bighead carp (*Aristichthys nobilis*), the relative gene expression of *GST rho* was higher than that of *GST pi* in muscle [42]. The higher expression of *GSTr* in muscle might lead to the more sensitive response to VA; however, this needs further characterization. In mammal, the expression of antioxidative enzymes genes (such as *CAT* and *CuZnSOD*) can be modulated by Nrf2 signaling [43]. The present study observed that dietary VA deficiency decreased mRNA levels of *Nrf2*, protein levels of total Nrf2, and nuclear Nrf2, while enhancing the mRNA levels of *Keap1a* in grass carp muscle. These results demonstrated that VA deficiency-decreased antioxidative capacity might partly be attributed to the down-regulated Nrf2 levels. However, the muscle *Keap1b* mRNA level was not influenced by VA. This is similar with our previous studies, which found that VA deficiency upregulated *Keap 1a* gene expressions rather than *Keap 1b* in intestine, head kidney, and spleen of grass carp [2,44], demonstrating that VA regulated antioxidant gene expressions mainly via Nrf2/Keap 1a signaling in fish with unknown mechanisms. Additionally, in vitro study showed that inhibition of TOR reduced total Nrf2 expression in human hepatic carcinoma cells [45]. Our current results indicated that dietary VA enhanced the mRNA levels of *TOR* and *S6K1*, and total protein and phosphorylation levels of TOR in muscle of grass carp, which showed similar change patterns as total Nrf2 and nuclear Nrf2 levels, implying that VA-activated Nrf2 signaling might partly be due to the activation of TOR in muscle.

4.3. Vitamin A Requirement for On-Growing Grass Carp

As demonstrated by the above data, VA deficiency could lead to the deteriorated muscle quality of on-growing grass carp probably via depressing antioxidant ability in muscle. Accordingly, it is valuable to estimate the VA requirements for grass carp based on the muscle quality indices. Based on muscle shear force, contents of ROS, the VA requirements for on-growing grass carp (262.02–996.67 g) were determined to be 2080 and 2244 IU/kg diet, respectively, which were higher than that based on growth (1929 IU/kg diet), suggesting that more VA might be required for improving fish muscle quality. It is consistent with the study that showed that on-growing grass carp had higher folic acid requirements based on flesh quality [46]. Meanwhile, the VA requirements for on-growing grass carp based on healthy indices of intestine, head kidney, and spleen were higher than that based on growth [16,44]. These results demonstrated that higher dietary intake of nutrients is often necessary to satisfy physiology functions other than growth in fish.

5. Conclusions

The present study demonstrated that dietary VA depressed deterioration of sensory appeal, nutrition value, as well as flavor quality of fish flesh probably through improving antioxidant capacity, as shown by increased activities and gene expressions of SOD, CAT, and GPx, which was closely related to Nrf2/Keap 1a (rather than Keap 1b) signaling in muscle. However, VA showed different impacts on GST and *GST* isoforms in muscle with unknown mechanisms. In addition, the dietary VA requirements for improving flesh quality of on-growing grass carp were determined to be 2080–2244 IU/kg, which was slightly higher than that based on growth (1929 IU/kg).

Author Contributions: P.W. and L.Z. completed the experiment and prepared the manuscript. W.J., Y.L., J.J., S.L., S.K., L.T. and W.T. helped analyzed the samples and data. X.Z. and L.F. designed the experiment and revised the manuscript. All authors have read and agreed to the published version of the manuscript.

Funding: This research was financially supported by the National Key R&D Program of China (2018YFD0900400, 2019YFD0900200), National Natural Science Foundation of China for Outstanding Youth Science Foundation (31922086), the Young Top-Notch Talent Support Program, China Agriculture Research System of MOF and MARA (CARS-45), and Sichuan Science and Technology Program (2019YFN0036).

Institutional Review Board Statement: All animal care and experimental procedures were approved by the Institutional Animal Care and Use Committee of Sichuan Agricultural University (SICAU-2016-098).

Informed Consent Statement: Not applicable.

Data Availability Statement: Data are contained within the article.

Conflicts of Interest: The authors declare no conflict of interest.

References

1. Różanowska, M.; Cantrell, A.; Edge, R.; Land, E.J.; Sarna, T.; Truscott, T.G. Pulse radiolysis study of the interaction of retinoids with peroxyl radicals. *Free Radic. Biol. Med.* **2005**, *39*, 1399–1405. [CrossRef] [PubMed]
2. Jiang, W.D.; Zhou, X.Q.; Zhang, L.; Liu, Y.; Wu, P.; Jiang, J.; Kuang, S.Y.; Tang, L.; Tang, W.N.; Zhang, Y.A.; et al. Vitamin A deficiency impairs intestinal physical barrier function of fish. *Fish Shellfish Immun.* **2019**, *87*, 546–558. [CrossRef] [PubMed]
3. Tilami, S.K.; Sampels, S. Nutritional value of fish: Lipids, proteins, vitamins, and minerals. *Rev. Fish Sci. Aquac.* **2017**, *26*, 243–253. [CrossRef]
4. Grigorakis, K. Compositional and organoleptic quality of farmed and wild gilthead sea bream (*Sparus aurata*) and sea bass (*Dicentrarchus labrax*) and factors affecting it: A review. *Aquaculture* **2007**, *272*, 55–75. [CrossRef]
5. Hughes, J.M.; Oiseth, S.K.; Purslow, P.P.; Warner, R.D. A structural approach to understanding the interactions between colour, water-holding capacity and tenderness. *Meat Sci.* **2014**, *98*, 520–532. [CrossRef]
6. Morachis-Valdez, G.; Dublán-García, O.; López-Martínez, L.X.; Galar-Martínez, M.; Saucedo-Vence, K.; Gómez-Oliván, L.M. Chronic exposure to pollutants in Madín Reservoir (Mexico) alters oxidative stress status and flesh quality in the common carp *Cyprinus carpio*. *Environ. Sci. Pollut. R.* **2015**, *22*, 9159–9172. [CrossRef] [PubMed]
7. Marti, S.; Realini, C.E.; Bach, A.; Pérez-Juan, M.; Devant, M. Effect of vitamin A restriction on performance and meat quality in finishing Holstein bulls and steers. *Meat Sci.* **2011**, *89*, 412–418. [CrossRef] [PubMed]
8. Daniel, M.J.; Dikeman, M.E.; Arnett, A.M.; Hunt, M.C. Effects of dietary vitamin A restriction during finishing on color display life, lipid oxidation, and sensory traits of longissimus and triceps brachii steaks from early and traditionally weaned steers. *Meat Sci.* **2009**, *81*, 15–21. [CrossRef]

9. Wang, Y.; Li, L.; Gou, Z.; Chen, F.; Fan, Q.; Lin, X.; Ye, J.; Zhang, C.; Jiang, S. Effects of maternal and dietary vitamin A on growth performance, meat quality, antioxidant status, and immune function of offspring broilers. *Poult. Sci.* **2020**, *99*, 3930–3940. [CrossRef]
10. Jennings, S.; Stentiford, G.D.; Leocadio, A.M.; Jeffery, K.R.; Metcalfe, J.D.; Katsiadaki, I.; Auchterlonie, N.A.; Mangi, S.C.; Pinnegar, J.K.; Ellis, T.; et al. Aquatic food security: Insights into challenges and solutions from an analysis of interactions between fisheries, aquaculture, food safety, human health, fish and human welfare, economy and environment. *Fish Fish.* **2016**, *17*, 893–938. [CrossRef]
11. Lund, M.N.; Heinonen, M.; Baron, C.P.; Estévez, M. Protein oxidation in muscle foods: A review. *Mol. Nutr. Food Res.* **2011**, *55*, 83–95. [CrossRef]
12. Falowo, A.B.; Fayemi, P.O.; Muchenje, V. Natural antioxidants against lipid–protein oxidative deterioration in meat and meat products: A review. *Food Res. Int.* **2014**, *64*, 171–181. [CrossRef]
13. Hu, C.; Chen, S.; Pan, C.; Huang, C. Effects of dietary vitamin A or β-carotene concentrations on growth of juvenile hybrid tilapia, *Oreochromis niloticus* × *O. aureus*. *Aquaculture* **2006**, *253*, 602–607. [CrossRef]
14. Tous, N.; Lizardo, R.; Theil, P.K.; Vilà, B.; Gispert, M.; Font-i-Furnols, M.; Esteve-Garcia, E. Effect of vitamin A depletion on fat deposition in finishing pigs, intramuscular fat content and gene expression in the longissimus muscle. *Livest Sci.* **2014**, *167*, 392–399. [CrossRef]
15. Yang, Q.; Ding, M.; Tan, B.; Dong, X.; Chi, S.; Zhang, S.; Liu, H. Effects of dietary vitamin A on growth, feed utilization, lipid metabolism enzyme activities, and fatty acid synthase and hepatic lipase mRNA expression levels in the liver of juvenile orange spotted grouper, *Epinephelus coioides*. *Aquaculture* **2017**, *479*, 501–507. [CrossRef]
16. Zhang, L.; Feng, L.; Jiang, W.; Liu, Y.; Wu, P.; Kuang, S.; Tang, L.; Tang, W.; Zhang, Y.; Zhou, X. Vitamin A deficiency suppresses fish immune function with differences in different intestinal segments: The role of transcriptional factor NF-κB and p38 mitogen-activated protein kinase signalling pathways. *Brit. J. Nutr.* **2017**, *117*, 67–82. [CrossRef]
17. Khan, M.A.; Jafri, A.K.; Chadha, N.K. Growth, reproductive performance, muscle and egg composition in grass carp, *Ctenopharyngodon idella* (Valenciennes), fed hydrilla or formulated diets with varying protein levels. *Aquac. Res.* **2004**, *35*, 1277–1285. [CrossRef]
18. Moren, M.; Opstad, I.; Berntssen, M.H.G.; Zambonino Infante, J.L.; Hamre, K. An optimum level of vitamin A supplements for Atlantic halibut (*Hippoglossus hippoglossus* L.) juveniles. *Aquaculture* **2004**, *235*, 587–599. [CrossRef]
19. NRC. *Nutrient Requirements of Fish and Shrimp*; The National Academies Press: Washington, DC, USA, 2011.
20. Chen, L.; Feng, L.; Jiang, W.; Jiang, J.; Wu, P.; Zhao, J.; Kuang, S.; Tang, L.; Tang, W.; Zhang, Y.; et al. Dietary riboflavin deficiency decreases immunity and antioxidant capacity, and changes tight junction proteins and related signaling molecules mRNA expression in the gills of young grass carp (*Ctenopharyngodon idella*). *Fish Shellfish Immun.* **2015**, *45*, 307–320. [CrossRef] [PubMed]
21. Brinker, A.; Reiter, R. Fish meal replacement by plant protein substitution and guar gum addition in trout feed, Part I: Effects on feed utilization and fish quality. *Aquaculture* **2011**, *310*, 350–360. [CrossRef]
22. AOAC. *Official Methods of Analysis*, 16th ed.; Association of Official Analytical Chemists International: Washington, DC, USA, 1998.
23. Carbonera, F.; Bonafe, E.G.; Martin, C.A.; Montanher, P.F.; Ribeiro, R.P.; Figueiredo, L.C.; Almeida, V.C.; Visentainer, J.V. Effect of dietary replacement of sunflower oil with perilla oil on the absolute fatty acid composition in Nile tilapia (GIFT). *Food Chem.* **2014**, *148*, 230–234. [CrossRef] [PubMed]
24. Bahuaud, D.; Mørkøre, T.; Østbye, T.K.; Veiseth-Kent, E.; Thomassen, M.S.; Ofstad, R. Muscle structure responses and lysosomal cathepsins B and L in farmed Atlantic salmon (*Salmo salar* L.) pre- and post-rigor fillets exposed to short and long-term crowding stress. *Food Chem.* **2010**, *118*, 602–615. [CrossRef]
25. Periago, M.J.; Ayala, M.D.; López-Albors, O.; Abdel, I.; Martínez, C.; García-Alcázar, A.; Ros, G.; Gil, F. Muscle cellularity and flesh quality of wild and farmed sea bass, *Dicentrarchus labrax* L. *Aquaculture* **2005**, *249*, 175–188. [CrossRef]
26. Hultmann, L.; Phu, T.M.; Tobiassen, T.; Aas-Hansen, Ø.; Rustad, T. Effects of pre-slaughter stress on proteolytic enzyme activities and muscle quality of farmed Atlantic cod (*Gadus morhua*). *Food Chem.* **2012**, *134*, 1399–1408. [CrossRef] [PubMed]
27. Elbarbary, N.S.; Ismail, E.A.R.; El-Naggar, A.R.; Hamouda, M.H.; El-Hamamsy, M. The effect of 12 weeks carnosine supplementation on renal functional integrity and oxidative stress in pediatric patients with diabetic nephropathy: A randomized placebo-controlled trial. *Pediatr. Diabetes* **2018**, *19*, 470–477. [CrossRef] [PubMed]
28. Wu, P.; Jiang, W.; Jiang, J.; Zhao, J.; Liu, Y.; Zhang, Y.; Zhou, X.; Feng, L. Dietary choline deficiency and excess induced intestinal inflammation and alteration of intestinal tight junction protein transcription potentially by modulating NF-κB, STAT and p38 MAPK signaling molecules in juvenile Jian carp. *Fish Shellfish Immun.* **2016**, *58*, 462–473. [CrossRef]
29. Livak, K.J.; Schmittgen, T.D. Analysis of relative gene expression data using real-time quantitative PCR and the $2^{-\triangle\triangle CT}$ method. *Methods* **2001**, *25*, 402–408. [CrossRef] [PubMed]
30. Wu, P.; Qu, B.; Feng, L.; Jiang, W.; Kuang, S.; Jiang, J.; Tang, L.; Zhou, X.; Liu, Y. Dietary histidine deficiency induced flesh quality loss associated with changes in muscle nutritive composition, antioxidant capacity, Nrf2 and TOR signaling molecules in on-growing grass carp (*Ctenopharyngodon idella*). *Aquaculture* **2020**, *526*, 735399. [CrossRef]

31. Jiang, W.; Wen, H.; Liu, Y.; Jiang, J.; Wu, P.; Zhao, J.; Kuang, S.; Tang, L.; Tang, W.; Zhang, Y.; et al. Enhanced muscle nutrient content and flesh quality, resulting from tryptophan, is associated with anti-oxidative damage referred to the Nrf2 and TOR signalling factors in young grass carp (*Ctenopharyngodon idella*): Avoid tryptophan deficiency or excess. *Food Chem.* **2016**, *199*, 210–219. [CrossRef]
32. Hu, K.; Zhang, J.; Feng, L.; Jiang, W.; Wu, P.; Liu, Y.; Jiang, J.; Zhou, X. Effect of dietary glutamine on growth performance, non-specific immunity, expression of cytokine genes, phosphorylation of target of rapamycin (TOR), and anti- oxidative system in spleen and head kidney of Jian carp (*Cyprinus carpio* var. Jian). *Fish Physiol. Biochem.* **2015**, *41*, 635–649. [CrossRef]
33. Liu, D.; Liang, L.; Xia, W.; Regenstein, J.M.; Zhou, P. Biochemical and physical changes of grass carp (*Ctenopharyngodon idella*) fillets stored at −3 and 0 °C. *Food Chem.* **2013**, *140*, 105–114. [CrossRef]
34. Johnston, I.A.; Li, X.; Vieira, V.L.A.; Nickell, D.; Dingwall, A.; Alderson, R.; Campbell, P.; Bickerdike, R. Muscle and flesh quality traits in wild and farmed Atlantic salmon. *Aquaculture* **2006**, *256*, 323–336. [CrossRef]
35. Aoki, T.; Ueno, R. Involvement of cathepsins B and L in the post-mortem autolysis of mackerel muscle. *Food Res. Int.* **1997**, *30*, 585–591. [CrossRef]
36. Kuah, M.; Jaya-Ram, A.; Shu-Chien, A.C. The capacity for long-chain polyunsaturated fatty acid synthesis in a carnivorous vertebrate: Functional characterisation and nutritional regulation of a Fads2 fatty acyl desaturase with Δ4 activity and an Elovl5 elongase in striped snakehead (*Channa striata*). *BBA Mol. Cell. Biol. Lipids* **2015**, *1851*, 248–260.
37. Tiruvalluru, M.; Ananthathmakula, P.; Ayyalasomayajula, V.; Nappanveettil, G.; Ayyagari, R.; Reddy, G.B. Vitamin A supplementation ameliorates obesity-associated retinal degeneration in WNIN/Ob rats. *Nutrition* **2013**, *29*, 298–304. [CrossRef]
38. Wang, J.; Jia, R.; Gong, H.; Celi, P.; Zhuo, Y.; Ding, X.; Bai, S.; Zeng, Q.; Yin, H.; Xu, S.; et al. The effect of oxidative stress on the chicken ovary: Involvement of microbiota and melatonin interventions. *Antioxidants* **2021**, *10*, 1422. [CrossRef] [PubMed]
39. Surai, P. Silymarin as a natural antioxidant: An overview of the current evidence and perspectives. *Antioxidants* **2015**, *4*, 204–247. [CrossRef]
40. Abele, D.; Puntarulo, S. Formation of reactive species and induction of antioxidant defence systems in polar and temperate marine invertebrates and fish. *Comp. Biochem. Phys. A* **2004**, *138*, 405–415. [CrossRef]
41. Kim, J.; Dahms, H.; Rhee, J.; Lee, Y.; Lee, J.; Han, K.; Lee, J. Expression profiles of seven glutathione S-transferase (GST) genes in cadmium-exposed river pufferfish (*Takifugu obscurus*). *Comp. Biochem. Phys. C* **2010**, *151*, 99–106. [CrossRef]
42. Li, G.; Xie, P.; Li, H.; Chen, J.; Hao, L.; Xiong, Q. Quantitative profiling of mRNA expression of glutathione S-transferase superfamily genes in various tissues of bighead carp (*Aristichthys nobilis*). *J. Biochem. Mol. Toxic.* **2010**, *24*, 250–259. [CrossRef]
43. Rajput, S.A.; Liang, S.J.; Wang, X.Q.; Yan, H.C. Lycopene protects intestinal epithelium from deoxynivalenol-induced oxidative damage via regulating Keap1/Nrf2 Signaling. *Antioxidants* **2021**, *10*, 1493. [CrossRef] [PubMed]
44. Jiang, W.; Zhang, L.; Feng, L.; Wu, P.; Liu, Y.; Jiang, J.; Kuang, S.; Tang, L.; Zhou, X. Inconsistently impairment of immune function and structural integrity of head kidney and spleen by vitamin A deficiency in grass carp (*Ctenopharyngodon idella*). *Fish Shellfish Immun.* **2020**, *99*, 243–256. [CrossRef] [PubMed]
45. Shay, K.P.; Michels, A.J.; Li, W.; Kong, A.N.; Hagen, T.M. Cap-independent Nrf2 translation is part of a lipoic acid-stimulated detoxification stress response. *Biochim. Biophys. Acta* **2012**, *1823*, 1102–1109. [CrossRef] [PubMed]
46. Shi, L.; Zhou, X.Q.; Jiang, W.D.; Wu, P.; Liu, Y.; Jiang, J.; Kuang, S.Y.; Tang, L.; Feng, L. The effect of dietary folic acid on flesh quality and muscle antioxidant status referring to Nrf2 signalling pathway in young grass carp (*Ctenopharyngodon idella*). *Aquacult. Nutr.* **2020**, *26*, 631–645. [CrossRef]

antioxidants

Article

Dietary Glutamine Inclusion Regulates Immune and Antioxidant System, as Well as Programmed Cell Death in Fish to Protect against *Flavobacterium columnare* Infection

Congrui Jiao [1,†], Jiahong Zou [1,†], Zhenwei Chen [1], Feifei Zheng [1], Zhen Xu [1], Yu-Hung Lin [2,*] and Qingchao Wang [1,*]

1 College of Fisheries, Huazhong Agricultural University, 1 Shizishan Street, Wuhan 430070, China; jcr-1998@webmail.hzau.edu.cn (C.J.); zjiahong@webmail.hzau.edu.cn (J.Z.); chenzhenwei@webmail.hzau.edu.cn (Z.C.); zhengfeifei@mail.hzau.edu.cn (F.Z.); zhenxu@mail.hzau.edu.cn (Z.X.)
2 Department of Aquaculture, National Pingtung University of Science and Technology, 1 Shuefu Road, Neipu, Pingtung 912, Taiwan
* Correspondence: yuhunglin@mail.npust.edu.tw (Y.-H.L.); qcwang@mail.hzau.edu.cn (Q.W.); Tel.: +886-7703-202 (Y.-H.L.); +86-872-821-13 (Q.W.)
† These authors contributed equally to this work.

Abstract: The susceptibility of animals to pathogenic infection is significantly affected by nutritional status. The present study took yellow catfish (*Pelteobagrus fulvidraco*) as a model to test the hypothesis that the protective roles of glutamine during bacterial infection are largely related to its regulation on the immune and antioxidant system, apoptosis and autophagy. Dietary glutamine supplementation significantly improved fish growth performance and feed utilization. After a challenge with *Flavobacterium columnare*, glutamine supplementation promoted *il-8* and *il-1β* expression via NF-κB signaling in the head kidney and spleen, but inhibited the over-inflammation in the gut and gills. Additionally, dietary glutamine inclusion also enhanced the systematic antioxidant capacity. Histological analysis showed the protective role of glutamine in gill structures. Further study indicated that glutamine alleviated apoptosis during bacterial infection, along with the reduced protein levels of caspase-3 and the reduced expression of apoptosis-related genes. Moreover, glutamine also showed an inhibitory role in autophagy which was due to the increased activation of the mTOR signaling pathway. Thus, our study for the first time illustrated the regulatory roles of glutamine in the fish immune and antioxidant system, and reported its inhibitory effects on fish apoptosis and autophagy during bacterial infection.

Keywords: antioxidant; apoptosis; autophagy; glutamine; immunity; mTOR signaling

Citation: Jiao, C.; Zou, J.; Chen, Z.; Zheng, F.; Xu, Z.; Lin, Y.-H.; Wang, Q. Dietary Glutamine Inclusion Regulates Immune and Antioxidant System, as Well as Programmed Cell Death in Fish to Protect against *Flavobacterium columnare* Infection. *Antioxidants* **2022**, *11*, 44. https://doi.org/10.3390/antiox11010044

Academic Editors: Min Xue, Junmin Zhang, Zhenyu Du, Jie Wang and Wei Si

Received: 10 November 2021
Accepted: 24 December 2021
Published: 26 December 2021

1. Introduction

Unlike mammals, teleosts live in aquatic environments which are more conducive to bacterial growth [1,2], therefore potential pathogens could enter the bodies of fish across their mucosal epithelial barriers including the gills, gastrointestinal system or skin lesions [3–5]. Bacterial infection is one of the most common causes for animal diseases, both in mammals and other animals [6]. During infection, animals would firstly activate the innate immune system to defend against the invaders, and some animal species, mainly vertebrates, could further activate the adaptive immune system to efficiently clearing invading pathogens with a prolonged infecting period [7]. In particular, fish represent the earliest bony vertebrate to develop both innate and adaptive immune responses during evolution [8]. Reactive oxygen species (ROS) would also be induced by bacterial infection, however, the over-production of O^{2-} would cause oxidative damage to proteins, nucleic acids, and lipids [9]. Fish, like other animals, develop a cellular antioxidant defenses system to function in multiple situations including ROS scavenging, oxidative stress protection, and attenuation of membrane lipid peroxidation [10]. The major front-line antioxidant enzymes, such as superoxide dismutase (SOD, neutralizes superoxide radicals to H_2O_2), catalase (CAT) and glutathione peroxidase (GPx, neutralizes H_2O_2 to water), and small non-protein antioxidants (scavenges all active oxygen species directly) work in a cascade to protect cells from oxidative stress [11]. Moreover, the programmed cell death including autophagy and apoptosis are also related to an early evolution of defense against bacterial infection, for example, successful autophagic contributes to the clearance of the invading microbe and cell survival [12] while apoptosis helps to preserve the whole organism or the species from the spread of infection [13]. Autophagy is a fundamental eukaryotic process with multiple cytoplasmic homeostatic roles, recently expanded to include unique stand-alone immunological functions and interactions with nearly all parts of the immune system [14]. Autophagy has been identified to function as the effector or regulatory functions downstream of systems sensing danger signals/alarmins, also known as damage-associated molecular patterns (DAMP), such as ATP and self-DNA-containing complexes [15]. The contribution of autophagy is complicated by the fact that autophagy can be either protective or harmful, depending on the biological context [16]. Besides autophagy, apoptosis could also be induced upon endogenous receptor/ligand systems on the surface of the pathogen-infected cell via the activation of several pro-apoptotic proteins, e.g., caspases, and the inactivation of anti-apoptotic proteins, e.g., NF-κB or MAP-kinases [17]. In particular, the induction of apoptosis in epithelial or endothelial cells might break the epithelia/endothelial cell barrier and permit the bacteria to reach the sub-mucosa [18].

The immune responses, antioxidant capacity and programmed cell death of animals would be significantly affected by their nutritional status [19]. Normally, functional nutrients could perform a substrate role in the initial development of the immune cells and during an actual immune response so that the responding cells can divide and synthesize effector molecules [20]. Nutrients may perform direct regulatory actions on the leukocytes that respond to infectious challenges, and also perform indirect effects during bacterial infection via the modulation of the endocrine system [21,22]. Moreover, antioxidant nutrients could act as the safeguard against the accumulation of ROS and their elimination from the system to ensure redox homeostasis [23] via preventing lipid peroxidation, stopping the oxidative chain reaction in membranes and lipoproteins [24], and promoting the synthesis of antioxidant enzymes [25]. Additionally, nutrient starvation and specific nutrients have also been reported to regulate the apoptosis and autophagy in multiple kinds of cells [26,27]. Among all nutrients, amino acids are traditionally considered to compose proteins, and recently have been reported to function as direct signals to activate several signaling pathways and further play multiple functions [28]. Glutamine (Gln) is an abundant amino acid in blood which also functions as the most important fuel for intestinal tissue to support gut protein synthesis [29,30], regulate antioxidant [31] and immune system [32,33], promote cell proliferation [34], and delay apoptosis [35,36]. Glutamine has also been reported to activate several cell signaling-related kinases to play multiple functions [37]. For example, glutamine enhances intestinal cells proliferation via activating MAPKs [38], promotes protein synthesis by activating mTOR signaling [39,40], and improves cell survival by regulating heat shock proteins in the intestine [41,42]. In particular, glutamine functions in maintaining gut integrity in both humans and terrestrial animals by serving as a major energy substrate for the rapid development of enterocytes [43,44]. Approximately 70% of glutamine is degraded by rat and pig small intestines during the first pass [45,46] and glutamine deprivation induces autophagy and alters the mTOR and MAPK signaling pathways in porcine intestinal epithelial cells [47].

In fish, glutamine has also been reported to promote growth, enhance antioxidant capacity, promote intestinal function, improve healthcare function and even muscle flavor [48,49], however, there is little information about the regulatory mechanism. Unlike the normally twice duplicated genome during the evolution of vertebrates, a third genome duplication occurs in teleosts, i.e., fish-specific genome duplication (FSGD), which makes the teleost a research model to illustrate novel mechanism [50,51]. Recent studies have identified the FSGD in yellow catfish (*Pelteobagrus fulvidraco*) [52] and also the regulatory roles of apoptosis and autophagy in lipid metabolism of yellow catfish, for example, the apoptosis signaling pathways is reported to mediate Met-induced changes of hepatic lipid deposition and metabolism [53] and autophagy is involved in FA-induced TG accumulation and lipotoxicity in yellow catfish [54]. However, the apoptosis and autophagy responses during bacterial infection are rarely evaluated in teleosts, including yellow catfish. The columnaris disease, induced by *Flavobacterium columnare* (*F. columnare*) infection [55], is common in yellow catfish which exhibits skin lesions, fin erosion and gill necrosis with a high degree of mortality [56]. During *F. columnare* infection, yellow catfish could activate both innate immunity, such as the increased release of pro-inflammatory cytokines and reactive oxygen species (ROS) and adaptive immunity such as the secretion of immunoglobulins [56,57]. However, unlike most teleosts, no IgT/IgZ which is the executor of fish mucosal immunity exists in the genome of yellow catfish, and its mucosal immunity is far from knowledgeable [57,58]. Thus, the yellow catfish is a good experimental model to study the regulatory mechanism of immune and antioxidant systems, apoptosis and autophagy in teleost after bacterial infection. Accordingly, glutamine was supplemented to the diet of the yellow catfish for an 8-week rearing experiment in the present study, which was then infected with *F. columnare* to systematically evaluate the regulatory mechanism. This study made the first attempt to evaluate the influence of dietary glutamine inclusion on the apoptosis and autophagy responses in fish during bacterial infection.

2. Materials and Methods

2.1. Fish Husbandry

Yellow catfish of approximately 4.0 g, bought from a fish farm in Wuhan (Hubei, China), were maintained in independent circular fiberglass tanks (80 cm in diameter, 100 cm in column height) at the wet lab of Huazhong Agricultural University. The temperature of all water tanks was maintained at 24 $\pm$ 1 °C, and the peripheral speed of water was 8 L/min. Fish were fed twice per day with a commercial diet. All animal proceeding procedures were approved by the Institutional Animal Care and Use Committee of Huazhong Agricultural University.

2.2. Experimental Diet Preparation and Feeding

The composition of the basal diet is shown in Table 1. In the experimental diet, glutamine (1%) was added to the basal diet. At the beginning of the fish rearing experiment, yellow catfish were randomly assigned to 10 fish tanks, with 30 fish per tank. Fish in five tanks fed with the basal diet were set as the control group, while fish in another five tanks fed with the experimental diet were set as the experimental group. All other feeding and domestication conditions were the same in the two groups. After 60 days of the rearing experiment, fish within each tank were weighed and the diet used in each tank was calculated. Then fish growth performance and feed utilization were calculated, including final body weight (FBW), weight gain rate (WGR), specific growth rate (SGR) and feed conversion ratio (FCR). The parameters were calculated as the following equations which were determined according to previous reported methods [56]:

Table 1. Formulation and proximate composition of the basal diet (% dry matter).

Ingredient	%	Ingredient	%
Fish meal	5	Vitamin premix [a]	1
Wheat gluten meal	8	Mineral premix [b]	1.5
Corn gluten meal	14	Monocalcium phosphate	1.5
Soybean meal	29	Choline chloride	0.5
Fish oil	2.5	Ethoxy quinoline	0.05
Soybean oil	2.5	Sodium alginate	2
Soy lecithin	1		

Table 1. *Cont.*

Ingredient	%	Ingredient	%
Wheat meal	31.45		
Proximate composition (%)			
Crude protein		39.62	
Crude lipid		7.49	

[a] Vitamin premix (mg/kg dry diet): vitamin A 10, vitamin D 0.05, vitamin E 400, vitamin K 40, vitamin B_1 50, vitamin B_2 200, niacin 500, vitamin B_6 50, biotin 5, folic acid 15, vitamin B_{12} 0.1, vitamin C 1000, inositol 2000, choline 5000. [b] Mineral premix (mg/kg dry diet): $FeSO_4 \cdot 7H_2O$ 372, $CuSO_4 \cdot 5H_2O$ 25, $ZnSO_4 \cdot 7H_2O$ 120, $MnSO_4 \cdot H_2O$ 5, $MgSO_4$ 2475, NaCl 1875, KH_2PO_4 1000, $Ca(H_2PO_4)_2$ 2500.

Weight gain rate (WGR, %) = 100 × [final body weight (g)-initial body weight (g)]/initial body weight (g);

Specific growth rate (SGR, %/day) = 100 × [ln (final body weight (g)) − ln (initial body weight (g))]/rearing period (days);

Feed conversion ratio (FCR) = dry feed intake (g)/[final body weight (g) − initial body weight (g)].

2.3. Bacterial Challenge Experiment

After the calculation of growth parameters, bacterial challenge test was conducted for fish in two groups. The water height of all 10 fish tanks was reduced to 20 cm, and then *F. columnare* medium was poured into tank at a final concentration of 5×10^5 CFU/mL which was kept for 2 h to perform bacterial infection [56]. After bacterial challenge experiment, fresh water was added to all tanks and fish were reared under the same conditions. Fish in the first two tanks per group were sampled on days 0, 1 and 30 after exposure. During sampling, fish were first anesthetized via MS222 exposure (1:10,000), and blood was drawn from the tail vein. Serum was obtained after centrifugation at 3000× *g* for 20 min at 4 °C and stored at −80 °C until further analyses. Then the head kidney (HK), spleen, gill, gut, and liver were removed. Partial gills were stored in paraformaldehyde, and other tissue samples (gill, gut, spleen, liver, HK, and serum) were immediately frozen in liquid nitrogen and stored at −80 °C before analysis. Fish in the other three tanks per group were not sampled and calculated at the end of 30 days to determine mortality.

2.4. Histopathological Structure

The gill samples were shaped and then fixed with paraformaldehyde for 24 h. Each tissue was cut into an appropriate size and set into the embedding box. Then tissues were transferred to gradient ethanol for dehydration, xylene for transparent treatment, and then paraffin wax for embedding. The embedded tissue was trimmed and then installed on a microtome for sectioning. The 5 μm-thick tissue sections were used for hematoxylin-eosin (H.E.) staining following the manufacturer's instructions. Stained glass slides were examined under an optical microscope (Olympus, DP72, Tokyo, Japan) equipped with a camera (Nikon E600, Nikon, Melville, NY, USA) and CellSens Standard Software (Olympus) to acquire images.

2.5. Enzyme Activity Assay

Total antioxidant capacity (T-AOC) and the activities of superoxide dismutase (SOD), catalase (CAT) and glutathione peroxidase (GPx) in serum were analyzed with commercially available assay kits (Nanjing Jiancheng Bioengineering Institute, Nanjing, China). T-AOC was assayed as the equal amount (mM) of trolox solution that could stop the oxidization of 2,2′-azino-bis(3-ethylbenzthiazoline-6-sulfonic acid (ABTS) to green $ABTS^+$ by reactive oxygen species (ROS). SOD activity (units per milliliter serum) was measured spectrophotochemically with each unit defined as the amount of enzyme necessary to inhibit the reduction of water-soluble tetrazolium salt by 50%. CAT activity (units per milliliter serum) was measured with one unit defined as the amount of activity required to transform 1 μmol of H_2O_2 per min. GPx activity was measured with one unit defined as the amount of GPx required to consumed 1 μmol of glutathione per min.

2.6. RNA Extraction and cDNA Preparation

Total RNA was extracted from gill, gut, head kidney, and spleen samples using Trizol (Invitrogen, Carlsbad, CA, USA) as described by the manufacturer. Briefly, samples were mechanically disrupted in 1 mL of Trizol using a homogenizer. Then, 200 mL of chloroform was added, and the suspension was centrifuged at 12,500 r/pm for 15 min. The clear upper phase was recovered, and mixed with an equal volume of isopropanol. The mixture was allowed to stand sufficiently to precipitate, and then centrifuged. The pellet was washed with ethanol, and added with appropriate DEPC-treated water to dissolve the precipitate. The spectrophotometer was used to determine the concentration and purity of RNA sample, controlling the its absorbance range with OD260/280 values between 1.8–2.2. RNA integrity was then checked by 1.0% agarose gel electrophoresis. DNase treatment was performed with 5× gDNA Digester Mix (Yeasen, China) according to the manufacturer's instructions and cDNA was synthesized from 1 μg of total RNA using 4× Hifair® III SuperMix plus (Yeasen, China). The synthesized cDNA was stored at −20 °C until use.

2.7. Quantitative Real-Time Polymerase Chain Reaction (PCR) Analysis

To assess the transcription levels of multiple genes, real-time quantitative PCR was performed on a 7500 Real-time PCR System (Applied Biosystems) using EvaGreen 2× qPCR Master Mix (Abm) with a 10 μL reaction volume containing 5 μL of 2× qPCR MasterMix, 1 μL of template (200 ng/μL), 0.15 μL of each forward and reverse primer (10 μM), and 3.7 μL of H_2O. The cycling conditions for all samples were: 95 °C for 30 s, then 40 cycles at 95 °C for 1 s and 58 °C for 10 s. Melt curve analysis was performed after the amplification cycle to confirm the accuracy of each amplicon. The expression of the target gene was calculated using the $2^{-\Delta\Delta Ct}$ method after normalized relative to the relative expression of ef-1α, which has been tested to be stable under the current experimental conditions. Table 2 lists the primer information used in this study.

Table 2. Primer used in the present study.

Gene Name AN	Accession No.	Forward Sequence	Reverse Sequence
		Inflammatory cytokines	
il-8	KY218792.1	CACCACGATGAAGGCTGCAACTC	TGTCCTTGGTTTCCTTCTGG
il-1β	MF770571.1	CGGCAGATGTGACCTGCACA	CAGAGTAAAAGCCAGCAGAAG
p65nfκb	KY751029.1	ACTACGTGGGTCATGCTCGG	TGCTGCAGGTTCCGTTCTCA
myd88	MH778540.1	GAGGTGTAAGAGGATGGTGGTT	TGTGGAGGGTCTGGTGTAGTCA
		Apoptosis-related genes	
caspase3	KY072821	TCTACGGCACAGATGGATCC	TGTGTGCCTTCTGACTCACT
caspase9	KY053837	TTCTGTCGAGGGGCATCTTT	AGGAACGGGTACAGGAACAG
apaf1	KY053839	ACCGCCAAATAGCAACCTG	CTGCTCCTCGTGCTCAACAT
p53	HQ419002.1	TGGGAAAACGAAGAGCAAAT	ATCGGAGGTGACAGGGACA
baxa	KY072819	TCGGAGACGAACTGGACAAC	TCGACAAGCAAAGTAGAAAAGC
bcl2	KY053838.1	TTTCACCGCCGTGATCG	CCAACTTGCTATGTTGTCCACC
		JAK-STAT signaling	
jak1	XM_027146298.1	CGGAACCTCTGAAAACAAGTC	TGTCCCCGAGAAAAGAGATAG
stat3	KP342389.1	ACTCCGGTTGCCAAATCACT	CCTCATTCCACAGAGCCAGTAT
stat5	KP342392.1	ATCACCAGACCACAGGCACC	CACCACGACAGGCAAAGACAG
		Autophagy	
becn1	KY062770	CTCAACTGGACCGCCTGAAGAAA	CACTCCACAGGAACGCTGGGTAAT
ulk1a	KY404999	GCGATTAAACAGGGCAAACTCTATCC	GCTGTGATGTTGTTCATTCGGTCC
atg5	KY062771	CAGAACCGTTTTATCTTCTCCTACCG	CGTCTACATCTTCAGCTTTCACGACTT
lc3β	KY062774	CCTGACCACGTCAACATGAGCGAACT	GGAAATGGCGGCAGACACGGAGA

Table 2. *Cont.*

Gene Name AN	Accession No.	Forward Sequence	Reverse Sequence
		Antioxidant enzymes	
cat	NW0208479561 1	TCTGTTCCCGTCCTTCATCC	ATATCCGTCAGGCAATCCAC
gpx	XR_003438442.1	ATCTACATTGGCTTGGAAAC	GAAAGTAGGGACTGAGGTGA
Cu/Zn-sododDsod	KT751173.1	GGCGGAGATGATGAAAGT	GAAAGGAAGCGGTGAAAC
Mn-sod	KT751172.1	TGGTGCTTGCTATGGTGA	GGCTTGAATCCCTTGCTG
ef1α	KR061492.1	GTCTGGAGATGCTGCCATTG	AGCCTTCTTCTCAACGCTCT

il-8, interleukin8; *il-1β*, interleukin 1 beta; *p65nfκb*, p65-nuclear factor kappa b; *myd88*, myeloid differentiation primary response 88; *caspase3*, cysteine-aspartic acid protease 3; *caspase9*, cysteine-aspartic acid protease 9; *apaf1*, apoptotic protease-activating factor 1; *p53*, tumor protein 53; *baxa*, *bcl2* associated X protein; *bcl2*, b-cell cll/lymphoma 2; *jak1*, janus kinase 1; *stat3*, signal transducer and activator of transcription 3; *stat5*, signal transducer and activator of transcription 5; *becn1*, *beclin1*; *ulk1a*, unc-51 like autophagy activating kinase 1a; *atg5*, autophagy related 5; *lc3β*, light chain 3 beta; *cat*, catalase; *gpx*, glutathione peroxidas; *Cu/Zn-sod*, Cu/Zn-superoxide dismutase; *Mn-sod*, Mn-superoxide dismutase; *ef-1α*, elongation factor 1 alpha.

2.8. TdT Mediated dUTP Nick-End Labeling (TUNEL) Assays

TdT-mediated dUTP nick-end labeling (TUNEL) assay was conducted to precisely detect apoptotic cells with the one-step TUNEL Apoptosis Assay kit (C1088, Beyotime, Nanjing, China) following the manufacturer's instructions. Briefly, gills were dewaxed in xylene for 5–10 min, and then changed to fresh xylene for another 5–10 min. Then, tissues were immersed into anhydrous ethanol for 5 min, 90% ethanol for 2 min, 70% ethanol for 2 min, and distilled water for 2 min. Tissues were treated with DNase-free proteinase K (Beyotime ST535, 20 μg/mL) at 20–37 °C for 15–30 min and then washed with PBS 3 times to remove excess proteinase K. Afterwards, the labelling reaction was performed using a labelling solution containing terminal deoxynucleotidyl transferase, buffer, and fluorescein dUTP at 37 °C for 60 min in a humidity chamber. Following incubation, excess labelling solution was washed off smears with PBS, and then cell smears were mounted with fluorescent microscopy mounting solution. Images were captured and analyzed using a CCD camera (Olympus CKX41, Tokyo, Japan).

2.9. Western Blotting Analysis

Tissues was homogenated with RIPA buffer which has been added with proteinase inhibitor and phosphorylated-protease inhibitor to extract protein. Protein concentration in all groups were calculated with an enhanced BCA Protein Assay Kit (P0010, Beyotime Tech, Shanghai, China) and then adjusted to same level. The same amounts of protein in each group were separated on SDS-PAGE, and then transferred to PVDF membranes. The membranes were blocked in 6% skim milk at room temperature for 2 h and then incubated at 4 °C with the relevant primary antibodies overnight. Then, membranes were incubated with the appropriate HRP-labeled secondary antibodies for 1 h at room temperature. Immunoreactive bands were visualized using ECL reagent under GE Amersham Imager 600 Imaging System (GE Healthcare, Boston, MA, USA) and protein levels were digitized using ImageQuant TL software (GE Healthcare). Primary antibodies for phospho-S6 (Ser 235) and LC3B-II were purchased from Cell Signaling Technology (Beverly, MA, USA). The primary antibody for caspase-3 was purchased from Abcam (Cambridge, UK). Primary antibody for β-actin was purchased from ABClonal Biotechnology (Wuhan, China).

2.10. Statistical Analysis

All statistical analyses were performed using SPSS 17.0. Data about fish growth performance and feed utilization was analyzed using student's *t*-test [56,59]. All gene-expression and histological calculation results were tested for normality using the Shapiro–Wilk-W's test, and normally distributed data were analyzed by factorial (two-way) analysis of variance (ANOVA) to determine the main effects of sampling days (D) and feeding treatment (T), and their interactions (D*T). When significant D*T were observed, data were analyzed by one-way ANOVA followed by Tukey's multiple range tests to inspect differences among all the data. When the significance is only with the main effects of D or T, the data were analyzed by the two-way ANOVA followed by Tukey's multiple range tests to assess the main effects of D or T only. Differences were considered significant when $p < 0.05$. All data were expressed as mean ± standard deviation of the mean (SD), except the specific statement.

3. Results

3.1. Dietary Glutamine Significantly Increased Growth Performance, Feed Utilization and the Disease Resistance of Yellow Catfish

In the present study, dietary glutamine inclusion showed a significant promoting effect on the growth performance of yellow catfish, with higher final body weight (Table 3). The weight gain rate and specific growth rate of yellow catfish in the glutamine group (530.44 ± 2.29% & 1.43 ± 0.01%/day, respectively) were significantly higher than those in the control group (507.74 ± 6.84% & 1.40 ± 0.01%/day, respectively), while the feed conversion rate in the glutamine group (0.94 ± 0.04) was significantly lower than that in the control group (1.07 ± 0.01).

Table 3. Effects of glutamine on weight gain rate, specific growth rate, feed conversion rate of yellow catfish, and its survival rate after challenge with *Flavobacterium columnare*. All data were analyzed using student's *t*-test [56,59].

	IBW (g)	FBW (g)	WGR (%)	SGR (%/day)	FCR	SR (%)
Con	4.61 ± 0.02	28.04 ± 0.30	507.74 ± 6.84	1.40 ± 0.01	1.07 ± 0.01	73 ± 4
Gln	4.66 ± 0.02	29.38 ± 0.09	530.44 ± 2.29	1.43 ± 0.00	0.94 ± 0.04	92 ± 4
p	0.159	0.012	0.035	0.036	0.042	0.033

In order to detect the disease resistance ability of yellow catfish after dietary glutamine supplementation, fish in two groups were challenged with *F. columnare*. After the immersed infection of *F. columnare*, yellow catfish fed with glutamine supplementation showed a significantly higher survival rate (92%) than fish fed with control diet (74%) (Table 3).

3.2. Glutamine Differentially Regulated Cytokines Expression in Multiple Fish Tissues after Bacterial Infection

The expression levels of cytokines along with genes involved in NF-κB/MyD88 signaling pathway were evaluated in both systematic (Figure 1a) and mucosal-associated immune tissues (Figure 1b). In the systematic immune organ, the expression of *il-8* and *il-1β* in the head kidney and spleen were significantly increased at 30 d after infection. Accordingly, the expression of *nf-κb* and *myd88* was also increased in the head kidney at 30 d. Moreover, compared to control group, dietary glutamine promoted the expression of *il-8* and *il-1β* in the head kidney, and also increased the expression of *nf-κb* and *myd88* at 1 d. In the spleen, interactions between sampling days (D) and dietary treatment (T) were found on the expression of *il-8* and *il-1β*. In the control group, *il-8* expression was increased at 30 d, and *il-1β* expression was increased at both 1d and 30 d. Dietary glutamine supplementation further promoted the expression of *il-8* and *il-1β* at 1 d and 30 d in the spleen. The expression of *nf-κb* and *myd88* significantly increased with prolonged infection period, while dietary glutamine further promoted the expression of *nf-κb* in the spleen.

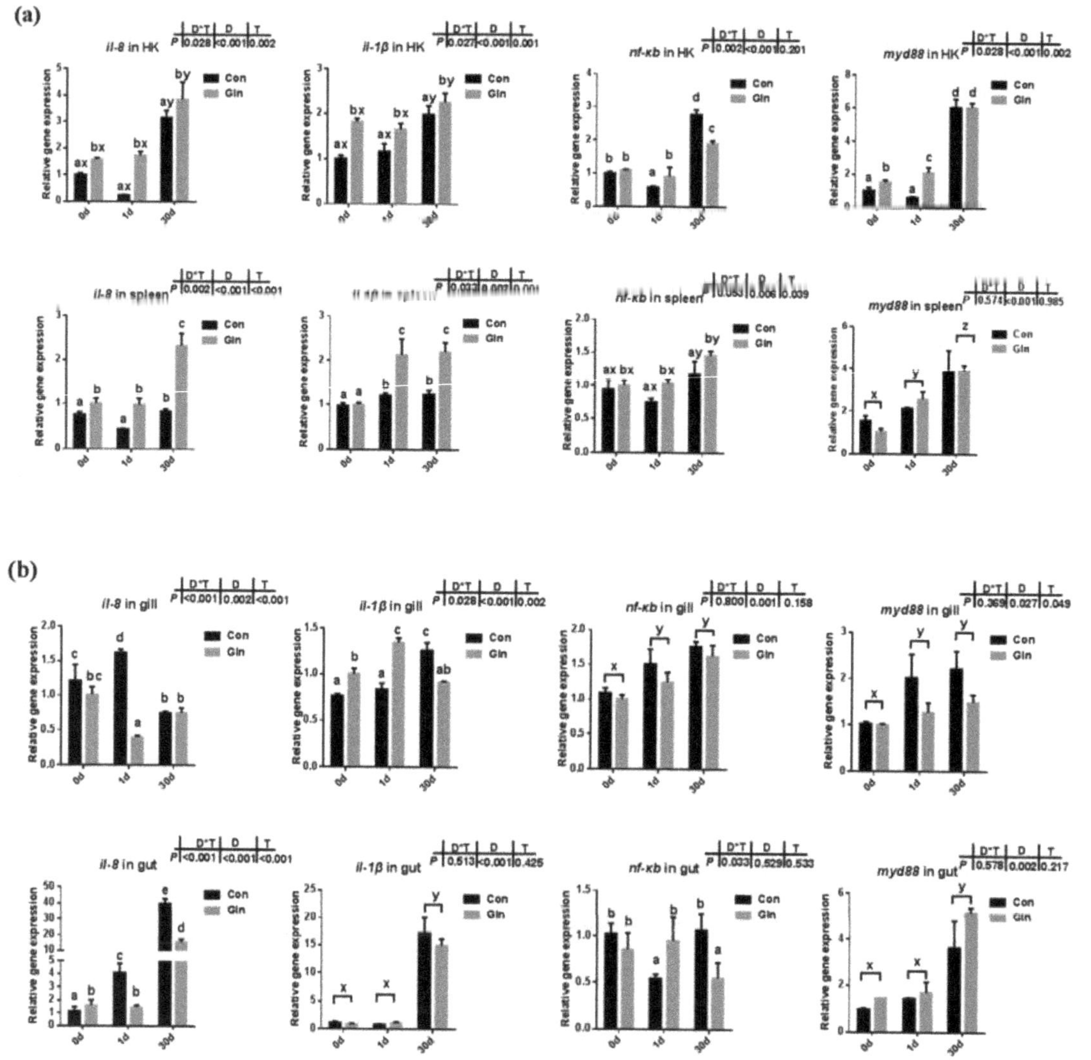

Figure 1. The effects of dietary glutamine on the inflammatory responses in systematic and mucosal immune tissues of yellow catfish during *F. columnare* infection. The relative mRNA expression levels of pro-inflammatory cytokines (*il-8*, *il-1β*) and genes involved in NF-κB signaling pathway (*nf-κb*, *myd88*) in the head kidney, spleen (**a**), and gill, gut (**b**) of yellow catfish fed with control diet (Con) and glutamine supplementation diet (Gln) at 0 d, 1 d and 30 d after *F. columnare* infection. All data was analyzed with two-way ANOVA and data are means ± SD (n = 3). When significant D*T were observed ($p < 0.05$), data were analyzed by one-way ANOVA followed by Tukey's multiple range tests to inspect differences among all the data, with "a–e" to indicate the significant difference. When the significance is only with the main effects of D or T, "a–e" and "x–z" were labeled to indicate the significant difference among groups in D or T, respectively.

In the mucosal-associated immune tissues, there was also a significant interactive effect of sampling days and dietary treatment were found on the expression of *il-8* ($p < 0.001$) and *il-1β* ($p = 0.002$) in the gill, and the expression of *il-8* in the gut ($p < 0.001$). In the control group, the expression of *il-8* was upregulated at 1 d in the gill and gut, but only upregulated at 30 d in the gut. Dietary glutamine significantly decreased the expression of *il-8* in the gill at 1 d and its expression in the gut at 1d and 30 d. In the control group, the expression of *il-1β* in the gill and gut significantly increased at 30 d, and the inhibitory effects of dietary glutamine was detected at 30 d. The expression of *nf-κb* in the gill and gut was significantly upregulated at 1d and 30 d compared to that at 0 d. Similarly, the expression of *myd88* in the gut was significantly upregulated at 30 d compared to that at 0 d, and dietary glutamine exhibited inhibitory effects on *myd88* expression in the gill at 30 d.

3.3. Glutamine Enhanced Fish Antioxidant Capacity against *F. columnare* Infection

Fish antioxidant capacity during *F. columnare* infection was also significantly affected by dietary glutamine inclusion. As shown in Figure 2a, dietary glutamine inclusion significantly enhanced serum CAT activity. Moreover, serum GPx activity at 1d and 30 d was also increased after dietary glutamine inclusion. Serum SOD activity was only affected by sampling days but not by dietary treatment. In all, the serum total antioxidant capacity (T-AOC) showed a significant increase at 30 d after glutamine inclusion.

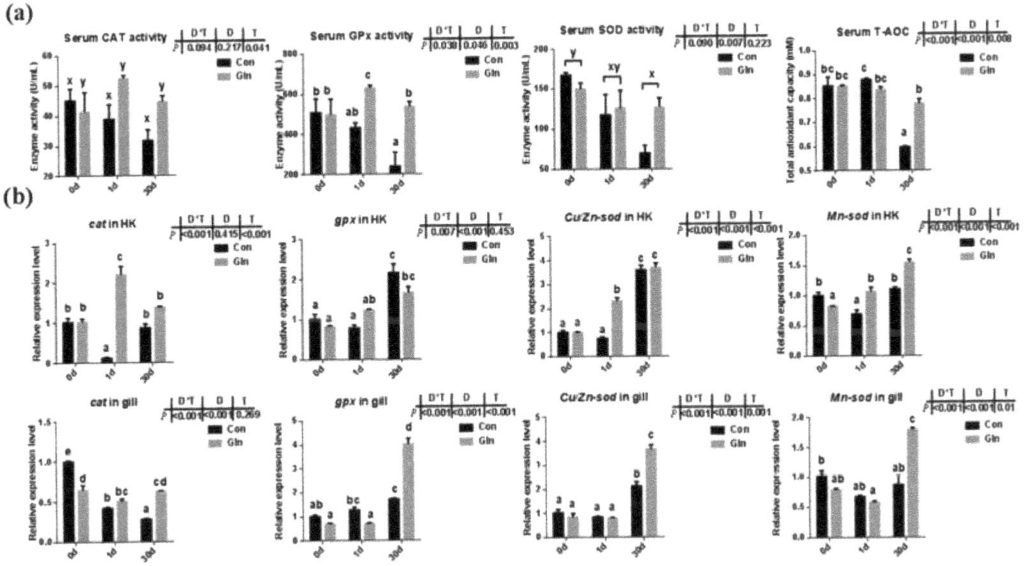

Figure 2. The effects of dietary glutamine on the antioxidant system of yellow catfish during *F. columnare* infection. (**a**) The enzyme activities of catalase (CAT), glutathione peroxidase (GPx), superoxide dismutase (SOD) along with total antioxidant capacity in serum of yellow catfish in both control group (Con) and glutamine supplementation group (Gln) at 0 d, 1 d and 30 d after *F. columnare* infection (n = 6). (**b**) The relative mRNA expression level of *cat*, *gpx*, *Cu/Zn-sod* and *Mn-sod* in head kidney and gill of yellow catfish in all treatments (n = 3). All data was analyzed with two-way analysis of variance (ANOVA) and data are means ± SD. When significant D*T were observed ($p < 0.05$), data were analyzed by one-way ANOVA followed by Tukey's multiple range tests to inspect differences among all the data, with "a–e" to indicate the significant difference. When the significance is only with the main effects of D or T, "a–e" and "x–z" were labeled to indicate the significant difference among groups in D or T, respectively.

3.4. Glutamine Protected Fish Gill Structures against *F. columnare* Infection

The gill plays an important role in the respiration, ion exchange and immune responses of fish, and the gill is the main infecting target tissue of *F. columnare*. In the present study, *F. columnare* infection significantly affected the histological structures of gill and the space between adjacent gill lamellae was occluded at 1 d (Figure 3). Moreover, the ratio of secondary lamellae length (SL) to secondary lamellae width (SW) and the ratio of SL to (SL and primary lamina length (PL)) were also significantly decreased after bacterial infection at both 1 d and 30 d. Dietary glutamine supplementation significantly alleviated the gill's structure induced by *F. columnare* infection, with an integrated structure of gill lamellar in glutamine supplementation group fish at 1d and 30 d. Moreover, the ratio of SL to SW and the ratio of SL to (SL+PL) were also significantly higher in the gill of yellow catfish with dietary glutamine supplementation than fish fed with the control diet at 30 d.

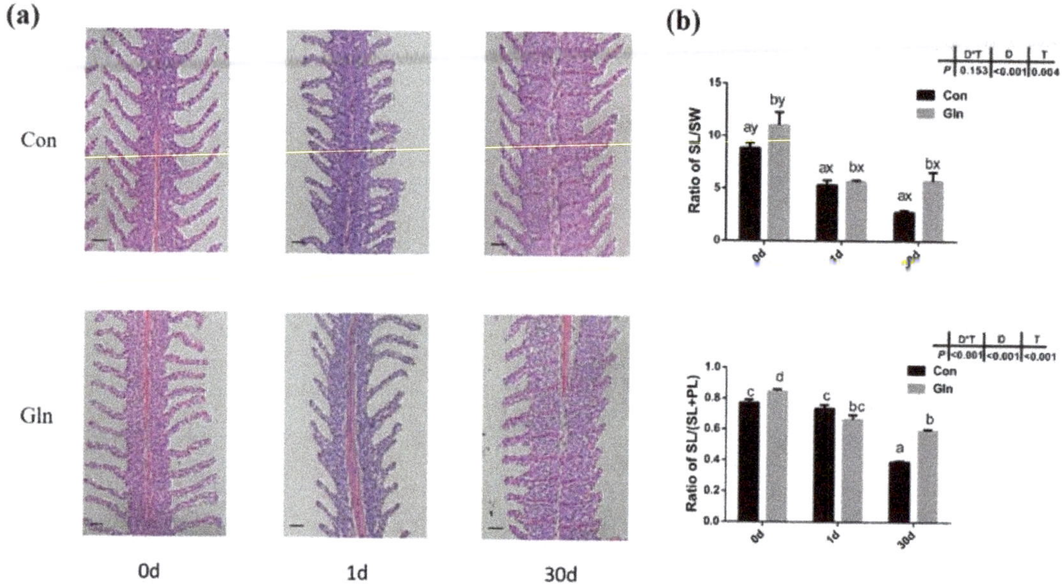

Figure 3. The effects of dietary glutamine on gill histological structures of yellow catfish during *F. columnare* infection. (**a**) The H.E. staining results of yellow catfish gill before and after *F. columnare* infection. Scale bars, 100 μm. (**b**) The ratio of secondary lamellae length (SL) to secondary lamellae width (SW) and the ratio of secondary lamellae length (SL) to the sum of primary lamellae length (PL) and SL. All data were analyzed with two-way ANOVA and data are means ± SD (n = 12). When significant D*T were observed ($p < 0.05$), data were analyzed by one-way ANOVA followed by Tukey's multiple range tests to inspect differences among all the data, with "a–e" to indicate the significant difference. When the significance is only with the main effects of D or T, "a–e" and "x–z" were labeled to indicate the significant difference among groups in D or T, respectively.

3.5. Glutamine Inhibited the Apoptosis of Fish Gill during Bacterial Infection

TUNEL analysis was adopted to determine the effects on the apoptosis in fish gills and results indicated that the apoptosis within fish gill significantly increased at 1d but then decreased at 30 d after bacterial infection (Figure 4a,b). Dietary glutamine supplementation showed a significant inhibitory role on the apoptosis of the gill. Considering that apoptosis was executed by caspase-family member proteins and regulated by different regulators, both the protein level and gene expression level of caspase-3 were evaluated, along with the expression of other related genes. As shown in Figure 4c,d, the protein levels of caspase-3 significantly increased at 30 d after bacterial infection, while dietary glutamine supplementation significantly decreased the caspase-3 protein level at both 1 d and 30 d. However, the ratio of cleaved caspase-3/caspase-3 was not increased but even decreased at 30 d after infection. Dietary glutamine increased the ratio of cleaved caspase-3/caspase-3 at 0 d but decreased this ratio after infection both at 1 d and 30 d. Additionally, the expression levels of *caspase-3*, *caspase-9*, *apaf1* and *baxa* were upregulated at 30 d in the control group

(Figure 5a). Dietary glutamine supplementation showed significant inhibitory roles on the expression of *caspase-3*, *caspase-9*, *apaf1* and *baxa* at 30 d, while it increased the expression of *caspase-3* at 1 d.

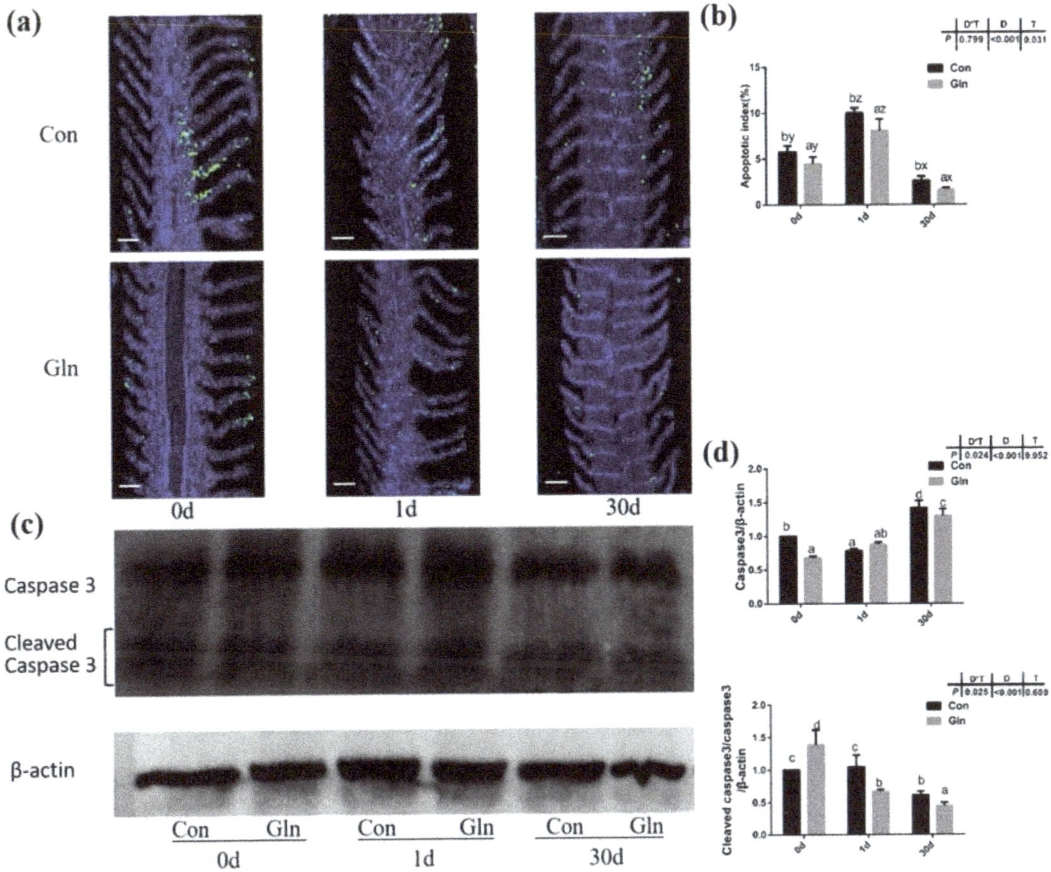

Figure 4. The effects of dietary glutamine on the gill apoptosis of yellow catfish during *F. columnare* infection. (**a**) TUNEL assay of apoptotic cells in the gill of yellow catfish in both the control group (Con) and glutamine supplementation group (Gln) at 0 d, 1 d and 30 d after *F. columnare* infection. Stained for apoptosis cells (green), nuclei are stained with DAPI (blue). Scale bar, 100 μm. (**b**) The apoptotic index of gill cells in yellow catfish among all treatments ($n = 12$). (**c**) The Western blotting analysis of caspase-3 and cleaved caspase-3 in yellow catfish among all six treatments. (**d**) The ratio of caspase-3 protein level/β-actin protein level and the ratio of cleaved caspase-3 protein level/caspase-3 protein level/β-actin protein level ($n = 6$). When significant D*T were observed ($p < 0.05$), data were analyzed by one-way ANOVA followed by Tukey's multiple range tests to inspect differences among all the data, with "a–e" to indicate the significant difference. When the significance is only with the main effects of D or T, "a–e" and "x–z" were labeled to indicate the significant difference among groups in D or T, respectively.

The apoptosis has been reported to be under regulation by JAK-STAT signaling pathway, and the expression levels of related genes were evaluated in the present study (Figure 5b). Results indicated that interactions between sampling days and dietary treatment were found on the expression of *jak1* ($p = 0.002$) and *stat5* ($p = 0.007$). In the control group, the expression levels of these genes were significantly higher at 30 d than at 0 d. Dietary glutamine supplementation significantly decreased the expression of *jak1* at 30 d, and *stat5* at 1 d and 30 d. However, the expression of *stat3* was only affected by different sampling days and it was significantly upregulated at 1 d.

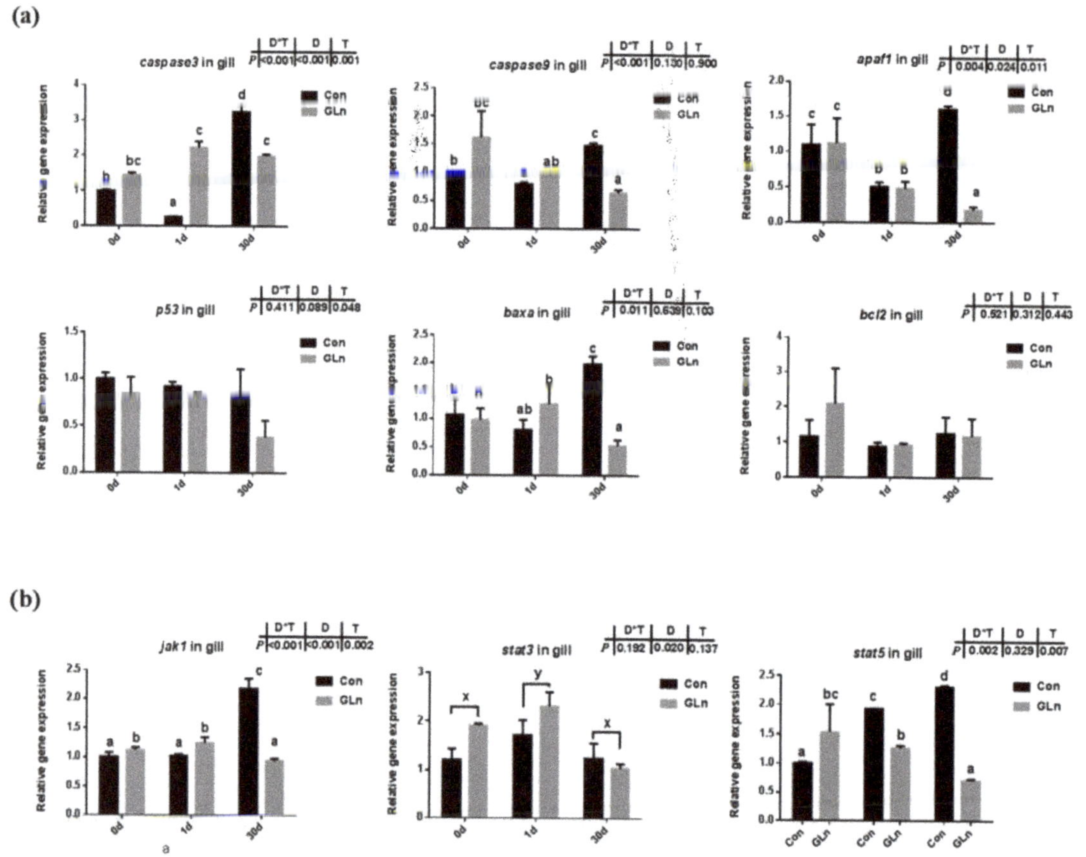

Figure 5. The effects of dietary glutamine on the apoptosis of yellow catfish during *F. columnare* infection along with the JAK-STAT signaling pathway. (**a**) The relative mRNA expression level of *caspase-3*, *caspase-9*, *apaf1*, *p53*, *baxa*, *bcl2* in gill of yellow catfish in both control group (Con) and glutamine supplementation group (Gln) at 0 d, 1 d and 30 d after *F. columnare* infection. (**b**) The relative mRNA expression level of *jak1*, *stat3*, *stat5* in gill of yellow catfish in all treatments. All data were analyzed with two-way ANOVA and data are means ± SD (n = 3). When significant D*T were observed ($p < 0.05$), data were analyzed by one-way ANOVA followed by Tukey's multiple range tests to inspect differences among all the data, with "a–e" to indicate the significant difference. When the significance is only with the main effects of D or T, "a–e" and "x–z" were labeled to indicate the significant difference among groups in D or T, respectively.

3.6. Glutamine Inhibited the Autophagy of Yellow Catfish via the Activation of mTOR Signaling

In order to determine the effects on autophagy, both the protein level of LC3B-II protein and the mRNA expression of related genes were evaluated. Results indicated that there was a significant interactive effect of sampling days and dietary treatment on the protein levels of LC3B-II (p = 0.002) and the ratio of LC3B-II/LC3B-I (p = 0.002) (Figure 6a,b). The protein level of LC3B-II significantly increased with prolonged infection period and highest level was detected at 30 d, however, the ratio of LC3B-II/LC3B-I showed the highest value at 1 d. Moreover, dietary glutamine significantly inhibited the protein levels of LC3B-II at 0 d and 30 d, and also significantly decreased the ratio of LC3B-II/LC3B-I at 1 d. Moreover, the mRNA expression levels of *becn1*, *ulk1a* and *atg5* were interactively affected by sampling days and feeding treatment. Dietary glutamine significantly decreased the expression of *becn1*, *ulk1a* and *atg5* at 30 d. The expression of *lc3β* was only affected by bacterial infection days but not feeding treatment, and its expression level was upregulated after bacterial infection.

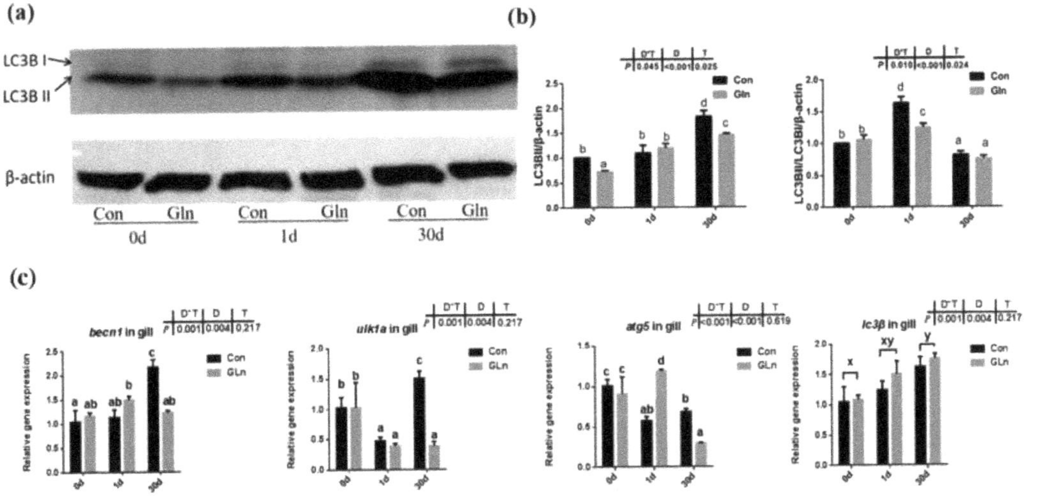

Figure 6. The effects of dietary glutamine on the autophagy of yellow catfish during *F. columnare* infection. (**a**) Western blotting analysis of LC3B I and LC3B II in yellow catfish in both control group (Con) and glutamine supplementation group (Gln) at 0 d, 1 d and 30 d after *F. columnare* infection. (**b**) The ratio of LC3B II protein level/β-actin protein level and the ratio of LC3B II/LC3B I/β-actin protein level in yellow catfish among all treatments ($n = 6$). (**c**) The mRNA expression of *becn1*, *ulk1a*, *atg5* and *lc3β* in the gill of yellow catfish among all treatments. All data were analyzed with two-way ANOVA and data are means ± SD ($n = 3$). When significant D*T were observed ($p < 0.05$), data were analyzed by one-way ANOVA followed by Tukey's multiple range tests to inspect differences among all the data, with "a–e" to indicate the significant difference. When the significance is only with the main effects of D or T, "a–e" and "x–z" were labeled to indicate the significant difference among groups in D or T, respectively.

Given that autophagy could be regulated by mTOR signaling pathway in mammals, the activation status of the mTOR signaling pathway in yellow catfish were evaluated by the phosphorylation of S6. Results showed that the phosphorylation level of S6 (p-S6) was significantly increased by dietary glutamine both at 0 d and 30 d (Figure 7), indicating the activation of mTOR signaling pathway. Thus glutamine might activate the mTOR signaling pathway to inhibit the autophagy responses during bacterial infection.

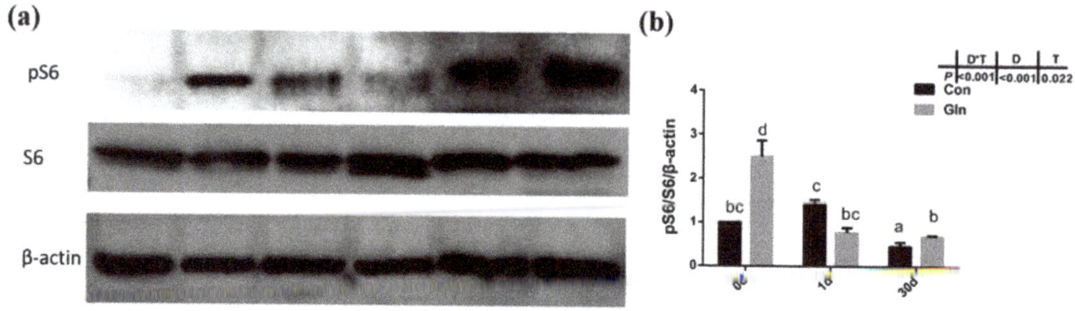

Figure 7. The effects of dietary glutamine on the mTOR signaling pathway activation of yellow catfish during *F. columnare* infection. (**a**) Western blotting analysis of pS6 and S6 in yellow catfish in both control group (Con) and glutamine supplementation group (Gln) at 0 d, 1 d and 30 d after *F. columnare* infection. (**b**) The ratio of pS6/S6/β-actin protein level in yellow catfish among all treatments. All data was analyzed with two-way ANOVA and data are means ± SD (n = 6). Mean values with different letters indicated significant differences among groups, $p < 0.05$.

4. Discussion

In the present study, after a 60-day feeding experiment, dietary glutamine supplementation significantly improved the growth performance of yellow catfish with higher final body weight (29.38 ± 0.09 g) than the control fish (28.04 ± 0.30 g) (Table 3). The increased body weight should be mainly attributed to higher protein synthesis [60], and protein synthesis in fish is mainly regulated by the mTOR signaling pathway [59]. In the present study, glutamine supplementation also significantly elevated the activation of the mTOR signaling pathway at the end of the feeding experiment (0 d), indicated by the increased phosphorylation of S6 (Figure 7). Thus glutamine activated TOR to promote fish protein synthesis, which resulted in the increased fish body weight. Additionally, dietary glutamine supplementation also improved feed utilization, which may be related to the enhanced nutrient-absorptive role, as glutamine has been reported to be one of the most important energy sources for enterocytes [46]. In order to study the protective effect of glutamine against bacterial infection, yellow catfish were infected with *F. columnare* via soaking, which is the pathogen responsible for columnaris disease [61]. The structure of the gill tissue, which is the target organ of *F. columnare*, in yellow catfish was significantly affected after *F. columnare* infection in the present study, showing occluded space between adjacent gill lamellae and decreased length of the secondary lamellae, which was similar to our previous study [56]. Glutamine has been reported to promote the histological structures of multiple mucosal tissues in both mammals and teleosts. In mammals, glutamine can positively affect gut health by supporting the gut microbiome and gut mucosal wall integrity [62], reduce mortality in rodents with sepsis [63], and attenuate lipopolysaccharide-induced acute lung injury via enhanced GSH synthesis [64,65]. In teleosts, earlier studies have also reported that dietary glutamine has a positive effect on improving the integrity of mucosal tissues such as gut in channel catfish [66] and hybrid striped bass (*Morone chrysops, Morone saxatilis*) [67]. Dietary nutrients such as valine [68], tryptophan [69] and isoleucine [70] affected the gill health status of young grass carp (*Ctenopharyngodon idella*) during normal feeding status while other nutrients such as vitamin E [71], phosphorus [72], pyridoxine [73], and choline [74] have also been reported to correlate with the incidence of rotten gill and gill morphology of grass carp during *F. columnare* infection. In the present study, dietary glutamine supplementation significantly alleviated the damage during bacterial infection, as the ratio of SL to SW and the ratio of SL to (SL+PL) were significantly increased with dietary glutamine supplementation at 30 d. All these results indicate that the addition of dietary glutamine helps yellow catfish to maintain the integrity of mucosal tissues and resist *F. columnare* infection, resulting in decreased mortality. Additionally, the protective roles of glutamine on fish gill structures significantly promote the growth performance of yellow catfish.

Teleosts were the first bony vertebrate to develop both innate and adaptive immunity, whose immune system contains both systemic immune organs (e.g., head kidney, spleen) and mucosal-associated immune tissues (e.g., skin, gill, gut) [75]. In the present study, the expression of *il-8* and *il-1β* were significantly upregulated in the head kidney and spleen at 30 d after infection, which was accompanied with the increased expression of *nf-κb* and *myd88*. Similar results were found in the gill and gut with the increased expressions of *il-8*, *il-1β*, *nf-κb* and *myd88* mainly at 30 d, indicating that *F. columnare* activates the NF-κB/Myd88 signaling pathway to stimulate the release of cytokines in both systematic and mucosal immune tissues, which was also consistent with our previous results [56]. The appropriate release of pro-inflammatory cytokines in immune organs is important for bacterial clearance and animal health [76]. Dietary glutamine supplementation promoted quicker inflammatory responses in the head-kidney and spleen at 1 d to fight against bacterial infection, indicated by the increased expression of *il-8* and *il-1β*. However, excess inflammation is also damaged to animal health, especially in the mucosal tissues that contained not only leukocytes but also more epithelial cells [77]. The limited inflammatory response and appropriate recovery of epithelial cells within those mucosal tissues are important to fish health. Here, dietary glutamine supplementation significantly decreased the expression of *il-8* in the gill at 1 d, and the expression of *il-1β* at 30 d, which contributed to the decreased mortality after bacterial infection. In mammals, treatment with glutamine before and after ischemia significantly attenuated the increases of TNF-α and CINC-1 levels in perfusate during ischemia-reperfusion induced acute lung injury ($p < 0.05$) [78]. Similarly, in the gut, the expression of *il-8* was significantly decreased by dietary glutamine supplementation at 1 d and 30 d, which indicated that the inflammation was significantly limited in a prolonged infection period. The inhibitory roles of glutamine on excess inflammation of gut are in agreement with previous reports in turbot which showed alleviation on the enteropathy induced by dietary soybean meal [79]. Glutamine has also been reported to be effective in protecting against H_2O_2-induced oxidative stress in carp intestinal epithelial cells [80]. In fact, the cellular antioxidant defense systems including antioxidant enzymes and small non-protein antioxidants in fish are also involved in the defense against bacterial infection. In the present study, the expression of representative antioxidant enzymes in the head kidney and gill were significantly decreased after dietary glutamine inclusion, indicating the lower oxidative stress. However, during *F. columnare* infection, the mRNA expression of *sod* (including *Mn-sod*, *Cu/Zn-sod*), *cat*, and *gpx* in the head kidney and gill were also significantly increased at 1 d and 30 d with dietary glutamine inclusion. Moreover, the serum enzyme activities of CAT and GPx were also significantly enhanced after dietary glutamine inclusion, resulting in the higher T-AOC in serum. The protective roles of glutamine in enhancing antioxidant capacity against various stresses has also been reported in other species such as mirror carp (*Cyprinus carpio* L.) [81], half-smooth tongue sole (*Cynoglossus semilaevis* Günther) [82], and sea cucumber (*Apostichopus japonicus* Selenka) [83].

As mentioned above, animals activate the immune and antioxidant system to execute the clearance of infecting microorganisms. However, if these immune responses fail to fight infection, the programmed cell death pathway can be activated to remove the infected cells from the organism [84]. Type I programmed cell death or apoptosis is critically important for the survival of multicellular organisms by getting rid of damaged or infected cells that may interfere with normal functions [85,86]. Considering the gill remains the main target organ of *F. columnare* infection [87] and the affected histological gill structures after infection, the apoptosis within fish gill was evaluated by a TUNEL assay in the present study. Bacterial infection significantly increased cell apoptosis in the gills of yellow catfish indicated by the TUNEL assay. Accompanied with this, both the protein level of caspase-3 and the mRNA expression of related genes including *caspase-3*, *caspase-9*, *apaf-1*, *p53* and *baxa* also significantly increased after infection. This was in accordance with previous studies in grass carp, which also detected the increased apoptosis in gill during *F. columnare* infection [73] and in the red blood cells during *Aeromonas hydrophila* infection [88]. In this study, dietary glutamine supplementation significantly reduced the ratio of apoptotic cells in the gill after bacterial infection, indicated by the TUNEL assay. In mammals, glutamine has also been found to have anti-apoptotic effects in rat intestinal epithelial (RIE-1) cell [89] and respiratory tissue (lung) [90]. In one former study, the metabolites of glutamine were reported to prevent hydroxyl radical-induced apoptosis through inhibiting mitochondria and calcium ion involved pathways in fish erythrocytes [91]. Thus it is reasonable that glutamine showed a protective role against apoptosis in fish gill in the present study. In order to further illustrate the regulatory mechanism of glutamine on apoptosis, the genes expression involved in the JAK-STAT signaling pathway were evaluated, which has been identified as one of the principal pathway responsible for apoptosis in mammals [92]. JAKs including JAK1, JAK2, JAK3, and TYK2, which function as the membrane receptors, can be activated upon preferential ligand binding and then recruits and phosphorylates STATs, preferentially STAT3 and STAT5, on their tyrosine residues [93]. Genes involved in the JAK-STAT signaling pathway have been identified in fish and also significantly affected by nutrients such as choline [74]. Our previous study also identified the regulatory role of the JAK-STAT signaling pathway on fish apoptosis during bacterial infection, which could be affected by herbal extracts [94]. In the present study, the expression of related genes including *jak1* and *stat5* were significantly increased at 30 d after *F. columnare* infection, while dietary glutamine supplementation significantly decreased their expression at 30 d. Thus glutamine might inhibit the JAK-STAT signaling pathway to decrease the apoptosis in fish gills after bacterial infection.

Besides apoptosis, autophagy, also called Type II programmed cell death, is a form of self-defense of cells against pathogenic micro-organisms, and has also been found to function during bacterial infection [95]. Autophagy has been reported to correlate with animal immune responses, as it can be used as a general factor to affect the basic functions of cells, and can also function as a special effector to regulate immune function [12]. Autophagy also plays an important role in teleost fish, and it has been reported to be involved in the lipid metabolism in gut epithelial cells of yellow catfish [96]. LC3B supported the extension and closure of phagocytes to form a double-membrane autophagosome, thereby enabling the autophagy mechanism towards maturity [97]. In the present study, both the protein level of LC3B-II and the ratio of LC3B-II/LC3B-I protein in yellow catfish was significantly increased after *F. columnare* infection. However, glutamine significantly inhibited the protein level of LC3B-II and the ratio of LC3B-II/LC3B-I protein, indicating the inhibitory effects on autophagy. Besides LC3B, other factors such as becn1, ulk1a and atg5 have also been reported to participate in the fish autophagy process [98]. In the present study, the expression of *ulk1a* and *atg5* were also significantly downregulated after glutamine supplementation at 30 d, confirming the inhibitory role of glutamine on the autophagy of yellow catfish. In order to illustrate the regulatory mechanism of glutamine on autophagy, the activation level of TOR signaling pathway was determined by the phosphorylation level of S6 (p-S6), which has been reported to function as a regulator in fish autophagy [99]. Dietary inclusion of glutamine significantly increased the phosphorylation level of S6 (p-S6), indicating the higher activation level of mTOR signaling pathway. Thus glutamine may activate the TOR signaling pathway, which functions in the inhibition of autophagy in yellow catfish after *F. columnare* infection. The protective roles of glutamine in the present study are similar to earlier reports on methionine-chelated Zn, which promotes anabolism by integrating the mTOR signal and autophagy pathway [100]. Taken together, this provides direct evidence that bacterial infection causes autophagy in the fish itself, and that glutamine suppresses autophagy via activating the TOR signaling pathway.

5. Conclusions

In conclusion, our results emphasized the effect of glutamine on the growth promotion and disease resistance of yellow catfish, and initially explored the functions of glutamine on fish somatic cell apoptosis and autophagy during bacterial infection. The discovery highlights the importance of functional nutrients in fish immune response, and may lay the foundation for the research of fish nutrition metabolism and anti-bacterial disease treatment.

Author Contributions: Q.W. designed and carried out the main context written. Y.-H.L. conducted the manuscript review. C.J., J.Z. and Z.C. conducted most experimental protocol. F.Z. and Z.X. supplied the relevant materials. All authors have read and agreed to the published version of the manuscript.

Funding: This article was funded by National Natural Science Foundation of China (Grant No. 31802317; 32172996).

Institutional Review Board Statement: All the fish care and experimental procedures have been reviewed and approved by the Animal Experiment Committee of Huazhong Agricultural University (permit number HZAUFI-2016-007).

Informed Consent Statement: Not applicable.

Data Availability Statement: Data are available from the corresponding author if requested.

Conflicts of Interest: The authors declare no conflict of interest.

Abbreviations

DAMP	damage-associated molecular patterns
FA	fatty acids
FBW	final body weight
FCR	feed conversion ratio
FSGD	fish specific genome duplication
Gln	glutamine
MAPK	mitogen-activated protein kinase
mTOR	mechanistic target of rapamycin
ROS	reactive oxygen species
SGR	specific growth rate
TAOC	total antioxidant capacity
TG	triglyceride
WGR	weight gain ratio

References

1. Pkala-Safińska, A. Contemporary threats of bacteribal infectioins in freshwater fish. *J. Vet. Res.* **2018**, *62*, 261–267. [CrossRef]
2. Yu, Y.; Wang, Q.; Huang, Z.; Ding, L.; Xu, Z. Immunoglobulins, mucosal immunity and vaccination in teleost fish. *Front. Immunol.* **2020**, *11*, 567941. [CrossRef] [PubMed]
3. Gomez, D.; Sunyer, J.O.; Salinas, I. The mucosal immune system of fish: The evolution of tolerating commensals while fighting pathogens. *Fish Shellfish Immunol.* **2013**, *35*, 1729–1739. [CrossRef]
4. Peatman, E.; Lange, M.; Zhao, H.; Beck, B.H. Physiology and immunology of mucosal barriers in catfish (*Ictalurus* spp.). *Tissue Barriers* **2015**, *3*, e1068907. [CrossRef]
5. Dash, S.; Das, S.K.; Samal, J.; Thatoi, H.N. Epidermal mucus, a major determinant in fish health: A review. *Iran. J. Vet. Res.* **2018**, *19*, 72–81. [CrossRef] [PubMed]
6. Swelum, A.A.; Elbestawy, A.R.; El-Saadony, M.T.; Hussein, E.O.S.; Alhotan, R.; Suliman, G.M.; Taha, A.E.; Ba-Awadh, H.; El-Tarabily, K.A.; Abd El-Hack, M.E. Ways to minimize bacterial infections, with special reference to *Escherichia coli*, to cope with the first-week mortality in chicks: An updated overview. *Poult. Sci.* **2021**, *100*, 101039. [CrossRef] [PubMed]
7. Boehm, T.; Swann, J.B. Origin and evolution of adaptive immunity. *Annu. Rev. Anim. Biosci.* **2014**, *2*, 259–283. [CrossRef]
8. Buchmann, K. Evolution of innate immunity: Clues from invertebrates via fish to mammals. *Front. Immunol.* **2014**, *5*, 459. [CrossRef]
9. Kurien, B.T.; Scofield, R.H. Free radical mediated peroxidative damage in systemic lupus erythematosus. *Life Sci.* **2003**, *73*, 1655–1666. [CrossRef]
10. Hermans, N.; Cos, P.; Maes, L.; De Bruyne, T.; Vanden Berghe, D.; Vlietinck, A.J.; Pieters, L. Challenges and pitfalls in antioxidant research. *Curr. Med. Chem.* **2007**, *14*, 417–430. [CrossRef]

11. Ighodaro, O.M.; Akinloye, O.A. First line defence antioxidants-superoxide dismutase (SOD), catalase (CAT) and glutathione peroxidase (GPX): Their fundamental role in the entire antioxidant defence grid. *Alex. J. Med.* **2018**, *54*, 287–293. [CrossRef]
12. Deretic, V. Autophagy in infection. *Curr. Opin. Cell Biol.* **2010**, *22*, 252–262. [CrossRef]
13. Jorgensen, I.; Rayamajhi, M.; Miao, E.A. Programmed cell death as a defence against infection. *Nat. Rev. Immunol.* **2017**, *17*, 151–164. [CrossRef] [PubMed]
14. Deretic, V. Autophagy: An emerging immunological paradigm. *J. Immunol.* **2012**, *189*, 15–20. [CrossRef] [PubMed]
15. Schaefer, L. Complexity of danger: The diverse nature of damage-associated molecular patterns. *J. Biol. Chem.* **2014**, *289*, 35237–35245. [CrossRef] [PubMed]
16. Mizushima, N.; Alerting, F.; Articles, C.D.; Mizushima, N. Review autophagy: Process and function. *Genes Dev.* **2007**, *21*, 2861–2873. [CrossRef]
17. Naderer, T.; Fulcher, M.C. Targeting apoptosis pathways in infections. *J. Leukoc. Biol.* **2018**, *103*, 275–285. [CrossRef]
18. Tribble, G.D.; Lamont, R.J. Bacterial invasion of epithelial cells and spreading in periodontal tissue. *Periodontology* **2010**, *52*, 68–83. [CrossRef]
19. Peter, K.; Judit, K.A. The interaction between nutrition and infection. *Clin. Infect. Dis.* **2008**, 1582–1588. [CrossRef]
20. Kedia-Mehta, N.; Finlay, D.K. Competition for nutrients and its role in controlling immune responses. *Nat. Commun.* **2019**, *10*, 2123. [CrossRef]
21. Calder, P.C.; Samantha, K. The immune system: A target for functional foods? *Br. J. Nutr.* **2002**, *88*, 165–177. [CrossRef]
22. Wu, D.; Lewis, E.D.; Pae, M.; Meydani, S.N. Nutritional modulation of immune function: Analysis of evidence, mechanisms and clinical relevance. *Front. Immunol.* **2019**, *9*, 3160. [CrossRef]
23. Delesderrier, E.; Curioni, C.; Omena, J.; Macedo, C.R.; Cople-Rodrigues, C.; Citelli, M. Antioxidant nutrients and hemolysis in sickle cell disease. *Clin. Chim. Acta* **2020**, *510*, 381–390. [CrossRef] [PubMed]
24. Rajendran, P.; Nandakumar, N.; Rengarajan, T.; Palaniswami, R.; Gnanadhas, E.N.; Lakshminarasaiah, U.; Gopas, J.; Nishigaki, I. Antioxidants and human diseases. *Clin. Chim. Acta* **2014**, *436*, 332–347. [CrossRef] [PubMed]
25. Liu, Y.; Hyde, A.S.; Simpson, M.A.; Barycki, J.J. Emerging regulatory paradigms in glutathione metabolism. *Adv. Cancer Res.* **2014**, *122*, 69–101. [CrossRef] [PubMed]
26. Lin, S.; Tian, H.; Lin, J.; Xu, C.; Yuan, Y.; Gao, S.; Song, C.; Lv, P.; Mei, X. Zinc promotes autophagy and inhibits apoptosis through AMPK/mTOR signaling pathway after spinal cord injury. *Neurosci. Lett.* **2020**, *736*, 135263. [CrossRef]
27. Sa-Nongdej, W.; Chongthammakun, S.; Songthaveesin, C. Nutrient starvation induces apoptosis and autophagy in C6 glioma stem-like cells. *Heliyon* **2021**, *26*, e06352. [CrossRef]
28. Li, P.; Yin, Y.L.; Li, D.; Kim, S.W.; Wu, G. Amino acids and immune function. *Br. J. Nutr.* **2007**, *98*, 237–252. [CrossRef]
29. Xi, P.; Jiang, Z.; Dai, Z.; Li, X.; Yao, K.; Zheng, C.; Lin, Y.; Wang, J.; Wu, G. Regulation of protein turnover by l-glutamine in porcine intestinal epithelial cells. *J. Nutr. Biochem.* **2012**, *23*, 1012–1017. [CrossRef]
30. Tomé, D. The roles of dietary glutamate in the intestine. *Ann. Nutr. Metab.* **2018**, *73*, 15–20. [CrossRef]
31. Pires, R.S.; Braga, P.G.S.; Santos, J.M.B.; Amaral, J.B.; Amirato, G.R.; Trettel, C.S.; Dos Santos, C.A.F.; Vaisberg, M.; Nali, L.H.S.; Vieira, R.P.; et al. l-Glutamine supplementation enhances glutathione peroxidase and paraoxonase-1 activities in HDL of exercising older individuals. *Exp. Gerontol.* **2021**, *156*, 111584. [CrossRef] [PubMed]
32. Cruzat, V.; Macedo Rogero, M.; Noel Keane, K.; Curi, R.; Newsholme, P. Glutamine: Metabolism and immune function, supplementation and clinical translation. *Nutrients* **2018**, *10*, 1564. [CrossRef]
33. Koufaris, C.; Nicolaidou, V.; Ma, J. Glutamine addiction in virus-infected mammalian cells: A target of the innate immune system? *Med. Hypotheses* **2021**, *153*, 110620. [CrossRef]
34. Kim, J.; Song, G.; Wu, G.; Gao, H.; Johnson, G.A.; Bazer, F.W. Arginine, leucine and glutamine stimulate proliferation of porcine trophectoderm cells through the mTOR-RPS6K-RPS6-EIF4EBP1 signal transduction pathway. *Biol. Reprod.* **2013**, *88*, 113. [CrossRef]
35. Evans, M.E.; Jones, D.P.; Ziegler, T.R. Glutamine prevents cytokine-induced apoptosis in human colonic epithelial cells. *J. Nutr.* **2003**, *133*, 3065–3071. [CrossRef] [PubMed]
36. Ban, K.; Kozar, R.A. Glutamine protects against apoptosis via downregulation of Sp3 in intestinal epithelial cells. *Am. J. Physiol. Gastrointest. Liver Physiol.* **2010**, *299*, 1344–1353. [CrossRef]
37. Marc Rhoads, J.; Wu, G. Glutamine, arginine and leucine signaling in the intestine. *Amino Acids* **2009**, *37*, 111–122. [CrossRef]
38. Rhoads, J.M.; Argenzio, R.A.; Chen, W.; Graves, L.M.; Licato, L.L.; Blikslager, A.T.; Smith, J.; Gatzy, J.; Brenner, D.A. Glutamine metabolism stimulates intestinal cell MAPKs by a cAMP-inhibitable, RAF-independent mechanism. *Gastroenterology* **2000**, *118*, 90–100. [CrossRef]
39. Yi, D.; Hou, Y.; Wang, L.; Ouyang, W.; Long, M.; Zhao, D.; Ding, B.; Liu, Y.; Wu, G. L-Glutamine enhances enterocyte growth via activation of the mTOR signaling pathway independently of AMPK. *Amino Acids* **2015**, *47*, 65–78. [CrossRef] [PubMed]
40. Meng, D.; Yang, Q.; Wang, H.; Melick, C.H.; Jewell, J.L. Glutamine and asparagine activate mTORC1 independently of Rag GTPases. *J. Biol. Chem.* **2020**, *295*, 2890–2899. [CrossRef]
41. Ogunlesi, F.; Cho, C.; Mcgrath-Morrow, S.A. The effect of glutamine on A549 cells exposed to moderate hyperoxia. *Biochim. Biophys. Acta* **2004**, *1688*, 112–120. [CrossRef] [PubMed]
42. Larson, S.D.; Jing, L.; Dai, H.C.; Evers, B.M. Molecular mechanisms contributing to glutamine-mediated intestinal cell survival. *Am. J. Physiol. Gastrointest. Liver Physiol.* **2007**, *293*, G1262. [CrossRef] [PubMed]

43. Kim, M.H.; Kim, H. The roles of glutamine in the intestine and its implication in intestinal diseases. *Int. J. Mol. Sci.* **2017**, *18*, 1051. [CrossRef] [PubMed]
44. Wang, B.; Wu, G.; Zhou, Z.; Dai, Z.; Sun, Y.; Ji, Y.; Li, W.; Wang, W.; Liu, C.; Han, F.; et al. Glutamine and intestinal barrier function. *Amino Acids* **2015**, *47*, 2143–2154. [CrossRef]
45. Reeds, P.J.; Burrin, D.G. Glutamine and the bowel. *J. Nutr.* **2001**, *131*, 2505S–2508S. [CrossRef]
46. Ji, F.J.; Wang, L.X.; Yang, H.S.; Hu, A.; Yin, Y.L. Review: The roles and functions of glutamine on intestinal health and performance of weaning pigs. *Animal* **2019**, *13*, 1–9. [CrossRef]
47. Zhu, Y.; Lin, G.; Dai, Z.; Zhou, T.; Li, T.; Yuan, T.; Wu, Z.; Wu, G.; Wang, J. L-Glutamine deprivation induces autophagy and alters the mTOR and MAPK signaling pathways in porcine intestinal epithelial cells. *Amino Acids* **2015**, *47*, 2185–2197. [CrossRef]
48. Coutinho, F.; Castro, C.; Rufino-Palomares, E.; Ordóñez-Grande, B.; Gallardo, M.A.; Oliva-Teles, A.; Peres, H. Dietary glutamine supplementation effects on amino acid metabolism, intestinal nutrient absorption capacity and antioxidant response of gilthead sea bream (*Sparus aurata*) juveniles. *Comp. Biochem. Physiol. A Mol. Integr. Physiol.* **2016**, *191*, 9–17. [CrossRef]
49. Ma, X.Z.; Feng, L.; Wu, P.; Liu, Y.; Kuang, S.Y.; Tang, L.; Zhou, X.Q.; Jiang, W.D. Enhancement of flavor and healthcare substances, mouthfeel parameters and collagen synthesis in the muscle of on-growing grass carp (*Ctenopharyngodon idella*) fed with graded levels of glutamine. *Aquaculture* **2020**, *528*, 735486. [CrossRef]
50. Sato, Y.; Hashiguchi, Y.; Nishida, M. Temporal pattern of loss/persistence of duplicate genes involved in signal transduction and metabolic pathways after teleost-specific genome duplication. *BMC. Evol. Biol.* **2009**, *9*, 127. [CrossRef]
51. Ravi, V.; Venkatesh, B. The divergent genomes of teleosts. *Annu. Rev. Anim. Biosci.* **2018**, *6*, 47–68. [CrossRef] [PubMed]
52. Gong, G.; Dan, C.; Xiao, S.; Guo, W.; Huang, P.; Xiong, Y. Chromosomal-level assembly of yellow catfish genome using third-generation DNA sequencing and Hi-C analysis. *Gigascience* **2018**, *7*, giy120. [CrossRef]
53. Song, Y.F.; Hogstrand, C.; Ling, S.C.; Chen, G.H.; Luo, Z. Creb-Pgc1α pathway modulates the interaction between lipid droplets and mitochondria and influences high fat diet-induced changes of lipid metabolism in the liver and isolated hepatocytes of yellow catfish. *J. Nutrl. Biochem.* **2020**, *80*, 108364. [CrossRef] [PubMed]
54. Wu, K.; Zhao, T.; Hogstrand, C.; Xu, Y.C.; Luo, Z. FXR-mediated inhibition of autophagy contributes to FA-induced TG accumulation and accordingly reduces FA-induced lipotoxicity. *Cell Commun. Signal* **2020**, *18*, 47. [CrossRef] [PubMed]
55. Evenhuis, J.P.; Mohammed, H.; Lapatra, S.E.; Welch, T.J.; Arias, C.R. Virulence and molecular variation of *Flavobacterium columnare* affecting rainbow trout in Idaho, USA. *Aquaculture* **2016**, *464*, 106–110. [CrossRef]
56. Wang, Q.; Shen, J.; Yan, Z.; Xiang, X.; Mu, R.; Zhu, P.; Yao, Y.; Zhu, F.; Chen, K.; Chi, S.; et al. Dietary *glycyrrhiza uralensis* extracts supplementation elevated growth performance, immune responses and disease resistance against *Flavobacterium columnare* in yellow catfish (*Pelteobagrus fulvidraco*). *Fish Shellfish Immunol.* **2020**, *97*, 153–164. [CrossRef]
57. Xu, J.; Zhang, X.; Luo, Y.; Wan, X.; Yao, Y.; Zhang, L.; Yu, Y.; Ai, T.; Wang, Q.; Xu, Z. IgM and IgD heavy chains of yellow catfish (*Pelteobagrus fulvidraco*): Molecular cloning, characterization and expression analysis in response to bacterial infection. *Fish Shellfish Immunol.* **2019**, *84*, 233–243. [CrossRef]
58. Xu, Z.; Takizawa, F.; Casadei, E.; Shibasaki, Y.; Sunyer, J.O. Specialization of mucosal immunoglobulins in pathogen control and microbiota homeostasis occurred early in vertebrate evolution. *Sci. Immunol.* **2020**, *5*, eaay3254. [CrossRef]
59. Wang, Q.; He, G.; Mai, K.; Xu, W.; Zhou, H.; Wang, X.; Mei, L. Chronic rapamycin treatment on the nutrient utilization and metabolism of juvenile turbot (*Psetta maxima*). *Sci. Rep.* **2016**, *6*, 28068. [CrossRef]
60. Li, X.; Zheng, S.; Jia, S.; Song, F.; Wu, G. Oxidation of energy substrates in tissues of largemouth bass (*Micropterus salmoides*). *Amino Acids* **2020**, *52*, 1017–1032. [CrossRef]
61. Declercq, A.M.; Haesebrouck, F.; Van den Broeck, W.; Bossier, P.; Decostere, A. Columnaris disease in fish: A review with emphasis on bacterium-host interactions. *Vet. Res.* **2013**, *44*, 27. [CrossRef]
62. Deters, B.J.; Saleem, M. The role of glutamine in supporting gut health and neuropsychiatric factors-ScienceDirect. *Food Sci. Human Wellness* **2021**, *10*, 149–154. [CrossRef]
63. Lai, Y.N.; Yeh, S.L.; Lin, M.T.; Shang, H.F.; Yeh, C.L.; Chen, W.J. Glutamine supplementation enhances mucosal immunity in rats with Gut-Derived sepsis. *Nutrition* **2004**, *20*, 286–291. [CrossRef]
64. Zhang, F.; Wang, X.; Pan, L.; Wang, W.; Li, N.; Li, J. Glutamine attenuates lipopolysaccharide-induced acute lung injury. *Nutrition* **2009**, *25*, 692–698. [CrossRef] [PubMed]
65. Vigeland, C.L.; Beggs, H.S.; Collins, S.L.; Chan-Li Yv Powell, J.D.; Doerschuk, C.M.; Horton, M.R. Inhibition of glutamine metabolism accelerates resolution of acute lung injury. *Physiol. Rep.* **2019**, *7*, e14019. [CrossRef] [PubMed]
66. Pohlenz, C.; Buentello, A.; Bakke, A.M.; Gatlin, D.M. Free dietary glutamine improves intestinal morphology and increases enterocyte migration rates, but has limited effects on plasma amino acid profile and growth performance of channel catfish *Ictalurus punctatus*. *Aquaculture* **2012**, *370*, 32–39. [CrossRef]
67. Cheng, Z.Y.; Gatlin, D.M.; Buentello, A. Dietary supplementation of arginine and/or glutamine influences growth performance, immune responses and intestinal morphology of hybrid striped bass (*Morone chrysops* × *Morone saxatilis*). *Aquaculture* **2012**, *s362*, 39–43. [CrossRef]
68. Feng, L.; Luo, J.B.; Jiang, W.D.; Liu, Y.; Wu, P.; Jiang, J.; Kuang, S.Y.; Tang, L.; Zhang, Y.A.; Zhou, X.Q. Changes in barrier health status of the gill for grass carp (*Ctenopharyngodon idella*) during valine deficiency: Regulation of tight junction protein transcript, antioxidant status and apoptosis-related gene expression. *Fish Shellfish Immunol.* **2015**, *45*, 239–249. [CrossRef] [PubMed]

69. Jiang, W.D.; Wen, H.L.; Liu, Y.; Jiang, J.; Kuang, S.Y.; Wu, P.; Zhao, J.; Tang, L.; Tang, W.N.; Zhang, Y.A.; et al. The tight junction protein transcript abundance changes and oxidative damage by tryptophan deficiency or excess are related to the modulation of the signalling molecules, NF-κB p65,TOR,caspase-(3,8,9) and Nrf2 mRNA levels, in the gill of young grass carp (*Ctenopharyngodon idellus*). *Fish Shellfish Immunol.* **2015**, *46*, 168–180. [CrossRef]
70. Feng, L.; Gan, L.; Jiang, W.D.; Wu, P.; Liu, Y.; Jiang, J.; Tang, L.; Kuang, S.Y.; Tang, W.N.; Zhang, Y.A.; et al. Gill structural integrity changes in fish deficient or excessive in dietary isoleucine: Towards the modulation of tight junction protein, inflammation, apoptosis and antioxidant defense via NF-κB, TOR and Nrf2 signaling pathways. *Fish Shellfish Immunol.* **2017**, *63*, 127–138. [CrossRef]
71. Xu, H.J.; Jiang, W.D.; Feng, L.; Liu, Y.; Wu, P.; Jiang, J.; Kuang, S.Y.; Tang, L.; Tang, W.N.; Zhang, Y.A.; et al. Vitamin E deficiency depressed gill immune response and physical barrier referring to NF-kB, TOR, Nrf2 and MLCK signalling in grass carp (*Ctenopharyngodon idella*) under infection of *Flavobacterium columnare*. *Aquaculture* **2018**, *481*, 13–27. [CrossRef]
72. Chen, K.; Zhou, X.Q.; Jiang, W.D.; Wu, P.; Liu, Y.; Jiang, J.; Kuang, S.Y.; Tang, L.; Tang, W.N.; Zhang, Y.A.; et al. Dietary phosphorus deficiency caused alteration of gill immune and physical barrier function in the grass carp (*Ctenopharyngodon idella*) after infection with *Flavobacterium columnare*. *Aquaculture* **2019**, *506*, 1–13. [CrossRef]
73. Zheng, X.; Feng, L.; Jiang, W.D.; Wu, P.; Liu, Y.; Kuang, S.Y.; Tang, L.; Zhou, X.Q. The regulatory effects of pyridoxine deficiency on the grass carp (*Ctenopharyngodon idella*) gill barriers immunity, apoptosis, antioxidant and tight junction challenged with *Flavobacterium columnar*. *Fish Shellfish Immunol.* **2020**, *105*, 209–223. [CrossRef] [PubMed]
74. Yuan, Z.H.; Jiang, W.D.; Feng, L.; Wu, P.; Liu, Y.; Kuang, S.Y.; Tang, L.; Zhou, X.Q. Dietary choline inhibited the gill apoptosis in association with the p38MAPK and JAK/STAT3 signalling pathways of juvenile grass carp (*Ctenopharyngodon idella*). *Aquaculture* **2020**, *529*, 735699. [CrossRef]
75. Salinas, I.; Zhang, Y.A.; Sunyer, J.O. Mucosal immunoglobulins and B cells of teleost fish. *Dev. Comp. Immunol.* **2011**, *35*, 1346–1365. [CrossRef]
76. Al-Banna, N.A.; Cyprian, F.; Albert, M.J. Cytokine responses in campylobacteriosis: Linking pathogenesis to immunity. *Cytokine Growth Factor Rev.* **2018**, *41*, 75–87. [CrossRef]
77. Kogut, M.H.; Genovese, K.J.; Swaggerty, C.L.; He, H.; Broom, L. Inflammatory phenotypes in the intestine of poultry: Not all inflammation is created equal. *Poultry Sci.* **2018**, *97*, 2339–2346. [CrossRef] [PubMed]
78. Peng, C.K.; Huang, K.L.; Wu, C.P.; Li, M.H.; Hu, Y.T.; Hsu, C.W.; Tsai, S.H.; Chu, S.J. Glutamine protects ischemia-reperfusion induced acute lung injury in isolated rat lungs. *Pulm Pharmacol. Ther.* **2011**, *24*, 153–161. [CrossRef]
79. Yang, L.; Chen, Z.; Dai, J.; Pei, Y.; Hu, H.; Ai, Q. The protective role of glutamine on enteropathy induced by high dose of soybean meal in turbot, *Scophthalmus maximus* L. *Aquaculture* **2018**, *497*, 510–519. [CrossRef]
80. Chen, J.; Zhou, X.Q.; Lin, F.; Liu, Y.; Jiang, J. Effects of glutamine on hydrogen peroxide-induced oxidative damage in intestinal epithelial cells of Jian carp (*Cyprinus carpio* var. *Jian*). *Aquaculture* **2009**, *288*, 285–289. [CrossRef]
81. Liu, J.; Mai, K.; Xu, W.; Zhang, Y.; Zhou, H.; Ai, Q. Effects of dietary glutamine on survival, growth performance, activities of digestive enzyme, antioxidant status and hypoxia stress resistance of half-smooth tongue sole (*Cynoglossus semilaevis* Günther) post larvae. *Aquaculture* **2015**, *446*, 48–56. [CrossRef]
82. Xu, H.; Zhu, Q.; Wang, C.; Zhao, Z.; Luo, L.; Wang, L.; Li, J.; Xu, Q. Effect of dietary Alanyl-glutamine supplementation on growth performance, development of intestinal tract, antioxidant status and plasma non-specific immunity of young mirror carp (*Cyprinus carpio* L.). *J. Northeast. Agric. Univ.* **2014**, *21*, 37–46. [CrossRef]
83. Yu, H.; Gao, Q.; Dong, S.; Lan, Y.; Ye, Z.; Wen, B. Regulation of dietary glutamine on the growth, intestinal function, immunity and antioxidant capacity of sea cucumber *Apostichopus japonicus* (Selenka). *Fish Shellfish Immunol.* **2016**, *50*, 56–65. [CrossRef]
84. Lamkanfi, M.; Dixit, V.M. Manipulation of host cell death pathways during microbial infections. *Cell Host Microbe* **2010**, *8*, 44–54. [CrossRef]
85. Labbé, K.; Saleh, M. Cell death in the host response to infection. *Cell Death Differ.* **2008**, *15*, 1339–1349. [CrossRef]
86. Portt, L.; Norman, G.; Clapp, C.; Greenwood, M.; Greenwood, M.T. Anti-apoptosis and cell survival: A review. *Biochimi. Biophys. Acta* **2011**, *1813*, 238–259. [CrossRef] [PubMed]
87. Kunttu, H.; Suomalainen, L.R.; Jokinen, E.I.; Valtonen, E.T. *Flavobacterium columnare* colony types: Connection to adhesion and virulence? *Microb. Pathog.* **2009**, *46*, 21–27. [CrossRef]
88. Lu, Z.; Yang, M.; Zhang, K.; Zhan, F.; Qin, Z. Aeromonas hydrophila infection activates death receptor apoptosis pathway in the red blood cells of grass carp (*Ctenopharyngodon idellus*). *Aquaculture* **2021**, *532*, 735956. [CrossRef]
89. Papaconstantinou, H.T.; Chung, D.H.; Zhang, W.; Ansari, N.H.; Hellmich, M.R.; Townsend, C.M., Jr.; Ko, T.C. Prevention of mucosal atrophy: Role of glutamine and caspases in apoptosis in intestinal epithelial cells. *J. Gastrointest. Surg.* **2000**, *4*, 416–423. [CrossRef]
90. Kwon, W.Y.; Suh, G.J.; Kim, K.S.; Jo, Y.H.; Lee, J.H.; Kim, K.; Jung, S.K. Glutamine attenuates acute lung injury by inhibition of high mobility group box protein-1 expression during sepsis. *Br. J. Nutr.* **2010**, *103*, 890–898. [CrossRef]
91. Li, H.; Jiang, W.; Liu, Y.; Jiang, J.; Zhang, Y.; Wu, P.; Zhao, J.; Duan, X.; Zhou, X.; Feng, L. The metabolites of glutamine prevent hydroxyl radical-induced apoptosis through inhibiting mitochondria and calcium ion involved pathways in fish erythrocytes. *Free Radic. Biol. Med.* **2016**, *92*, 126–140. [CrossRef]
92. Brouwer, I.J.; Out-Luiting, J.; Vermeer, M.H.; Tensen, C.P. Cucurbitacin E and I target the JAK/STAT pathway and induce apoptosis in Sézary cells. *Biochem. Biophys. Rep.* **2020**, *24*, 100832. [CrossRef] [PubMed]

93. Kiu, H.; Nicholson, S.E. Biology and significance of the JAK/STAT signalling pathways. *Growth Factors* **2012**, *30*, 88–106. [CrossRef] [PubMed]
94. Song, Z.; Jiao, C.; Chen, B. Dietary Acanthopanax senticosus extracts modulated the inflammatory and apoptotic responses of yellow catfish to protect against *Edwardsiella ictaluri* infection. *Aquac. Res.* **2021**, *52*, 5078–5092. [CrossRef]
95. Deretic, V.; Levine, B. Autophagy, immunity and microbial adaptations. *Cell Host Microbe* **2009**, *3*, 527–549. [CrossRef] [PubMed]
96. Ling, S.C.; Wu, K.; Zhang, D.G.; Luo, Z. Endoplasmic reticulum stress–mediated autophagy and apoptosis alleviate dietary fat-induced triglyceride accumulation in the intestine and in isolated intestinal epithelial cells of yellow catfish. *J. Nutr.* **2019**, *149*, 1732–1741. [CrossRef]
97. Fujita, N.; Hayashi-Nishino, M.; Fukumoto, H.; Omori, H.; Yamamoto, A.; Noda, T.; Yoshimori, T. An Atg4B mutant hampers the lipidation of LC3 paralogues and causes defects in autophagosome closure. *Mol. Biol. Cell* **2008**, *19*, 4651–4659. [CrossRef]
98. Wei, C.C.; Zhi, L.; Hogstrand, C.; Xu, Y.H.; Song, Y.F. Zinc reduces hepatic lipid deposition and activates lipophagy via Zn^{2+}/MTF-1/PPARα and Ca^{2+}/CaMKKβ/AMPK pathways. *FASEB J.* **2018**, *32*, 6666–6680. [CrossRef]
99. Neufeld, T.P. TOR-dependent control of autophagy: Biting the hand that feeds. *Curr. Opin. Cell Biol.* **2010**, *22*, 157–168. [CrossRef]
100. Wu, K.; Chen, G.H.; Hogstrand, C.; Ling, S.C.; Wu, L.X.; Luo, Z. Methionine-chelated Zn promotes anabolism by integrating mTOR signal and autophagy pathway in juvenile yellow catfish. *J. Trace Elem. Med. Biol.* **2021**, *65*, 126732. [CrossRef]

antioxidants

Article

The Protective Effect of Taurine on Oxidized Fish-Oil-Induced Liver Oxidative Stress and Intestinal Barrier-Function Impairment in Juvenile *Ictalurus punctatus*

Yong Shi [1,2,†], Yi Hu [1,2,†], Ziqin Wang [1,2], Jiancheng Zhou [3], Junzhi Zhang [1,2], Huan Zhong [1,2], Guihong Fu [1,2] and Lei Zhong [1,2,*]

1 Hunan Research Center of Engineering Technology for Utilization of Distinctive Aquatic Resource, Hunan Agricultural University, Changsha 410128, China; shiyong@stu.hunau.edu.cn (Y.S.); huyi0231@hunau.edu.cn (Y.H.); wangziqin202109@163.com (Z.W.); zhjun123@hunau.edu.cn (J.Z.); zhonghuan@hunau.edu.cn (H.Z.); fuguihong@hunau.edu.cn (G.F.)
2 College of Animal Science and Technology, Hunan Agricultural University, Changsha 410128, China
3 Wuhan Dabeinong Aquatic Science and Technology Co. Ltd., Wuhan 430000, China; jiancheng1949@163.com
* Correspondence: zhonglei-5@163.com or zhonglei@hunau.edu.cn; Tel.: +86-135-4853-3428
† These two authors contributed to this work equally.

Abstract: Dietary lipids provide energy for growth and development and provide fatty acids necessary for normal structure and biological function. However, oxidized lipids cause oxidative stress and intestinal damage. An 8-week feeding trial with fresh fish oil (FFO, control group), oxidized fish oil (OFO), and taurine-supplemented diets (OFOT, OFO + 0.2% of taurine) was conducted to evaluate the protective effect of taurine on oxidized fish-oil-induced liver oxidative stress and intestine impairment in juvenile *Ictalurus punctatus*. The results showed that (1) Growth performance was significantly lower in fish fed OFO than in those fed other diets, whereas the opposite occurred in the hepatosomatic index. (2) OFO-feeding significantly increased lipid deposition compared with the FFO group. The addition of taurine ameliorated the OFO-induced increase in lipid vacuolization in the liver, significantly upregulated *lpl* mRNA expression, and downregulated *fas* and *srebp1* mRNA expression. (3) OFO-feeding significantly reduced oxidative damage of liver. Compared with the OFO group, the OFOT group remarkably upregulated antioxidant enzyme mRNA expression through the Nrf2-Keap1 signaling pathway based on the transcriptional expression. (4) OFO diets induced intestinal physical and immune barrier damage. Compared with the OFO group, OFOT diets remarkably downregulated *il-1β*, *il-6*, *tnf-α*, and *il-8* mRNA expression and upregulated *tgf-β* mRNA expression through the NF-κB signaling pathway. Besides, the addition of taurine to OFO diets significantly upregulated *zo-2* and *zo-1* mRNA expression, and downregulated *claudin-15* and *claudin-12* mRNA expression. In conclusion, oxidized-fish-oil diets can cause negative physiological health effects in *Ictalurus punctatus*, while adding taurine can increase growth and antioxidant ability, reduce lipid deposition, and improve intestinal health.

Keywords: channel catfish; oxidative damage; immune response; intestinal health; signaling pathway

Citation: Shi, Y.; Hu, Y.; Wang, Z.; Zhou, J.; Zhang, J.; Zhong, H.; Fu, G.; Zhong, L. The Protective Effect of Taurine on Oxidized Fish-Oil-Induced Liver Oxidative Stress and Intestinal Barrier-Function Impairment in Juvenile *Ictalurus punctatus*. *Antioxidants* **2021**, *10*, 1690. https://doi.org/10.3390/antiox10111690

Academic Editors: Min Xue, Junmin Zhang, Zhenyu Du, Jie Wang and Wei Si

Received: 23 September 2021
Accepted: 24 October 2021
Published: 26 October 2021

1. Introduction

It is well known that, as one of the important nutrients of aquatic animals, dietary lipids not only provide energy for growth and development in fish, but also provide the essential fatty-acid and fat-soluble vitamins that maintain normal structure and biological function [1]. At present, the main lipid sources in aquatic feed are fish oil and soybean oil. Compared with soybean oil, fish oil has a high content of unsaturated fatty acids (HUFAs), such as eicosapentaenoic acid (EPA) and docosahexaenoic acid (DHA), and a good feeding attraction effect, so it is the best lipid source for aquatic animals [2,3]. However, EPA and DHA are easily oxidized during the storage and processing of fish oil and feed, producing harmful substances such as lipid hydroperoxides, ketones, aldehydes, and acids [4,5]. Studies have also reported that feeding an oxidized-fish-oil diet has been found to decrease growth performance and cause oxidative stress of tilapia (*Oreochromis niloticus*) [6], Wuchang bream (*Megalobrama amblycephala*) [7], and orange spotted grouper (*Epinephelus coioides*) [5], disrupt lipid metabolism of *Rhynchocypris lagowski* Dybowski [8], and induce intestinal injury of *Megalobrama amblycephala* [9] and rice field eel (*Monopterus albus*) [10]. Therefore, exploring effective dietary strategies is imperative to alleviate the negative effects of oxidized-fish-oil diets on aquatic animals.

Taurine is a type of non-protein amino acid in the form of a free amino acid, which has a wide range of physiological functions, such as calcium homeostasis, osmotic regulation, membrane stability, and antioxidant and anti-inflammatory functions [11,12]. Many studies have also demonstrated the versatility of taurine in aquatic animals. Previous studies in our laboratory indicated that taurine supplementation in a low-fish-meal diet increased growth performance and immunity function and enhanced anti-stress ability in black carp (*Mylopharyngodon piceus*) [13] and rice field eel [14]. Similar studies have found, in other aquatic animals, that taurine can increase growth performance, enhance antioxidant ability, improve intestinal health, and reduce lipid deposition of seabass (*Dicentrarchus labrax*) [15], grass carp [16], and California yellowtail (*Seriola dorsalis*) [17]. However, there is no report on whether taurine can alleviate the negative effects caused by oxidized-fish-oil diets in aquatic animals. Therefore, we have carried out related research.

Channel catfish (*Ictalurus punctatus*), which belongs to catfish family (*Siluriformes*), is an important freshwater aquaculture fish in China. Because of its delicious meat, high nutritional value, and fast growth, it is welcomed by producers and consumers. In 2018, the production of channel catfish exceeded 390,000 tons in the world, an increase of 3.34% over the previous year [18]. Channel catfish has a high requirement for feed freshness, and feed mildew, deterioration, and oxidation will have a negative impact on growth and health. Therefore, this study aimed to investigate whether taurine can alleviate the negative effects of lipid deposition, oxidative stress, and intestinal damage induced by oxidized-fish-oil diets in juvenile channel catfish. It is of great significance to explore the side-effects of oxidized fat ingestion on the growth and health of aquatic animals for the study of fish nutrition and health, and to provide solutions for practical production and a theoretical basis for the application of taurine.

2. Materials and Methods

2.1. Preparation of Oxidized Fish Oil

Oxidized fish oil was prepared by constant temperature water bath aeration. The detailed steps are as follows: fill the beaker with fresh fish oil, place it in a constant temperature water bath at 50 °C, insert the air pump snorkel into the container and aerate it for five days. The peroxide value of the fish oil was monitored daily until it reached 897.4 meq/kg. The peroxide value of fresh fish oil was 9.2 meq/kg.

2.2. Experimental Diets

Three isonitrogenous and isolipid diets were designed in this experiment—fresh fish oil (FFO, control group) diet, oxidized fish oil (OFO) diet, and OFO diet with 0.2% taurine (OFOT) (Table 1). The ingredients were finely ground, sieved (0.25 mm), mixed and supplemented with fish oil and soybean oil. A 10% volume of water of the weight of the ingredients was added. After mixing, pellets were squeezed (1.0 and 1.5 mm in size) and then dried naturally in the shade. The experimental diets were then stored at −20 °C until use.

Table 1. Composition and nutrient levels of basic diet (%, dry matter).

Ingredients	FFO	OFO	OFOT
Fish meal	10.00	10.00	10.00
Soybean meal	28.00	28.00	28.00
Rapeseed meal	20.00	20.00	20.00
Rice bran	3.00	3.00	3.00
Wheat flour	25.26	25.26	25.06
Chicken meal	9.00	9.00	9.00
Fish oil	2.00	0.00	0.00
Oxidized fish oil	0.00	2.00	2.00
Premix [1]	1.00	1.00	1.00
Choline	0.20	0.20	0.20
$Ca(H_2PO_4)_2$	1.50	1.50	1.50
Mold inhibitor	0.03	0.03	0.03
Antioxidants	0.01	0.01	0.01
Taurine	0.00	0.00	0.20
	Approximate composition (%) [2]		
Crude protein	36.27	36.14	36.31
Crude lipid	5.47	5.52	5.43
Crude ash	6.57	6.49	6.53
POV (meq/kg)	3.7	21.4	21.2

[1] Provided by MGO Ter Bio-Tech (Qingdao, Shandong, China). Vitamin and mineral premix composition (mg/kg diet): KCl 200 mg, KI (1%) 60 mg, $CoCl_2·6H_2O$ (1%) 50 mg, $CuSO_4·5H_2O$ 30 mg, $FeSO_4·H_2O$ 400 mg, $ZnSO_4·H_2O$ 400 mg, $MnSO_4·H_2O$ 150 mg, $Na_2SeO_3·5H_2O$ (1%) 65 mg, $MgSO_4·H_2O$ 2000 mg, zeolite power 3645.85 mg, VB1 12 mg, riboflavin 12 mg, VB6 8 mg, VB12 0.05 mg, VK3 8 mg, inositol 100 mg, pantothenic acid 40 mg, niacin acid 50 mg, folic acid 5 mg, biotin 0.8 mg, VA 25 mg, VD 35 mg, VE 50 mg, VC 100 mg, ethoxyquin 150 mg, flour 2434.15 mg. [2] Crude protein, crude lipid, ash, and POV were measured values. The detection method was referenced to previous studies [19].

2.3. *Experimental Animals and Feeding Experiment*

Channel catfish were purchased from a fine seed farm (Wuhan, Hubei, China), and the breeding experiment was carried out in the recirculating aquaculture system of Wuhan Dabeinong Aquatic Science and Technology Co., Ltd. (Wuhan, Hubei, China). During the acclimatization period, the FFO group was fed until the channel catfish showed obvious feeding behavior, and then the fish were fasted for 24 h. Channel catfish fingerlings (average weight 6.00 ± 0.01 g) were randomly distributed into 12 breeding barrels (diameter 1.0 m, water depth 0.8 m, indoor) with three replicates in each treatment group, each containing 35 fish per replicate. During the 8-week the feeding trial, the channel catfish were manually fed three times per day (8 a.m., 12 p.m., and 5 p.m.) at 3%–5% of their body weight. The water temperature was maintained at 27.32 ± 0.17 °C, dissolved oxygen was more than 6.5 mg/L, and ammonia and nitrate were less than 0.2 mg/L.

2.4. *Sample Collection*

All experiments followed the regulations of Hunan Agricultural University for laboratory animal protection. After the experiment, growth performance was calculated after 24 h of fasting. The fish were anesthetized with MS-222 (100 mg/L, Sigma Aldrich Co. LLC., St. Louis, MO, USA) before sampling [19]. Three fish were taken from each breeding barrel for tail vein blood collection, which was collected in a 2 mL centrifuge tube and placed at 4 °C for 12 h. After that, the supernatant was centrifuged and stored at −80 °C. Three fish from each breeding barrel were quickly dissected on ice, and the liver, intestine, and skin (backside and abdomen) tissues were removed in enzyme-free centrifuge tube (1.5 mL) and put it in liquid nitrogen, then stored at −80 °C.

2.5. *Determination of Growth Parameters*

The weight gain rate (WGR), feed conversion ratio (FCR), survival rate (SR), condition factor (CF), hepatosomatic index (HSI), and viserosomatic index (VSI) were calculated, as follows:

$$\text{Weight gain rate (WGR, \%)} = \text{(final body weight} - \text{initial body weight)/initial body weight} \times 100 \quad (1)$$

$$\text{Feed conversion rate (FCR)} = \text{total amount of the feed consumed/(final body weight-initial body weight)} \quad (2)$$

$$\text{Survival rate (SR, \%)} = \text{final number of fish/initial number of fish} \times 100 \quad (3)$$

$$\text{Condition factor (CF, g/cm}^3\text{)} = 100 \times \text{whole body weight/(body length)}^3 \quad (4)$$

$$\text{Hepatosomatic index (HSI, \%)} = \text{liver weight/whole body weight} \times 100 \quad (5)$$

$$\text{Viserosomatic index (VSI, \%)} = \text{visceral weight/whole body weight} \times 100 \quad (6)$$

2.6. Skin Pigment and Body Color Analysis

Three fish were randomly selected from each breeding barrel and tested on the backside and abdomen of each fish with a chromometer (model: 601, Beijing, China) to obtain L*, a*, and b* values. "L*" is brightness: 0–100 from black to white; "a*" is red-green: red is represented as a positive value, green is represented as a negative value; and "b*" is yellow-blue: yellow is represented as a positive value, blue is represented as a negative value. The activities of carotenoids, lutein, and tyrosinase in the backside and abdomen skin of each fish were assessed by Elisa kits (Meimian, Jiangsu, China).

2.7. Biochemical Index Analysis

The levels of total cholesterol (TC), triacylglycerol (TG), immunoglobulin M (IgM), complement 3 (C3), complement 4 (C4), alanine aminotransferase (ALT), and aspartate aminotransferase (AST) in serum were assayed by using a commercial kit (Nanjing Jiancheng Bioengineering Institute, Nanjing, China).

The glutathione (GSH), superoxide dismutase (SOD), glutathione peroxidase (GPx), malondialdehyde (MDA), glutathione reductase (GR), and total antioxidant capacity (T-AOC) levels in the liver were assayed by using a commercial kit (Nanjing Jiancheng Bioengineering Institute, Nanjing, China).

2.8. Histological Analysis

There were three replicates in each treatment group, and three fish were taken from each replicate. Liver and intestine were fixed in paraformaldehyde and embedded with paraffin wax. According to the previous experimental method, the steps of hematoxylin-eosin (H&E) staining were as follows: eight-micron tissue was taken on the glass slide with a slicer, the tissue was stained with hematoxylin, and the results were observed under an electron microscope [19]. Liver histological measurements covered 50 cells and the nuclei of the analysed tissues collected from each individual.

2.9. Real-Time Polymerase Chain Reaction

Total RNA from the liver and intestine was extracted using TRIzol reagent (Invitrogen, Carlsbad, CA, USA) and the quality was assessed according to Shi et al. [20]. First-strand cDNA was synthesized and RT-qPCR analysis of mRNA was performed according to a previous report [20]. The amplification efficiency was between 0.95 and 1.10, as calculated by the formula $E = 10^{(-1/\text{slope})} - 1$. Primer sequences are shown in Table 2. With *gapdh* as the reference, the calculation is carried out according to the $E = 2^{-\Delta\Delta CT}$ formula [21].

2.10. Statistical Analysis

All data were compared by one-way analysis of variance (ANOVA), and differences between the means were tested by Duncan's multiple-range test. All results are reported as the "mean ± S.E.", and all statistical analyses were performed using SPSS 24.0 (New York, NY, USA). Differences were considered significant at $p < 0.05$ ($p < 0.05$) [22].

Table 2. Primers used for mRNA quantitative real-time PCR.

Gene	Forward Sequences (5′→3′)	Reverse Sequences (5′→3′)	Accession No.
fas	CTGCTGTCTGAGGGCGTAA	CGATGGCGATGAGGTTCT	NM_001200193.1
lpl	AGCAACATTACCCAACCTCAGC	CCAGCTACATGAGCACCCAAA	KF693235.1
srebp1	GTTGCGGAAGGCGATTGA	GCAGTGGGCTGTTGGGTTC	XM_017480901.1
sod	GACTTGGGCAAAGGTGGAAA	CACTCAGCAATGCCTATCACG	NM_001200992.1
gpx1	TCTGAGGCACGACCACCA	GCGTCTTTCCCGTTCACAT	NM_001200741.1
gpx8	TCACTTCACCGTGTTGGCTT	CCCTCAGCACTCACCAGAAA	XM_017466944.1
gr	GGATGTGAAGGATAAGCGAAAC	TTCGGCAACACGGGTATG	GU588318.1
keap1	CGGCAAGCATCTCAGTCG	TGCTCGGGTCCAACTGC	XM_017482237.1
nrf2	GGTCCACGCCTACCAACAA	CAGGGAGGAATGGAGGGAT	XM_017470076.1
zo-1	TACCAAACCGTGGATACAAACC	CTTCTATGGGTGGAGGAGGC	XM_017458510.1
zo-2	GAGGTCAAAGGGCAGCAAA	GAAATCTTCGGGCAGGTCA	XM_017488926.1
claudin-12	GCTGGGATGTTCCTCTTGATAG	AGAGCGGCGAACTCAAGG	XM_017453476.1
claudin-15	GTGGTTCTCGGCACATTCG	CAAGCCCTGTAGGATGAAGAAG	XM_017471911.1
occludin	GCATCGGTAGCGGGTCAT	GACTTGGTTGAGTTCTGCCTTG	XM_017451558.1
tnf-α	CGCCAGCGGTAAACACG	CCGTTGAATGTCCGAAAGG	XM_017464718.1
il-1β	CTGAAGGGTGGAAACAAGGAT	GGAGTCACCAGTGCCGTTT	AJ586102.1
il-6	GAAGATTGATACTCCGCTCCTG	GATTAAATGTAACAGCCTGGTGG	XM_017455306.1
il-8	TCCAAGTGCCTCCTGTTCAA	CCCTTCTTCCCTTGGACTTTAT	KP701473.1
il-10	GCAGGCTTACGAAAGGGTTA	CGGCGTATGAAGAACGAAGT	XM_017450800.1
tgf-β1	GGAACGGCTGAGTGGGTCT	TGCTTACTGAGGCGGCTATG	XM_017483625.1
tgf-β2	TGAAGCGGTCAGCGAATG	CTCACTCTTGTTTGGGATGATGTA	XM_017476217.1
tgf-β3	TCGGTGCCCTGTCCTATTG	GCGGAGAACGAGGCTTACA	XM_017476492.1
nf-κb	CTCAGCCCATCTACGACAACA	CGTCAGGTTCGTATCGCAGT	KF572025.1
gapdh	TGTCCGTTTGGAGAAGCCT	ATCAGGTCACAGACACGGTTG	NM_001201199.1

3. Results

3.1. Growth Performance

As shown in Table 3, there was no significant change ($p > 0.05$) in VSI, CF, SR, or FCR of channel catfish among treatment groups. Compared with the FFO group, WGR and final weight of the OFO group were significantly reduced, while adding taurine significantly increased ($p < 0.05$). HSI in the OFO group was significantly higher than that in the FFO group ($p < 0.05$). Compared with the OFO group, HSI in the OFOT group significantly decreased ($p < 0.05$), and there was no significant difference from the FFO group ($p > 0.05$).

Table 3. Effects of dietary taurine on the growth performance of channel catfish (*Ictalurus punctatus*) fed oxidized-fish-oil diets.

	FFO	OFO	OFOT	*p*-Value
Initial weight (g)	6.00 ± 0.01	6.00 ± 0.01	5.99 ± 0.00	0.702
Final weight (g)	26.20 ± 0.14 ^{b}	23.43 ± 0.72 ^{a}	25.62 ± 0.40 ^{b}	0.036
WGR	336.53 ± 1.38 ^{b}	290.49 ± 11.93 ^{a}	327.27 ± 6.55 ^{b}	0.034
SR	97.14 ± 2.86	94.29 ± 2.86	93.33 ± 6.67	0.829
FCR	1.16 ± 0.04	1.35 ± 0.09	1.20 ± 0.08	0.223
HSI	2.26 ± 0.05 ^{a}	2.74 ± 0.12 ^{b}	2.46 ± 0.07 ^{a}	0.003
VSI	13.20 ± 0.66	13.67 ± 0.47	13.28 ± 0.30	0.781
CF	1.52 ± 0.04	1.57 ± 0.03	1.49 ± 0.01	0.216

Note: Data indicate the mean values of three replicate cages per treatment (three fish per replicate breeding barrel). Mean values with different superscripts in a row are significantly different (one-way ANOVA, $p < 0.05$). Weight gain rate (WGR, %) = (final body weight − initial body weight)/initial body weight × 100; survival rate (SR, %) = final number of fish/initial number of fish × 100; feed conversion rate (FCR) = total amount of the feed consumed/(final body weight − initial body weight); hepatosomatic index (HSI, %) = liver weight/whole body weight × 100; viserosomatic index (VSI, %) = visceral weight/whole body weight × 100; condition factor (CF, g/cm^3) = 100 × whole body weight/(body length)^3.

3.2. Skin Pigment and Body Color

There was no significant change ($p > 0.05$) in L* of backside and b* of abdomen among treatment groups (Table 4). Compared with the FFO group, carotenoids of backside and L* of abdomen in the OFO group were significantly decreased ($p < 0.05$), lutein of backside and abdomen were significantly increased ($p < 0.05$). Compared with the OFO group, carotenoids, tyrosinase and a* of backside and tyrosinase and a* of abdomen in the OFOT group markedly increased ($p < 0.05$).

Table 4. Effects of dietary taurine on skin pigment and body color of channel catfish (*Ictalurus punctatus*) fed oxidized-fish oil diets.

	FFO	OFO	OFOT	*p*-Value
		Backside		
Carotenoids (μg/mL)	16.00 ± 0.47 [b]	14.04 ± 0.32 [a]	18.42 ± 0.57 [c]	0.002
Lutein (pg/mL)	1746.8 ± 84.5 [a]	2331.8 ± 26.0 [b]	2243.5 ± 17.3 [b]	<0.001
Tyrosinase (ng/mL)	4986.3 ± 43.3 [a]	4699.7 ± 87.4 [a]	5583.0 ± 120.0 [b]	0.001
L*	51.14 ± 0.49	53.24 ± 1.25	47.64 ± 2.87	0.125
a*	−5.10 ± 0.06 [a]	−5.44 ± 0.08 [a]	−4.54 ± 0.29 [b]	0.008
b*	1.60 ± 0.23 [a]	2.49 ± 0.15 [ab]	2.07 ± 0.25 [b]	0.031
		Abdomen		
Carotenoids (μg/mL)	13.18 ± 0.23 [a]	13.61 ± 0.07 [ab]	14.91 ± 0.64 [b]	0.049
Lutein (pg/mL)	2216.0 ± 21.7 [b]	2391.8 ± 22.4 [c]	2083.5 ± 40.1 [a]	0.001
Tyrosinase (ng/mL)	4448.0 ± 31.8 [a]	4309.7 ± 57.0 [a]	4693.0 ± 75.5 [b]	0.009
L*	82.41 ± 0.20 [b]	81.52 ± 0.25 [a]	81.55 ± 0.14 [a]	0.011
a*	−3.47 ± 0.06 [ab]	−3.55 ± 0.06 [a]	−3.31 ± 0.03 [b]	0.017
b*	8.19 ± 0.05	8.67 ± 0.12	8.01 ± 0.42	0.200

Note: Data indicate the mean values of three replicate cages per treatment (three fish per replicate breeding barrel). Mean values with different superscripts in a row are significantly different (one-way ANOVA, $p < 0.05$).

3.3. Lipid Deposition and Histological Structure of Liver

The TG and TC contents in the OFO group were remarkably increased in comparison with the FFO group, while supplementation with 0.2% taurine significantly reduced the TG and TC contents ($p < 0.05$), there was no significant difference ($p > 0.05$) between the OFOT and FFO groups (Figure 1A,B). In comparison of the OFO and FFO groups, *fas* and *srebp1* mRNA expression in the liver of the channel catfish were significantly upregulated ($p < 0.05$), and *lpl* mRNA expression was significantly downregulated ($p < 0.05$) (Figure 1C–E). The OFOT group significantly upregulated *lpl* mRNA expression, and significantly downregulated *fas* and *srebp1* mRNA expression compared with the OFO group ($p < 0.05$). As showed in Figure 2 and Table 5, the OFO group remarkably decreased the size of nuclei, and increased the size of hepatocytes ($p < 0.05$) compared with the FFO group. The size of nuclei in the FFOT group was significantly increased compared with the OFO group, whereas the opposite result was observed for the size of hepatocytes ($p < 0.05$). Therefore, the fish fed OFO diets showed more hepatic lipid vacuolization than those fed FFO or OFOT.

3.4. Serum Immune Indices

The OFO group significantly reduced IgM, C4, and C3 contents, while the supplementation of taurine remarkably increased ($p < 0.05$) these immune indices compared with the OFO group (Table 6). In addition, compared with the FFO group, AST and ALT activities in the OFO group were significantly increased ($p < 0.05$), while adding 0.2% taurine remarkably decreased the activities of AST and ALT ($p < 0.05$), there was no significant difference ($p > 0.05$) between the OFOT and FFO treatment.

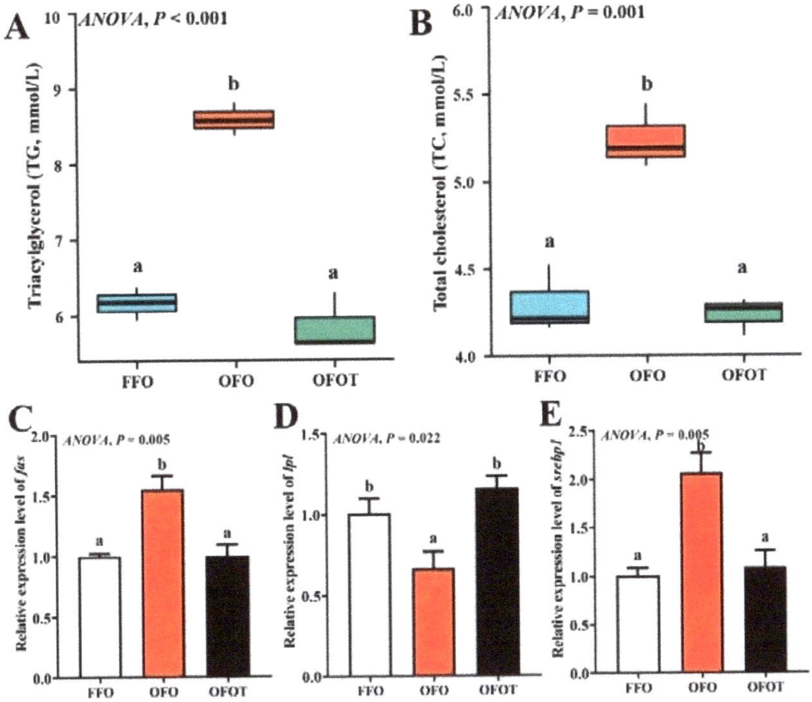

Figure 1. Serum biochemical indices and lipid-metabolism-related gene expression of the liver in channel catfish (*Ictalurus punctatus*) fed the diets. (**A**) Triacylglycerol, TG; (**B**) total cholesterol, TC; and (**C**–**E**) lipid-metabolism-related genes (*fas*, *acc*, and *srebp1*). Data indicate the mean values of three replicate cages per treatment (three fish per replicate breeding barrel). Significance was evaluated by one-way ANOVA ($p < 0.05$) followed by Duncan's multiple range tests. Values marked with different letters are significantly different between the treatment groups.

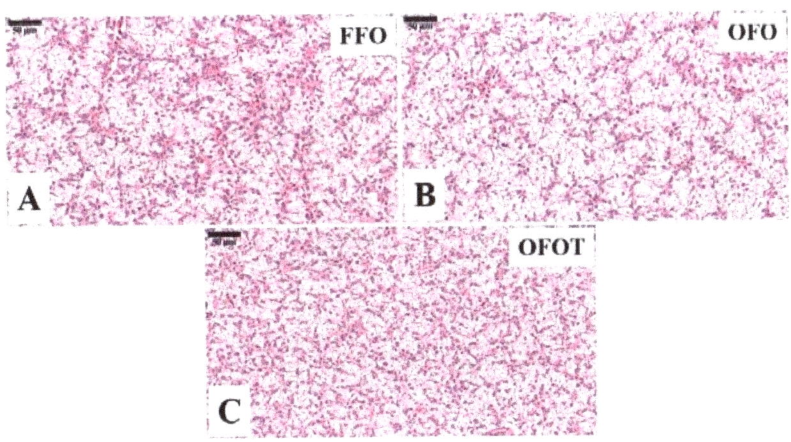

Figure 2. Histological characteristics of the liver in channel catfish (*Ictalurus punctatus*) fed the diets (H&E stain, magnification 400×). (**A**) Fresh fish oil (FFO) group; (**B**) oxidized fish oil (OFO) group; and (**C**) OFO diet with 0.2% taurine (OFOT) group.

Table 5. Effects of dietary taurine on the morphometrics of the liver of channel catfish (*Ictalurus punctatus*) fed oxidized-fish-oil diets.

	FFO	OFO	OFOT	*p*-Value
Size of nuclei (μm)	6.57 ± 0.02	5.35 ± 0.03	6.61 ± 0.02	<0.001
Size of hepatocytes (μm)	17.03 ± 0.05	21.66 ± 0.26	17.02 ± 0.05	<0.001

Note: Data indicate the mean values of three replicate cages per treatment (three fish per replicate breeding barrel).

Table 6. Effects of dietary taurine on serum immune indices of channel catfish (*Ictalurus punctatus*) fed oxidized-fish-oil diets.

	FFO	OFO	OFOT	*p*-Value
C3 (g/L)	1.04 ± 0.03 [c]	0.72 ± 0.02 [a]	0.87 ± 0.03 [b]	<0.001
C4 (g/L)	0.65 ± 0.02 [b]	0.5 ± 0.02 [a]	0.69 ± 0.03 [b]	<0.001
IgM (g/L)	1.42 ± 0.08 [b]	1.01 ± 0.03 [a]	1.37 ± 0.07 [b]	0.003
AST (U/L)	33.02 ± 6.31 [a]	56.45 ± 0.97 [b]	35.2 ± 0.42 [a]	0.008
ALT (U/L)	8.65 ± 0.58 [a]	13.33 ± 0.77 [b]	7.57 ± 0.62 [a]	<0.001

Note: Data indicate the mean values of three replicate cages per treatment (three fish per replicate breeding barrel). Mean values with different superscripts in a row are significantly different (one-way ANOVA, $p < 0.05$). C3, complement 3; C4, complement 4; IgM, immunoglobulin M; AST, aspartate aminotransferase; and ALT, alanine aminotransferase.

3.5. Antioxidant Indices in the Liver

As shown in Figure 3, the MDA content in the OFO group was significantly increased ($p < 0.05$) compared with that of the FFO group. The content of MDA of fish fed the 0.2% taurine supplementation diets was remarkably lower than values in fish fed OFO diets. In addition, the OFO treatment significantly decreased the levels of SOD, GPx, GR, GSH, and T-AOC, while the supplementation of taurine markedly increased ($p < 0.05$) these antioxidant indices compared with the OFO group.

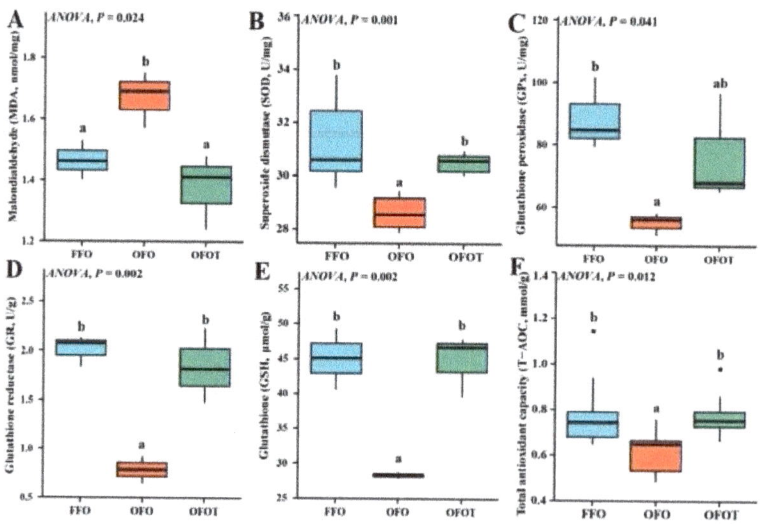

Figure 3. Liver antioxidant indices of channel catfish (*Ictalurus punctatus*) subject to different treatment. (**A**) Malondialdehyde, MDA; (**B**) Superoxide dismutase, SOD; (**C**) Glutathione peroxidase, GPx; (**D**) Glutathione reductase, GR; (**E**) Glutathione, GSH; (**F**) Total antioxidant capacity, T-AOC. Data indicate the mean values of three replicate cages per treatment (three fish per replicate breeding barrel). Significance was evaluated by one-way ANOVA ($p < 0.05$) followed by Duncan's multiple range tests. Values marked with different letters are significantly different between the treatment groups.

3.6. Antioxidant-Related Gene Expression in the Liver

As shown in Figure 4, there was no significant difference in *sod* and *gpx8* gene expression among the treatment groups. The OFO group remarkably downregulated *gpx1*, *gr*, and *nrf2* mRNA expression, and upregulated *keap1* mRNA expression ($p < 0.05$) compared with the FFO group. *gpx1*, *gr*, and *nrf2* mRNA expression in the FFOT group were significantly upregulated compared with the OFO group, whereas the opposite result was observed for the mRNA expression level of *keap1* ($p < 0.05$)

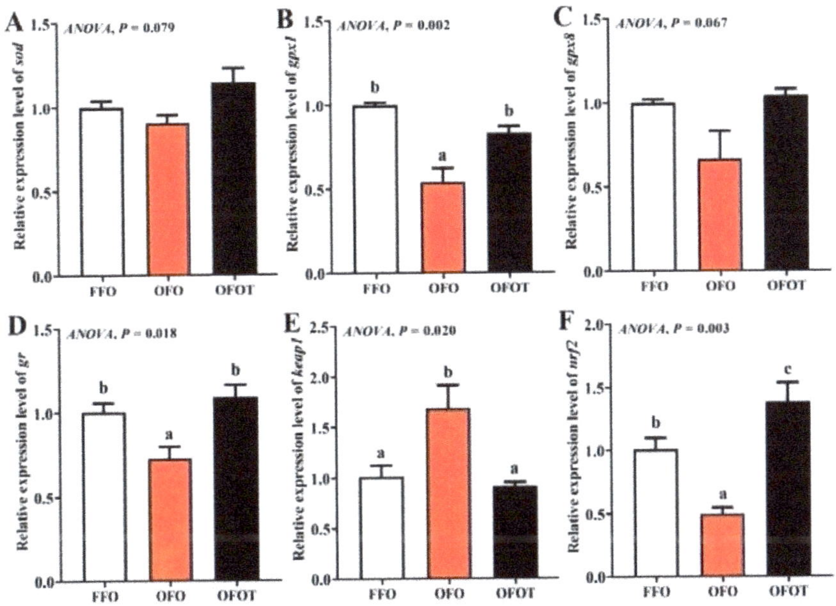

Figure 4. Effects of dietary taurine on liver antioxidant-related genes expression of channel catfish (*Ictalurus punctatus*) fed oxidized-fish-oil diets. (**A**) *sod*; (**B**) *gpx1*; (**C**) *gpx8*; (**D**) *gr*; (**E**) *keap1*; (**F**) *nrf2*. Data indicate the mean values of three replicate cages per treatment (three fish per replicate breeding barrel). Significance was evaluated by one-way ANOVA ($p < 0.05$) followed by Duncan's multiple range tests. Values marked with different letters are significantly different between the treatment groups.

3.7. Histological Structure in the Intestine

As shown in Figure 5, through the H&E staining analysis of the intestine, the feeding of oxidized-fish-oil diets significantly reduced the goblet cell quantity, villi length, and muscular thickness of the intestine. Compared with the OFO group, the OFOT group significantly increased the goblet cell quantity, villi length, and muscular thickness of intestine.

3.8. Intestinal Physical-Barrier-Related Gene Expression

As showed in Figure 6, there was no significant change in *occludin* mRNA expression among the treatment groups. In comparison of the OFO and FFO groups, *claudin-12* and *claudin-15* mRNA expression in the intestine of the channel catfish were markedly upregulated, and *zo-2* and *zo-1* mRNA expression were significantly downregulated ($p < 0.05$). Adding taurine remarkably upregulated *zo-1* and *zo-2* mRNA expression, and downregulated *claudin-15* and *claudin-12* mRNA expression compared with the OFO group ($p < 0.05$).

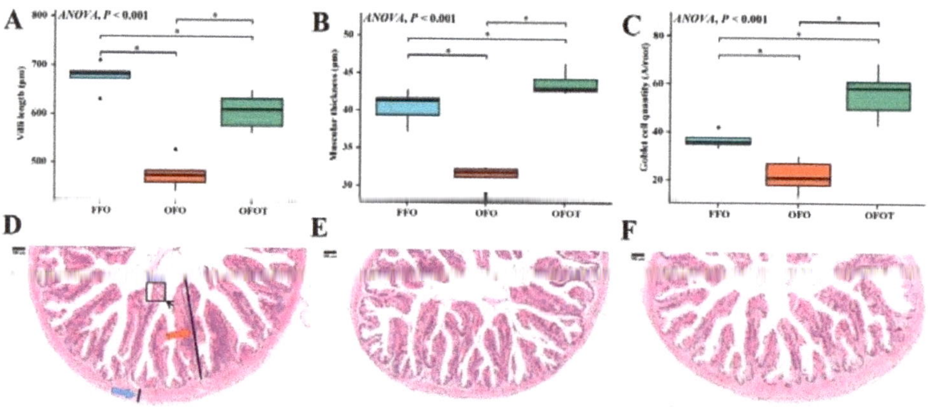

Figure 5. Effects of dietary taurine on intestinal morphology of channel catfish (*Ictalurus punctatus*) fed oxidized-fish-oil diets (magnification 40×). (**A**) Villi length; (**B**) muscular thickness; (**C**) goblet cell quantity; (**D**) FFO group; (**E**) OFO group; and (**F**) OFOT group. The red arrow indicates the villi length and the blue arrow indicates the muscular thickness, respectively. Significance was evaluated by one-way ANOVA ($p < 0.05$) followed by Duncan's multiple range tests. * $p < 0.05$.

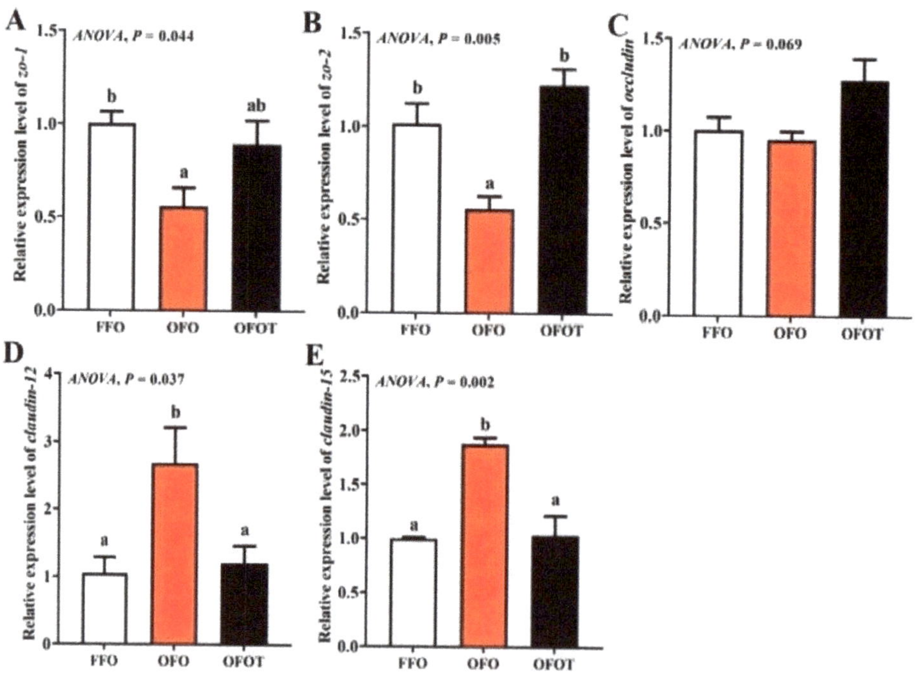

Figure 6. Effects of dietary taurine on intestinal physical-barrier-related genes expression of channel catfish (*Ictalurus punctatus*) fed oxidized-fish-oil diets. (**A**) *zo-1*; (**B**) *zo-2*; (**C**) *occludin*; (**D**) *claudin-12*; (**E**) *claudin-15*. Data indicate the mean values of three replicate cages per treatment (three fish per replicate breeding barrel). Significance was evaluated by one-way ANOVA ($p < 0.05$) followed by Duncan's multiple range tests. Values marked with different letters are significantly different between the treatment groups.

3.9. Intestinal Immune-Barrier-Related Gene Expression

As showed in Figure 7, the OFO group markedly downregulated *tgf-β1*, *tgf-β2*, and *tgf-β3* mRNA transcription levels, and upregulated *tnf-α*, *nf-κb*, *il-1β*, *il-6*, and *il-8* mRNA transcription levels compared with the FFO group ($p < 0.05$). *tgf-β1*, *tgf-β2*, and *tgf-β3* mRNA transcription levels in the OFOT group were markedly upregulated compared with the OFO group, whereas the opposite result was observed for *tnf-α*, *nf-κb*, *il-1β*, *il-6*, and *il-8* mRNA transcription levels ($p < 0.05$).

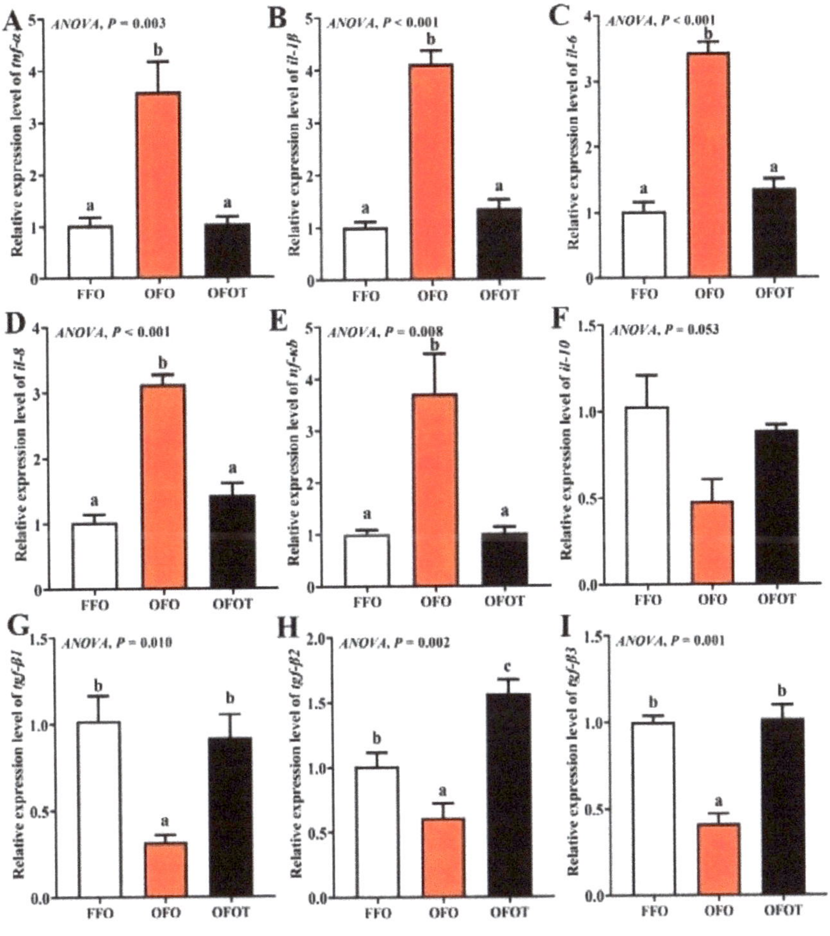

Figure 7. Effects of dietary taurine on intestinal immune-barrier-related genes expression of channel catfish (*Ictalurus punctatus*) fed oxidized-fish-oil diets. (**A**) *tnf-α*; (**B**) *il-1β*; (**C**) *il-6*; (**D**) *il-8*; (**E**) *nf-κb*; (**F**) *il-10*; (**G**) *tgf-β1*; (**H**) *tgf-β2*; (**I**) *tgf-β3*. Data indicate the mean values of three replicate cages per treatment (three fish per replicate breeding barrel). Significance was evaluated by one-way ANOVA ($p < 0.05$) followed by Duncan's multiple range tests. Values marked with different letters are significantly different between the treatment groups.

As shown in Figure 8, correlation analyses shown that *nf-κb* mRNA transcription level was negatively correlated with *tnf-α*, *il-1β*, *il-6*, and *il-8* mRNA transcription levels ($p < 0.05$), and positively correlated with *tgf-β1*, *tgf-β2*, and *tgf-β3* mRNA transcription levels ($p < 0.05$).

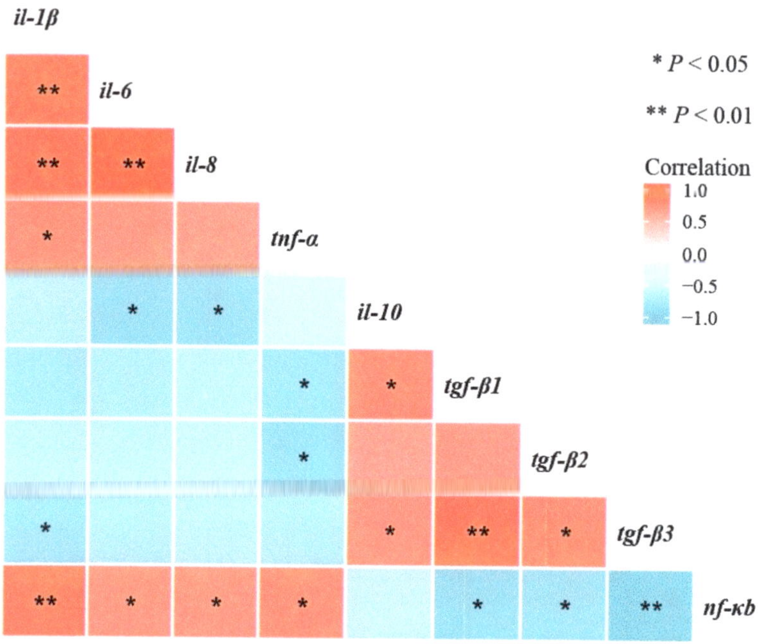

Figure 8. Correlative analysis of intestinal immune-barrier-related gene expression was performed using the R Programming Language. * $p < 0.05$, ** $p < 0.01$.

4. Discussion

A fresh fish oil can provide the HUFAs needed during the growth of fish [23]. HUFAs are prone to oxidative rancidity and have a negative impact on fish. In this study, oxidized-fish-oil diets significantly reduced the growth performance of channel catfish, which is similar to results obtained in juvenile hybrid grouper (♀*Epinephelus fuscoguttatus* × ♂*Epinephelus lanceolatus*) [24], farmed tilapia [6], orange spotted grouper [5], and yellow catfish (*Pelteobagrus fulvidraco*) [25]. One of the reasons is that toxic and harmful substances such as lipid hydroperoxides, ketones, aldehydes, and acids are produced after oxidation of fish oil, which induces oxidative stress, leads to inflammatory response, and then inhibits growth [5]. Another reason is that oxidized fish oil has reduced nutritional value compared with non-oxidized fish oil [26]. Taurine has been widely used in aquatic feeds. Adding an appropriate amount of taurine to diets can obviously increase growth performance of yellowtail kingfish (*Seriola lalandi*) [27], turbot [28], and tiger puffer (*Takifugu rubripes*) [29]. Experimental results also showed that the addition of 0.2% taurine to the oxidized-fish-oil diet obviously increased the growth performance of channel catfish, and there was no significant difference from the FFO group. There are two main reasons why taurine promotes fish growth: first, taurine has a good attractant effect [30]; second, taurine may alleviate the negative effects caused by oxidized-fish-oil diets, such as lipid deposition, oxidative damage, and inflammatory response.

Long-term feeding of oxidized-fish-oil diets can lead to the lipid deposition of liver [31]. Based on H&E staining, liver fat vacuolation is usually expressed as the size of hepatocytes and their nuclei [32,33]. In this study, oxidized-fish-oil diets led to lipid deposition in the liver of channel catfish, which was supported by the phenomenon of increased lipid vacuolization in the liver (such as smaller nuclei and larger hepatocytes), HSI, serum TC, and TG contents. Similar studies have been found in yellow catfish [25], loach (*Misgurnus anguillicaudatus*) [34] and largemouth bass (*Micropterus salmoides*) [35]. Further studies showed that oxidized-fish-oil diets resulted in liver lipid deposition due to the upregulation of the expression of lipid synthesis gene (*fas*) and the downregulation of the expression of lipolysis gene (*lpl*). Sterol-regulatory element binding protein 1 (*srebp1*) is mainly involved in the activation of enzymes related to lipid synthesis, and can promote lipid synthesis by targeting the expression of fatty-acid-catalyzing enzymes such as *fas* [36,37]. The present study showed that feeding oxidized-fish-oil diets significantly upregulated *srebp1* mRNA expression, indicating that oxidized-fish-oil diets can induce lipid deposition by regulating the mRNA expression of lipid synthesis and lipolysis. Taurine has a good function of reducing lipid deposition. It has been found in broiler chickens that taurine can reduce blood lipid content [38]. There are also studies in aquatic animals that have found that taurine can promote lipolysis of European seabass [15], white seabream (*Diplodus sargus*) [39], and Persian sturgeon (*Acipenser persicus*) [40]. In this study, the addition of taurine to oxidized-fish-oil diets remarkably reduced lipid vacuolization in the liver, HSI, serum TC, and TG contents. Furthermore, taurine remarkably downregulated the transcriptional levels of *fas* and *srebp1* in the liver, and upregulated the transcriptional level of *lpl*, indicating that taurine alleviated lipid deposition induced by oxidized-fish-oil diets. Studies have speculated that taurine has a good lipolysis effect, which may be related to the AMPK/SIRT1 signaling pathway [38]. Previous studies confirmed that activation of AMPK can inhibit the activities of FAS and ACC, thereby reducing the concentration of malonyl-CoA and enhancing CPT1 activity, thus increasing lipid catabolism and reducing lipid deposition [41,42]. However, the mechanism of taurine alleviating lipid deposition needs further study.

For animals, the oxidation diet is one of important exogenous factors leading to oxidative stress. Long-term feeding of oxidized-fish-oil diets can induce the production of reactive oxygen species in mitochondria, and excessive reactive oxygen species (ROS) can lead to tissue oxidative damage [43–45]. Malondialdehyde (MDA) is the final decomposition product of lipid peroxidation caused by ROS, and its content reflects the degree of peroxidation [46]. In the process of ROS removal, CAT breaks down hydrogen peroxide into oxygen and water, and SOD and GPx also play an important role, which can decrease hydrogen peroxide [47,48]. The present study showed that oxidized-fish-oil-diet feeding led to markedly a higher the content of MDA and lower the levels of CAT, SOD, GPx, GR, and T-AOC in the liver than in the FFO group. As is well known, the increase of serum AST and ALT activities is one of the important markers of liver injury [49]. Besides, the present study has showed that oxidized-fish-oil diets significantly increased serum AST and ALT activities of channel catfish, indicating that oxidized-fish-oil diets leads to oxidative stress and damage in the liver. Similar studies have found in other aquatic animals that oxidized-fish-oil diets significantly decreased antioxidant enzyme activities and increased AST and ALT activities of Wuchang bream [7] and tilapia (*Oreochromis niloticus*) [6]. Some studies have shown that taurine is a powerful antioxidant, mainly due to its stable biofilm and direct scavenging ability of ROS [50]. Furthermore, taurine can also improve antioxidant capacity by increasing the activity of antioxidant enzymes [51]. The results of this experiment also showed that the addition of taurine to the oxidized-fish-oil diet remarkably promoted CAT, GPx, GR, SOD, and T-AOC levels, whereas the opposite result was observed for the MDA level. The antioxidant capacity of taurine is related to its role as a precursor of glutathione [52], and taurine can also enhance the regeneration of glutathione from glutathione disulfide [53].

Antioxidant enzyme activity is regulated by the nrf2/keap1 signaling pathway [54]. *keap1* inhibits the expression of antioxidant genes by inhibiting the nuclear translocation of *nrf2* [55]. The present study showed that oxidized-fish-oil diets remarkably downregulated the transcriptional levels of *nrf2*, *gr*, and *gpx1* in the liver, while the transcriptional levels of *keap1* were reversed. These results were consistent with the results of antioxidant enzyme activities, indicating that long-term feeding of oxidized-fish-oil diets can reduce the antioxidant capacity of channel catfish. Previous studies in pufferfish (*Takifugu obscurus*) [56] and yellow catfish (*Pelteobagrus fulvidraco*) [57] have found that when fish are under oxidative stress, dietary taurine can increase the expression levels of antioxidant enzyme genes in the liver, thus improving antioxidant capacity. In this study, the addition of taurine to oxidized-fish-oil diets remarkably upregulated *nrf2*, *gr*, and *gpx1* mRNA expression, whereas the opposite result was observed for the transcriptional level of *keap1*. These results were consistent with the results of antioxidant enzyme activities, indicating that taurine can relieve peroxidation injury of channel catfish caused by oxidized-fish-oil diets. Similar studies have confirmed that taurine can remarkably improve antioxidant ability in juvenile turbot by regulating the *nrf2/keap1* signaling pathway [58]. Therefore, we speculate that taurine regulates the activity of antioxidant enzymes through the *nrf2/keap1* signaling pathway, thereby enhancing the ability of fish to resist oxidative stress.

Immune-active substances such as immunoglobulin and complement factor in serum play an important role in animal immune response [59]. Fish mainly rely on the non-specific immune system to respond to external environmental stimuli and pathogen invasion [60]. As a protein response system, the complement system is mainly responsible for destroying or removing pathogenic microorganisms, and is an important part of the non-specific immunity [61]. The present study showed that oxidized-fish-oil diets significantly decreased immune function, which was supported by the phenomenon of decreased serum C3, C4, and IgM contents. Some studies have indicated that dietary supplementation of taurine can alleviate acute ammonia poisoning of yellow catfish by increasing the content of total immunoglobulin in serum [57]. In addition, our previous study revealed that taurine can improve serum C3 and C4 levels in rice field eel to alleviate the immune response induced by high-fat diets [62]. The results of this study showed that the dietary supplementation of taurine to a oxidized-fish-oil diet increased serum C3, C4, and IgM contents, which indicated that taurine can improve immune function of channel catfish. Similar experimental results were found in Chinese mitten crab (*Eriocheir sinensis*) [63] and yellow catfish [64].

Further research has shown that taurine can enhance immune function though controlling intestinal inflammatory response [65]. Intestinal inflammatory response is mainly regulated by cytokines, including anti-inflammatory cytokines (including *tgf-β* and *il-10*) and pro-inflammatory cytokines (including *tnf-α il-1β*, *il-6*, and *il-8*) [66]. The present study showed that the mRNA transcription levels of *tnf-α il-1β*, *il-6*, and *il-8* were remarkably upregulated when channel catfish fed oxidized-fish-oil diets, whereas the opposite result was observed for the mRNA transcription levels of *tgf-β1*, *tgf-β2*, and *tgf-β3*. A similar study has been conducted in *Rhynchocypris lagowski*, which showed that oxidized-fish-oil diets lead to high expression of pro-inflammatory cytokines (*tnf-α*, *il-1β*, and *il-8*) and low expression of anti-inflammatory cytokines (*il-10* and *tgf-β*) [67]. Finding how to alleviate the inflammatory reaction caused by an oxidized-fish-oil diet is very important to improving the utilization rate of aquatic feed. Previous studies have reported that adding taurine significantly downregulated the expression levels of anti-inflammatory cytokines in grass carp [16] and yellow catfish [57]. The results of this experiment also shown that the addition of taurine to a oxidized-fish-oil diet dramatically downregulated *tnf-α il-1β*, *il-6*, and *il-8* mRNA expression in the liver, and upregulated *tgf-β1*, *tgf-β2*, and *tgf-β3* mRNA expression, indicating that taurine can reduce inflammatory response in the intestine induced by oxidized-fish-oil diets.

Cytokine expression in inflammatory response is regulated by various signaling pathways, among which nuclear transcription factor-κB (NF-κB), as an important signaling factor, plays an important role in inflammatory response [68]. In this study, long-term feeding of oxidized-fish-oil diets remarkably upregulated the transcriptional level of *nf-κb*. However, the addition of taurine to oxidized-fish-oil diets reversed this trend. Furthermore, correlation analyses showed that the mRNA expression level of *nf-κb* was negatively correlated with the mRNA expression levels of *tnf-α*, *il-1β*, *il-6*, and *il-8*, and positively correlated with the mRNA expression levels of *tgf-β1*, *tgf-β2*, and *tgf-β3*, which suggested that taurine inhibited the NF-κB signaling pathway to protecting oxidized fish-oil-induced inflammation response in channel catfish.

Intestinal physical-barrier function is an indispensable part of intestinal health of aquatic animals [69]. Generally speaking, the muscular thickness and villi length in the intestine are important criteria to measure the efficiency of digestion and absorption [70]. Goblet cells on intestinal villi, as typical mucous cells, play an important role in regulating the integrity of intestinal epithelial cells and the immune response to foreign antigens [71]. The present study showed that oxidized-fish-oil diets significantly reduced the villi length, goblet cell quantity, and muscular thickness of intestine. However, the addition of taurine to oxidized-fish-oil diets reversed this trend, indicating taurine can maintain the structural integrity of the intestine. An important component of the intestinal physical barrier is tight junction protein. Studies have reported that tight junction proteins are closely related to the integrity of intestinal structure, and the upregulation of transmembrane protein-related genes (including *occluding*, *zo-1*, and *zo-2*) can maintain the structural integrity of intestinal epithelial cells, while the upregulation of cytoplasmic protein-related genes (including *claudin-12* and *claudin-15*) can damage the structural integrity of intestinal epithelial cells [54,72]. This present study found that oxidized-fish-oil diets substantially downregulated intestinal *zo-1* and *zo-2* mRNA transcriptional levels of channel catfish, and upregulated *claudin-12* and *claudin-15* mRNA transcriptional levels, indicating that oxidized-fish-oil diets may increase the intestinal barrier structure damage caused by intercellular space by regulating tight junction protein genes. Taurine has been reported to enhance intestinal morphology and barrier function [73]. The results of this experiment also showed that the addition of taurine to a oxidized-fish-oil diet remarkably upregulated intestinal *zo-1* and *zo-2* mRNA transcriptional levels, and downregulated *claudin-12* and *claudin-15* mRNA transcriptional levels, indicating that taurine can repair the intestinal physical barrier damage induced by oxidized-fish-oil diets. However, the specific regulatory mechanism needs to be further studied.

5. Conclusions

The present study indicated that oxidized-fish-oil diets have a negative effect on growth performance, lipid metabolism, antioxidant ability, and intestinal health in channel catfish. However, addition of taurine to a oxidized-fish oil diet can increase growth performance of channel catfish. Taurine reduced lipid deposition in the liver through promoting the transcription factors of lipid metabolism including *srebp1*, *lpl*, and *fas*. In addition, our findings revealed that the supplementation of taurine alleviated oxidized fish-oil-induced oxidative damage of the liver through the Nrf2-Keap1 signaling pathway based on the transcriptional expression, and then significantly improved the activity of antioxidant enzymes. Furthermore, the current study revealed that the supplementation of taurine alleviated inflammatory response in the intestine through the NF-κB signaling pathway based on the transcriptional expression.

Author Contributions: Conceptualization, L.Z. and Y.H.; writing—original draft preparation, Y.S. and Y.H.; writing—review and editing, L.Z.; methodology, Y.S.; software, Z.W.; data curation, Y.S. and Z.W.; visualization, Y.S.; supervision, Y.H.; validation, Y.S., Y.H., and L.Z.; formal analysis, Y.S. and J.Z. (Junzhi Zhang); investigation, H.Z. and G.F.; resources, J.Z. (Jiancheng Zhou); project administration, L.Z. and Y.H. All authors have read and agreed to the published version of the manuscript.

Funding: This research was funded by National Natural Science Foundation of China, grant numbers 32172985 and 32172986, the National Key R&D Program of China, grant number 2019YFD0900200, and the Hunan Provincial Natural Science Foundation, grant number 2020JJ4044.

Institutional Review Board Statement: This study was approved by the Hunan Agricultural University Animal Care and Use Committee (430317, 26 April 2020).

Informed Consent Statement: Not applicable.

Data Availability Statement: All data generated or analyzed during this study are included in this published article.

Acknowledgments: We thank Changbao Che, Bo Zhu, Bin Geng, and Xiaoli Cao for their help in diet production and sampling.

Conflicts of Interest: The authors declare no conflict of interest. Although Dr. Jiancheng Zhou is from the Wuhan Dabeinong Aquatic Science and Technology Co. Ltd., the company had no role in the design of the study in the collection, analyses, or interpretation of data; in the writing of the manuscript; or in the decision to publish the results.

References

1. Sargent, J.; Bell, G.; McEvoy, L.; Tocher, D.; Estevez, A. Recent developments in the essential fatty acid nutrition of fish. *Aquaculture* **1999**, *177*, 191–199. [CrossRef]
2. Zhou, L.; Han, D.; Zhu, X.; Yang, Y.; Jin, J.; Xie, S. Effects of total replacement of fish oil by pork lard or rapeseed oil and recovery by a fish oil finishing diet on growth, health and fish quality of gibel carp (*Carassius auratus gibelio*). *Aquac. Res.* **2016**, *47*, 2961–2975. [CrossRef]
3. Gray, J.I. Measurement of lipid oxidation: A review. *J. Am. Oil Chem. Soc.* **1978**, *55*, 539–546. [CrossRef]
4. Chen, Y.J.; Liu, Y.J.; Yang, H.J.; Yuan, Y.; Liu, F.J.; Tian, L.X.; Liang, G.Y.; Yuan, R.M. Effect of dietary oxidized fish oil on growth performance, body composition, antioxidant defence chanism and liver histology of juvenile largemouth bass *Micropterus salmoides*. *Aquac. Nutr.* **2012**, *18*, 321–331. [CrossRef]
5. Liu, D.; Chi, S.; Tan, B.; Dong, X.; Yang, Q.; Liu, H.; Zhang, S.; Han, F.; He, Y. Effects of fish oil with difference oxidation degree on growth performance and expression abundance of antioxidant and fat metabolism genes in orange spotted grouper, *Epinephelus coioides*. *Aquac. Res.* **2019**, *50*, 188–197. [CrossRef]
6. Yu, L.J.; Wen, H.; Jiang, M.; Wu, F.; Tian, J.; Lu, X.; Xiao, J.; Liu, W. Effects of ferulic acid on growth performance, immunity, and antioxidant status in genetically improved farmed tilapia (*Oreochromis niloticus*) fed oxidized fish oil. *Aquac. Nutr.* **2020**, *26*, 1431–1442. [CrossRef]
7. Song, C.; Liu, B.; Xu, P.; Ge, X.; Zhang, H. Emodin ameliorates metabolic and antioxidant capacity inhibited by dietary oxidized fish oil through PPARs and Nrf2-Keap1 signaling in Wuchang bream (*Megalobrama amblycephala*). *Fish Shellfish Immunol.* **2019**, *94*, 842–851.
8. Yu, T.; Wang, Q.J.; Chen, X.M.; Chen, Y.K.; Ghonimy, A.; Zhang, D.M.; Wang, G.Q. Effect of dietary L-carnitine supplementation on growth performance and lipid metabolism in *Rhynchocypris lagowski* Dybowski fed oxidized fish oil. *Aquac. Res.* **2020**, *51*, 3698–3710. [CrossRef]
9. Song, C.; Liu, B.; Xu, P.; Xie, J.; Ge, X.; Zhou, Q.; Sun, C.; Zhang, H.; Shan, F.; Yang, Z. Oxidized fish oil injury stress in *Megalobrama amblycephala*: Evaluated by growth, intestinal physiology, and transcriptome-based PI3K-Akt/NF-κB/TCR inflammatory signaling. *Fish Shellfish Immunol.* **2018**, *81*, 446–455. [CrossRef]
10. Peng, M.; Luo, H.; Kumar, V.; Kajbaf, K.; Hu, Y.; Yang, G. Dysbiosis of intestinal microbiota induced by dietary oxidized fish oil and recovery of diet-induced dysbiosis via taurine supplementation in rice field eel (*Monopterus albus*). *Aquaculture* **2019**, *512*, 734288. [CrossRef]
11. Huxtable, R. Physiological actions of taurine. *Physiol. Rev.* **1992**, *72*, 101–142. [CrossRef] [PubMed]
12. Oliveira, M.W.; Minotto, J.B.; de Oliveira, M.R.; Zanotto-Filho, A.; Behr, G.A.; Rocha, R.F.; Moreira, J.C.F.; Klamt, F. Scavenging and antioxidant potential of physiological taurine concentrations against different reactive oxygen/nitrogen species. *Pharmacol. Rep.* **2010**, *62*, 185–193. [CrossRef]
13. Zhang, J.Z.; Hu, Y.; Ai, Q.H.; Mao, P.; Tian, Q.Q.; Zhong, L.; Xiao, T.Y.; Chu, W.Y. Effect of dietary taurine supplementation on growth performance, digestive enzyme activities and antioxidant status of juvenile black carp (*Mylopharyngodon piceus*) fed with low fish meal diet. *Aquac. Res.* **2018**, *49*, 3187–3195. [CrossRef]
14. Hu, Y.; Yang, G.; Li, Z.; Hu, Y.; Zhong, L.; Zhou, Q.; Peng, M. Effect of dietary taurine supplementation on growth, digestive enzyme, immunity and resistant to dry stress of rice field eel (*Monopterus albus*) fed low fish meal diets. *Aquac. Res.* **2018**, *49*, 2108–2118.
15. Martins, N.; Diógenes, A.F.; Magalhães, R.; Matas, I.; Oliva-Teles, A.; Peres, H. Dietary taurine supplementation affects lipid metabolism and improves the oxidative status of European seabass (*Dicentrarchus labrax*) juveniles. *Aquaculture* **2021**, *531*, 735820. [CrossRef]
16. Yan, L.C.; Feng, L.; Jiang, W.D.; Wu, P.; Liu, Y.; Jiang, J.; Tang, L.; Tang, W.N.; Zhang, Y.A.; Yang, J.; et al. Dietary taurine supplementation to a plant protein source-based diet improved the growth and intestinal immune function of young grass carp (*Ctenopharyngodon idella*). *Aquac. Nutr.* **2019**, *25*, 873–896. [CrossRef]
17. Garcia-Organista, A.A.; Mata-Sotres, J.A.; Viana, M.T.; Rombenso, A.N. The effects of high dietary methionine and taurine are not equal in terms of growth and lipid metabolism of juvenile California Yellowtail (*Seriola dorsalis*). *Aquaculture* **2019**, *512*, 734304. [CrossRef]
18. FAO. *Fishery and Aquaculture Statistics 2018/FAO Annuaire*; Food and Agriculture Organization of the United Nations: Rome, Italy, 2020.
19. Shi, Y.; Zhong, L.; Ma, X.; Liu, Y.; Tang, T.; Hu, Y. Effect of replacing fishmeal with stickwater hydrolysate on the growth, serum biochemical indexes, immune indexes, intestinal histology and microbiota of rice field eel (*Monopterus albus*). *Aquac. Rep.* **2019**, *15*, 100223. [CrossRef]

20. Shi, Y.; Zhong, L.; Zhang, J.Z.; Ma, X.K.; Zhong, H.; Peng, M.; He, H.; Hu, Y. Substitution of fish meal with krill meal in rice field eel (*Monopterus albus*) diets: Effects on growth, immunity, muscle textural quality, and expression of myogenic regulation factors. *Anim. Feed Sci. Technol.* **2021**, *280*, 115047. [CrossRef]
21. Livak, K.J.; Schmittgen, T.D. Schmittgen, Analysis of relative gene expression data using Real-Time Quantitative PCR and the $2^{-\triangle\triangle CT}$ Method. *Methods* **2001**, *25*, 402–408. [CrossRef]
22. Zar, J.H. Biostatistical analysis. *Q. Rev. Biol.* **1999**, *18*, 797–799.
23. Izquierdo, M.S.; Obach, A.; Arantzamendi, L.; Montero, D.; Robaina, L.; Rosenlund, G. Dietary lipid sources for seabream and seabass: Growth performance, tissue composition and flesh quality. *Aquac. Nutr.* **2003**, *9*, 397–407.
24. Long, S.; Dong, X.; Tan, B.; Zhang, S.; Xie, S.; Yang, Q.; Chi, S.; Liu, H.; Deng, J.; Yang, Y.; et al. Growth performance, antioxidant ability, biochemical index in serum, liver histology and hepatic metabolomics analysis of juvenile hybrid grouper (♀*Epinephelus fuscoguttatus* × ♂*Epinephelus lanceolatus*) fed with oxidized fish oil. *Aquaculture* **2021**, *545*, 737261.
25. Zhang, D.G.; Zhao, T.; Hogstrand, C.; Ye, H.M.; Xu, X.J.; Luo, Z. Oxidized fish oils increased lipid deposition via oxidative stress-mediated mitochondrial dysfunction and the CREB1-Bcl2-Beclin1 pathway in the liver tissues and hepatocytes of yellow catfish. *Food Chem.* **2021**, *360*, 129814. [CrossRef] [PubMed]
26. Dong, G.F.; Huang, F.; Zhu, X.M.; Zhang, L.; Mei, M.X.; Hu, Q.W.; Liu, H.Y. Nutriphysiological and cytological responses of juvenile channel catfish (*Ictalurus punctatus*) to dietary oxidized fish oil. *Aquac. Nutr.* **2012**, *18*, 673–684. [CrossRef]
27. Candebat, C.L.; Booth, M.; Codabaccus, M.B.; Pirozzi, I. Dietary methionine spares the requirement for taurine in juvenile Yellowtail Kingfish (*Seriola lalandi*). *Aquaculture* **2020**, *522*, 735090. [CrossRef]
28. Zhang, Y.; Wei, Z.; Liu, G.; Deng, K.; Yang, M.; Pan, M.; Gu, Z.; Liu, D.; Zhang, W.; Mai, K. Synergistic effects of dietary carbohydrate and taurine on growth performance, digestive enzyme activities and glucose metabolism in juvenile turbot *Scophthalmus maximus* L. *Aquaculture* **2019**, *499*, 32–41. [CrossRef]
29. Wei, Y.; Zhang, Q.; Xu, H.; Liang, M. Taurine requirement and metabolism response of tiger puffer *Takifugu rubripes* to graded taurine supplementation. *Aquaculture* **2020**, *524*, 735237. [CrossRef]
30. Qi, G.; Ai, Q.; Mai, K.; Xu, W.; Liufu, Z.; Yun, B.; Zhou, H. Effects of dietary taurine supplementation to a casein-based diet on growth performance and taurine distribution in two sizes of juvenile turbot (*Scophthalmus maximus* L.). *Aquaculture* **2012**, *358–359*, 122–128. [CrossRef]
31. Chen, K.; Ye, Y.; Cai, C.; Wu, P.; Huang, Y.; Wu, T.; Lin, X.; Luo, Q.; Zhang, B.; Xiao, P. Damage of oxidized fish oil on the structure and function of hepatopancreas of grass carp (*Ctenopharyngodon idellus*). *Acta Hydrobiol. Sin.* **2016**, *40*, 793–803.
32. Prusińska, M.; Nowosad, J.; Jarmołowicz, S.; Mikiewicz, M.; Duda, A.; Wiszniewski, G.; Sikora, M.; Biegaj, M.; Samselska, A.; Arciuch-Rutkowska, M.; et al. Effect of feeding barbel larvae (*Barbus barbus* (L., 1758)) Artemia sp. nauplii enriched with PUFAs on their growth and survival rate, blood composition, alimentary tract histological structure and body chemical composition. *Aquac. Rep.* **2020**, *18*, 100492.
33. Stejskal, V.; Gebauer, T.; Sebesta, R.; Nowosad, J.; Sikora, M.; Biegaj, M.; Kucharczyk, D. Effect of feeding strategy on survival, growth, intestine development, and liver status of maraena whitefish *Coregonus maraena* larvae. *J. World Aquac. Soc.* **2021**, *52*, 829–842. [CrossRef]
34. Zhang, Y.; Li, Y.; Liang, X.; Cao, X.; Huang, L.; Yan, J.; Wei, Y.; Gao, J. Hepatic transcriptome analysis and identification of differentially expressed genes response to dietary oxidized fish oil in loach *Misgurnus anguillicaudatus*. *PLoS ONE* **2017**, *12*, e0172386. [CrossRef] [PubMed]
35. Xie, S.; Yin, P.; Tian, L.; Liu, Y.; Niu, J. Lipid metabolism and plasma metabolomics of juvenile largemouth bass *Micropterus salmoides* were affected by dietary oxidized fish oil. *Aquaculture* **2020**, *522*, 735158. [CrossRef]
36. Zhang, L.; Li, H.X.; Pan, W.S.; Khan, F.U.; Qian, C.; Qi-Li, F.R.; Xu, X. Novel hepatoprotective role of *Leonurine hydrochloride* against experimental non-alcoholic steatohepatitis mediated via AMPK/SREBP1 signaling pathway. *Biomed. Pharmacother.* **2019**, *110*, 571–581. [CrossRef] [PubMed]
37. McPherson, R.; Gauthier, A. Molecular regulation of SREBP function: The Insig-SCAP connection and isoform-specific modulation of lipid synthesis. *Biochem. Cell Biol.* **2004**, *82*, 201–211. [CrossRef] [PubMed]
38. Han, H.L.; Zhang, J.F.; Yan, E.F.; Shen, M.M.; Wu, J.M.; Gan, Z.D.; Wei, C.H.; Zhang, L.L.; Wang, T. Effects of taurine on growth performance, antioxidant capacity, and lipid metabolism in broiler chickens. *Poult. Sci.* **2020**, *99*, 5707–5717. [CrossRef] [PubMed]
39. Magalhães, R.; Martins, N.; Martins, S.; Lopes, T.; Diáz-Rosales, P.; Pousão-Ferreira, P.; Oliva-Tele-S, A.; Peres, H. Is dietary taurine required for white seabream (*Diplodus sargus*) juveniles? *Aquaculture* **2018**, *502*, 296–302. [CrossRef]
40. Hoseini, S.M.; Hosseini, S.A.; Eskandari, S.; Amirahmadi, M. Effect of dietary taurine and methionine supplementation on growth performance, body composition, taurine retention and lipid status of Persian sturgeon, *Acipenser persicus* (Borodin, 1897), fed with plant-based diet. *Aquac. Nutr.* **2018**, *24*, 324–331. [CrossRef]
41. Tan, Y.; Kim, J.; Cheng, J.; Ong, M.; Lao, W.G.; Jin, X.L.; Lin, Y.G.; Xiao, L.; Zhu, X.Q.; Qu, X.Q. Green tea polyphenols ameliorate non-alcoholic fatty liver disease through upregulating AMPK activation in high fat fed Zucker fatty rats. *World J. Gastroenterol.* **2017**, *23*, 3805–3814. [CrossRef]
42. Kim, S.J.; Tang, T.; Abbott, M.; Viscarra, J.A.; Wang, Y.; Sul, H.S. AMPK phosphorylates desnutrin/ATGL and hormone-sensitive lipase to regulate lipolysis and fatty acid oxidation within adipose tissue. *Mol. Cell. Biol.* **2016**, *36*, 1961–1976. [CrossRef] [PubMed]

43. Rahal, A.; Kumar, A.; Singh, V.; Yadav, B.; Tiwari, R.; Chakraborty, S.; Ghama, K. Oxidative stress, prooxidants, and antioxidants: The interplay. *Biomed. Res. Int.* **2014**, *2014*, 761264. [CrossRef] [PubMed]
44. Giorgio, M.; Trinei, M.; Migliaccio, E.; Pelicci, P.G. Hydrogen peroxide: A metabolic by-product or a common mediator of ageing signals? *Nat. Rev. Mol. Cell Biol.* **2007**, *8*, 722–728. [CrossRef] [PubMed]
45. Bondia-Pons, I.; Ryan, L.; Martinez, J.A. Oxidative stress and inflammation interactions in human obesity. *J. Physiol. Biochem.* **2012**, *68*, 701–711.
46. Kanner, J.; Lapidot, T. The stomach as a bioreactor: Dietary lipid peroxidation in the gastric fluid and the effects of plant-derived antioxidants. *Free Radic. Biol. Med.* **2001**, *31*, 1388–1395. [CrossRef]
47. Dinu, D.; Marinescu, D.; Munteanu, M.C.; Staicu, A.C.; Costache, M.; Dinischiotu, A. Modulatory effects of deltamethrin on antioxidant defense mechanisms and lipid peroxidation in *Carassius auratus* gibelio liver and intestine. *Arch. Environ. Contam. Toxicol.* **2010**, *58*, 757–764. [CrossRef] [PubMed]
48. Ramalingam, M.; Kim, S.J. Reactive oxygen/nitrogen species and their functional correlations in neurodegenerative diseases. *J. Neural Transm.* **2012**, *119*, 891–910.
49. Mehra, L.; Hasija, Y.; Mittal, G. Therapeutic potential of alpha-ketoglutarate against acetaminophen-induced hepatotoxicity in rats. *J. Pharm. Bioallied Sci.* **2016**, *8*, 296–299.
50. Aly, H.A.; Khafagy, R.M. Taurine reverses endosulfan-induced oxidative stress and apoptosis in adult rat testis. *Food Chem. Toxicol.* **2014**, *64*, 1–9.
51. Salze, G.P.; Davis, D.A. Taurine: A critical nutrient for future fish feeds. *Aquaculture* **2015**, *437*, 215–229. [CrossRef]
52. Hayes, J.; Tipton, K.F.; Bianchi, L.; Corte, L.D. Complexities in the neurotoxic actions of 6-hydroxydopamine in relation to the cytoprotective properties of taurine. *Brain Res. Bull.* [illegible] [CrossRef]
53. Morales, A.E.; Pérez-Jiménez, A.; Hidalgo, M.C.; Abellán, E.; Cardenete, G. Oxidative stress and antioxidant defenses after prolonged starvation in Dentex dentex liver. *Comp. Biochem. Physiol. C Toxicol. Pharmacol.* **2004**, *139*, 153–161. [CrossRef]
54. Shi, Y.; Zhong, L.; Liu, Y.L.; Zhang, J.Z.; Lv, Z.; Li, Y.; Hu, Y. Effects of dietary andrographolide levels on growth performance, antioxidant capacity, intestinal immune function and microbioma of rice field eel (*Monopterus albus*). *Animals* **2020**, *10*, 1744. [CrossRef] [PubMed]
55. Jung, K.A.; Kwak, M.K. The Nrf2 system as a potential target for the development of indirect antioxidants. *Molecules* **2010**, *15*, 7266–7291. [CrossRef] [PubMed]
56. Cheng, C.H.; Guo, Z.X.; Wang, A.L. The protective effects of taurine on oxidative stress, cytoplasmic free-Ca^{2+} and apoptosis of pufferfish (*Takifugu obscurus*) under low temperature stress. *Fish Shellfish Immunol.* **2018**, *77*, 457–464. [CrossRef]
57. Zhang, M.; Li, M.; Wang, R.; Qian, Y. Effects of acute ammonia toxicity on oxidative stress, immune response and apoptosis of juvenile yellow catfish *Pelteobagrus fulvidraco* and the mitigation of exogenous taurine. *Fish Shellfish Immunol.* **2018**, *79*, 313–320. [CrossRef]
58. Zhang, Y.; Wei, Z.; Yang, M.; Liu, D.; Pan, M.; Wu, C.; Zhang, W.; Mai, K. Dietary taurine modulates hepatic oxidative status, ER stress and inflammation in juvenile turbot (*Scophthalmus maximus* L.) fed high carbohydrate diets. *Fish Shellfish Immunol.* **2021**, *109*, 1–11. [CrossRef]
59. Ni, P.J.; Jiang, W.D.; Wu, P.; Liu, Y.; Kuang, S.Y.; Tang, L.; Tang, W.N.; Zhang, Y.A.; Zhou, X.Q.; Feng, L. Dietary low or excess levels of lipids reduced growth performance, and impaired immune function and structure of head kidney, spleen and skin in young grass carp (*Ctenopharyngodon idella*) under the infection of Aeromonas hydrophila. *Fish Shellfish Immunol.* **2016**, *55*, 28–47. [CrossRef] [PubMed]
60. Ai, Q.H.; Mai, K.S. Advance on nutritional immunity of fish. *Acta Hydrobiol. Sin.* **2007**, *31*, 425–430.
61. Mori, K.; Nakanishi, T.; Suzuki, T.; Oohara, I. Defense mechanisms in invertebrates and fish. *Tanpakushitsu Kakusan Koso* **1989**, *34*, 214–223.
62. Shi, Y.; Zhong, L.; Zhong, H.; Zhang, J.; Che, C.; Fu, G.; Hu, Y.; Mai, K. Taurine supplements in high-fat diets improve survival of juvenile *Monopterus albus* by reducing lipid deposition and intestinal damage. *Aquaculture* **2022**, *547*, 737431. [CrossRef]
63. Dong, J.; Cheng, R.; Yang, Y.; Zhao, Y.; Wu, G.; Zhang, R.; Zhu, X.; Li, L.; Li, X. Effects of dietary taurine on growth, non-specific immunity, anti-oxidative properties and gut immunity in the Chinese mitten crab *Eriocheir sinensis*. *Fish Shellfish Immunol.* **2018**, *82*, 212–219. [CrossRef]
64. Li, M.; Lai, H.; Li, Q.; Gong, S.; Wang, R. Effects of dietary taurine on growth, immunity and hyperammonemia in juvenile yellow catfish *Pelteobagrus fulvidraco* fed all-plant protein diets. *Aquaculture* **2016**, *450*, 349–355. [CrossRef]
65. Kim, C.; Cha, Y.N. Taurine chloramine produced from taurine under inflammation provides anti-inflammatory and cytoprotective effects. *Amino Acids.* **2014**, *46*, 89–100. [CrossRef]
66. Rymuszka, A.; Adaszek, Ł. Pro- and anti-inflammatory cytokine expression in carp blood and head kidney leukocytes exposed to cyanotoxin stress–an in vitro study. *Fish Shellfish Immunol.* **2012**, *33*, 382–388. [CrossRef]
67. Zhang, D.M.; Guo, Z.X.; Zhao, Y.L.; Wang, Q.J.; Gao, Y.S.; Yu, T.; Chen, Y.K.; Chen, X.M.; Wang, G.Q. L-carnitine regulated Nrf2/Keap1 activation in vitro and in vivo and protected oxidized fish oil-induced inflammation response by inhibiting the NF-κB signaling pathway in *Rhynchocypris lagowski* Dybowski. *Fish Shellfish Immunol.* **2019**, *93*, 1100–1110. [CrossRef]
68. Sen, R.; Baltimore, D. Multiple nuclear factors interact with the immunoglobulin enhancer sequences. *Cell* **1986**, *46*, 705–716. [CrossRef]

69. Cario, E.; Gerken, G.; Podolsky, D.K. Toll-like receptor 2 controls mucosal inflammation by regulating epithelial barrier function. *Gastroenterology* **2007**, *132*, 1359–1374. [CrossRef]
70. Pirarat, N.; Pinpimai, K.; Endo, M.; Katagiri, T.; Ponpornpisit, A.; Chansue, N.; Maita, M. Modulation of intestinal morphology and immunity in nile tilapia (*Oreochromis niloticus*) by *Lactobacillus rhamnosus* GG. *Res. Vet. Sci.* **2011**, *91*, e92–e97. [CrossRef]
71. Farhangi, M.; Carter, C.G. Growth, physiological and immunological responses of rainbow trout (*Oncorhynchus mykiss*) to different dietary inclusion levels of dehuded lupin (*Lupinus angustifolius*). *Aquac. Res.* **2002**, *32*, 329–340. [CrossRef]
72. Xu, H.J.; Jiang, W.D.; Feng, L.; Liu, Y.; Wu, P.; Jiang, J.; Kuang, S.Y.; Tang, L.; Tang, W.N.; Zhang, Y.A.; et al. Dietary vitamin C deficiency depresses the growth, head kidney and spleen immunity and structural integrity by regulating NF-κB, TOR, Nrf2, apoptosis and MLCK signaling in young grass carp (*Ctenopharyngodon idella*). *Fish Shellfish Immunol.* **2016**, *52*, 111–138. [CrossRef] [PubMed]
73. Xu, M.; Che, L.; Gao, K.; Wang, L.; Yang, X.; Wen, X.; Jiang, Z.; Wu, D. Effects of dietary taurine supplementation to gilts during late gestation and lactation on offspring growth and oxidative stress. *Animals* **2019**, *9*, 220. [CrossRef] [PubMed]

antioxidants

Article

Lactobacillus plantarum Ameliorates High-Carbohydrate Diet-Induced Hepatic Lipid Accumulation and Oxidative Stress by Upregulating Uridine Synthesis

Rong Xu, Tong Wang, Fei-Fei Ding, Nan-Nan Zhou, Fang Qiao, Li-Qiao Chen, Zhen-Yu Du * and Mei-Ling Zhang *

School of Life Sciences, East China Normal University, Shanghai 200241, China;
52181300029@stu.ecnu.edu.cn (R.X.); 52211300035@stu.ecnu.edu.cn (T.W.); 52201300060@stu.ecnu.edu.cn (F.-F.D.); 51201300089@stu.ecnu.edu.cn (N.-N.Z.); fqiao@bio.ecnu.edu.cn (F.Q.); lqchen@bio.ecnu.edu.cn (L.-Q.C.)
* Correspondence: zydu@bio.ecnu.edu.cn (Z.-Y.D.); mlzhang@bio.ecnu.edu.cn (M.-L.Z.)

Abstract: The overconsumption of carbohydrates induces oxidative stress and lipid accumulation in the liver, which can be alleviated by modulation of intestinal microbiota; however, the underlying mechanism remains unclear. Here, we demonstrated that a strain affiliated with *Lactobacillus plantarum* (designed as MR1) efficiently attenuated lipid deposition, oxidative stress, as well as inflammatory response, which are caused by high-carbohydrate diet (HC) in fish with poor utilization ability of carbohydrates. Serum untargeted metabolome analysis indicated that pyrimidine metabolism was the significantly changed pathway among the groups. In addition, the content of serum uridine was significantly decreased in the HC group compared with the control group, while it increased by supplementation with *L. plantarum* MR1. Further analysis showed that addition of *L. plantarum* MR1 reshaped the composition of gut microbiota and increased the content of intestinal acetate. In vitro experiment showed that sodium acetate could induce the synthesis of uridine in hepatocytes. Furthermore, we proved that uridine could directly ameliorate oxidative stress and decrease liver lipid accumulation in the hepatocytes. In conclusion, this study indicated that probiotic *L. plantarum* MR1 ameliorated high-carbohydrate diet-induced hepatic lipid accumulation and oxidative stress by increasing the circulating uridine, suggesting that intestinal microbiota can regulate the metabolism of nucleotides to maintain host physiological homeostasis.

Keywords: high-carbohydrate diet; *Lactobacillus plantarum*; oxidative stress; acetate; uridine

Citation: Xu, R.; Wang, T.; Ding, F.-F.; Zhou, N.-N.; Qiao, F.; Chen, L.-Q.; Du, Z.-Y.; Zhang, M.-L. *Lactobacillus plantarum* Ameliorates High-Carbohydrate Diet-Induced Hepatic Lipid Accumulation and Oxidative Stress by Upregulating Uridine Synthesis. *Antioxidants* **2022**, *11*, 1238. https://doi.org/10.3390/antiox11071238

Academic Editor: Han Moshage

Received: 23 May 2022
Accepted: 21 June 2022
Published: 24 June 2022

1. Introduction

Over intake of high-carbohydrate diet promotes the oxidative stress and subsequently induces the inflammatory response in animals [1,2]. Moreover, it will increase the de novo lipogenesis. Oxidative stress and lipid accumulation contribute to metabolic syndrome [3,4], including fatty liver [5], high glucose level, and impaired insulin function in animals [6]. Postprandial oxidative stress is regarded as a secondary response to postprandial high glucose and triglyceride levels [7], and oxidative stress may be the mechanistic link between lipid metabolism and related complications [8]. Therefore, the study to attenuate lipid accumulation and oxidative stress induced by high-carbohydrate diet has attracted increased attention.

Carbohydrates are important energy sources for fish; however, fish are natural glucose intolerant with poor utilization ability of carbohydrates [9,10]. High dietary carbohydrates cause negative effects on the growth performance and liver health, and easily trigger oxidative stress in fish. For example, it has been reported that high-carbohydrate diet induces de novo fatty acid synthesis, insulin resistance, and fatty liver disease in blunt snout bream (*Megalobrama amblycephala*) [11]. High-carbohydrate diet affects the growth performance, hepatic glucose metabolism, and antioxidant capacity of largemouth bass

(*Micropterus salmoides*) [12]. Fish has become an important model organism to study the host physiological mechanisms [13,14], particularly for glucose metabolism [15].

Gut microbiota has been considered as an important organ with properties involving host metabolism and physiological functions [16]. It has been found that gut microbiota dysbiosis is strongly correlated with hepatic lipid metabolism disorder or oxidative stress [17,18], while gut microbiota homeostasis restoration can prevent diet-induced fatty liver diseases, glucose metabolism disorder or other metabolic syndromes [19]. Therefore, intestinal microbiota indeed acts as a mediator in the lipid metabolism dysregulation and oxidative stress development [20,21]. It has been found that *Bifidobacterium lactis HN019* has a beneficial effect on the regulation of inflammatory and oxidative biomarkers in subjects with the metabolic syndrome [22]. In addition, *Lactobacillus plantarum* can exert a beneficial effect in high-fat/high-fructose diet-induced fatty liver disease in rats [23], and attenuate oxidative stress by regulating the composition of gut microbiota in D-galactose-exposed mice [24,25]. However, the mechanisms by which microbes regulate lipid accumulation or activate host antioxidant responses still require further investigation. A recent study found that *Lactobacillus rhamnosus* GG activated hepatic nuclear factor erythroid-2-related factor 2 (Nrf2) signal by producing 5-methoxyindoleacetic acid in Drosophila melanogaster and mice, suggesting that bacteria-derived molecules can protect against liver oxidative injury [26].

The short-chain fatty acids (SCFAs) produced by bacterial fermentation of dietary fiber in the gut play critical roles in glucose and lipid metabolism or oxidative stress modulation [27]. For example, butyrate has been reported to exert an antioxidant effect by activating Nrf2 and H3K9/14 acetylation via G-protein coupled receptor (GPR) 109A in bovine mammary epithelial cells [28]. Propionate induces intestinal oxidative stress resulting from Sod2 propionylation in zebrafish fed with high-fat diet (HFD) [29]. Acetate, the most abundant SCFAs in the gut, can regulate lipid metabolism, improve glucose intolerance and insulin resistance, and inhibit the inflammatory response by activating GPRs [27,30]. Accordingly, except for acting as signal molecules, SCFAs can be used as a substrate to synthesize some precursor substances, and then participate in lipid metabolism. However, the related mechanisms need to be further elucidated [31,32].

Nile tilapia (*Oreochromis niloticus*) is the second most farming fish species worldwide, which is also an important model for the physiological study of fish with available genomic information [33,34]. In this study, a strain affiliated with *L. plantarum* (MR1) was isolated from the intestine of healthy Nile tilapia. Three diets, including a normal control diet (NC), a high-carbohydrate diet (HC), and a high-carbohydrate diet supplemented with *L. plantarum* MR1 (HCL), were used to feed the fish for 10 weeks. Then, the effect of *L. plantarum* MR1 on the growth performance, liver lipid metabolism, inflammatory response, and oxidative stress in the HC-fed Nile tilapia was analyzed. The possible mechanism was identified by serum untargeted metabolome, gut microbiome analysis, and in vitro experiment. The present study revealed the method by which intestinal microbiota regulates the metabolism of nucleotides to maintain host physiological homeostasis.

2. Materials and Methods

2.1. Animal Ethics

Experiments were conducted under the Guidance of the Care and Use of Laboratory Animals in China. This study was approved by the Committee on the Ethics of Animal Experiments of East China Normal University (F20190101).

2.2. Animal Experiments

A strain affiliated with *L. plantarum*, designed as MR1, was purified from the gut of healthy Nile tilapia (additional details in the Supplementary Methods).

Nile tilapia juveniles were obtained from Yueqiang aquafarm (Guangzhou, China). All fish were acclimated to 28 °C and fed with a commercial diet twice per day for 2 weeks. Following the acclimation, 270 similar-sized fish at 1.66 (± 0.05) g were randomly distributed into three groups (three replicates for each group, thirty fish per replicate) and fed with a normal control diet (NC), a high-carbohydrate diet (HC) or a high-carbohydrate diet supplemented with *L. plantarum* MR1 (HCL) for 10 weeks. All fish were fed twice per day at 5% of body weight. The formulations of the diets were listed in Table S1. The total weight of fish in each tank was recorded every 2 weeks. For additional details on sampling collection, see Supplementary Methods.

2.3. Body Composition Assessment

Nine fish from each group were collected for the body composition assay. The whole fish were dried in an electric oven at 105 °C to a constant weight. The dried fish were ground to mince for determination of total protein analysis using a semi-automatic Kjeldahl System (Kjeltec 8200, FOSS, Hillerod, Denmark) and for total lipid content detection by a classic methanol–chloroform method. The ash was determined by incineration in a muffle furnace at 550 °C to a constant weight.

2.4. Biochemical Analysis

Serum samples were directly used for the biochemical detection. Liver samples from fish were homogenized with 1 × PBS to obtain a 10% (*w/v*) liver homogenate, centrifuged at 4 °C (3500× *g* for 10 min) for further experiments. Aspartate aminotransferase (AST, C010-2-1), alanine aminotransferase (ALT, C009-2-1), triglyceride (TG, A110-1-1), non-esterified free fatty acids (NEFA, A042-2-1), superoxide dismutase (SOD, A001-3-2), malondialdehyde (MDA, A003-1-2), and reduced glutathione (GSH, A006-2-1) were detected using biochemical assay kits, in accordance with the manufacturer's instructions (Nanjing Jiancheng Bioengineering Institute, Nanjing, China).

2.5. Histological Analysis

Liver tissues were fixed in 4% paraformaldehyde followed by gradient ethanol dehydration and xylene transparency procedure. In addition, the tissues were embedded in paraffin, which were then sliced into 5 μm for hematoxylin and eosin (H&E) staining. For oil red O staining, liver tissues were immediately frozen at −80 °C using the OCT embedding agent (G6059-110ML, Servicebio, Wuhan, China). Approximately 5–10 μm frozen slices were stained with oil red O (ORO) for 10 min in the dark, and then immersed in 60% isopropanol for a few seconds. The frozen slices were counterstained with hematoxylin to visualize lipid droplets. The histological features were observed and captured under a Nikon Eclipse Ti-SR inverted microscope (Nikon, Japan). Quantification and statistical analysis were conducted using Image J v.1.8.0 (National Institutes of Health, Bethesda, MD, USA).

2.6. 16S rRNA Amplicon Sequencing

Intestinal bacterial genomic DNA was purified using an E.Z.N.A.® Soil DNA Kit (D5625, Omega, Norcross, GA, USA), in accordance with the manufacturer's instructions. The V3-V4 region of bacterial 16S rRNA gene was amplified by PCR and sequenced on an Illumina MiSeq PE300 platform/NovaSeq PE250 platform (Illumina, San Diego, CA, USA). For additional details, see Supplementary Methods.

2.7. Measurement of SCFAs

Quantification of bacterial and intestinal SCFAs (including acetate, propionate, and butyrate) was conducted using the previously described methods [35]. For additional details, see Supplementary Methods.

2.8. Mass Spectrometry

Serum untargeted metabolome was analyzed by liquid chromatography–mass spectrometry (LC–MS). Chromatographic separation of the metabolites was performed on a Thermo UHPLC system equipped with an ACQUITY UPLC HSS T3 (100 × 2.1 mm i.d.,

1.8 μm, Waltham, MA, USA) column. For additional details on the detection of all the samples, see Supplementary Methods.

2.9. Cell Culture

The primary hepatocytes were generated from healthy Nile tilapia, as previously described [36]. Briefly, the liver tissues were washed three times using 1 × PBS and digested with DMEM containing 0.1% collagenase IV (17104019, Gibco, Suwanee, GA, USA), then filtered through a 70 μm-nylon mesh. The hepatocytes were harvested by centrifuging at 850× *g* for 10 min. Red blood cells were removed using a red cell lysis buffer (RT122-02, Tiangen, Beijing, China), and the remaining cells were incubated in DMEM with 10% FBS and 1% penicillin–streptomycin at 28 °C. To explore whether acetate acted as the substrate for uridine synthesis, 20 mM sodium acetate (NaAc) (S5636, Sigma-Aldrich, St. Louis, MO, USA) was added to the hepatocytes for 24 h to detect the concentration of uridine. For additional details on the effects of uridine in oleic acid (OA) and hydrogen peroxide (H_2O_2)-treated hepatocytes, see Supplementary Methods.

2.10. Uridine and Acetyl-CoA Detection by HPLC

Uridine and acetyl-CoA were detected by high performance liquid chromatography (HPLC) (LC-20AT, Shimadzu, Takamatsu, Kagawa, Japan) with a tandem double plunger. The separation was achieved on the Shim-pack GIST 5 μm C18 column (4.6 × 250 μm, 5 μm particle size, Shimadazu®, Takamatsu, Kagawa, Japan). For additional details on the contents of serum uridine and liver acetyl-CoA analysis, see Supplementary Methods.

2.11. Real-Time Quantitative PCR (RT-qPCR)

Real-time quantitative PCR (RT-qPCR) was performed as previously described [35]. Briefly, total RNA was extracted, and cDNAs were transcribed using RNAs as templates. The cDNAs were amplified by PCR with ChamQ Universal SYBR qPCR Master Mix (Q711, Vazyme, Nanjing, China) using the primers listed in Table S2. Real-time quantitative PCR (RT-qPCR) analyses were performed using CFX Connect (Bio-rad, Richmond, CA, USA). The *β-actin* and *ef-1α* were used as house-keeping genes, and the relative gene expression values were quantified using the $2^{-\Delta\Delta CT}$ method.

2.12. Western Blot Analysis

Tissues or cells were lysed, and total protein contents were extracted and quantified. Proteins were subjected to SDS-PAGE and transferred to NC membrane, then blocked with 5% BSA for 1 h at room temperature. The membrane was incubated with primary antibodies overnight at 4 °C. Following the secondary antibody incubation, the protein bands were visualized in the Odyssey CLx Imager (Li-Cor Biotechnology, Lincoln, NE, USA) and quantified using Image J v.1.8.0. GAPDH was served as a reference protein. The antibodies used in this study were listed in Table S3.

2.13. Statistical Analysis

Statistical analysis of all data was performed using GraphPad Prism 7.0. The results of biological assays are presented as mean ± standard error of the mean (SEM). Datasets were assessed using one-way analysis of variance (ANOVA) with Tukey's adjustment. Differences between two groups were analyzed by Student's *t*-test. In the Figures: *, $p < 0.05$; **, $p < 0.01$; ***, $p < 0.001$.

3. Results

3.1. *L. plantarum* MR1 Promotes Growth Performance and Decreases Lipid Accumulation of Nile Tilapia

At the end of the feeding experiment, the growth performance showed that final body weight (FBW) and weight gain rate (WGR) (Table 1) were significantly increased in the HC group compared with the NC group, and the WGR was further increased by addition of *L. plantarum* MR1 (Table 1). The body composition indexes showed that the visceral somatic index (VSI) and hepatosomatic index (HSI) were markedly increased in the HC group compared with the NC group; however, they were decreased in the HCL group when compared with the HC group (Table 1). Although there was no significant difference, the mesenteric fat index (MFI) was higher in the HC group than the NC group, and it had a declining tendency in the HCL group, compared with the HC group (Table 1). Compared with the HC-fed fish, the carcass index (CI) was significantly increased in *L. plantarum* MR1-treated fish (Table 1). The total lipid contents of the whole fish were clearly increased in the HC-fed fish, but significantly decreased by *L. plantarum* MR1 administration without influencing the total protein contents (Table 1). These results suggested that *L. plantarum* MR1 could promote the growth performance and decrease the total lipid contents in Nile tilapia.

Table 1. Effect of *L. plantarum* MR1 on growth performance parameters, body composition indexes, and body composition contents of Nile tilapia.

Groups	NC	HC	HCL
	Growth performance parameters		
IBW (g)	1.64 ± 0.05	1.67 ± 0.05	1.68 ± 0.05
FBW (g)	20.16 ± 0.60	22.01 ± 2.41 #	25.98 ± 2.22
WGR (%)	1089.00 ± 48.15	1250.77 ± 32.05 #	1391.08 ± 20.04 *
	Body composition indexes (%)		
VSI (%)	12.31 ± 1.14	14.44 ± 1.45 ##	11.78 ± 0.78 ***
HSI (%)	2.85 ± 0.25	3.66 ± 0.58 #	2.93 ± 0.41 *
MFI (%)	0.59 ± 0.16	0.71 ± 0.39	0.52 ± 0.09
CI (%)	48.77 ± 1.61	47.12 ± 2.93	49.43 ± 1.73 *
	Body composition contents (%)		
Moisture	72.31 ± 1.04	71.49 ± 1.21	72.42 ± 0.66
Total lipid	6.55 ± 0.75	8.23 ± 0.64 ##	6.92 ± 0.55 *
Total protein	14.67 ± 0.29	13.94 ± 0.58	14.16 ± 0.45
Ash	2.66 ± 0.47	3.09 ± 0.22	3.34 ± 0.27

IBW: Initial body weight (g); FBW: Final body weight (g); WGR: Weight gain rate (%); VSI: Visceral somatic index (%); HSI: Hepatosomatic index (%); MFI: Mesenteric fat index (%); CI: Carcass index (%). Statistics were analyzed by one-way ANOVA with Tukey's adjustment and presented as mean ± standard error of the mean (SEM) (#, NC v.s. HC; *, HC v.s. HCL, #, $p < 0.05$; ##, $p < 0.01$; *, $p < 0.05$; ***, $p < 0.001$).

3.2. *L. plantarum MR1 Reduces HC-Induced Liver Lipid Deposition of Nile Tilapia*

Compared with the NC group, the levels of serum aspartate aminotransferase (AST) and alanine aminotransferase (ALT) (markers of liver injury) were considerably higher in the HC group; however, they were clearly reduced by supplementation with *L. plantarum* MR1 (Figure 1a,b). The hematoxylin and eosin (H&E) staining displayed that the hepatic vacuolation was significantly increased in the HC group when compared with the NC group; however, it was decreased by addition of *L. plantarum* MR1 (Figure 1c,d). Moreover, liver oil red O (ORO) staining showed that HC-fed fish had the highest lipid accumulation, which was significantly prevented by addition of *L. plantarum* MR1 (Figure 1e,f). Further analysis also indicated that the contents of triglyceride (TG) and non-esterified free fatty acids (NEFA) in liver and serum were significantly decreased in the fish supplemented with *L. plantarum* MR1 (Figure 1g–j), suggesting that *L. plantarum* MR1 could reduce the HC-induced liver lipid deposition in Nile tilapia.

To investigate the mechanism by which *L. plantarum* MR1 reduces the liver lipid accumulation, the relative expression of genes related to lipid metabolism were detected. The expression levels of genes related to lipid synthesis, such as *fas* and *ppar*γ, were significantly higher in the HC group than those in the NC group, and the expression levels of *fas*, *acc*α, *dgat2*, and *ppar*γ, were considerably lower in the HCL group relative to the HC group (Figure 1k). The genes associated with lipid catabolism, such as *atgl*, *fatp1*, and *hsl*, were downregulated in the HC group compared with the NC group; however, addition of *L. plantarum* MR1 significantly increased the transcript levels of *atgl*, *fatp1*, *cpt1a*, and *hsl* (Figure 1l). To define whether the energy expenditure was increased, the protein expression of AMP-activated protein kinase alpha (AMPKα), a key protein in energy signaling pathway, was analyzed, and a significant increase in phosphorylated-AMPKα (p-AMPKα) was observed in *L. plantarum* MR1-treated group (Figure 1m,n), suggesting that addition of *L. plantarum* MR1 reduced lipid deposition by activating the AMPKα signaling pathway in Nile tilapia.

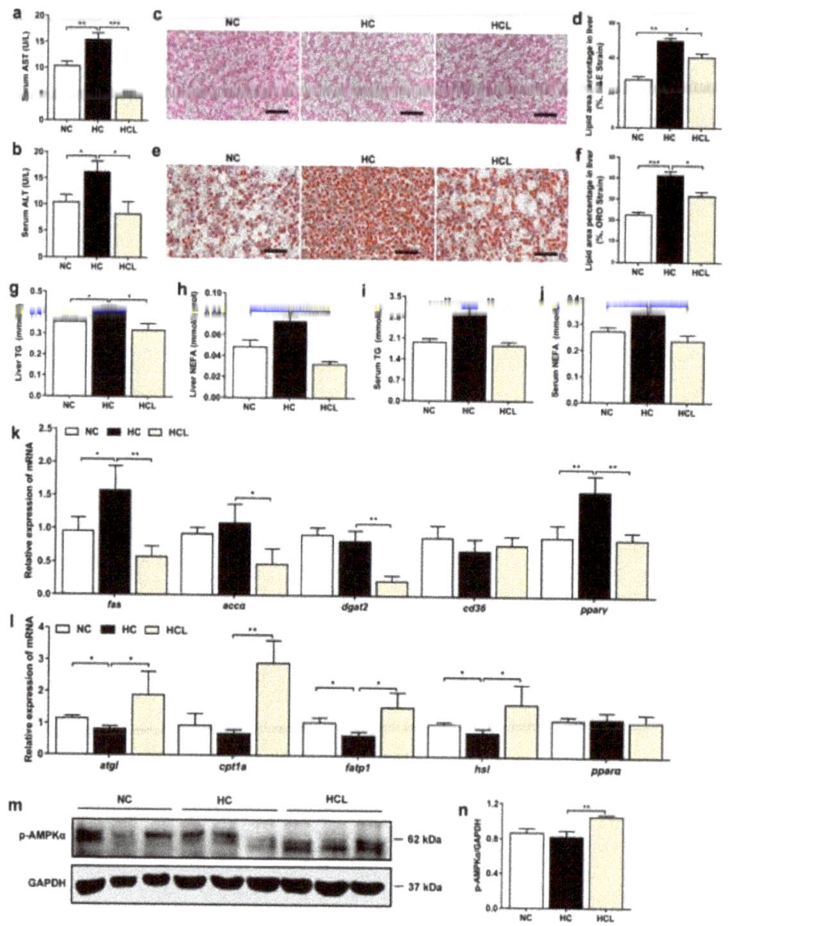

Figure 1. *L. plantarum* MR1 improved liver health in Nile tilapia fed with high-carbohydrate diet. (**a**,**b**) Serum contents of aspartate aminotransferase (AST) (**a**) and alanine aminotransferase (ALT) (**b**) ($n = 6$). (**c**–**f**) Examination of the liver condition by hematoxylin and eosin (H&E) staining (**c**,**d**) and oil red O (ORO) staining (**e**,**f**). Scale bar, 50 μm ($n = 3$ slices). (**g**–**j**) Detection of the liver and serum triglyceride (TG) and non-esterified free fatty acid (NEFA) levels ($n = 6$). (**k**) Relative expression of lipid synthesis genes in the liver ($n = 6$). (**l**) Relative expression of lipid catabolism genes in the liver ($n = 6$). (**m**) Protein level of phosphorylated-AMP-activated protein kinase alpha (p-AMPKα) energy homeostasis regulator in the liver ($n = 3$). (**n**) The protein level of p-AMPKα was quantified and normalized to the loading control (GAPDH) ($n = 3$). Statistics were analyzed by one-way ANOVA with Tukey's adjustment and presented as mean ± SEM (*, $p < 0.05$; **, $p < 0.01$; ***, $p < 0.001$).

Lipid accumulation in the liver normally causes inflammation. The inflammatory cytokines were significantly higher in the HC group than the NC group; however, they were notably lowered in *L. plantarum* MR1-treated group (Figure S1a). In addition, the inflammatory marker proteins, phosphorylated-nuclear factor-kappa B (p-NF-κB), and cleaved interleukin-1 beta (IL-1β), were clearly decreased in the HCL group (Figure S1b,c), indicating that *L. plantarum* MR1 could suppress the HC-induced inflammatory response.

3.3. *L. plantarum MR1 Attenuates the HC-Induced Oxidative Stress of Nile Tilapia*

As lipid accumulation may trigger the oxidative stress in the liver, the antioxidant effect of *L. plantarum* MR1 was evaluated. The activity of superoxide dismutase (SOD) was elevated by addition of *L. plantarum* MR1 compared with the HC group (Figure 2a,b). The lipid peroxidation product malondialdehyde (MDA), an oxidative damage marker, was significantly increased in the HC-treated fish compared with the NC-treated fish; however, it was notably reduced by *L. plantarum* MR1 treatment (Figure 2c,d). The contents of reduced glutathione (GSH) were significantly decreased in serum or liver in the HC-fed fish compared with the NC-fed fish; however, they were markedly increased in *L. plantarum* MR1-treated fish (Figure 2e,f). As a master regulator of antioxidative responses, the protein expression of Nrf2 was significantly decreased in the HC group, but increased in the HCL group (Figure 2g,h). The expression levels of genes in Nrf2 signaling pathway, including *nqo1* and *ho1*, were downregulated in the HC group compared with the NC group; however, the expression levels of *nrf2* and *ho1* were upregulated in *L. plantarum* MR1-treated group (Figure 2i). Considered together, these results demonstrated that addition of *L. plantarum* MR1 could attenuate the HC-induced oxidative stress by activating the Nrf2 signaling pathway in Nile tilapia.

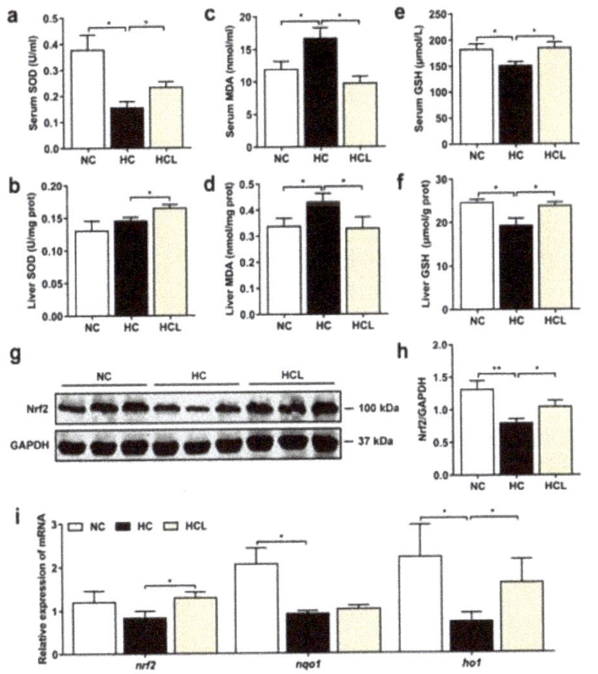

Figure 2. *L. plantarum* MR1 relieved the HC-induced oxidative stress in Nile tilapia. (**a**,**b**) The activities of superoxide dismutase (SOD) in the serum (**a**) and liver (**b**) (n = 6). (**c**,**d**) The contents of malondialdehyde (MDA) in the serum (**c**) and liver (**d**) (n = 6). (**e**,**f**) The contents of reduced glutathione (GSH) in the serum (**e**) and liver (**f**) (n = 6). (**g**,**h**) The protein expression and quantification of Nrf2 in the liver (n = 6). (**i**) The relative mRNA expression of genes for Nrf2 signaling pathway in the liver (n = 6). Statistics were analyzed by one-way ANOVA with Tukey's adjustment and presented as mean ± SEM (*, $p < 0.05$; **, $p < 0.01$).

3.4. *L. plantarum* MR1 Regulates the Nucleotide Metabolism of Nile Tilapia

The serum untargeted metabolome analysis was performed to identify the composition of metabolites among the treatments. The results showed that the metabolism process was obviously changed among the groups (NC v.s. HC, 66.1%; HC v.s. HCL, 64.2%) (Figure 3a). Further analysis showed that the change in pyrimidine metabolism was shared by the comparison between NC v.s. HC and HC v.s. HCL (Figure 3b). The heatmap displayed that the metabolite uridine, a pyrimidine nucleotide, was markedly decreased in the HC group, but clearly increased in the HCL group (Figure 3c,d). Moreover, the uridine concentration in the serum was confirmed by high performance liquid chromatography (HPLC), and the result was consistent with the metabolome analysis (Figure 3e). Considered together, these data demonstrated that supplementation of *L. plantarum* MR1 could increase the contents of uridine, which were decreased in the serum of HC-fed Nile tilapia.

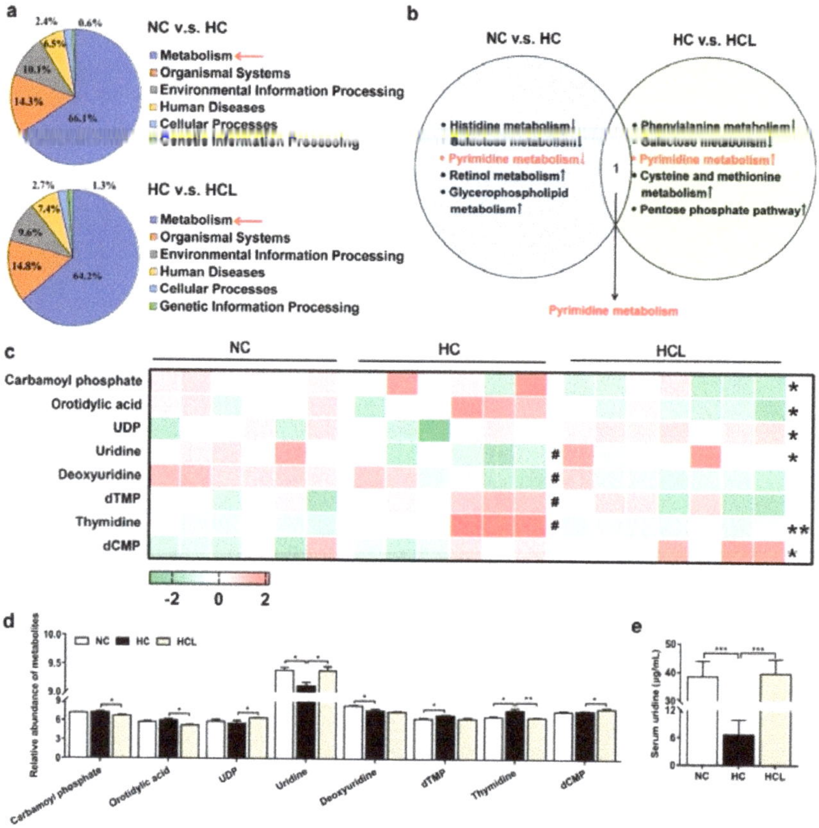

Figure 3. *L. plantarum* MR1 modulated the metabolic patterns in Nile tilapia. (**a**) Identification of differentially biological processes percentage in the HC-treated fish compared with the NC-fed fish or in *L. plantarum* MR1-treated fish compared with the HC-treated fish by liquid chromatography–mass spectrometry (LC–MS) analysis. (**b**) Annotation of the identical significant altered metabolism pathway in the three groups. (**c**) Heat map of selected differentially changed metabolites involved in pyrimidine metabolism (#, NC v.s. HC; *, HC v.s. HCL, #, $p < 0.05$, *, $p < 0.05$, **, $p < 0.01$). (**d**) Histogram statistics of differential metabolites in the heat map. (**e**) The contents of serum uridine detection by high performance liquid chromatography (HPLC); $n = 6$ in NC and HC groups and $n = 7$ in HCL group. Statistics were analyzed by one-way ANOVA with Tukey's adjustment and presented as mean ± SEM (*, $p < 0.05$; **, $p < 0.01$; ***, $p < 0.001$).

3.5. Uridine Alleviates the OA-Induced Lipid Accumulation in the Primary Hepatocytes of Nile Tilapia

To explore the effect of uridine in reducing liver lipid deposition, 250 μM of oleic acid (OA) was added to the primary hepatocytes of Nile tilapia to construct the lipid accumulation model. Uridine at the concentration of 25 μM (OA + UL), 125 μM (OA + UM) or 250 μM (OA + UH) was added to the OA-treated cells. The cell proliferation rate was significantly decreased in the OA group when compared with the Ctrl group; however, it was notably increased by addition of uridine, indicating that no cytotoxicity was caused by uridine (Figure 4a). The contents of TG in the cells were markedly increased in the OA-treated group compared with the Ctrl group; however, they were dramatically reduced by addition of uridine (Figure 4b). Compared with the Ctrl group, the lipid droplets were significantly increased in the OA group, but were clearly diminished by addition of uridine in the hepatocytes, in accordance with BODIPY 493/503 staining (Figure 4c). The relative expression of genes related to lipid synthesis, such as *fas*, *accα*, *dgat2*, and *pparγ*, were significantly upregulated in the OA-treated group relative to the Ctrl group, but were downregulated by supplementation with uridine (Figure 4d). The relative expression of genes associated with lipid catabolism, such as *atgl* and *cpt1a*, was lower in the OA-treated group than the Ctrl group, while the expression of *atgl*, *cpt1a*, and *pparα*, was higher in the uridine administration group (Figure 4e). Considered together, our data suggested that uridine could ameliorate the OA-induced lipid droplets accumulation by increasing lipid catabolism and decreasing lipid synthesis.

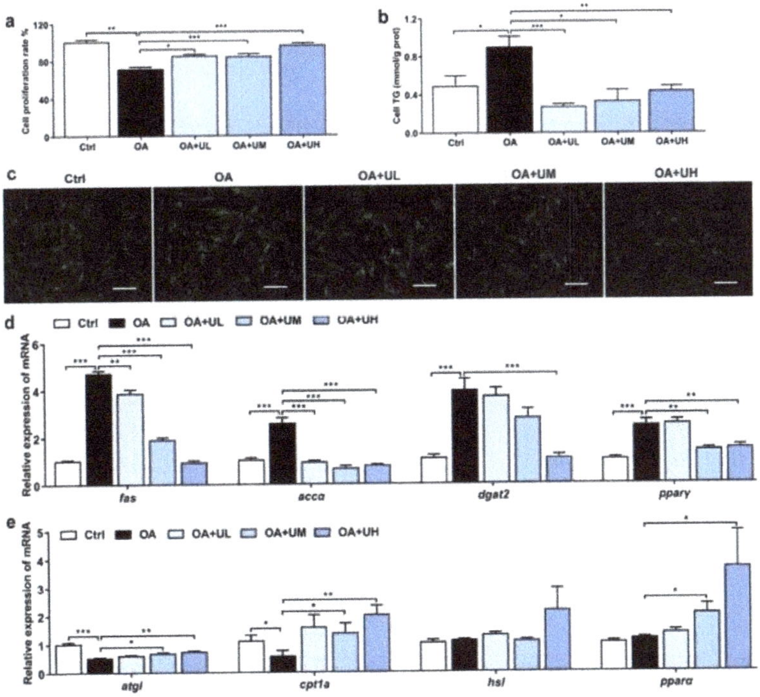

Figure 4. Uridine reduced the OA-induced accumulation of lipid droplets in the primary hepatocytes of Nile tilapia. (**a**) The proliferation of cells that are analyzed by CCK-8 in oleic acid (OA)-induced cell model (n = 10). (**b**) The contents of TG in the hepatocytes (n = 6). (**c**) Lipid droplets staining with BODIPY 493/503 in the hepatocytes. Scale bar, 100 μm (n = 6). (**d**,**e**) Gene expression of lipid synthesis (**d**) and catabolism (**e**) in the hepatocytes (n = 6). Statistics were analyzed by one-way ANOVA with Tukey's adjustment and presented as mean ± SEM (*, $p < 0.05$; **, $p < 0.01$; ***, $p < 0.001$).

Based on our previous assumption, we further evaluated the influence of uridine on inflammatory response. Consistently, the expression of p-NF-κB and cleaved IL-1β was remarkably increased in the OA-treated group compared with the Ctrl group, but notably suppressed by the uridine treatment (Figure S2).

3.6. *Uridine Ameliorates Oxidative Stress in the Primary Hepatocytes of Nile Tilapia*

We further investigated whether uridine could attenuate OA-induced oxidative injury in the hepatocytes. The data showed that the activity of SOD was considerably lower in the OA group than the Ctrl group; however, it was significantly increased in the OA + UH group (Figure 5a). Compared with the Ctrl group, the contents of MDA were markedly increased in the OA group, but notably decreased by uridine administration (Figure 5b). The contents of GSH were decreased in the OA-treated cells, compared with the Ctrl cells; however, they were clearly increased in the OA + UM and OA + UH treatment groups when compared with the OA group (Figure 5c). Furthermore, the protein level of Nrf2 was decreased in the OA-treated group, but increased by uridine treatment (Figure 5d,e). The gene expression levels of *nqo1* and *ho1*, which were regulated by Nrf2 signaling pathway, were lower in the OA group relative to the Ctrl group, while the relative expression of *nrf2*, *nqo1*, and *ho1*, were increased in the uridine treatment group (Figure 5f). These data suggested that uridine could ameliorate the OA-induced oxidative stress by enhancing the antioxidative biomarkers and activating the Nrf2 signaling pathway.

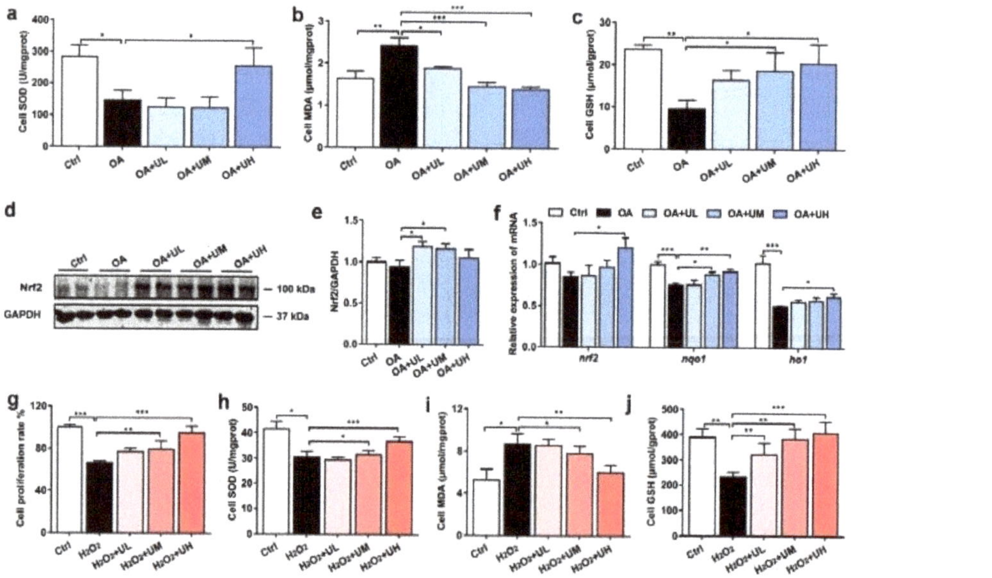

Figure 5. Uridine inhibited oxidative stress in the primary hepatocytes of Nile tilapia. (**a**) The activity of SOD in the primary hepatocytes ($n = 6$). (**b**) The contents of MDA in the primary hepatocytes ($n = 6$). (**c**) The contents of GSH in the primary hepatocytes ($n = 6$). (**d**) The protein expression of Nrf2 in the hepatocytes ($n = 4$). (**e**) Quantitation of Nrf2 levels normalized to GAPDH levels from blots shown in (**d**) ($n = 4$). (**f**) The gene expression of Nrf2 signaling pathway in the hepatocytes ($n = 6$). (**g**) The cells proliferation analyzed by CCK-8 in hydrogen peroxide (H_2O_2)-induced cell model ($n = 10$). (**h**) The activity of SOD in the primary hepatocytes ($n = 6$). (**i**) The contents of MDA in the primary hepatocytes ($n = 6$). (**j**) The contents of GSH in the primary hepatocytes ($n = 6$). Statistics were analyzed by one-way ANOVA with Tukey's adjustment and presented as mean ± SEM (*, $p < 0.05$; **, $p < 0.01$; ***, $p < 0.001$).

In addition, different concentrations of uridine at 25 μM (H_2O_2 + UL), 125 μM (H_2O_2 + UM), and 250 μM (H_2O_2 + UH) were added to the 250 μM hydrogen peroxide (H_2O_2)-induced oxidative injury model in hepatocytes to define whether uridine could directly inhibit oxidative stress. We observed that the proliferation rate of cells was significantly decreased by H_2O_2 treatment, but clearly increased by uridine (Figure 5g). The activity of SOD was considerably lower in the H_2O_2 group than the Ctrl group; however, it was significantly increased in the H_2O_2 + UM and H_2O_2 + UH group (Figure 5h). Compared with the Ctrl group, the contents of MDA were markedly increased in the H_2O_2 group, but notably decreased by uridine administration (Figure 5i). The contents of GSH were decreased in the H_2O_2-treated cells; however, they were clearly increased in uridine-treated cells (Figure 5j). These results indicated that uridine could attenuate oxidative stress directly.

3.7. *L. plantarum MR1 Supplementation Alters the Gut Microbiota Composition of Nile Tilapia*

To reveal the relationship between the addition of *L. plantarum* MR1 and the content of serum uridine, the intestinal bacterial composition and intestinal microbiota derived metabolites were detected. The composition of gut microbiota was characterized by 16S rRNA amplicon pyrosequencing. The principal coordinate analysis (PCoA) displayed that addition of *L. plantarum* MR1 changed the composition of gut microbiota (Figure 6a). The abundance of Firmicutes was increased in the HC group in contrast with the NC group and further increased in the HCL group (Figure 6b). The abundance of *Lactobacillus* belonging to Firmicutes was slightly increased in the HC group relative to the NC group, but markedly enriched by *L. plantarum* MR1 supplementation (Figure 6c). The bacteria metabolites were detected, and compared with the NC group. The intestinal acetate was significantly decreased in the HC group, but dramatically elevated in the HCL group (Figure 6d).

A previous study has reported that sodium acetate (NaAc) can be metabolized to acetyl-CoA, and then act as a substrate to promote the synthesis of uridine in endothelial cells [37]; however, whether intestinal microbiota-derived acetate could increase the uridine synthesis remains unknown. The gene expression levels of *acss1* and *acss2*, which were responsible for the synthesis of acetyl-CoA from acetate, were detected in the liver. Compared with the NC group, *ascc1* was significantly downregulated in the HC group, and the presence of *L. plantarum* MR1 increased the transcript levels of *ascc1* and *ascc2* (Figure 6e). The contents of acetyl-CoA in the liver were lower in the HC group than the NC group, but distinctly increased in the HCL group (Figure 6f). To detect whether acetate could directly promote the synthesis of uridine, 20 mM of NaAc was incubated with the primary hepatocytes of Nile tilapia, and the concentration of uridine was detected. The contents of uridine were significantly increased by the addition of NaAc (Figure 6g). The gene expression levels of *acss1* and *acss2* were both stimulated by the NaAc treatment in the hepatocytes (Figure 6h). Considered together, these data suggested that *L. plantarum* MR1 altered the composition of gut microbiota and increased the abundance of *Lactobacillus*. Moreover, gut microbiota-derived acetate could act as a substrate for the uridine synthesis in the liver of Nile tilapia.

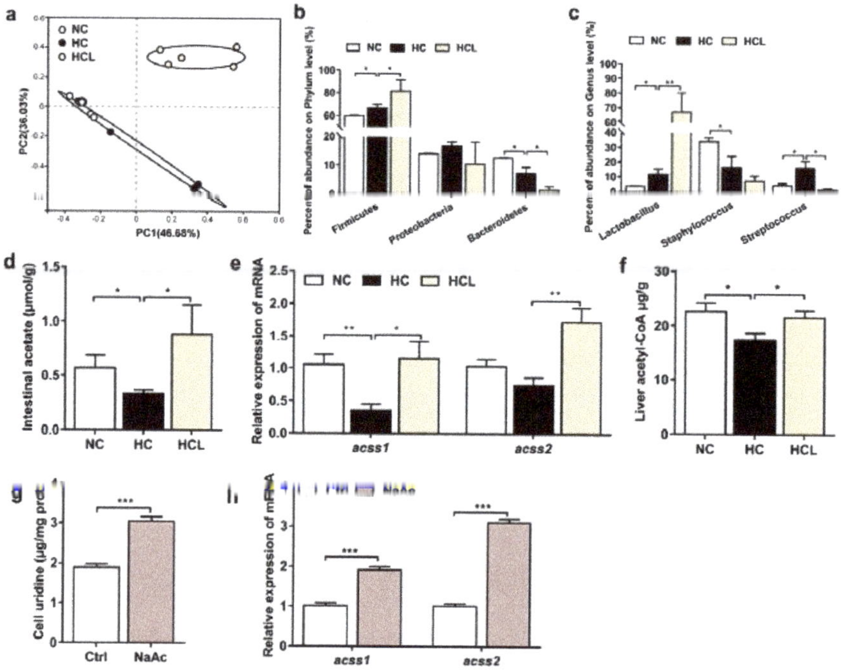

Figure 6. *L. plantarum* MR1 altered the gut microbiota composition. *L. plantarum* MR1-produced acetate promoted the uridine synthesis of Nile tilapia. (**a**) Principal coordinate analysis (PCoA) of the intestinal bacterial community. (**b**) Abundance of the gut bacteria at the phylum levels. (**c**) Abundance of the gut bacteria at the Genus levels. (**a–c**) $n = 6$ in NC and HC groups and $n = 5$ in HCL group. (**d**) Intestinal short-chain fatty acids (SCFAs) detection in Nile tilapia ($n = 6$). (**e**) Gene expression of *acss1* and *acss2* in the liver ($n = 6$). (**f**) Detection of the concentration of acetyl-CoA in the liver by HPLC ($n = 6$). (**g**) Analysis of uridine contents in the sodium acetate (NaAc)-treated hepatocytes ($n = 6$). (**h**) Gene expression of *acss1* and *acss2* in the hepatocytes ($n = 6$). Statistics were analyzed by one-way ANOVA with Tukey's adjustment (**a–f**) or Student's *t*-test (**g**,**h**), and presented as mean ± SEM (*, $p < 0.05$; **, $p < 0.01$; ***, $p < 0.001$).

4. Discussion

Host-microbiota mutualism is an integral part for the maintenance of health status. In addition, it is increasingly clear that the interaction between the host and intestinal microbiota has an impact on all tissues, in addition to the gut [38]. Beyond the fact that the components of bacteria have a direct influence on the host, intestinal microbiota can also metabolize dietary components to produce metabolites that may enter the circulation system to influence the host [39]. In the present study, we demonstrated that addition of *L. plantarum* MR1 altered the composition of gut microbiota and increased the contents of intestinal acetate to promote the production of uridine. We further detected higher contents of uridine in the NaAc-treated primary hepatocytes of the fish. The data suggested that the intestinal microbiota-derived acetate may promote the synthesis of uridine in the liver. Increased uridine may be responsible for the *L. plantarum* MR1 mediated beneficial effect, including protection against oxidative stress and lipid accumulation in the liver.

SCFAs are the most abundant dietary metabolites produced by gut microbiota. In addition, multiple studies indicated that they can regulate lipid and glucose metabolism in the host [40]. However, whether the microbiota and its metabolites can regulate the nucleotides synthesis of the host, as well as the mechanisms, are still unclear. It has been found that gut microbiome alteration is closely related to the change in the metabolism of nucleotides. For example, individual bat foraging on birds (by the great evening bat) had a higher abundance of Firmicutes and the Firmicutes–Bacteroidetes ratio, which may be associated with the metabolism of carbohydrates and nucleotides; however, the conclusion was drawn from the potential function prediction based on the sequencing data [41]. *L. plantarum* belongs to the phylum of Firmicutes, and we found a higher abundance of Firmicutes in the HCL group. It is well known that collapsing data to "phylum" level has generated a decade-long debate over controversial results of the relationship between the bacteria composition and host metabolic characteristics [42]. Therefore, elucidating the bacteria function will provide more solid evidence to understand the microbiota–host communications. We hypothesized that intestinal microbiota-derived acetate could be metabolized into acetyl-CoA to induce the synthesis of uridine in the liver. It has been reported that the loss of carnitine palmitoyl transferase 1 (CPT1) reduced acetyl-CoA, which originates from fatty acid and impaired de novo nucleotide synthesis, while supplementation with NaAc facilitated nucleotide synthesis in the endothelial cells [37]. Liver is generally considered as the major organ for the synthesis of uridine, which can enter the systemic circulation through the portal vein [43]. In the present study, we demonstrated that addition of sodium acetate in the primary hepatocytes of Nile tilapia could promote the uridine synthesis in vitro, and uridine may be responsible for exerting the antioxidant function and alleviating lipid accumulation directly. To our knowledge, this is the first study to show that intestinal bacteria-derived acetate contributed to the synthesis of uridine.

Uridine is the main pyrimidine nucleoside in the plasma of human and rodent. It is critical for RNA synthesis, glycogen synthesis, and many other cellular processes [44]. Moreover, it has been reported that uridine significantly decreases the accumulation of white adipose tissue and liver lipid by modifying gut microbiota composition in mice [45]. There are two possible mechanisms for the relationship between uridine and lipid metabolism. It has been found that uridine may alter the ratios of NAD+/NADH and NADP+/NADPH in the liver and modulate the protein acetylation profile to regulate lipid metabolism [46]. Furthermore, it has been found that inhibition of dihydroorotate dehydrogenase (DHODH) may cause micro-vesicular steatosis, which can be alleviated by the treatment with uridine; however, uridine has no effect on DHODH activity in vitro [46]. The exact regulatory mechanisms of uridine on glucose and lipid metabolism still remain unclear [44]. In the present study, the activation of AMPKα was increased in the HCL group, suggesting that uridine may reduce lipid deposition by activating the AMPKα signaling pathway [47].

More recently, uridine has been suggested to prevent the oxidative damage of lungs, which are caused by the coronavirus disease 2019 (COVID-19). The possible reason for this is the fact that uridine acts as the substrate of uridine-5′-diphosphate (UDP), which can prevent the excessive production of hydrogen peroxides [48]. Consistently, we proved that uridine could directly diminish H_2O_2-induced oxidative stress in the primary hepatocytes of Nile tilapia, suggesting that uridine had a resistance effect on oxidative stress. Furthermore, uridine could reduce lipid accumulation and oxidative stress of OA-induced damage in the hepatocytes.

5. Conclusions

In conclusion, these data demonstrate that addition of *L. plantarum* MR1 in high-carbohydrate diet alters the composition of intestinal microbiota, and thus increases the intestinal acetate which promotes the synthesis of uridine. Uridine exerts an antioxidant effect and decreases the hepatic lipid accumulation. Our study deepened the knowledge of gut microbiota and its metabolites, which are responsible for the alleviation of oxidative stress and lipid deposition. Overall, this suggests that gut microbiota-derived metabolites can act as a precursor substance for the metabolism of nucleotides to maintain the liver health in fish fed with high-carbohydrate diet.

Supplementary Materials: The following supporting information can be downloaded at: https://www.mdpi.com/article/10.3390/antiox11071238/s1. Supplementary Methods, Table S1: Formulations of the diets; Table S2: Primers used for RT-qPCR expression analysis; Table S3: The antibodies information for immunofluorescence and Western blotting assay; Figure S1: *L. plantarum* MR1 inhibited lipid accumulation-induced liver inflammation in Nile tilapia; Figure S2: Uridine suppressed the OA-induced inflammation in the primary hepatocytes of Nile tilapia.

Author Contributions: Conceptualization and design, M.-L.Z. and Z.-Y.D.; feeding of experiments, R.X. and T.W.; contribution to the collection of samples, T.W., F.-F.D. and N.-N.Z.; provision of essential materials and technical supports, M.-L.Z., Z.-Y.D., F.Q. and L.-Q.C.; data analysis and interpretation, M.-L.Z., Z.-Y.D. and R.X.; writing—review and editing, R.X., M.-L.Z. and Z.-Y.D. All authors contributed experimental assistance and intellectual input to this study. All authors have read and agreed to the published version of the manuscript.

Funding: This work was supported by the National Key Research and Development Program (grant number: 2018YFD0900400) and National Natural Science Foundation of China (grant number: 31972798).

Institutional Review Board Statement: The animal study protocol was approved by the Ethics of Animal Experiments of East China Normal University (F20190101).

Informed Consent Statement: Not applicable.

Data Availability Statement: The 16S rRNA gene sequence of L. plantarum MR1 was submitted to NCBI GenBank (https://www.ncbi.nlm.nih.gov/genbank/, accessed on 22 May 2022) with the accession number: OM535884. The microbiota raw data were deposited into the NCBI Sequence Read Archive (SRA) database (https://www.ncbi.nlm.nih.gov/sra, accessed on 22 May 2022) with accession number: PRJNA615286 and PRJNA805534. The metabolomic data were deposited and available at MetaboLights (https://www.ebi.ac.uk/metabolights/, accessed on 22 May 2022) under accession number: MTBLS4262. Data is contained within the article and supplementary materials.

Acknowledgments: We thank Renata C. Matos and François Leulier from institute de Génomique Fonctionnelle de Lyon in France for careful reading of the manuscript and professional suggestions.

Conflicts of Interest: The authors declare no conflict of interest.

Abbreviations

HC	high-carbohydrate diet
L. plantarum	lactobacillus plantarum
H&E	hematoxylin and eosin
ORO	oil red O
SCFAs	short-chain fatty acids
GPR	G-protein coupled receptor
FBW	final body weight
WGR	weight gain rate
VSI	visceral somatic index
HSI	hepatosomatic index
MFI	mesenteric fat index
CI	carcass index
AST	aspartate aminotransferase
ALT	alanine aminotransferase
PCoA	principal coordinate analysis
TG	triglyceride
NEFA	non-esterified free fatty acids
SOD	superoxide dismutase
MDA	malondialdehyde
GSH	reduced glutathione
H_2O_2	hydrogen peroxide
OA	oleic acid
NaAc	sodium acetate

References

1. Tan, B.L.; Norhaizan, M.E.; Liew, W.P. Nutrients and Oxidative Stress: Friend or Foe? *Oxid Med. Cell Longev.* **2018**, *2018*, 9719584. [CrossRef] [PubMed]
2. Fernández-Sánchez, A.; Madrigal-Santillán, E.; Bautista, M.; Esquivel-Soto, J.; Morales-González, A.; Esquivel-Chirino, C.; Durante-Montiel, I.; Sánchez-Rivera, G.; Valadez-Vega, C.; Morales-González, J.A. Inflammation, oxidative stress, and obesity. *Int. J. Mol. Sci.* **2011**, *12*, 3117–3132. [CrossRef] [PubMed]
3. Chiu, S.; Mulligan, K.; Schwarz, J.M. Dietary carbohydrates and fatty liver disease: De novo lipogenesis. *Curr. Opin. Clin. Nutr. Metab. Care* **2018**, *21*, 277–282. [CrossRef]
4. Semiane, N.; Foufelle, F.; Ferré, P.; Hainault, I.; Ameddah, S.; Mallek, A.; Khalkhal, A.; Dahmani, Y. High carbohydrate diet induces nonalcoholic steato-hepatitis (NASH) in a desert gerbil. *C. R. Biol.* **2017**, *340*, 25–36. [CrossRef]
5. Neuschwander-Tetri, B.A. Carbohydrate intake and nonalcoholic fatty liver disease. *Curr. Opin. Clin. Nutr. Metab. Care* **2013**, *16*, 446–452. [CrossRef]
6. Hall, K.D.; Chung, S.T. Low-carbohydrate diets for the treatment of obesity and type 2 diabetes. *Curr. Opin. Clin. Nutr. Metab. Care* **2018**, *21*, 308–312. [CrossRef]
7. Ceriello, A.; Genovese, S. Atherogenicity of postprandial hyperglycemia and lipotoxicity. *Rev. Endocr. Metab. Disord.* **2016**, *17*, 111–116. [CrossRef]
8. Savini, I.; Catani, M.V.; Evangelista, D.; Gasperi, V.; Avigliano, L. Obesity-associated oxidative stress: Strategies finalized to improve redox state. *Int. J. Mol. Sci.* **2013**, *14*, 10497–10538. [CrossRef]
9. Hemre, G.I.; Mommsen, T.P.; Krogdahl, Å. Carbohydrates in fish nutrition: Effects on growth, glucose metabolism and hepatic enzymes. *Aquac. Nutr.* **2002**, *8*, 175–194. [CrossRef]
10. Kamalam, B.S.; Medale, F.; Panserat, S. Utilisation of dietary carbohydrates in farmed fishes: New insights on influencing factors, biological limitations and future strategies. *Aquaculture* **2017**, *467*, 3–27. [CrossRef]
11. Prisingkorn, W.; Prathomya, P.; Jakovlić, I.; Liu, H.; Zhao, Y.H.; Wang, W.M. Transcriptomics, metabolomics and histology indicate that high-carbohydrate diet negatively affects the liver health of blunt snout bream (Megalobrama amblycephala). *BMC Genom.* **2017**, *18*, 856. [CrossRef] [PubMed]
12. Ma, H.J.; Mou, M.M.; Pu, D.C.; Lin, S.M.; Chen, Y.J.; Luo, L. Effect of dietary starch level on growth, metabolism enzyme and oxidative status of juvenile largemouth bass, Micropterus salmoides. *Aquaculture* **2019**, *498*, 482–487. [CrossRef]
13. Teame, T.; Zhang, Z.; Ran, C.; Zhang, H.; Yang, Y.; Ding, Q.; Xie, M.; Gao, C.; Ye, Y.; Duan, M.; et al. The use of zebrafish (Danio rerio) as biomedical models. *Anim. Front.* **2019**, *9*, 68–77. [CrossRef] [PubMed]
14. López Nadal, A.; Ikeda-Ohtsubo, W.; Sipkema, D.; Peggs, D.; McGurk, C.; Forlenza, M.; Wiegertjes, G.F.; Brugman, S. Feed, Microbiota, and Gut Immunity: Using the Zebrafish Model to Understand Fish Health. *Front. Immunol.* **2020**, *11*, 114. [CrossRef] [PubMed]
15. Riddle, M.R.; Aspiras, A.C.; Gaudenz, K.; Peuß, R.; Sung, J.Y.; Martineau, B.; Peavey, M.; Box, A.C.; Tabin, J.A.; McGaugh, S.; et al. Insulin resistance in cavefish as an adaptation to a nutrient-limited environment. *Nature* **2018**, *555*, 647–651. [CrossRef]
16. Fan, Y.; Pedersen, O. Gut microbiota in human metabolic health and disease. *Nat. Rev. Microbiol.* **2021**, *19*, 55–71. [CrossRef]
17. Sun, X.; Seidman, J.S.; Zhao, P.; Troutman, T.D.; Spann, N.J.; Que, X.; Zhou, F.; Liao, Z.; Pasillas, M.; Yang, X.; et al. Neutralization of Oxidized Phospholipids Ameliorates Non-alcoholic Steatohepatitis. *Cell Metab.* **2020**, *31*, 189–206.e8. [CrossRef]
18. Borrelli, A.; Bonelli, P.; Tuccillo, F.M.; Goldfine, I.D.; Evans, J.L.; Buonaguro, F.M.; Mancini, A. Role of gut microbiota and oxidative stress in the progression of non-alcoholic fatty liver disease to hepatocarcinoma: Current and innovative therapeutic approaches. *Redox Biol.* **2018**, *15*, 467–479. [CrossRef]
19. Leung, C.; Rivera, L.; Furness, J.B.; Angus, P.W. The role of the gut microbiota in NAFLD. *Nat. Rev. Gastroenterol. Hepatol.* **2016**, *13*, 412–425. [CrossRef]
20. Puri, P.; Sanyal, A.J. The Intestinal Microbiome in Nonalcoholic Fatty Liver Disease. *Clin. Liver Dis.* **2018**, *22*, 121–132. [CrossRef]
21. Tomasello, G.; Mazzola, M.; Leone, A.; Sinagra, E.; Zummo, G.; Farina, F.; Damiani, P.; Cappello, F.; Gerges Geagea, A.; Jurjus, A.; et al. Nutrition, oxidative stress and intestinal dysbiosis: Influence of diet on gut microbiota in inflammatory bowel diseases. *Biomed. Pap.* **2016**, *160*, 461–466. [CrossRef] [PubMed]
22. Bernini, L.J.; Simão, A.N.C.; de Souza, C.H.B.; Alfieri, D.F.; Segura, L.G.; Costa, G.N.; Dichi, I. Effect of Bifidobacterium lactis HN019 on inflammatory markers and oxidative stress in subjects with and without the metabolic syndrome. *Br. J. Nutr.* **2018**, *120*, 645–652. [CrossRef] [PubMed]
23. Park, E.J.; Lee, Y.S.; Kim, S.M.; Park, G.S.; Lee, Y.H.; Jeong, D.Y.; Kang, J.; Lee, H.J. Beneficial Effects of Lactobacillus plantarum Strains on Non-Alcoholic Fatty Liver Disease in High Fat/High Fructose Diet-Fed Rats. *Nutrients* **2020**, *12*, 542. [CrossRef]
24. Gao, Y.; Liu, Y.; Ma, F.; Sun, M.; Song, Y.; Xu, D.; Mu, G.; Tuo, Y. Lactobacillus plantarum Y44 alleviates oxidative stress by regulating gut microbiota and colonic barrier function in Balb/C mice with subcutaneous d-galactose injection. *Food Funct.* **2021**, *12*, 373–386. [CrossRef] [PubMed]
25. Zhao, J.; Tian, F.; Yan, S.; Zhai, Q.; Zhang, H.; Chen, W. Lactobacillus plantarum CCFM10 alleviating oxidative stress and restoring the gut microbiota in d-galactose-induced aging mice. *Food Funct.* **2018**, *9*, 917–924. [CrossRef]
26. Saeedi, B.J.; Liu, K.H.; Owens, J.A.; Hunter-Chang, S.; Camacho, M.C.; Eboka, R.U.; Chandrasekharan, B.; Baker, N.F.; Darby, T.M.; Robinson, B.S.; et al. Gut-Resident Lactobacilli Activate Hepatic Nrf2 and Protect Against Oxidative Liver Injury. *Cell Metab.* **2020**, *31*, 956–968. [CrossRef]

27. He, J.; Zhang, P.; Shen, L.; Niu, L.; Tan, Y.; Chen, L.; Zhao, Y.; Bai, L.; Hao, X.; Li, X.; et al. Short-Chain Fatty Acids and Their Association with Signalling Pathways in Inflammation, Glucose and Lipid Metabolism. *Int. J. Mol. Sci.* **2020**, *21*, 6356. [CrossRef] [PubMed]
28. Guo, W.; Liu, J.; Sun, J.; Gong, Q.; Ma, H.; Kan, X.; Cao, Y.; Wang, J.; Fu, S. Butyrate alleviates oxidative stress by regulating NRF2 nuclear accumulation and H3K9/14 acetylation via GPR109A in bovine mammary epithelial cells and mammary glands. *Free Radic. Biol. Med.* **2020**, *152*, 728–742. [CrossRef]
29. Ding, Q.; Zhang, Z.; Li, Y.; Liu, H.; Hao, Q.; Yang, Y.; Ringø, E.; Olsen, R.E.; Clarke, J.L.; Ran, C.; et al. Propionate induces intestinal oxidative stress via Sod2 propionylation in zebrafish. *iScience* **2021**, *24*, 102515. [CrossRef]
30. Hernández, M.A.G.; Canfora, E.E.; Jocken, J.W.E.; Blaak, E.E. The Short-Chain Fatty Acid Acetate in Body Weight Control and Insulin Sensitivity. *Nutrients* **2019**, *11*, 1943. [CrossRef]
31. den Besten, G.; van Eunen, K.; Groen, A.K.; Venema, K.; Reijngoud, D.J.; Bakker, B.M. The role of short-chain fatty acids in the interplay between diet, gut microbiota, and host energy metabolism. *J. Lipid Res.* **2013**, *54*, 2325–2340. [CrossRef]
32. Morrison, D.J.; Preston, T. Formation of short chain fatty acids by the gut microbiota and their impact on human metabolism. *Gut Microbes* **2016**, *7*, 189–200. [CrossRef] [PubMed]
33. Wang, M.; Lu, M. Tilapia polyculture: A global review. *Aquac. Res.* **2016**, *47*, 2363–2374. [CrossRef]
34. Brawand, D.; Wagner, C.E.; Li, Y.I.; Malinsky, M.; Keller, I.; Fan, S.; Simakov, O.; Ng, A.Y.; Lim, Z.W.; Bezault, E.; et al. The genomic substrate for adaptive radiation in African cichlid fish. *Nature* **2014**, *513*, 375–381. [CrossRef] [PubMed]
35. Xu, R.; Li, M.; Wang, T.; Zhao, Y.W.; Shan, C.J.; Qiao, F.; Chen, L.Q.; Zhang, W.B.; Du, Z.Y.; Zhang, M.L. Bacillus amyloliquefaciens ameliorates high-carbohydrate diet-induced metabolic phenotypes by restoration of intestinal acetate-producing bacteria in Nile Tilapia. *Br. J. Nutr.* **2021**, *127*, 653–666. [CrossRef]
36. Jiao, J.G.; Liu, Y.; Zhang, H.; Li, L.Y.; Qiao, F.; Chen, L.Q.; Zhang, M.L.; Du, Z.Y. Metabolism of linoleic and linolenic acids in hepatocytes of two freshwater fish with different n-3 or n-6 fatty acid requirements. *Aquaculture* **2020**, *515*, 734595. [CrossRef]
37. Schoors, S.; Bruning, U.; Missiaen, R.; Queiroz, K.C.; Borgers, G.; Elia, I.; Zecchin, A.; Cantelmo, A.R.; Christen, S.; Goveia, J.; et al. Fatty acid carbon is essential for dNTP synthesis in endothelial cells. *Nature* **2015**, *520*, 192–197. [CrossRef]
38. Trompette, A.; Gollwitzer, E.S.; Yadava, K.; Sichelstiel, A.K.; Sprenger, N.; Ngom-Bru, C.; Blanchard, C.; Junt, T.; Nicod, L.P.; Harris, N.L.; et al. Gut microbiota metabolism of dietary fiber influences allergic airway disease and hematopoiesis. *Nat. Med.* **2014**, *20*, 159–166. [CrossRef]
39. Mardinoglu, A.; Wu, H.; Bjornson, E.; Zhang, C.; Hakkarainen, A.; Räsänen, S.M.; Lee, S.; Mancina, R.M.; Bergentall, M.; Pietiläinen, K.H.; et al. An Integrated Understanding of the Rapid Metabolic Benefits of a Carbohydrate-Restricted Diet on Hepatic Steatosis in Humans. *Cell Metab.* **2018**, *27*, 559–571. [CrossRef]
40. Agus, A.; Clément, K.; Sokol, H. Gut microbiota-derived metabolites as central regulators in metabolic disorders. *Gut* **2021**, *70*, 1174–1182. [CrossRef]
41. Bagheri, M.; Shah, R.D.; Mosley, J.D.; Ferguson, J.F. A metabolome and microbiome wide association study of healthy eating index points to the mechanisms linking dietary pattern and metabolic status. *Eur. J. Nutr.* **2021**, *60*, 4413–4427. [CrossRef] [PubMed]
42. Wu, G.; Zhao, N.; Zhang, C.; Lam, Y.Y.; Zhao, L. Guild-based analysis for understanding gut microbiome in human health and diseases. *Genome Med.* **2021**, *13*, 22. [CrossRef] [PubMed]
43. Deng, Y.; Wang, Z.V.; Gordillo, R.; An, Y.; Zhang, C.; Liang, Q.; Yoshino, J.; Cautivo, K.M.; De Brabander, J.; Elmquist, J.K.; et al. An adipo-biliary-uridine axis that regulates energy homeostasis. *Science* **2017**, *355*, eaaf5375. [CrossRef]
44. Zhang, Y.; Guo, S.; Xie, C.; Fang, J. Uridine Metabolism and Its Role in Glucose, Lipid, and Amino Acid Homeostasis. *BioMed Res. Int.* **2020**, *2020*, 7091718. [CrossRef] [PubMed]
45. Liu, Y.; Xie, C.; Zhai, Z.; Deng, Z.Y.; De Jonge, H.R.; Wu, X.; Ruan, Z. Uridine attenuates obesity, ameliorates hepatic lipid accumulation and modifies the gut microbiota composition in mice fed with a high-fat diet. *Food Funct.* **2021**, *12*, 1829–1840. [CrossRef]
46. Le, T.T.; Ziemba, A.; Urasaki, Y.; Hayes, E.; Brotman, S.; Pizzorno, G. Disruption of uridine homeostasis links liver pyrimidine metabolism to lipid accumulation. *J. Lipid Res.* **2013**, *54*, 1044–1057. [CrossRef]
47. Ran, C.; Xie, M.; Li, J.; Xie, Y.; Ding, Q.; Li, Y.; Zhou, W.; Yang, Y.; Zhang, Z.; Olsen, R.E.; et al. Dietary Nucleotides Alleviate Hepatic Lipid Deposition via Exogenous AMP-Mediated AMPK Activation in Zebrafish. *J. Nutr.* **2021**, *151*, 2986–2996. [CrossRef]
48. Mironova, G.D.; Belosludtseva, N.V.; Ananyan, M.A. Prospects for the use of regulators of oxidative stress in the comprehensive treatment of the novel Coronavirus Disease 2019 (COVID-19) and its complications. *Eur. Rev. Med. Pharmacol. Sci.* **2020**, *24*, 8585–8591. [CrossRef]

antioxidants

MDPI

Article

Drinking Water Supplemented with Acidifiers Improves the Growth Performance of Weaned Pigs and Potentially Regulates Antioxidant Capacity, Immunity, and Gastrointestinal Microbiota Diversity

Qing-Lei Xu [1,†], Chang Liu [1,†], Xiao-Jian Mo [2], Meng Chen [1], Xian-Le Zhao [1], Ming-Zheng Liu [1], Shu-Bai Wang [3], Bo Zhou [1,*] and Cheng-Xin Zhao [2,*]

[1] College of Animal Science and Technology, Nanjing Agricultural University, Nanjing 210095, China; 2019205004@njau.edu.cn (Q.-L.X.); 2019805121@njau.edu.cn (C.L.); 2020805124@stu.njau.edu.cn (M.C.); 2020805123@stu.njau.edu.cn (X.-L.Z.); 2020205018@stu.njau.edu.cn (M.-Z.L.)

[2] Yantai Jinhai Pharmaceutical Co., Ltd., Yantai 265323, China; moxiaojian@jinhaiyaoye.com

[3] College of Animal Science and Technology, Qingdao Agricultural University, Qingdao 266109, China; wangshubai@qau.edu.cn

* Correspondence: zhoubo@njau.edu.cn (B.Z.); zhaochengxin@jinhaiyaoye.com (C.-X.Z.)

† These authors contributed equally to this work.

Citation: Xu, Q.-L.; Liu, C.; Mo, X.-J.; Chen, M.; Zhao, X.-L.; Liu, M.-Z.; Wang, S.-B.; Zhou, B.; Zhao, C.-X. Drinking Water Supplemented with Acidifiers Improves the Growth Performance of Weaned Pigs and Potentially Regulates Antioxidant Capacity, Immunity, and Gastrointestinal Microbiota Diversity. *Antioxidants* **2022**, *11*, 809. https://doi.org/10.3390/antiox11050809

Academic Editors: Min Xue, Junmin Zhang, Zhenyu Du, Jie Wang and Wei Si

Received: 25 March 2022
Accepted: 20 April 2022
Published: 21 April 2022

Abstract: This study evaluated the potential effects of adding acidifiers to the drinking water on the growth performance, complete blood count, antioxidant indicators, and diversity of gastrointestinal microbiota for weaned pigs. A total of 400 weaned pigs were randomly divided into four treatments. Pigs were fed the same basal diet and given either water (no acidifier was added, control) or water plus blends of different formulas of acidifiers (acidifier A1, A2, or A3) for 35 days. On d 18 and 35 of the experimental period, 64 pigs (four pigs per pen) were randomly selected to collect blood for a CBC test (n = 128) and an antioxidant indicators test (n = 128); 24 pigs (six pigs per group) were randomly selected to collect fresh feces (n = 48) from the rectum for 16S rRNA gene sequencing. Compared to the control, supplementing the drinking water with acidifiers improved the growth performance and survival rate of weaned pigs. Acidifier groups also increased serum catalase (CAT) and total antioxidant capacity (T-AOC) activities, while also displaying a decreased malondialdehyde (MDA) concentration compared to the control. The relative abundance of *Firmicutes* in the acidifier A1 group was greater than that in the control group ($p < 0.05$) on d 35; the relative abundance of *Lactobacillus* in the acidifier A1 group was greater than that in the control group ($p < 0.05$) on d 18 and 35. The microbial species *Subdoligranulum* or *Ruminococcaceae_UCG-005* had significantly positive correlations with ADG and ADFI or with serum antioxidant indicators, respectively. These findings suggest that supplementing the drinking water with an acidifier has a potential as an antioxidant, which was reflected in the improvement of growth performance, immunity, antioxidant capacity, and intestinal flora.

Keywords: weaned pigs; acidifier; antioxidants; animal feed; intestinal flora

1. Introduction

Segregated early weaning (SEW) improved the fertility of sows and reduced the transmission of diseases from sows to piglets, which improved the production efficiency and provide economic benefits to the pig industry [1,2]. However, due to immature digestive organs, insufficient secretion of gastric acid and digestive enzymes, and low immunity of early weaned pigs, they often display early weaning stress syndrome, which is characterized by digestive dysfunction, growth retardation, and diarrhea [3,4]. After weaning, pigs are vulnerable to the invasion of pathogens, resulting in a dysfunction of the intestinal flora and immune system due to the changes of environment, feed, and other factors [5].

In the past, antibiotics had played an important role in the prevention and treatment of a pig's diseases and the reduction of weaning stress [6,7]. However, long-term abuse of antibiotics affected the diversity of intestinal microbiota in newborn piglets and beneficial bacterial colonization, which led to subsequent intestinal diseases [8–10]. In addition, long-term abuse of antibiotics not only produces antibiotic-resistant bacteria, but also causes antibiotic residues in animal-derived food products and the environment [11,12]. Following the ban of antibiotic growth promoters, new alternatives are required to solve these problems.

As non-toxic, pollution-free, and non-resistance functional feed additives, acidifiers have played important roles in the improvement of animal growth, immunity, and intestinal health [13]. Supplementation with acidifiers improved the activity of pepsin and enhanced the digestion and absorption of nutrients in the gastrointestinal tract of weaned pigs [14]. In addition, acidifiers increased feed intake by inducing and stimulating taste buds, thus promoting the growth performance of pigs [8,15]. Acidifiers also improved the antioxidant capacity of piglets [16,17] and reduced the diarrhea caused by weaning stress [18]. Acidifiers decreased the pH value of the gastrointestinal tract [19], changed the composition of intestinal microorganisms, and inhibited the harmful microorganisms which are sensitive to low pH (such as *Enterobacteriaceae*) [20]. Therefore, acidifiers are increasingly being added to feed as an antibiotic-free growth promoter to replace antibiotics.

Previous studies investigated acidifiers as feed additives [21,22]. However, when the feed contains more protein or minerals with high buffer capacity, the effect of acidifiers may be weakened. At the same time, the acidifier was neutralized by the alkaline substances in the feed, thus losing the acidification effect [23]. The liquid acidifier can be added to the drinking water through the dosing device of the drinking water pipe, which is more convenient to add when needed. Moreover, acidified drinking water had the same effect as adding it to the feed [24,25].

Common drinking water compound organic acidifiers contained formic acid, acetic acid, and propionic acid [26]. As an important component of an organic acidifier, lactic acid decreased intestinal *Salmonella* and improved intestinal health [27]. Stevioside was widely used as a sweetener in feed additives to improve the feed intake of livestock and poultry [28,29]. Therefore, the compound organic acidifier containing formic acid, acetic acid, and propionic acid was selected as one treatment. On this basis, lactic acid was added as another treatment. In order to improve the taste of the drinking water, stevioside was added as the third treatment, and water without any additions was set as the control group. The effectiveness of acidified drinking water, containing a large number of organic acids, is largely based on the pH being lowered to a level of 4.0, at which *Enterobacteriaceae* cannot multiply [30]. Therefore, the acidified drinking water of each treatment group contains different proportions of organic acids, so as to adjust the pH value to 4.0. The present experiment aimed to evaluate the effects of acidified drinking water on the growth performance, immunity parameters, antioxidant capacity, and intestinal microflora of weaned pigs, and then compare which types of acidifiers are more suitable to be added to the drinking water of weaned pigs as an antibiotic substitute.

2. Materials and Methods

2.1. Pigs, Experimental Design, and Housing

This study was approved by the Nanjing Agricultural University Animal Care and Use Committee (SYXK Su 2017-0027). In this experiment, a total of 400 weaned pigs at 28 d of age were randomly divided into four groups: (1) acidifiers A1 group (continuous supply of acidified drinking water with 19% formic acid +19% acetic acid +3.5% propionic acid +15% lactic acid, pH = 4); (2) acidifiers A2 group (continuous supply of acidified drinking water with 22% formic acid +16% acetic acid +4% propionic acid, pH = 4); (3) acidifiers A3 group (continuous supply of acidified drinking water with 22% formic acid +16% acetic acid +4% propionic acid +0.5% stevioside, pH = 4); (4) control group (continuous supply of non-treated water, pH = 7). There were four replicates (pens) in each group, and 25 pigs per pen; the acidifiers were used for 8 h every day and the pigs were free to drink. A ratio of acidifier: water = 1:500 was used in this study.

All pens were equipped with slatted floors, stainless-steel vibrating feeders, and cup drinking bowls. The temperature was automatically controlled between 22 to 27 °C by air-exhaust fans, evaporative cooling pads, and hot blast heaters. A water-powered proportional dosing pump (DOSATRON, INTERNATIONAL, Bordeaux, France) was used to supplement the liquid acidifiers into the drinking water for 35 d before the end of the nursery period. According to the standard of NRC (2012), a corn-soybean meal basal diet was formulated to meet the nutritional needs of pigs (Table S1). Feed and water were provided ad libitum. The experimental pigs were immunized according to the specified immunization procedure.

2.2. Growth Performance and Diarrhea Rate

Pigs were weighed at d 0 and 35 of the experiment and the data was used to calculate the average daily gain (ADG). Average daily feed intake (ADFI) was recorded to calculate feed-to-gain ratio (F:G). The survival rates of pig were determined at the end of the experiment.

The diarrhea scores of pigs were recorded from 08:00 to 09:00 every day according to the criteria: 1-solid, well-formed feces; 2-loose and shapeless feces; 3-runny feces; and 4-watery diarrhea [31]. Diarrheal symptoms and mortality, if any, were recorded daily for each labelled pig during the trial period. The percentage of diarrhea occurrence in the total number of pigs in a pen (diarrhea severity), diarrhea duration (d), and feces score were calculated.

2.3. Complete Blood Count (CBC) Test

On d 18 and 35 of the experimental period, 64 pigs (four pigs per pen) were randomly selected to collect 5 mL of blood via jugular venipuncture. The blood was then divided equally into two parts: one for a CBC test (n = 128) and the other for the determination of antioxidant indicators (n = 128). A total of 23 CBC indicators of the whole blood samples were determined using animal specialty automatic physiological analysis instruments (Mindray, BC-5000 Vet, Shenzhen Mindray Biomedical Electronics Co., Ltd., Shenzhen, China) within several hours. The serum samples were obtained by centrifuging (4000× g for 10 min) at 4 °C, and stored at −80 °C until analysis.

2.4. Antioxidant Indicators Measurements

All the serum samples (n = 128) were used for antioxidant activity analysis. Antioxidant indicators, such as malondialdehyde (MDA), total superoxide dismutase (SOD), glutathione (GSH), glutathione peroxidase (GSH-Px), catalase (CAT), and total antioxidant capacity (T-AOC), were determined using ELISA Kits (Shanghai Langdun Biotechnology Co., Ltd., Shanghai, China) with assay sensitivities of 99.0% according to the manufacturer's instructions. Briefly, we use a competitive method to detect the content of MDA, SOD, GSH, GSH-Px, and CAT in the sample. The samples were added to the ELISA plates that were pre-coated with antibodies, then biotin-labeled antigens were added, and the plates were incubated at 37 °C for 30 min. The two compete with antibodies to form immune complexes. The unbound biotin-labeled antigens are removed by rinsing in PBS with 0.15% Tween 20 (PBST), then horseradish peroxidase avidin (Avidin-HRP) is added, and the plates are incubated at 37 °C for 30 min. Avidin-HRP binds to biotin-labeled antigens, and, after washing, the bound HRP enzyme catalyzes tetramethylbenzidine (TMB) into a blue dye and then into yellow under the action of acid. We measured the absorbance (OD value) of each well at a wavelength of 450 nm within 10 min and calculated the test samples according to the standard curve.

T-AOC was measured using a ferric reducing ability of power (FRAP) assay. Briefly, antioxidants react with the ferric tripyridyl triazine (Fe^{3+}-TPTZ) complex and convert it into ferrous tripyridyl triazine (Fe^{2+}-TPTZ) under acidic conditions. The absorbance of T-AOC was measured at 593 nm. The reaction absorbance for MDA, SOD, GSH, GSH-Px, CAT, and T-AOC were measured using a microplate reader (Tecan, Austria GmbH, Grödig, Austria). The sensitivities of MDA, SOD, GSH, GSH-Px, and CAT were 0.15 nmol/mL, 0.1 ng/mL, 10 μg/mL, 0.3 ng/mL, and 0.2 ng/mL, respectively. Each sample was tested three times. The intra- and inter-assay coefficients of variation were less than 10%.

2.5. DNA Extraction and 16S rRNA Gene Sequencing

On d 18 and 35 of the experiment period, 24 pigs (six pigs per group) were randomly selected to collect fresh feces from the rectum. The feces samples (n = 48) were immediately frozen in liquid nitrogen until DNA extraction. Total bacterial genomic DNA was extracted using a stool DNA kit (Omega, Bio-Tek Inc., Norcross, GA, USA) according to the manufacturer's protocol. The hypervariable region V3–V4 of the microbial 16S rRNA gene (accession numbers: SAMN26994143 to SAMN26994190) was amplified by PCR with indices and adaptors-linked universal primers (F: 5′-ACTCCTACGGGAGGCAGCAG-3′; R: 5′-GGACTACHVGGGTWT-CTAAT-3′). The PCR products were confirmed with 2% agarose gel electrophoresis, purified with AMPure XT beads (Beckman Coulter Genomics, Danvers, MA, USA), and then quantified by an Invitrogen Qubit 4.0 fluorometer (Invitrogen, Thermo Fisher Scientific, Waltham, MA, USA). The amplicon pools were prepared for sequencing, and the quantity of the amplicon library was assessed on the Agilent 2100 bioanalyzer (Agilent Technologies, Palo Alto, CA, USA) and the Library Quantification Kit for Illumina (Kapa Biosciences, Woburn, MA, USA). Amplicon libraries were sequenced on the Illumina MiSeq platform (Illumina, San Diego, CA, USA) for paired-end sequencing of 2 × 300 bp reads according to the manufacturer's recommendations. The raw paired-end reads were truncated by removing the barcode and primer sequence. Quality filtering on the raw tags was performed to obtain high-quality clean tags using QIIME2 software [32]. The high-quality clean tags were clustered into operational taxonomic units (OTUs) with a similarity threshold of 0.97. Representative sequences were selected for each OUT, and the RDP classifier was used to further classify OTUs with representative sequences at an 0.80 confidence level [33]. Principal coordinate analysis (PCoA) plots were generated according to the unweighted UniFrac distance metrics [34]. The number of observed species and the indices of Chao 1 (species richness), as well as Shannon and Simpson (diversity), were calculated to estimate alpha diversity using QIIME 2 [32].

2.6. Statistical Analyses

Data of growth performances, the CBC test, antioxidant indicators, and fecal microbiota were analyzed using the GLIMMIX procedure of SAS 9.4 (SAS Institute Inc., Cary, NC, USA) with treatment, gender, time, and their interaction as fixed effects, and pen as a random effect according to the completely randomized design. Appropriate post-test comparisons for means were made for multiple groups using the Bonferroni Multiple Comparisons Test. A partial correlation analysis between the gut microbiota and antioxidant activity and growth performance indicators was carried out using R statistical software. The results are expressed as means ± standard error of the mean (SEM). Differences were considered significant at $p < 0.05$, while differences were considered to show a tendency at $0.05 \leq p < 0.10$.

3. Results

3.1. Growth Performance

No serious adverse events were observed during the whole experiment period. The effects of acidifiers on the growth performances of weaned pigs are presented in Table 1. The FBW in the acidifier groups showed an increased trend compared with that in the control group ($0.05 < p < 0.10$). The ADG of the A1 group was greater than that of the control group ($p < 0.05$). The ADFI of the A1 group was also greater than that of the control group ($p < 0.05$). However, no significant difference was found in the F:G between groups ($p > 0.05$).

Table 1. Effects of acidifiers on the growth performances of weaned pigs [1].

Items	Treatment [2]				SEM [3]	*p*-Value
	Control	A1	A2	A3		
IBW, kg	9.20 [c]	10.41 [a]	10.02 [b]	9.05 [c]	0.14	0.002
FBW, kg	24.81 [b]	27.16 [a]	26.01 [a,b]	25.38 [a,b]	0.42	0.091
ADG, g	445.29 [b]	512.25 [a]	478.81 [a,b]	462.26 [b]	12.85	0.040
ADFI, g	645.40 [b]	743.80 [a]	696.20 [a,b]	653.50 [b]	11.35	0.024
F:G	1.45	1.45	1.46	1.41	0.009	0.130

IBW, Initial body weight; FBW, Final body weight; ADG, average daily gain; ADFI, average daily feed intake; G:F, gain-to-feed ratio. [a,b,c] Different superscript within a row indicate a significant difference ($p < 0.05$). [1] Covariance analysis was used to correct the effect of different initial weights on average daily gain. The initial weight was corrected to 9.7026 kg. [2] A1, acidifiers A1 group, continuous supply of acidified drinking water with 19% formic acid +19% acetic acid +3.5% propionic acid +15% lactic acid; A2, acidifiers A2 group, continuous supply of acidified drinking water with 22% formic acid +16% acetic acid +4% propionic acid; A3, acidifiers A3 group, continuous supply of acidified drinking water with 22% formic acid +16% acetic acid +4% propionic acid +0.5% stevioside; control, continuous supply of non-treated water. [3] SEM means standard error of the means ($n = 100$).

3.2. Diarrhea and Survival Rate

The effects of acidifiers on the diarrhea and survival rate of weaned pigs are shown in Table 2. There was no significant difference in diarrhea rates between the acidifier treatment groups and the control group during the experimental period ($p > 0.05$). Nonetheless, the diarrhea rates of pigs in the A1, A2, and A3 groups were decreased by 7.5%, 2.9%, and 6.5% compared with the control group during the entire experimental period, respectively. Although there were also no significant differences in the survival rate of pigs between the A1, A3, and control groups ($p > 0.05$), there was a significant improvement ($p = 0.02$) on the survival rate of pigs between the A2 group and the control group in the Multiple Comparisons test. In summary, drinking water supplementation with acidifiers had a greater survival rate in the A2 group and numerically lower diarrhea rates than the control group at the end of the trial period.

Table 2. Effects of acidifiers on the diarrhea and survival rate of weaned pigs.

Items	Treatment [1]				SEM [2]	*p*-Value
	Control	A1	A2	A3		
			Diarrhea rate, %			
d 1 to 18	3.11	2.72	2.95	2.78	0.30	0.802
d 19 to 35	3.00	2.94	3.00	2.94	0.23	0.995
d 1 to 35	3.06	2.83	2.97	2.86	0.25	0.905
			Survival rate, %			
d 1 to 35	93.00 [b]	96.00 [a,b]	99.00 [a]	95.00 [a,b]	1.56	0.107

[a,b] Different superscripts within a row indicate a significant difference ($p < 0.05$). [1] A1, acidifiers A1 group, continuous supply of acidified drinking water with 19% formic acid +19% acetic acid +3.5% propionic acid +15% lactic acid; A2, acidifiers A2 group, continuous supply of acidified drinking water with 22% formic acid +16% acetic acid +4% propionic acid; A3, acidifiers A3 group, continuous supply of acidified drinking water with 22% formic acid +16% acetic acid +4% propionic acid +0.5% stevioside; control, continuous supply of non-treated water. [2] SEM means standard error of the means ($n = 100$).

3.3. CBC Test Indicators

The effects of acidifiers on the CBC test indicators of weaned pigs are shown in Table 3. On d 18, Neu in the A1 and A2 groups was less than that in the A3 group ($p < 0.05$); Neu%, HCT, and MCV in the A1 and A2 groups were less than those in the control group ($p < 0.05$); MCH, MCHC, and RDW-SD in the A1 and A2 groups were greater than those in the control group ($p < 0.05$); Mon % in the A1 group was greater than those in the other three groups ($p < 0.05$); EOS, RBC, MPV, and PDW in the A1 group were less than those in the control group ($p < 0.05$); RDW-CV in the A1 group was greater than those in control group ($p < 0.05$); LYM% in the A2 group was greater than that in the control group ($p < 0.05$). There was no significant difference in other blood routine indices among the four groups ($p > 0.05$).

Table 3. Effects of acidifiers on the complete blood count (CBC) test of weaned pigs.

Items [1]	Treatment [2]				SEM [3]	*p* Value
	Control	A1	A2	A3		
			d18			
WBC, 10^9/L	21.17	18.56	17.92	20.50	2.09	0.653
Neu, 10^9/L	6.59 ^{a,b}	2.74 ^{b}	2.96 ^{b}	8.65 ^{a}	1.39	0.007
LYM, 10^9/L	12.34	12.50	12.98	10.18	1.41	0.507
Mon, 10^9/L	1.79 ^{a,b}	3.12 ^{a}	1.84 ^{a,b}	1.26 ^{b}	0.55	0.109
Eos, 10^9/L	0.37 ^{a}	0.15 ^{b}	0.13 ^{b}	0.31 ^{a,b}	0.07	0.057
Bas, 10^9/L	0.09	0.06	0.07	0.10	0.02	0.107
Neu, %	31.52 ^{a}	13.93 ^{b}	14.08 ^{b}	37.08 ^{a}	2.48	<0.001
LYM, %	57.66 ^{b,c}	68.46 ^{a,b}	73.65 ^{a}	55.00 ^{c}	4.42	0.010
Mon, %	8.69 ^{c}	15.48 ^{a}	11.46 ^{b}	6.01 ^{c}	1.18	<0.001
Eos, %	1.77 ^{a}	0.89 ^{b,c}	0.56 ^{c}	1.50 ^{a,b}	0.27	0.008
Bas, %	0.36 ^{a,b}	0.26 ^{b}	0.26 ^{b}	0.41 ^{a}	0.05	0.042
RBC, 10^9/L	5.30 ^{a}	4.35 ^{b}	4.53 ^{b}	5.63 ^{a}	0.25	<0.001
HGB, g/L	100.56	103.30	103.01	104.92	5.45	0.366
HCT, %	30.57 ^{a}	19.80 ^{b}	21.58 ^{b}	31.89 ^{a}	1.40	<0.001
MCV, fL	57.66 ^{a}	45.50 ^{b}	46.79 ^{b}	56.18 ^{a}	0.85	0.000
MCH, pg	18.96 ^{b}	22.20 ^{a}	21.83 ^{a}	18.53 ^{b}	0.58	<0.001
MCHC, g/L	329.06 ^{c}	477.67 ^{a}	462.67 ^{b}	330.13 ^{c}	12.99	0.000
RDW-CV, %	19.41 ^{c}	36.23 ^{a}	33.28 ^{b}	19.23 ^{c}	0.59	0.000
RDW-SD, fL	38.99 ^{b}	57.04 ^{a}	54.21 ^{a}	37.59 ^{b}	1.01	0.000
PLT, 10^9/L	247.19	371.25	384.98	212.00	65.61	0.003
MPV, fL	9.16 ^{a}	7.99 ^{b}	8.28 ^{b}	9.25 ^{a}	0.29	0.001
PDW, %	15.53 ^{a}	14.95 ^{c}	15.04 ^{b,c}	15.43 ^{a,b}	0.16	0.023
PCT, %	0.23 ^{a,b}	0.18 ^{b}	0.27 ^{a}	0.20 ^{a,b}	0.04	0.106
			d35			
WBC, 10^9/L	26.02 ^{a,b}	29.77 ^{a}	31.40 ^{a}	22.60 ^{b}	2.09	0.017
Neu, 10^9/L	9.39	12.46	11.76	8.85	1.39	0.194
LYM, 10^9/L	12.93	11.91	13.77	9.90	1.41	0.246
Mon, 10^9/L	2.98 ^{b}	4.40 ^{a,b}	3.76 ^{a}	3.04 ^{b}	0.55	0.034
Eos, 10^9/L	0.58 ^{b}	0.87 ^{a}	0.87 ^{a}	0.73 ^{a,b}	0.07	0.019
Bas, 10^9/L	0.15 ^{a}	0.13 ^{a,b}	0.15 ^{a}	0.09 ^{b}	0.02	0.061
Neu, %	34.43	40.69	37.79	37.59	1.36	0.761
LYM, %	51.39	41.92	44.96	44.99	4.42	0.489
Mon, %	11.49	14.03	12.08	13.74	1.18	0.360
Eos, %	2.19 ^{b}	2.96 ^{a}	2.76 ^{a,b}	3.34 ^{a}	0.27	0.026
Bas, %	0.51 ^{a}	0.41 ^{a,b}	0.45 ^{a,b}	0.34 ^{b}	0.02	0.078
RBC, 10^9/L	5.78 ^{a}	6.03 ^{a}	5.91 ^{a}	5.10 ^{b}	0.10	0.006
HGB, g/L	105.31 ^{a}	104.31 ^{a}	104.75 ^{a}	92.81 ^{b}	5.45	0.098
HCT, %	34.68 ^{a}	36.34 ^{a}	34.51 ^{a}	29.23 ^{b}	1.40	0.003
MCV, fL	59.99	60.33	58.54	59.71	0.85	0.473
MCH, pg	18.24	17.36	17.76	18.29	0.58	0.626
MCHC, g/L	304.75	287.81	303.94	307.12	12.99	0.709
RDW-CV, %	20.13	20.62	20.86	20.07	0.59	0.740
RDW-SD, fL	42.36	43.53	43.13	41.84	1.01	0.646
PLT, 10^9/L	227.00	305.00	346.25	256.62	65.61	0.588

Table 3. *Cont.*

Items [1]	Treatment [2]				SEM [3]	p-Value
	Control	A1	A2	A3		
MPV, fL	9.03 [b]	10.07 [a]	10.68 [a]	9.99 [a]	0.29	0.001
PDW, %	16.28	16.13	15.84	16.19	0.16	0.229
PCT, %	0.22 [b]	0.32 [a,b]	0.37 [a]	0.28 [a,b]	0.04	0.048

[a,b,c] Different superscripts within a row indicate a significant difference ($p < 0.05$). [1] WBC, white blood cell; Neu, neutrophil count; LYM, lymphocyte; Mon, monocyte; Eos, eosinophilic; Bas, basophil; RBC, red blood cell; HGB, hemoglobin; HCT, hematocrit; MCV, mean corpuscular volume; MCH, mean corpuscular hemoglobin; MCHC, mean corpuscular hemoglobin concentration; RDW-CV, red cell distribution width-coefficient of variation; RDW-SD, red cell distribution width-standard deviation; PLT, platelets; MPV, mean platelet volume; PDW, platelet distribution width; PCT, thrombocytocrit. [2] A1, acidifiers A1 group, continuous supply of acidified drinking water with 19% formic acid +19% acetic acid +3.5% propionic acid +15% lactic acid; A2, acidifiers A2 group, continuous supply of acidified drinking water with 22% formic acid +16% acetic acid +4% propionic acid; A3, acidifiers A3 group, continuous supply of acidified drinking water with 22% formic acid +16% acetic acid +4% propionic acid +0.5% stevioside; control, continuous supply of non-treated water. [3] SEM means standard error of the means ($n = 16$).

Furthermore, on d 35, Mon of the A2 group was greater than that of the control group ($p < 0.05$); EOS in the A1 and A2 groups was greater than that in the control group ($p < 0.05$); EOS% in the A1 and A3 groups was greater than that in the control group ($p < 0.05$); PCT% in the A2 group was greater than that in the control group ($p < 0.05$); RBC and HCT in the A3 group were less than those in the control group ($p < 0.05$); MPV of the three acidifier groups was greater than that of the control group ($p < 0.05$). There were no significant differences in other blood routine indices among the four groups ($p < 0.05$).

3.4. Antioxidant Capacity Level

The effects of acidifiers on the level of antioxidant capacity of weaned pigs are shown in Table 4. The level of SOD in the A2 group was greater than that of the control group on d 18 ($p < 0.05$). The activities of CAT in the A1 and A2 groups were greater than the control group on d 18 ($p < 0.05$). The activities of GSH in the A1 and A3 groups were greater than the control on d 18 ($p < 0.05$). The level of T-AOC in the acidifier groups was greater than that of the control group on d 18 ($p < 0.05$), whereas the level of MDA in the acidifier supplementation groups was significantly lower than that in the control group on d 18. The activity of GSH in the acidifier groups was greater than that of the control group on d 35 ($p < 0.05$). The level of T-AOC in the A1 group was greater than that of the other three groups on d 35 ($p < 0.05$), whereas no significant difference was found in the activity of MDA between groups on d 35 ($p > 0.05$). Interestingly, the level of GSH-Px showed a tendency to increase in the A1 group when compared to control group on d 35 ($0.05 < p < 0.10$).

Table 4. Effects of acidifiers on antioxidant capacity in pigs.

Items [1]	Treatment [2]				SEM [3]	p-Value
	Control	A1	A2	A3		
			d18			
SOD, U/mL	141.76 [b]	123.87 [c]	162.43 [a]	140.32 [b]	3.51	<0.001
CAT, U/mL	6.28 [b]	7.70 [a]	8.25 [a]	5.37 [b]	0.37	< 0.001
GSH, μg/mL	162.03 [b]	218.24 [a]	150.95 [b]	215.37 [a]	10.14	<0.001
GSH-Px, U/mL	656.06	653.43	547.75	689.38	50.77	0.227
T-AOC, U/mL	4.08 [c]	5.12 [b]	5.74 [a]	5.83 [a]	0.19	<0.001
MDA, nmol/mL	4.69 [a]	2.67 [b]	2.75 [b]	2.70 [b]	0.19	<0.001
			d35			
SOD, U/mL	118.29 [a,b]	127.58 [a]	115.74 [b]	123.22 [a,b]	3.51	0.087

Table 4. *Cont.*

Items [1]	Treatment [2]				SEM [3]	*p*-Value
	Control	A1	A2	A3		
CAT, U/mL	13.89 ^{b}	14.80 ^{a,b}	14.64 ^{a,b}	14.97 ^{a}	0.37	0.188
GSH, μg/mL	195.46 ^{c}	236.22 ^{b}	281.37 ^{a}	227.00 ^{b}	10.14	<0.001
GSH-Px, U/mL	436.38 ^{b}	583.96 ^{a}	432.92 ^{b}	551.07 ^{a,b}	50.77	0.076
T-AOC, U/mL	4.23 ^{b}	5.44 ^{a}	4.60 ^{b}	4.41 ^{b}	0.19	<0.001
MDA, nmol/mL	2.61	2.27	2.45	2.58	0.19	0.563

^{a,b,c} Different superscripts within a row indicate a significant difference ($p < 0.05$). [1] SOD, superoxide dismutase; CAT, catalase, GSH, glutathione, GSH-Px, glutathione peroxidase; T-AOC, total oxidative capacity; MDA, malonaldehyde. [2] A1, acidifiers A1 group, continuous supply of acidified drinking water with 19% formic acid +19% acetic acid +3.5% propionic acid +15% lactic acid; A2, acidifiers A2 group, continuous supply of acidified drinking water with 22% formic acid +16% acetic acid +4% propionic acid; A3, acidifiers A3 group, continuous supply of acidified drinking water with 22% formic acid +16% acetic acid +4% propionic acid +0.5% stevioside; control, continuous supply of non-treated water. [3] SEM means standard error of the means ($n = 16$).

3.5. Microbiological Analysis of Gastrointestinal Contents

3.5.1. Alpha Diversity Analysis

There were 56,867 to 79,927 valid reads obtained from each sample (File S1). Alpha diversity analysis including Observed species, Good's coverage, Shannon, Simpson, and Chao1 indices are presented in Figure 1. The Shannon index of the A2 group was less than that of the A3 group on d 18 ($p < 0.05$). The observed OTUs (species) and Chao1 indices of the A3 group were less than those of the control group on d 35 ($p < 0.05$). The Simpson of the A1 group was less than that of the control group ($p < 0.05$), whereas the Good's coverage index of the A1 and A3 groups was greater than that of the control group on d 35 ($p < 0.05$).

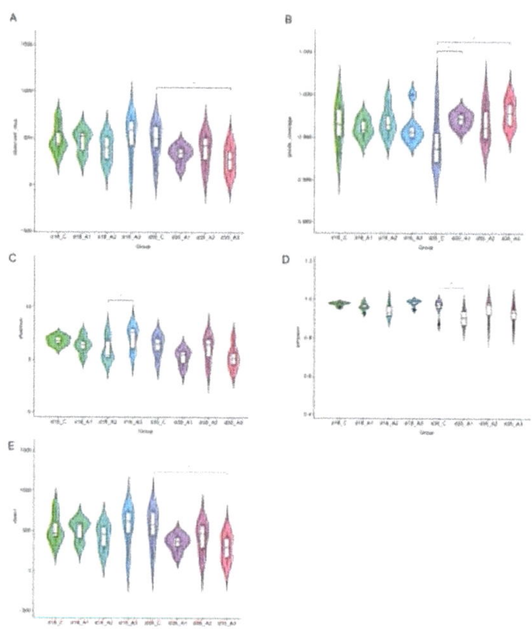

Figure 1. Comparison of the Alpha diversity index in the four groups on d 18 and d 35. Observed OTUs (observed species) (**A**). Good's coverage (**B**). Simpson indices (**C**). Shannon indices (**D**). Chao1 indices (**E**). Labeled with * indicates a significant difference, $p < 0.05$; Mean values are based on six pigs (six pigs per group).

3.5.2. Analysis of Group Differences of Intestinal Flora at the Phylum Level

The composition of intestinal flora at the phylum level is presented in Figure 2A and File S2. The most abundant phylum across all groups was *Firmicutes* at the phylum level, followed by *Bacteroidetes*, *Actinobacteria*, and *Proteobacteria*. The relative abundance of *Firmicutes* in intestinal flora on d 18 and d 35 were 75.86% and 84.12%, respectively. Next, we compared the differences of intestinal flora at the phylum level between treatments. The relative abundance of *Firmicutes* in the A1 and A2 groups was greater than that in the A3 group on d 18 ($p < 0.05$) (Figure 3A). The relative abundance of *Firmicutes* in the A1 group was greater than that in the control group on d 35 ($p < 0.05$) (Figure 3B). The relative abundance of *Proteobacteria* in the A2 group was less than that in the A3 group on d 35 ($p < 0.05$) (Figure 3H).

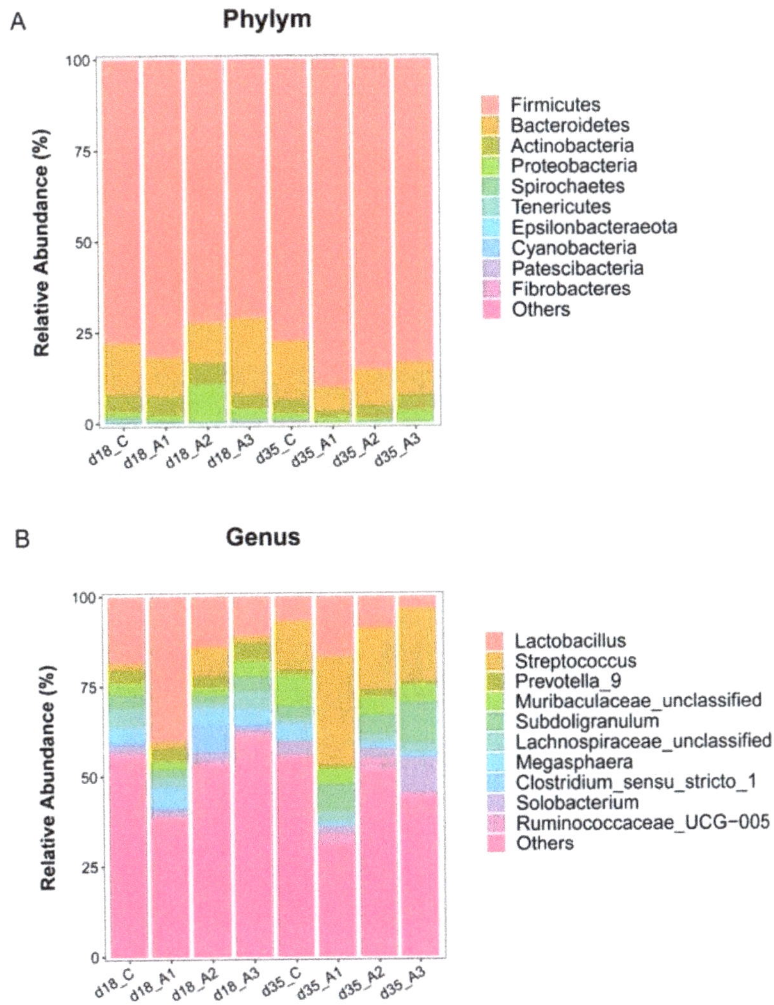

Figure 2. Gut microbiota composition of experimental group pigs and control group pigs: The relative abundance of the top 10 intestinal flora at the phylum level (**A**); the relative abundance of the top 10 intestinal flora at the genus level (**B**).

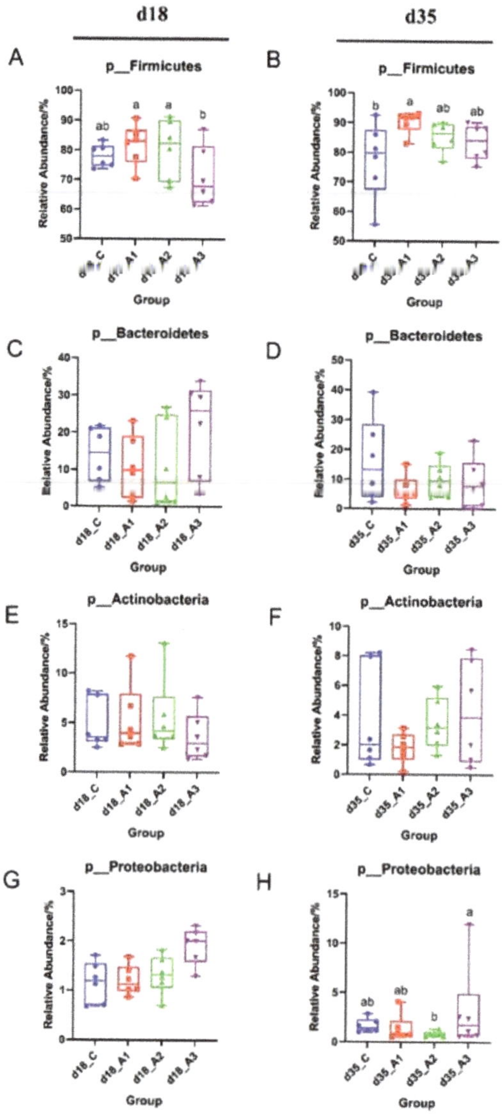

Figure 3. Comparison of dominant intestinal flora at the phylum level. Relative abundances of *Firmicutes* (**A**), *Bacteroidetes* (**C**), *Actinobacteria* (**E**), and *Proteobacteria* (**G**) among the four groups on d 18. Relative abundances of *Firmicutes* (**B**), *Bacteroidetes* (**D**), *Actinobacteria* (**F**), and *Proteobacteria* (**H**) among the four groups on d 35. Labeled means with different superscript letters are significantly different, $p < 0.05$. Mean values are based on six pigs (six pigs per group).

3.5.3. Analysis of Group Differences of Intestinal Flora at the Genus Level

The composition of intestinal flora at the genus level in each group is presented in Figure 2B and File S3. The most abundant genus across all groups was *Lactobacillus*, accounting for 20.99% and 8.94% on d 18 and 35, respectively. The genus level analysis showed that the relative abundance of *Lactobacillus* in the A1 group was greater than that of the other three groups ($p < 0.05$); the relative abundance of *Lachnospiraceae_unclassified* in the A2 group was less than that of the control group or the A3 group on d 18 ($p < 0.05$)

(Figure 4A). The relative abundance of *Lactobacillus* in the A1 group was greater than that of the control group and the A3 group on d 35 ($p < 0.05$); the relative abundance of *Streptococcus* in the A1 group was greater than that of the control group on d 35 ($p < 0.05$); the relative abundance of *Subdoligranulum* in the A3 group was greater than that of the control group on d 35 ($p < 0.05$); the relative abundance of *Solobacterium* in the A3 group was greater than that of other three groups on d 35 ($p < 0.05$); the relative abundance of *Ruminococcaceae_UCG-005* in the A2 group was greater than that of the control group and the A3 group on d 35 ($p < 0.05$) (Figure 4B).

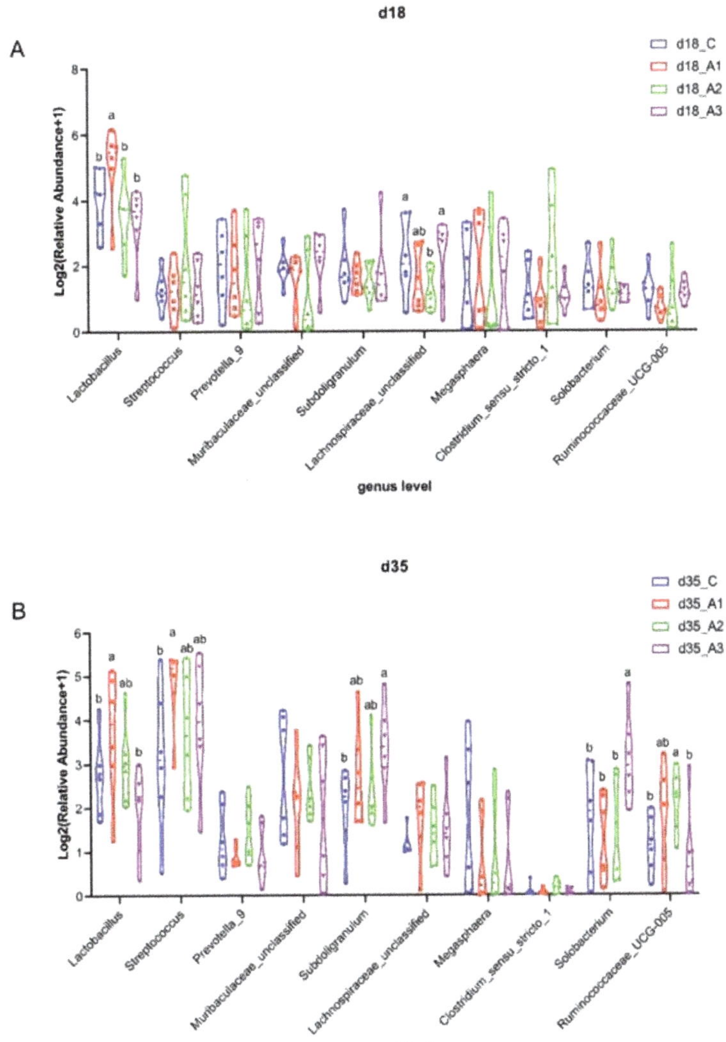

Figure 4. Comparison of dominant intestinal flora at the genus level. Relative abundances of dominant intestinal flora at the genus level among the four groups on d18 (**A**). Relative abundances of dominant intestinal flora at the genus level among the four groups on d18 (**B**). Labeled means with different superscript letters are significantly different, $p < 0.05$. Mean values are based on six pigs (six pigs per group).

3.6. Partial Correlation Analyses between the Differential Microbial Species and Measured Parameters

The results of partial correlation analyses are presented in Figure 5. The results showed that *Streptococcus* had a significant positive correlation with serum CAT level (R = 0.998, $p < 0.05$). *Subdoligranulum* had a significant positive correlation with ADG and ADFI (R = 0.92, $p < 0.05$), while having a negative correlation with diarrhea rate (R = −0.88, $p < 0.05$). *Ruminococcaceae_UCG-005* was significantly positively correlated with serum CAT, GSH, GSH-Px, and T-AOC level (R = 0.99–1.00, $p < 0.05$), while negatively correlated with MDA concentration (R = −1.00, $p < 0.05$). There were no significant associations found in other microbiota ($p > 0.05$).

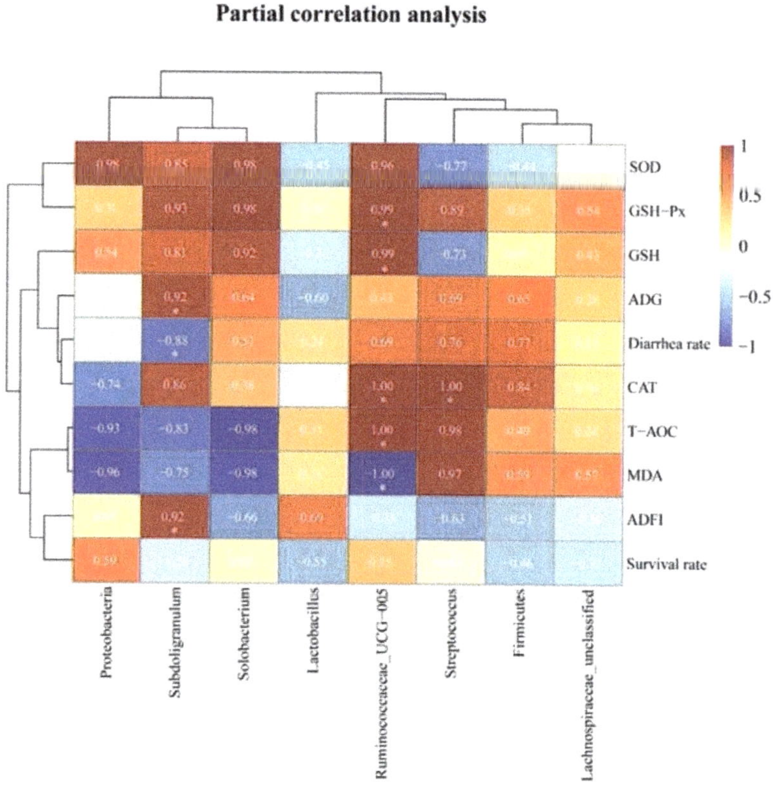

Figure 5. The partial correlation analyses between the differential microbial species and measured parameters. ADG, average daily gain; ADFI, Average daily feed intake; SOD, total superoxide dismutase; CAT, catalase; GSH, glutathione; GSH-Px, glutathione peroxidase; T-AOC, total antioxidant capacity; MDA, malondialdehyde; The red represents positive correlation, blue represents negative correlation, and the depth of the color represents the degree of correlation, respectively (Labeled with * indicate significantly different, $p < 0.05$).

4. Discussion

The digestive system of weaned pigs is immature, and the secretion of gastric acid and digestive enzymes is insufficient, resulting in the inability to activate digestive enzymes such as pepsin, making them unable to effectively digest nutrients such as protein and starch in the diet [35]. Therefore, improving the digestive capacity and environment of pigs plays a vital role in promoting the growth performance of pigs. Supplementation with a microencapsulated blend of organic acids improved the feed intake, average daily gain, and weight gain rate of weaned pigs [36]. Yang et al. [37] also found that adding the mixture of essential oils and organic acids to the diet improved the final weight and daily gain of weaned pigs. Several studies have reported the beneficial effects of compound organic acids (formic acid, acetic acid, propionic acid, and lactic acid) in swine feed [38]. However, due to the type and dose of dietary organic acids, coating or mixing mode, dietary composition, and other factors, the response of dietary organic acids to weaned pigs is different. Compared with supplementation in feed, drinking water supplementation with acidifiers is more convenient and makes it easier to control the dosage in pig farms. In addition, the acidifiers can disinfect drinking water and inhibit pathogenic bacteria. Meanwhile, few studies have investigated the effects of different organic acid combinations supplied via the water in the post-weaning period of piglets. Here, we present, for the first time, the synergistic effect of an organic acid formula (containing formic, acetic, and propionic acids) with or without lactic acid in pigs. In our present study, the ADG and ADFI of the A1 group was greater than that of the control group, while there was no significant difference between the A2 and A3 groups and the control group. The organic acids combination of the A1 group contains lactic acid compared with the A2 and A3 groups. A study showed that the addition of compound organic acids containing formic acid and lactic acid significantly reduced the *Salmonella* seroprevalence compared with the addition of formic acid alone in the feed of fattening pigs [39]. It has been previously reported that lactic acid reduced the pH value of the gastric juices and inhibited the reproduction of enterotoxin *Escherichia coli*, which is more effective than other organic acids in improving the growth performance of pigs [40], which is basically consistent with our results. The effect of compound organic acid as an alternative to antibiotics is better than a single organic acid in weaned pigs [41]. This indicates that the synergistic blend of formic acid, acetic acid, propionic acid, and lactic acid could improve intestinal health, appetite, and feed intake, so as to improve the utilization rate of feed and make the weight gain of weaned pigs significant.

Moreover, piglets are subjected to nutritional, environmental, and psychological stress during the weaning period, which leads to post-weaning diarrhea syndrome (PWDS) [42]. When the degree of diarrhea is mild, it causes malnutrition and affects the growth and development of piglets [43]. When it is serious, it causes dehydration and even death in piglets [44]. A previous study found that dietary supplementation with an acidifier inhibited the reproduction of ETEC F4 and ETEC K99, and reduced the severity and duration of diarrhea in weaned pigs [45]. In the present study, the diarrhea rate of pigs in the experimental groups were numerically less than that in the control group, but the difference was not significant. This might be due to the addition of the acidifier to the drinking water having less of an effect on piglet diarrhea than adding the acidifier to feed. On the other hand, the weaned pigs were raised in a modern nursery and had an acclimation period in the nursery before the experiment, which might also be one of the reasons why the diarrhea rate was not significantly different between the acidifier group and the control group. Interestingly, there was significant improvement on the survival rate of pigs between the A2 group and the control group in the Multiple Comparisons, and the survival rate of other acidifier groups was also numerically greater than that of the control group. Most pigs that died were small and weak during the whole experimental period. The reestablishment of social hierarchy after mixing causes weak pigs to be at a disadvantage in competing for feed, while social competition has no effect on drinking water behavior [46,47]. It suggests that adding acidifiers to the drinking water can improve the survival rate of weak piglets.

When animals are healthy, there is a dynamic balance between the production of free radicals and the ability of the antioxidant system to scavenge free radicals in the body [48]. Post-weaning piglets produced too many oxides and free radicals, resulting in an imbalance of the redox potential and oxidative stress damage, which affects the immune response and growth performance of piglets [49,50]. Therefore, it is very important to improve the antioxidant stress ability of weaned pigs by eliminating free radicals and regulating the balance and stability of redox potential through nutritional intervention. Oxidative stress was determined by detecting the activities of glutathione peroxidase (GPX) and superoxide dismutase (SOD) in piglets [51]. In addition, genes related to oxidative stress, including catalase (CAT), lactate dehydrogenase (LDHA), glutathione peroxidase 2 (GPX2), and superoxide dismutase 3 (SOD3), often changed during the weaning transition period of piglets [52]. Organic acid-based feed additives improved the levels of GSH and ferric-reducing ability potential (FRAP) in the ileum, and had a significant antioxidant effect [53]. Furthermore, supplementing the water plus organic acid blends to the basal diet of piglets had significantly increased serum T-AOC activities [54]. As an immunoassay technique, ELISA has been widely used to detect and quantify proteins, antibodies, or hormones [55]. An ELISA-based competition assay has the advantages of high sensitivity, simple operation, and affordability [56], exhibiting greater effectiveness than the HPLC-FLD method [57]. The FRAP assay (Ferric Reducing Ability of Plasma), a simple method to determine the total antioxidant capacity, has been applied to detect the T-AOC of animal serum [58]. The FRAP assay has the advantages of being inexpensive, has a simple reagent preparation, and high reproducibility [59]. In the present experiment, the concentrations of MDA, SOD, GSH, GSH-Px, and CAT were measured by ELISA-based competition assays. The results showed that adding acidifiers to the drinking water significantly increased the T-AOC level of weaned pigs, which was consistent with the above research results. This indicated that adding acidifiers to the drinking water effectively enhanced the antioxidant capacity and reduced oxidative stress in weaning pigs.

The stability of intestinal flora has long been known to play a vital role to maintain the metabolic health of the host [60]. If the microflora was disordered, it reduced the immunity of the animals and caused a variety of diseases [61]. The structure of the intestinal flora of adult animals is generally considered to be stable. Meanwhile, the bacteria in piglets mainly comes from the mother from newborn to pre-weaning, and the intestinal micro ecosystem remains relatively balanced [62]. However, the food of weaned pigs is changed from liquid breast milk to solid feed, coupled with the changes of environmental factors such as humidity, temperature, and population transformation, the harmful flora in the intestine of piglets increases, which destroys the balance of intestinal flora and affects the growth and development of the host [63]. Therefore, it is of great significance to find feed additives that improve the intestinal microflora of weaned pigs and maintain the intestinal health of animals. Acidifiers reduced the pH of the gastrointestinal tract and created an acidic environment, which might be helpful in improving the gastrointestinal environment of piglets [64]. Observed species, Chao1, Shannon, and Simpson indices are indicators reflecting alpha diversity [65]. In this study, the 16S rRNA gene sequencing of the intestinal contents of piglets showed that the Observed species and Chao1 index of the A3 group were less than those of the control group on d 35. The Simpson index of the A1 group was less than that of the control group on d 35. The above results showed that the number of microbial species in the feces of weaned pigs decreased significantly after adding acidifiers to the drinking water for 35 d. This suggests that acidifiers might regulate the pH value of the gastrointestinal tract and inhibit the growth of harmful bacteria in the gastrointestinal tract of weaned pigs, so as to improve the structure of the gastrointestinal flora of piglets.

In addition to its efforts on microbial richness and diversity, this study had indicated that supplementing the drinking water with acidifiers also exerted an effect on the abundance of intestinal flora in piglets. In this experiment, *Firmicutes* and *Bacteroidetes* were the dominant bacteria in the intestinal microorganisms of piglets, and their relative abundance was high, which was basically consistent with previous experimental results [66,67]. The relative abundance of *Firmicutes* in the A1 group was the greatest on d 18 and 35, and also greater than that of the control group on d 35, which indicated that acidifier 1 could improve the relative abundance of *Firmicutes* in piglets. *Firmicutes* was reported to be related to the weight gain of weaned pigs [68]; however, there was no significant correlation between *Firmicutes* and the daily weight gain of weaned pigs in this study. It might be that the increase of *Firmicutes* abundance improved antioxidant capacity and indirectly improved growth performance, which might be the reason for the significant increase of ADG in the A1 group. In addition, Li et al. [69] found that when the compound acidifier was composed of formic acid, acetic acid, propionic acid, and medium chain fatty acids were added to the diet of growing pigs, the content of *Bacteroides* in the cecum of growing pigs decreased by 8.8%. In this study, the relative abundance of *Bacteroidetes* in the three experimental groups were less than that of the control group on d 35, which is similar to the results of the above research.

All bacteria belonging to *Proteobacteria* are Gram-negative bacteria, including *Salmonella*, *Escherichia coli*, and other pathogenic bacteria [70], which seriously affect the health of animals [71]. In this study, the relative abundance of *Proteobacteria* in the A2 group was the lowest among the four groups on d 35, indicating that acidifier 2 had the trend of reducing the relative abundance of *Proteobacteria* in pigs. *Lactobacillus* belongs to Gram-positive bacteria, which regulates the balance of intestinal flora of animals, enhances immunity, and improves feed digestibility [72]. Studies have shown that acidifiers also increased the abundance of *Lactobacillus* in the feces of weaned pigs [73]. The relative abundance of *Lactobacillus* in the A1 group was greater than that of the other three groups on d 18; the relative abundance of *Lactobacillus* in the A1 group was greater than that of the control group and A3 group on d 35. This might be because acidifier 1 contains lactic acid, which increased the relative abundance of *Lactobacillus* in the intestine of piglets, allowing *Lactobacillus* to became the dominant flora, thus improving the micro ecological balance of the intestinal environment. Therefore, adding acidifier 1 to the drinking water improved the intestinal flora structure of weaned pigs, so as to regulated the balance of gastrointestinal flora.

Partial correlation is used to calculate the strength of the relationship between two variables while accounting for the effects of one or more other variables. The partial correlation analyses showed that the relative abundance of *Subdoligranulum* was positively correlated with ADG and ADFI, while negatively correlated with diarrhea rate. The relative abundance of *Subdoligranulum* was lower in the colon of Ningxiang pigs (a fatty-type Chinese Indigenous pig breed) compared to that of Large White pigs [74]. *Subdoligranulum* has a positive effect on fecal microbiota transplantation in the treatment of necrotizing enterocolitis by affecting the production of butyrate [75]. In the present study, the relative abundance of *Subdoligranulum* in the A3 group was greater than that of the control group. The relative abundances of *Subdoligranulum* in the A1 and A2 groups were also numerically greater than that of the control group on d 35. We speculated that the addition of the acidifier to drinking water improves the abundance of *Subdoligranulum*, which contributes to improve intestinal health, increase feed intake, and daily weight gain, while decreasing diarrhea rate. The gut is a target site of reactive oxygen species (ROS) accumulation and consequent oxidative stress [76]. The partial correlation analysis showed that *Ruminococcaceae_UCG-005* was positively correlated with serum CAT, GSH, GSH-Px, and T-AOC levels, whereas it was negatively correlated with MDA concentration. Feeding gallic acid increased catalase and the total antioxidant capacity levels, while decreasing malondialdehyde concentrations and increasing the relative abundances of *Ruminococcaceae_UCG-005* in preweaning calves [77]. Rodents drinking silicon-containing water (BT) with antioxidant activity increased plasma H_2O_2 scavenging activity and glutathione peroxidase activity, while significantly increasing the abundance of *Ruminococcaceae_UCG-005*, which is basically consistent with the present study [78]. Our study showed that the relative abundance of *Ruminococcaceae_UCG-005* in the A2 group was significantly greater than that of the control group, and the A1 group was also numerically greater than the control group on d 35. It indicates that *Ruminococcaceae_UCG-005* may play an antioxidant role when piglets are subjected to weaning stress. Altogether, the disruption of gut microbial composition could be the major underlying factor inducing the decline in the antioxidant capacity of weaning-challenged piglets. These results contribute to the new understanding of the acidifier-enhanced antioxidant capacity, at least in part, due to alterations in the gut microbiota in weaned pigs.

5. Conclusions

In conclusion, the current findings suggested that adding acidifier 1 (19% formic acid +19% acetic acid +3.5% propionic acid +15% lactic acid) to the drinking water of weaned piglets significantly improved the ADG and ADFI of weaned pigs. In addition, we also found that adding acidifiers to the drinking water significantly improved the total antioxidant capacity of serum in weaned pigs. The addition of acidifier 1 to the drinking water increased the relative abundance of *Firmicutes* and *Lactobacillus* in the intestines of pigs, and improved the community structure of intestinal flora of pigs. These results also demonstrated the potential of acidifiers as an alternative to antibiotics in promoting the growth of weaned pigs, improving the ability of antioxidant capacity, and improving the intestinal microflora.

Supplementary Materials: The following supporting information can be downloaded at: https://www.mdpi.com/article/10.3390/antiox11050809/s1, Table S1: Experimental diet composition and nutrients (as-fed basis, %); File S1: Statistics of clean data; File S2: Phylum-level composition of bacteria (%); File S3: Genus level composition of bacteria (%).

Author Contributions: Conceptualization, B.Z. and C.-X.Z.; methodology, C.L.; software, Q.-L.X.; validation, [illegible]; formal analysis, B.Z.; investigation, Q.-L.X. and C.L.; resources, X.-J.M.; data curation, Q.-L.X. and C.L.; writing—original draft preparation, Q.-L.X. and C.L.; writing—review and editing, B.Z. and S.-B.W.; visualization, M.-Z.L.; supervision, B.Z. and C.-X.Z.; project administration, B.Z.; funding acquisition, B.Z. All authors have read and agreed to the published version of the manuscript.

Funding: This work was supported by the "National Natural Science Foundation of China" (grant number 31672465).

Institutional Review Board Statement: This study was conducted in Huaduo Animal Husbandry Technology Co., Ltd., Xuzhou City, Jiangsu, China. This study was approved by the Nanjing Agricultural University Animal Care and Use Committee (protocol code SYXK Su 2017-0027). All procedures involving piglet feeding, management, and animal welfare strictly followed the Nanjing Agricultural University experimental guidelines during the experiment.

Informed Consent Statement: Not applicable.

Data Availability Statement: Data are contained within the article.

Acknowledgments: This study was supported by the National Natural Science Foundation of China (31672465) and we thank the products from Jinhai Pharmaceutical Co., Ltd. Sequencing services that were provided by Lianchuan Biotechnology Co., Ltd. (Hangzhou, China).

Conflicts of Interest: The authors declare the following competing interest(s): Xiao-Jian Mo and Cheng-Xin Zhao are co-authors in this manuscript. They are employees at Yantai Jinhai Pharmaceutical Co., Ltd., which provided the acidifiers for this work. Cheng-Xin Zhao played a major role in the design of the study and supervised the published results. Xiao-Jian Mo prepared the acidifiers for this study.

References

1. Koketsu, Y.; Tani, S.; Iida, R. Factors for improving reproductive performance of sows and herd productivity in commercial breeding herds. *Porc. Health Manag.* **2017**, *3*, 1. [CrossRef] [PubMed]
2. Mueller, N.J.; Kuwaki, K.; Knosalla, C.; Dor, F.J.; Gollackner, B.; Wilkinson, R.A.; Arn, S.; Sachs, D.H.; Cooper, D.K.; Fishman, J.A. Early weaning of piglets fails to exclude porcine lymphotropic herpesvirus. *Xenotransplantation* **2005**, *12*, 59–62. [CrossRef] [PubMed]
3. Campbell, J.M.; Crenshaw, J.D.; Polo, J. The biological stress of early weaned piglets. *J. Anim. Sci. Biotechnol.* **2013**, *4*, 19. [CrossRef] [PubMed]
4. Moeser, A.J.; Pohl, C.S.; Rajput, M. Weaning stress and gastrointestinal barrier development: Implications for lifelong gut health in pigs. *Anim. Nutr.* **2017**, *3*, 313–321. [CrossRef]
5. Xu, J.; Li, Y.; Yang, Z.; Li, C.; Liang, H.; Wu, Z.; Pu, W. Yeast Probiotics Shape the Gut Microbiome and Improve the Health of Early-Weaned Piglets. *Front. Microbiol.* **2018**, *9*, 2011. [CrossRef]
6. Hung, D.Y.; Cheng, Y.H.; Chen, W.J.; Hua, K.F.; Pietruszka, A.; Dybus, A.; Lin, C.S.; Yu, Y.H. Bacillus licheniformis-Fermented Products Reduce Diarrhea Incidence and Alter the Fecal Microbiota Community in Weaning Piglets. *Animals* **2019**, *9*, 1145. [CrossRef]

7. Jensen, M.L.; Thymann, T.; Cilieborg, M.S.; Lykke, M.; Molbak, L.; Jensen, B.B.; Schmidt, M.; Kelly, D.; Mulder, I.; Burrin, D.G.; et al. Antibiotics modulate intestinal immunity and prevent necrotizing enterocolitis in preterm neonatal piglets. *Am. J. Physiol. Gastrointest. Liver Physiol.* **2014**, *306*, G59–G71. [CrossRef]
8. Lourenco, J.M.; Hampton, R.S.; Johnson, H.M.; Callaway, T.R.; Rothrock, M.J., Jr.; Azain, M.J. The Effects of Feeding Antibiotic on the Intestinal Microbiota of Weanling Pigs. *Front. Vet. Sci.* **2021**, *8*, 601394. [CrossRef]
9. Wang, M.; Huang, H.; Hu, Y.; Huang, J.; Yang, H.; Wang, L.; Chen, S.; Chen, C.; He, S. Effects of dietary microencapsulated tannic acid supplementation on the growth performance, intestinal morphology, and intestinal microbiota in weaning piglets. *J. Anim. Sci.* **2020**, *98*, skaa112. [CrossRef]
10. Yan, H.; Yu, B.; Degroote, J.; Spranghers, T.; Van Noten, N.; Majdeddin, M.; Van Poucke, M.; Peelman, L.; De Vrieze, J.; Boon, N.; et al. Antibiotic affects the gut microbiota composition and expression of genes related to lipid metabolism and myofiber types in skeletal muscle of piglets. *BMC Vet. Res.* **2020**, *16*, 392. [CrossRef]
11. Manyi-Loh, C.; Mamphweli, S.; Meyer, E.; Okoh, A. Antibiotic Use in Agriculture and Its Consequential Resistance in Environmental Sources: Potential Public Health Implications. *Molecules* **2018**, *23*, 795. [CrossRef]
12. Tian, M.; He, X.; Feng, Y.; Wang, W.; Chen, H.; Gong, M.; Liu, D.; Clarke, J.L.; van Eerde, A. Pollution by Antibiotics and Antimicrobial Resistance in LiveStock and Poultry Manure in China, and Countermeasures. *Antibiotics* **2021**, *10*, 539. [CrossRef]
13. Suiryanrayna, M.V.; Ramana, J.V. A review of the effects of dietary organic acids fed to swine. *J. Anim. Sci. Biotechnol.* **2015**, *6*, 45. [CrossRef]
14. Dibner, J.J.; Buttin, P. Use of organic acids as a model to study the impact of gut microflora on nutrition and metabolism. *J. Appl. Poult. Res.* **2002**, *11*, 453–463. [CrossRef]
15. Liu, Y.; Espinosa, C.D.; Abelilla, J.J.; Casas, G.A.; Lagos, L.V.; Lee, S.A.; Kwon, W.B.; Mathai, J.K.; Navarro, D.; Jaworski, N.W.; et al. Non-antibiotic feed additives in diets for pigs: A review. *Anim. Nutr.* **2018**, *4*, 113–125. [CrossRef]
16. Chen, J.L.; Li, Y.; Yu, B.; Chen, D.W.; Mao, X.B.; Zheng, P.; Luo, J.Q.; He, J. Dietary chlorogenic acid improves growth performance of weaned pigs through maintaining antioxidant capacity and intestinal digestion and absorption function. *J. Anim. Sci.* **2018**, *96*, 1108–1118. [CrossRef]
17. Zhang, Y.; Wang, Y.; Chen, D.; Yu, B.; Zheng, P.; Mao, X.; Luo, Y.; Li, Y.; He, J. Dietary chlorogenic acid supplementation affects gut morphology, antioxidant capacity and intestinal selected bacterial populations in weaned piglets. *Food Funct.* **2018**, *9*, 4968–4978. [CrossRef]
18. Yu, J.; Song, Y.; Yu, B.; He, J.; Zheng, P.; Mao, X.; Huang, Z.; Luo, Y.; Luo, J.; Yan, H.; et al. Tannic acid prevents post-weaning diarrhea by improving intestinal barrier integrity and function in weaned piglets. *J. Anim. Sci. Biotechnol.* **2020**, *11*, 87. [CrossRef]
19. Canibe, N.; Hojberg, O.; Hojsgaard, S.; Jensen, B.B. Feed physical form and formic acid addition to the feed affect the gastrointestinal ecology and growth performance of growing pigs. *J. Anim. Sci.* **2005**, *83*, 1287–1302. [CrossRef]
20. Matsumoto, H.; Miyagawa, M.; Yin, Y.; Oosumi, T. Effects of organic acid, Enterococcus faecalis strain EC-12 and sugar cane extract in feed against enterotoxigenic Escherichia coli-induced diarrhea in pigs. *Amb. Express* **2021**, *11*, 68. [CrossRef]
21. Dahmer, P.L.; Jones, C.K. The Impacts of Commercial Dietary Acidifiers on Growth Performance of Nursery Pigs. *J. Anim. Sci.* **2021**, *99*, 87. [CrossRef]
22. Mudarra, R.A.; Tsai, T.C.C.; Bottoms, K.; Shieh, T.S.; Bradly, C.; Maxwell, C.V. Effect of Adding Bioactive Peptide in Combination of Pharmaceutical Zinc Oxide or Organic Acids on Growth Performance, Hematology Profile, and Nutrient Digestibility in Nursery Pigs. *J. Anim. Sci.* **2021**, *99*, 22. [CrossRef]
23. Marden, J.P.; Julien, C.; Monteils, V.; Auclair, E.; Moncoulon, R.; Bayourthe, C. How does live yeast differ from sodium bicarbonate to stabilize ruminal pH in high-yielding dairy cows? *J. Dairy Sci.* **2008**, *91*, 3528–3535. [CrossRef]
24. Lingbeek, M.M.; Borewicz, K.; Febery, E.; Han, Y.; Doelman, J.; van Kuijk, S.J.A. Short-chain fatty acid administration via water acidifier improves feed efficiency and modulates fecal microbiota in weaned piglets. *J. Anim. Sci.* **2021**, *99*, skab307. [CrossRef]
25. Mustafa, A.; Bai, S.P.; Zeng, Q.F.; Ding, X.M.; Wang, J.P.; Xuan, Y.; Su, Z.W.; Zhang, K.Y. Effect of organic acids on growth performance, intestinal morphology, and immunity of broiler chickens with and without coccidial challenge. *AMB Express* **2021**, *11*, 140. [CrossRef]
26. De Busser, E.V.; Dewulf, J.; Nollet, N.; Houf, K.; Schwarzer, K.; De Sadeleer, L.; De Zutter, L.; Maes, D. Effect of organic acids in drinking water during the last 2 weeks prior to slaughter on Salmonella shedding by slaughter pigs and contamination of carcasses. *Zoonoses Public Health* **2009**, *56*, 129–136. [CrossRef]
27. Kuley, E.; Ozyurt, G.; Ozogul, I.; Boga, M.; Akyol, I.; Rocha, J.M.; Ozogul, F. The Role of Selected Lactic Acid Bacteria on Organic Acid Accumulation during Wet and Spray-Dried Fish-based Silages. Contributions to the Winning Combination of Microbial Food Safety and Environmental Sustainability. *Microorganisms* **2020**, *8*, 172. [CrossRef]
28. Jiang, J.; Qi, L.; Lv, Z.; Wei, Q.; Shi, F. Dietary stevioside supplementation increases feed intake by altering the hypothalamic transcriptome profile and gut microbiota in broiler chickens. *J. Sci. Food. Agric.* **2021**, *101*, 2156–2167. [CrossRef]
29. Wu, X.; Yang, P.; Sifa, D.; Wen, Z. Effect of dietary stevioside supplementation on growth performance, nutrient digestibility, serum parameters, and intestinal microflora in broilers. *Food Funct.* **2019**, *10*, 2340–2346. [CrossRef]
30. Ogunade, I.M.; Jiang, Y.; Kim, D.H.; Cervantes, A.A.P.; Arriola, K.G.; Vyas, D.; Weinberg, Z.G.; Jeong, K.C.; Adesogan, A.T. Fate of Escherichia coli O157:H7 and bacterial diversity in corn silage contaminated with the pathogen and treated with chemical or microbial additives. *J. Dairy Sci.* **2017**, *100*, 1780–1794. [CrossRef]
31. Pedersen, K.S.; Toft, N. Intra- and inter-observer agreement when using a descriptive classification scale for clinical assessment of faecal consistency in growing pigs. *Prev. Vet. Med.* **2011**, *98*, 288–291. [CrossRef] [PubMed]

32. Bolyen, E.; Rideout, J.R.; Dillon, M.R.; Bokulich, N.; Abnet, C.C.; Al-Ghalith, G.A.; Alexander, H.; Alm, E.J.; Arumugam, M.; Asnicar, F.; et al. Reproducible, interactive, scalable and extensible microbiome data science using QIIME 2. *Nat. Biotechnol.* **2019**, *37*, 852–857. [CrossRef] [PubMed]
33. Cole, J.R.; Wang, Q.; Fish, J.A.; Chai, B.; McGarrell, D.M.; Sun, Y.; Brown, C.T.; Porras-Alfaro, A.; Kuske, C.R.; Tiedje, J.M. Ribosomal Database Project: Data and tools for high throughput rRNA analysis. *Nucleic Acids Res.* **2014**, *42*, D633–D642. [CrossRef] [PubMed]
34. Lozupone, C.; Knight, R. UniFrac: A new phylogenetic method for comparing microbial communities. *Appl. Environ. Microbiol.* **2005**, *71*, 8228–8235. [CrossRef]
35. Nowak, P.; Zaworska-Zakrzewska, A.; Frankiewicz, A.; Kasprowicz-Potocka, M. The Effects and Mechanisms of Acids on the Health of Piglets and Weaners—A Review. *Ann. Anim. Sci.* **2021**, *21*, 433–455. [CrossRef]
36. Gong, J.; Yu, H.; Liu, T.; Li, M.; Si, W.; de Lange, C.F.M.; Dewey, C. Characterization of ileal bacterial microbiota in newly-weaned pigs in response to feeding lincomycin, organic acids or herbal extract. *Livest. Sci.* **2008**, *116*, 318–322. [CrossRef]
37. Yang, C.; Zhang, L.; Cao, G.; Feng, J.; Yue, M.; Xu, Y.; Dai, B.; Han, Q.; Guo, X. Effects of dietary supplementation with essential oils and organic acids on the growth performance, immune system, fecal volatile fatty acids, and microflora community in weaned piglets. *J. Anim. Sci.* **2019**, *97*, 133–143. [CrossRef]
38. Michiels, J.; Missotten, J.; Rasschaert, G.; Dierick, N.; Heyndrickx, M.; De Smet, S. Effect of organic acids on Salmonella colonization and shedding in weaned piglets in a seeder model. *J. Food Prot.* **2012**, *75*, 1974–1983. [CrossRef]
39. Creus, E.; Perez, J.F.; Peralta, B.; Baucells, F.; Mateu, E. Effect of acidified feed on the prevalence of Salmonella in market-age pigs. *Zoonoses Public Health* **2007**, *54*, 314–319. [CrossRef]
40. Tsiloyiannis, V.K.; Kyriakis, S.C.; Vlemmas, J.; Sarris, K. The effect of organic acids on the control of porcine post-weaning diarrhoea. *Res. Vet. Sci.* **2001**, *70*, 287–293. [CrossRef]
41. Ahmed, S.T.; Hwang, J.A.; Hoon, J.; Mun, H.S.; Yang, C.J. Comparison of Single and Blend Acidifiers as Alternative to Antibiotics on Growth Performance, Fecal Microflora, and Humoral Immunity in Weaned Piglets. *Asian Austral. J. Anim.* **2014**, *27*, 93–100. [CrossRef]
42. Mou, Q.; Yang, H.S.; Yin, Y.L.; Huang, P.F. Amino Acids Influencing Intestinal Development and Health of the Piglets. *Animals* **2019**, *9*, 302. [CrossRef]
43. Lu, X.; Zhang, M.; Zhao, L.; Ge, K.; Wang, Z.; Jun, L.; Ren, F. Growth Performance and Post-Weaning Diarrhea in Piglets Fed a Diet Supplemented with Probiotic Complexes. *J. Microbiol. Biotechnol.* **2018**, *28*, 1791–1799. [CrossRef]
44. Bertolini, F.; Harding, J.C.; Mote, B.; Ladinig, A.; Plastow, G.S.; Rothschild, M.F. Genomic investigation of piglet resilience following porcine epidemic diarrhea outbreaks. *Anim. Genet.* **2017**, *48*, 228–232. [CrossRef]
45. Bosi, P.; Sarli, G.; Casini, L.; De Filippi, S.; Trevisi, P.; Mazzoni, M.; Merialdi, G. The influence of fat protection of calcium formate on growth and intestinal defence in Escherichia coli K88-challenged weanling pigs. *Anim. Feed Sci. Technol.* **2007**, *139*, 170–185. [CrossRef]
46. Andersen, H.M.; Dybkjaer, L.; Herskin, M.S. Growing pigs' drinking behaviour: Number of visits, duration, water intake and diurnal variation. *Animal* **2014**, *8*, 1881–1888. [CrossRef]
47. Tong, X.; Shen, C.; Chen, R.; Gao, S.; Liu, X.; Schinckel, A.P.; Zhou, B. Reestablishment of Social Hierarchies in Weaned Pigs after Mixing. *Animals* **2019**, *10*, 36. [CrossRef]
48. Da Costa, R.M.; Rodrigues, D.; Pereira, C.A.; Silva, J.F.; Alves, J.V.; Lobato, N.S.; Tostes, R.C. Nrf2 as a Potential Mediator of Cardiovascular Risk in Metabolic Diseases. *Front. Pharmacol.* **2019**, *10*, 382. [CrossRef]
49. Feng, Y.; An, Z.; Chen, H.; He, X.; Wang, W.; Li, X.; Zhang, H.; Li, F.; Liu, D. Ulva prolifera Extract Alleviates Intestinal Oxidative Stress via Nrf2 Signaling in Weaned Piglets Challenged With Hydrogen Peroxide. *Front. Immunol.* **2020**, *11*, 599735. [CrossRef]
50. Xiong, X.; Tan, B.; Song, M.; Ji, P.; Kim, K.; Yin, Y.; Liu, Y. Nutritional Intervention for the Intestinal Development and Health of Weaned Pigs. *Front. Vet. Sci.* **2019**, *6*, 46. [CrossRef]
51. Novais, A.K.; Martel-Kennes, Y.; Roy, C.; Deschene, K.; Beaulieu, S.; Bergeron, N.; Laforest, J.P.; Lessard, M.; Matte, J.J.; Lapointe, J. Tissue-specific profiling reveals modulation of cellular and mitochondrial oxidative stress in normal- and low-birthweight piglets throughout the peri-weaning period. *Animal* **2020**, *14*, 1014–1024. [CrossRef]
52. Meng, Q.; Luo, Z.; Cao, C.; Sun, S.; Ma, Q.; Li, Z.; Shi, B.; Shan, A. Weaning Alters Intestinal Gene Expression Involved in Nutrient Metabolism by Shaping Gut Microbiota in Pigs. *Front. Microbiol.* **2020**, *11*, 694. [CrossRef] [PubMed]
53. Jimenez, M.J.; Berrios, R.; Stelzhammer, S.; Bracarense, A. Ingestion of organic acids and cinnamaldehyde improves tissue homeostasis of piglets exposed to enterotoxic Escherichia coli (ETEC). *J. Anim. Sci.* **2020**, *98*, skaa012. [CrossRef] [PubMed]
54. Xiang, X.D.; Deng, Z.C.; Wang, Y.W.; Sun, H.; Wang, L.; Han, Y.M.; Wu, Y.Y.; Liu, J.G.; Sun, L.H. Organic Acids Improve Growth Performance with Potential Regulation of Redox Homeostasis, Immunity, and Microflora in Intestines of Weaned Piglets. *Antioxidants* **2021**, *10*, 1665. [CrossRef] [PubMed]
55. Balintova, J.; Welter, M.; Marx, A. Antibody-nucleotide conjugate as a substrate for DNA polymerases. *Chem. Sci.* **2018**, *9*, 7122–7125. [CrossRef]
56. Liu, J.C.; Yang, Y.H.; Zhu, Q.; Wang, Z.H.; Hu, G.J.; Shi, H.C.; Zhou, X.H. ELISA-Based Method for Variant-Independent Detection of Total Microcystins and Nodularins via a Multi-immunogen Approach. *Environ. Sci. Technol.* **2021**, *55*, 12984–12993. [CrossRef] [PubMed]
57. Qiu, X.; Ma, J.; Li, P.; Geng, Z.; Sun, C.; Wang, D.; Xu, W. Development of indirect competitive ELISA for determination of dehydroabietic acid in duck skin and comparison with the HPLC method. *Poult. Sci.* **2020**, *99*, 3280–3285. [CrossRef] [PubMed]
58. Rubio, C.P.; Martinez-Subiela, S.; Hernandez-Ruiz, J.; Tvarijonaviciute, A.; Ceron, J.J. Analytical validation of an automated assay for ferric-reducing ability of plasma in dog serum. *J. Vet. Diagn. Investig.* **2017**, *29*, 574–578. [CrossRef]

59. Benzie, I.F.F.; Strain, J.J. The ferric reducing ability of plasma (FRAP) as a measure of "antioxidant power": The FRAP assay. *Anal. Biochem.* **1996**, *239*, 70–76. [CrossRef]
60. Guevarra, R.B.; Lee, J.H.; Lee, S.H.; Seok, M.J.; Kim, D.W.; Kang, B.N.; Johnson, T.J.; Isaacson, R.E.; Kim, H.B. Piglet gut microbial shifts early in life: Causes and effects. *J. Anim. Sci. Biotechnol.* **2019**, *10*, 1. [CrossRef]
61. Fan, Y.; Pedersen, O. Gut microbiota in human metabolic health and disease. *Nat. Rev. Microbiol.* **2021**, *19*, 55–71. [CrossRef]
62. Beaumont, M.; Cauquil, L.; Bertide, A.; Ahn, I.; Barilly, C.; Gil, L.; Canlet, C.; Zemb, O.; Pascal, G.; Samson, A.; et al. Gut Microbiota-Derived Metabolite Signature in Suckling and Weaned Piglets. *J. Proteome Res.* **2021**, *20*, 982–994. [CrossRef]
63. Sun, X.; Cui, Y.; Su, Y.; Gao, Z.; Diao, X.; Li, J.; Zhu, X.; Li, D.; Li, Z.; Wang, C.; et al. Dietary Fiber Ameliorates Lipopolysaccharide-Induced Intestinal Barrier Function Damage in Piglets by Modulation of Intestinal Microbiome. *Msystems* **2021**, *6*, e01374-20. [CrossRef]
64. De Busser, E.V.; Dewulf, J.; Zutter, L.D.; Haesebrouck, F.; Callens, J.; Meyns, T.; Maes, W.; Maes, D. Effect of administration of organic acids in drinking water on faecal shedding of E. coli, performance parameters and health in nursery pigs. *Vet. J.* **2011**, *188*, 184–188. [CrossRef]
65. Nikolova, V.L.; Smith, M.R.B.; Hall, L.J.; Cleare, A.J.; Stone, J.M.; Young, A.H. Perturbations in Gut Microbiota Composition in Psychiatric Disorders: A Review and Meta-analysis. *JAMA Psychiatry* **2021**, *78*, 1343–1354. [CrossRef]
66. Arfken, A.M.; Frey, J.F.; Ramsay, T.G.; Summers, K.L. Yeasts of Burden: Exploring the Mycobiome-Bacteriome of the Piglet GI Tract. *Front. Microbiol.* **2019**, *10*, 1–14. [CrossRef]
67. Dhakal, S.; Wang, L.; Antony, L.; Rank, J.; Bernardo, P.; Ghimire, S.; Bondra, K.; Siems, C.; Lakshmanappa, Y.S.; Renu, S.; et al. Amish (Rural) vs. non-Amish (Urban) Infant Fecal Microbiotas Are Highly Diverse and Their Transplantation Lead to Differences in Mucosal Immune Maturation in a Humanized Germfree Piglet Model. *Front. Immunol.* **2019**, *10*, 1509. [CrossRef]
68. Ding, X.; Lan, W.; Liu, G.; Ni, H.; Gu, J.D. Exploring possible associations of the intestine bacterial microbiome with the pre-weaned weight gaining performance of piglets in intensive pig production. *Sci. Rep.* **2019**, *9*, 15534. [CrossRef]
69. Li, M.; Long, S.; Wang, Q.; Zhang, L.; Hu, J.; Yang, J.; Cheng, Z.; Piao, X. Mixed organic acids improve nutrients digestibility, volatile fatty acids composition and intestinal microbiota in growing-finishing pigs fed high-fiber diet. *Asian-Australas J. Anim. Sci.* **2019**, *32*, 856–864. [CrossRef]
70. Vaz-Moreira, I.; Nunes, O.C.; Manaia, C.M. Ubiquitous and persistent Proteobacteria and other Gram-negative bacteria in drinking water. *Sci. Total Environ.* **2017**, *586*, 1141–1149. [CrossRef]
71. Shin, N.R.; Whon, T.W.; Bae, J.W. Proteobacteria: Microbial signature of dysbiosis in gut microbiota. *Trends Biotechnol.* **2015**, *33*, 496–503. [CrossRef] [PubMed]
72. Naghmouchi, K.; Belguesmia, Y.; Bendali, F.; Spano, G.; Seal, B.S.; Drider, D. Lactobacillus fermentum: A bacterial species with potential for food preservation and biomedical applications. *Crit. Rev. Food Sci. Nutr.* **2020**, *60*, 3387–3399. [CrossRef] [PubMed]
73. Lan, R.X.; Kim, I. Effects of organic acid and medium chain fatty acid blends on the performance of sows and their piglets. *Anim. Sci. J.* **2018**, *89*, 1673–1679. [CrossRef]
74. Lei, L.F.; Wang, Z.B.; Li, J.Z.; Yang, H.S.; Yin, Y.L.; Tan, B.; Chen, J.S. Comparative Microbial Profiles of Colonic Digesta between Ningxiang Pig and Large White Pig. *Animals* **2021**, *11*, 1862. [CrossRef] [PubMed]
75. Lin, H.; Guo, Q.; Ran, Y.; Lin, L.; Chen, P.; He, J.; Chen, Y.; Wen, J. Multiomics Study Reveals Enterococcus and Subdoligranulum Are Beneficial to Necrotizing Enterocolitis. *Front. Microbiol.* **2021**, *12*, 752102. [CrossRef]
76. Vaccaro, A.; Dor, Y.K.; Nambara, K.; Pollina, E.A.; Lin, C.D.; Greenberg, M.E.; Rogulja, D. Sleep Loss Can Cause Death through Accumulation of Reactive Oxygen Species in the Gut. *Cell* **2020**, *181*, 1307–1328.e15. [CrossRef]
77. Xu, H.J.; Zhang, Q.Y.; Wang, L.H.; Zhang, C.R.; Li, Y.; Zhang, Y.G. Growth performance, digestibility, blood metabolites, ruminal fermentation, and bacterial communities in response to the inclusion of gallic acid in the starter feed of preweaning dairy calves. *J. Dairy Sci.* **2022**, *105*, 3078–3089. [CrossRef]
78. Wu, W.Y.; Chou, P.L.; Yang, J.C.; Chien, C.T. Silicon-containing water intake confers antioxidant effect, gastrointestinal protection, and gut microbiota modulation in the rodents. *PLoS ONE* **2021**, *16*, e0248508. [CrossRef]

antioxidants

MDPI

Article

Effect of Supplementing Different Levels of L-Glutamine on Holstein Calves during Weaning

Shuo Wang [1,†], Fuwei Wang [2,†], Fanlin Kong [1], Zhijun Cao [1], Wei Wang [1], Hongjian Yang [1], Yajing Wang [1], Yanliang Bi [3,*] and Shengli Li [1,*]

1 State Key Laboratory of Animal Nutrition, Beijing Engineering Technology Research Center of Raw Milk Quality and Safety Control, College of Animal Science and Technology, China Agricultural University, Beijing 100193, China; b20213040351@cau.edu.cn (S.W.); fanlinkong@cau.edu.cn (F.K.); caozhijun@cau.edu.cn (Z.C.); wei.wang@cau.edu.cn (W.W.); yang_hongjian@cau.edu.cn (H.Y.); yajingwang@cau.edu.cn (Y.W.)
2 Beijing Sunlon Livestock Development Co., Ltd., Beijing 100076, China; wang_fuwei@foxmail.com
3 Key Laboratory of Feed Biotechnology of the Ministry of Agriculture and Rural Affairs, Institute of Feed Research, Chinese Academy of Agricultural Sciences, Beijing 100081, China
* Correspondence: biyanliang@caas.cn (Y.B.); lishengli@cau.edu.cn (S.L.)
† These authors contributed equally to this work.

Abstract: Weaning stress affects the health and performance of calves. L-glutamine (L-Gln) is commonly used as a functional antioxidant and energy supplement in the body. However, dietary L-Gln supplementation improving weaning stress of calves is unclear. Thus, we aimed to explore the effects of L-Gln (provided by rumen-protected L-Gln) on calves during weaning. Seventy-five Holstein calves (54.0 ± 2.68 kg; 42 ± 2.1 d of age) were assigned to five groups: no supplementation and L-Gln with 1%, 2%, 3%, and 4% dry matter daily intake (DMI) supplementation groups, respectively. The experiment lasted for 28 days (42–70 d of age of calves), and the calves were weaned at 15 d of experiment. DMI and body weekly weight of all calves were recorded. Blood samples of nine healthy calves with similar body weight were collected from each group at 0, 7, 14, 16, 18, 21, and 28 d of experiment for detecting serum L-Gln, glucose, insulin, urea nitrogen, D-lactate, cortisol, haptoglobin, interleukin-8, immunoglobulin (Ig) G, IgA, IgM, total antioxidant capacity, superoxide dismutase, glutathione peroxidase, catalase, and malondialdehyde. At the end of the experiment, six healthy calves with similar body weight from each group were selected for slaughter and morphological analysis of small intestine tissue. The results showed that the L-Gln supplementation in the diets improved the negative effects of sudden weaning in calves. Furthermore, compared to the higher-level L-Gln supple-mentation (3 and 4% of DMI) groups, the dietary lower-level L-Gln supplementation (1 and 2% of DMI) had higher average daily gain, glutathione peroxidase and IgG concentration, and villus height/crypt depth of the duodenum and jejunum, as well as lower cortisol, haptoglobin, and interleukin-8 concentration of weaned calves. These results provided effective reference for relieving the negative effects of calves during weaning.

Keywords: weaning stress; L-Gln; calves; rumen-protected

Citation: Wang, S.; Wang, F.; Kong, F.; Cao, Z.; Wang, W.; Yang, H.; Wang, Y.; Bi, Y.; Li, S. Effect of Supplementing Different Levels of L-Glutamine on Holstein Calves during Weaning. *Antioxidants* **2022**, *11*, 542. https://doi.org/10.3390/antiox11030542

Academic Editors: Peter F. Surai and Stanley Omaye

Received: 14 January 2022
Accepted: 10 March 2022
Published: 12 March 2022

1. Introduction

In large-scale dairy farms, calves weaning as early as possible is necessary without negative effects. Not only does it increase the amount of milk available for human consumption, but it also improves dairy management and increases cow reproductive efficiency [1,2]. However, early weaning can result in significant physical and psychological damage to calves, producing weaning stress [3]. It is well known that weaning stress can reduce feed intake and damage the small intestinal structure of calves, which affects their productive capacity in adulthood [4–6].

Weaning stress is associated with significant changes in amino acid metabolism [7]. In particular, L-glutamine (L-Gln) metabolism after weaning is more intense than that before weaning. L-Gln, an energy source and precursor of endogenous substance (amino acid, nucleic acid, etc.) synthesis in the small intestinal epithelium, is the most abundant non-essential amino acid in mammals [5,8]. In addition, L-Gln is involved in the synthesis of glutathione, which indirectly protects cells from free-radical damage [9]. Therefore, L-Gln is commonly used as a functional antioxidant and energy supplement in the body [10,11]. L-Gln can effectively relieve weaning stress of piglets by increasing feed intake, improving immunity and antioxidant capacity, and increasing the length of small intestinal villus [12–14]. This response may be due to the involvement of L-Gln and other compounds in the adaptation and regeneration of intestinal tissues under stress conditions [7,15]. Furthermore, adding L-Gln to piglet diets was found to prevent intestinal atrophy during weaning, because L-Gln can stimulate lymphocyte regeneration and macrophage activity without affecting the normal intestinal structure [16,17]. Some studies also showed that L-Gln added at 1% of milk DM in high-dose milk benefitted the growth of small intestine [18], and addition of L-Gln at 1% of milk replacer DM in high-dose milk replacer enhanced the dry matter intake of calves [19]. Additionally, a recent study reported that supplementing L-Gln with 2% of DMI in 9 L/d milk weaned completely 3 d earlier than calves without L-Gln [20]. However, there are few studies on the effects of L-Gln on weaned calves. Although Hu et al. found that intravenous injection of low or medium concentrations of L-Gln in weaned calves significantly improved their immune capacity using parenteral nutrition technology [21], L-Gln injected directly into the bloodstream can be preferentially utilised by the liver and kidney rather than small intestine. Based on these results, dietary L-Gln supplementation that can effectively reduce negative effects of weaning in calves is unclear. Rumen-protected technology is widely used in ruminant nutrition to protect nutrients through the rumen, allowing the digestion and utilisation of nutrients in the small intestine. Therefore, rumen-protected L-Gln (RPG), as an excellent supply of L-Gln, provides L-Gln directly to the small intestine of calves.

In this study, we hypothesized that L-Gln supplementation in diet could effectively relieve weaning stress in a dose-dependent manner, which is probably due to a dose of L-Gln improving calves' immunity, antioxidant capacity, and intestinal morphology. Based on the hypotheses, this study was conducted to investigate the effects of L-Gln on the growth performance, immune function, antioxidant capacity, and small intestinal structure of weaned calves by adding RPG to a starter diet before and after weaning. Simultaneously, four different levels of L-Gln were used to determine the optimal supplemental level of L-Gln for weaned calves.

2. Materials and Methods

2.1. Animal and Management

A total of 75 male and healthy Holstein calves with similar body weight (BW) and age (54.0 ± 2.68 kg; 42 ± 2.1 d of age) were selected for the experiment. Each calf was housed in an individual pen (3 × 2 × 1.5-m^3; length × width × height) with sand bedding. The pens were cleaned before 0800 h every day to ensure the health and hygiene of the calves. All calves had the same feeding pattern. Briefly, all calves were drenched with a total of 6 L of colostrum, with 4 L drenched within 1 h and 2 L drenched 6 h after birth. Next, the calves were provided commercial milk replacer (Grober Nutrition, Cambridge, ON, Canada; water content: 4.02%, Crude protein: 22%, ether extract: 17%, lactose: 45%, and L-Gln: 0.468%) 2 times a day. The milk replacer was reconstituted as an emulsion (12.5%, w/v) using cooled (50–60 °C) boiled water and fed to the calves when the temperature cooled to 39 °C. Meanwhile, in the first week of life, the calves were fed 6 L of emulsion per day; 7 L of emulsion per day in the second week of life; and 8 L of emulsion per day from the third week to the weaning day (56 days of age). Along with the emulsion, a commercial pelleted starter feed (Lian Ying Co., Ltd.; water content: 9.35%, crude protein: 22.41%, crude ash: 6.49%, crude fibre: 8.45%, calcium: 0.97%, phosphorus: 0.63%, L-Gln: 0.116%)

and clean water (39 °C) were provided ad libitum from 3 d of age. Subsequently, the calves were weaned at 56 days (d) of age (15 d of experiment) and fed only the starter feed after weaning.

2.2. Experimental Design

These 75 calves, according to BW and age, were assigned to one of five groups: supplementing L-Gln at 0 (CON), 1% (1%Gln), 2% (2%Gln), 3% (3%Gln), or 4% (4%Gln) of DMI; each group consisted of 15 calves. The experiment lasted 4 weeks (42 d to 70 d of age of calves); all calves, expect the CON group, were provided L-Gln with a commercial RPG, which was provided by Wansheng Biological Co., Ltd. (Zhongwei, Ningxia, China) and contained 50% L-Gln. In situ ruminal degradability of RPG was 25.3% on average, and in vitro small intestinal degradability of RPG was 87.1% on average in our previous study (unpublished). Three Holstein bull with rumen fistula were selected for the in situ ruminal degradability experiment under 1.3 times the maintenance energy level. The starter feed intake of the calves was measured continuously for 3 d before the experiment, and the added amount of RPG was determined according to the DMI of each calf. In addition, we adjusted the amount of RPG per week according to the change in the levels of DMI in each calf. Except for the Con group, each calf was first fed a mixture of 200 g starter feed and RPG every day, and then remaining starter feed was fed when the RPG was exhausted.

2.3. Feed Intake and Growth Performance Measurement

To calculate the DMI (milk DMI + starter feed DMI; g/d) of each calf, the remaining and added amounts of starter feed were recorded at 0800 h every day, and body weights of the calves were measured every 7 d thereafter before the morning feeding during the experimental period. The average daily gain (ADG) of the calves was calculated every 7 d. Feed efficiency was calculated as ADG/DMI. The content of L-Gln in the milk replacer and starter feed were detected via ultra-performance liquid chromatography-tandem mass spectrometry (UPLC-MS/MS). The UPLC with an Acquity BEH C_{18} column (50 mm × 2.1 mm, 1.7 μm particle size) (Waters, Milford, MA, USA), and the UPLC system was coupled to a Micromass Xevo TQ-S triple quadrupole mass spectrometer (Waters, Manchester, UK) fitted with an electrospray ionization source in negative mode. The referenced detection method was described by Xiao et al. [22].

2.4. Blood Sample Collection and Measurement

Nine healthy calves with similar body weight were selected from each group to collect blood samples from the jugular vein before the morning feeding on 0, 7, 14, 16, 18, 21, and 28 d of the experiment. Each blood sample was collected using 10 mL non-anticoagulant tubes. Next, blood samples in non-anticoagulant tubes were centrifuged at 1500× *g* for 30 min at 4 °C (Tiangen OSE-MP25, Beijing, China) to obtain serum, which was then stored in 1.5 mL centrifuge tubes in liquid nitrogen for further analyses.

The concentrations of serum immunoglobulin (Ig) G, IgA, IgM, D-lactate, cortisol, haptoglobin (HP), and interleukin-8 (IL-8) were determined using respective ELISA kits (Nanjing Jian Cheng Bioengineering Institute, Nanjing, China). Serum glucose (Glu), insulin (INS), and urea nitrogen (SUN) concentrations were determined using an automatic biochemical analyser (Kehua-zy KHB-1280, Shanghai, China). Total antioxidant capacity (T-AOC), superoxide dismutase (SOD), glutathione peroxidase (GSH-PX), malondialdehyde (MDA), and catalase (CAT) activity and L-Gln concentration in the serum were determined using commercial kits (Nanjing Jian Cheng Bioengineering Institute, Nanjing, China).

2.5. Small Intestine Sampling, Processing, and Morphological Analysis

At the end of the experiment, six healthy calves with similar body weight (the calves from whom blood samples had been obtained) were slaughtered from each group. Next, we collected the entire small intestine of the calves, from the pyloric sphincter to the ileocecal valve. The duodenum, jejunum, and ileum samples were collected 6 cm posterior to the pylorus, the middle part of the jejunum, and 6 cm posterior to the jejunum-ileum junction by an animal anatomist. All tissue samples were immediately washed with saline to remove the chyme of the samples then preserved in 10% buffered formalin for analysis of intestinal morphology. After fixation in the buffered formalin, the samples were trimmed to maintain an integrated cross section. Next, the sections were trimmed from each sample by a transversal cut through the villi and dehydrated overnight using graded alcohol (50%, 75%, 85%, 95%, and 100% alcohol) at ambient temperature. The samples were cleared in a mixture of xylene: ethyl alcohol = 1:1) and then impregnated with Histosec paraffin pastilles (Merck Ltd., Auckland, New Zealand). Subsequently, the samples were embedded (KD-BM, Kedi Ltd., Jinghua, China) and cut using a paraffin slicing machine (RM2016, Laika Ltd., Shanghai, China). The 5 μm-thick sections were stained with haematoxylin-eosin until they were no longer faded. Morphometric analysis included the evaluation of villus height and width, crypt depth, villus height/crypt depth (V/C). Measurements were taken from 10 villi per section and 2 sections per calf. The measurements were performed by an investigator blinded to the treatment allocation of the calves using the image processing Pannoramic Viewer software (v 1.15.3, 3DHISTECH Ltd., Budapest, Pest, Hungary); the measurement criteria are described in Figure 1.

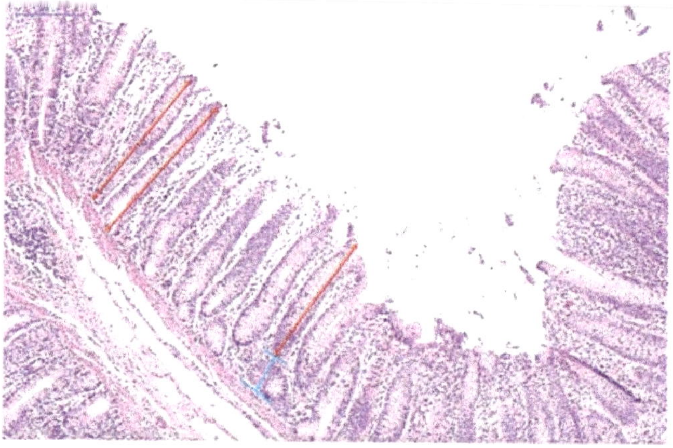

Figure 1. Representative histological pictures of the small intestine, magnified 80×. Red lines represent villus height; blue lines represent crypt depth.

2.6. Statistical Analyses

Statistical analyses were performed using R software (v. 4.0.3, https://www.r-project.org/, accessed on 25 June 2021). The statistical power for the serum samples in this study was >0.8 with a significance level of 0.05 using G*Power software (v. 3.1.9.6, https://g-power.apponic.com, accessed on 25 June 2021). Before analysis, all data were checked for normality using shapiro.test function of the R software, and the data fitting non-normality was log-transformed. Next, indicators of feed intake, growth performance, serum immunity, and serum antioxidants were analysed via a mixed linear model, as follows:

$$Y_{ijkl} = \mu + T_i + D_j + TD_{ij} + Calf_k + e_{ijkl}$$

where Y_{ijkl} is the dependent variable; μ is the average experimental value; T_i is the fixed effect of L-Gln supplementation or not; D_j designates the repeated effect; TD_{ij} is the interaction effect of T_i and D_j; $Calf_k$ is the random effect of calf; and e_{ijkl} is the error term. Another mixed linear model was used to analyse the intestinal tissue samples; the relevant model is as follows:

$$Y_{ijk} = \mu + T_i + Calf_j + e_{ijk}$$

where Y_{ijk} is the dependent variable; μ is the average experimental value; T_i is the fixed effect of L-Gln supplementation or not; $Calf_j$ is the random effect of calf; and e_{ijk} is the error term. The values of all data were reported as least squares means, and polynomial orthogonal contrasts in R software were used to test linear and quadratic responses. Differences of $p \leq 0.05$ were considered significant, and those of $0.05 < p \leq 0.10$ were considered to show a tendency of difference.

3. Results

3.1. Feed Intake and Growth Performance

With increasing quantity of L-Gln added to the diet, the body weight of the calves in the 4-week experimental period had a quadratically increasing tendency, whereas BW at 1-, 2-, and 3-week experimental periods was not affected (Table 1). Meanwhile, L-Gln supplementation significantly boosted the average daily gain, and increasing the quantity of L-Gln quadratically increased the ADG during the pre-weaning period and total experimental period. We also found that L-Gln supplementation significantly enhanced the DMI, and the DMI had a linear increase during the pre-weaning period and total experimental period. L-Gln intake was significantly affected by L-Gln supplementation, and it showed a linear increase with increasing supplemental levels of L-Gln. In addition, the DMI during the total experimental period showed a quadratically increasing tendency. L-Gln supplementation during the post-weaning experimental period significantly affected the feed efficiency, and the feed efficiency showed a quadratically increasing tendency with increasing supplemental levels of L-Gln.

Table 1. Growth performance and feed intake of calves during different treatments (N = 15 per group).

Items	Treatment					SEM	*p*-Value			
	CON	1%Gln	2%Gln	3%Gln	4%Gln		T	T × D	Linear	Quadratic
BW, kg										
0	57.8	58.0	58.1	56.9	57.7	0.28	0.86		0.64	0.93
1-wk	62.3	63.6	63	61.6	63.3	0.31	0.92		0.99	0.97
2-wk	67.1	68.8	68.6	66.3	68.3	0.32	0.73		0.97	0.64
3-wk	68.6	72.0	72.1	68.3	71.7	0.45	0.21		0.53	0.45
4-wk	72.3	76.6	77.2	72.5	75.3	0.45	0.07		0.62	0.09
ADG, kg/d										
Pre-weaning	0.67	0.77	0.75	0.68	0.76	0.01	0.36	0.29	0.40	0.42
Post-weaning	0.37	0.56	0.61	0.44	0.50	0.02	<0.01	0.12	0.29	0.01
Overall	0.52	0.67	0.68	0.56	0.63	0.01	<0.01	0.61	0.21	0.02
DMI, g/d										
Pre-weaning	1183	1372	1355	1293	1391	14.5	<0.01	0.92	0.01	0.17
Post-weaning	1645	1864	1822	1685	1821	16.3	0.22	0.56	0.17	0.21
Overall	1414	1618	1588	1489	1606	13.3	<0.01	0.42	0.01	0.09
L-Gln intake [1], g/d										
Pre-weaning	4.72	18.48	31.86	43.53	60.39	2.35	<0.01	0.77	<0.01	0.17
Post-weaning	0.32	19.00	36.78	50.87	73.19	3.05	<0.01	0.19	<0.01	0.23
Overall	2.52	18.74	34.32	47.20	66.79	2.69	<0.01	0.39	<0.01	0.14
Feed efficiency										
Pre-weaning	0.57	0.56	0.55	0.53	0.55	0.02	0.78	0.95	0.98	0.96
Post-weaning	0.22	0.30	0.34	0.26	0.28	0.03	0.03	0.31	0.37	0.08
Overall	0.37	0.42	0.42	0.38	0.39	0.02	0.26	0.76	0.63	0.21

[1] L-Gln intake (g/d) = L-Gln intake of milk replacer (milk replacer intake(mL) × 0.125(w/v, g/mL) × 0.00468) + L-Gln intake of starter feed (feed intake(g/d) × 0.00116× 0.15 (the rumen bypass efficiency of feed protein was 15% in NRC2001 [23])) + L-Gln supplementation (0 or 1% or 2% or 3% or 4% of DMI); BW, body weight; ADG, average daily gain; DMI, dry matter intake; Feed efficiency was calculated as ADG/DMI; T, treatment; D, day.

3.2. Serum L-Gln, Glu, Urea Nitrogen, and Insulin

We found that the concentrations of serum L-Gln and Glu were not different among the five groups during the entire experimental period (Table 2). Interestingly, L-Gln supplementation at 16, 18, and 28 d of experiment and during post-weaning and the whole experiment period significantly enhanced the concentration of SUN. Additionally, we also found that the SUN concentrations showed linear increases with increasing supplemental levels of L-Gln during post-weaning and the whole experiment period; at 16 and 18 d of the experiment, the concentration of SUN showed linear and quadratic increases with increasing supplemental levels of L-Gln. Furthermore, L-Gln supplementation significantly boost INS concentration, and the INS concentration presented linear increasing with increasing L-Gln levels during the pre-weaning period.

Table 2. Serum Gln, Glu, BUN, and INS indicators of calves among different groups (N = 9 per group).

Items	Treatment					SEM	*p*-Value			
	CON	1%Gln	2%Gln	3%Gln	4%Gln		T	T × D	Linear	Quadratic
L-Gln, mmol/L										
0 d	0.67	0.66	0.68	0.65	0.68	0.02	0.77		0.81	0.57
7 d	0.71	0.76	0.75	0.70	0.74	0.02	0.20		0.37	0.63
14 d	0.72	0.74	0.76	0.73	0.75	0.03	0.93		0.59	0.89
16 d	0.75	0.75	0.77	0.77	0.78	0.02	0.64		0.56	0.79
18 d	0.74	0.78	0.77	0.81	0.79	0.03	0.16		[illegible]	[illegible]
21 d	0.80	0.81	0.74	0.80	0.81	0.04	0.64		0.18	0.59
28 d	0.82	0.84	0.81	0.81	0.84	0.04	0.97		0.73	0.63
Pre-weaning	0.70	0.72	0.73	0.69	0.72	0.02	0.39	0.58	0.66	0.29
Post-weaning	0.78	0.80	0.77	0.80	0.81	0.03	0.67	0.69	0.19	0.72
Overall	0.74	0.76	0.75	0.75	0.77	0.01	0.72	0.44	0.97	0.33
Glu, mmol/L										
0 d	6.12	6.16	6.12	6.16	6.14	0.18	1.00		0.98	0.86
7 d	6.18	6.06	5.99	6.04	5.51	0.23	0.29		0.83	0.44
14 d	5.60	5.52	5.68	5.39	5.32	0.12	0.17		0.46	0.45
16 d	4.20	4.36	4.34	4.31	4.32	0.17	0.97		0.61	0.69
18 d	5.23	5.44	4.93	5.09	5.11	0.18	0.35		0.27	0.26
21 d	5.43	5.34	5.20	5.23	5.42	0.14	0.13		0.29	0.43
28 d	5.44	5.33	5.28	5.19	5.22	0.20	0.91		0.57	0.63
Pre-weaning	5.97	5.91	5.93	5.86	5.66	0.15	0.71	0.31	0.51	0.82
Post-weaning	5.08	5.12	4.94	4.96	5.02	0.09	0.33	0.57	0.64	0.46
Overall	5.46	5.46	5.36	5.34	5.29	0.08	0.62	0.39	0.71	0.82
SUN, mmol/L										
0 d	4.54	4.29	4.32	4.30	4.42	0.51	1.00		0.74	0.83
7 d	4.18	4.75	4.41	5.31	5.47	0.79	0.71		0.91	0.31
14 d	2.43	2.69	2.73	2.58	2.82	0.33	0.93		0.74	0.70
16 d	3.31	5.47	5.35	4.38	5.09	0.44	<0.01		0.04	0.05
18 d	8.50	9.67	10.54	12.02	11.51	0.58	<0.01		0.02	0.04
21 d	12.56	12.71	11.99	13.00	12.94	0.79	0.91		0.68	0.45
28 d	8.31	8.85	9.81	10.37	11.18	0.64	0.02		0.31	0.23
Pre-weaning	3.72	3.91	3.82	4.06	4.24	0.53	0.42	0.99	0.26	0.62
Post-weaning	8.17	9.18	9.42	9.94	10.26	0.62	<0.01	0.58	0.02	0.13
Overall	6.26	6.92	7.02	7.42	7.68	0.33	0.04	0.44	0.04	0.14
INS, mIU/L										
0 d	19.49	22.34	19.86	20.58	21.30	1.98	0.84		1.00	0.30
7 d	15.79	18.29	19.32	17.49	18.47	3.45	0.96		0.59	0.86
14 d	14.58	17.51	20.71	15.01	17.69	2.68	0.54		0.28	0.78
16 d	8.73	8.93	10.01	6.50	7.85	1.66	0.65		0.81	0.40
18 d	12.73	18.13	17.08	10.92	13.75	2.11	0.13		0.38	0.87
21 d	14.18	15.56	15.21	13.58	13.98	1.53	0.88		0.73	0.99
28 d	4.76	6.01	5.09	5.54	6.27	0.94	0.77		0.98	0.26
Pre-weaning	16.62	19.38	19.96	17.69	19.15	1.92	0.04	0.72	0.03	0.17
Post-weaning	10.10	12.16	11.85	9.14	10.46	1.73	0.74	0.55	0.48	0.53
Overall	12.89	15.25	15.33	12.80	14.19	0.99	0.15	0.81	0.23	0.19

Glu, glucose; SUN, serum urea nitrogen; INS, insulin; T, treatment; D, day.

3.3. Stress-Related Hormone Indicators

To assess the stress status of calves in each group, we measured the concentrations of serum cortisol, HP, IL-8, and D-lactate (Table 3). Serum cortisol concentrations in L-Gln supplementation groups were significantly higher than that in the CON group at 16 and 21 d of experiment, and during the post-weaning period. Meanwhile, they showed a significant increasing tendency compared to the CON group. Furthermore, we also found cortisol concentration quadratically decreased as L-Gln inclusion in the diet increased at 16 and 21 d of experiment and during the post-weaning and the whole experiment period. Although L-Gln supplementation groups had no effect on the concentration of serum HP and IL-8, we found that HP and IL-8 concentrations showed a quadratic decrease with increasing L-Gln levels in the diets at 16 and 21 d of experiment and during the post-weaning period. At 21 d of experiment, L-Gln supplementation groups significantly decreased serum D-lactate concentration, and D-lactate level showed a quadratic decrease with increasing L-Gln levels in the diets. In addition, we also found that D-lactate showed a quadratically decreased tendency with increasing L-Gln levels in the diets during the post-weaning and whole experiment period.

Table 3. Serum stress indicators of calves among different groups (N = 9 per group).

Items	Treatment					SEM	*p*-Value			
	CON	1%Gln	2%Gln	3%Gln	4%Gln		T	T × D	Linear	Quadratic
D-lactate, μmol/L										
0 d	6.18	6.16	6.30	6.25	6.32	0.05	0.63		0.24	0.86
14 d	6.09	5.98	6.12	6.02	6.13	0.05	0.71		0.78	0.65
16 d	7.52	7.21	7.47	7.82	7.16	0.10	0.83		0.78	0.8
21 d	6.72	6.17	6.03	6.07	6.41	0.07	<0.01		0.12	<0.01
28 d	6.10	6.13	6.00	6.17	6.17	0.06	0.25		0.66	0.61
Pre-weaning	6.14	6.07	6.21	6.14	6.23	0.04	0.97	0.99	0.88	0.79
Post-weaning	6.78	6.50	6.50	6.69	6.58	0.06	0.18	0.58	0.23	0.09
Overall	6.52	6.33	6.38	6.47	6.44	0.03	0.15	0.63	0.14	0.06
Cortisol, μg/dL										
0 d	3.09	3.01	3.22	3.06	2.99	0.11	0.94		0.85	0.72
14 d	3.16	3.19	3.39	3.16	3.18	0.11	0.92		1.00	0.64
16 d	6.72	5.58	5.79	5.85	6.08	0.12	<0.01		0.18	0.01
21 d	5.28	4.29	4.33	4.84	5.13	0.12	<0.01		0.74	<0.01
28 d	3.30	3.20	3.20	3.29	3.33	0.10	0.79		0.85	0.68
Pre-weaning	3.13	3.10	3.31	3.11	3.09	0.10	0.92	0.84	0.77	0.59
Post-weaning	5.10	4.36	4.44	4.66	4.85	0.09	<0.01	0.92	<0.01	<0.01
Overall	4.31	3.85	3.99	4.04	4.14	0.07	0.07	0.36	0.17	0.02
HP, ng/mL										
0 d	0.53	0.52	0.54	0.57	0.51	0.05	0.92		0.77	0.88
14 d	0.54	0.51	0.55	0.52	0.51	0.04	0.88		0.82	0.75
16 d	0.81	0.65	0.67	0.79	0.83	0.05	0.22		0.64	0.03
21 d	0.73	0.62	0.63	0.76	0.74	0.03	0.37		0.83	0.05
28 d	0.56	0.58	0.54	0.61	0.59	0.05	0.83		0.79	0.62
Pre-weaning	0.54	0.52	0.55	0.55	0.51	0.04	0.91	0.92	0.80	0.78
Post-weaning	0.70	0.62	0.61	0.72	0.72	0.04	0.29	0.84	0.63	0.04
Overall	0.63	0.58	0.59	0.65	0.64	0.03	0.34	0.96	0.52	0.17
IL-8, pg/mL										
0 d	50.87	46.07	49.84	47.13	47.92	3.35	0.22		0.50	0.66
14 d	44.14	45.94	47.21	44.01	46.99	2.26	0.35		0.72	0.40
16 d	64.17	49.94	52.88	65.93	64.24	3.13	0.16		0.87	0.03
21 d	58.25	47.46	46.62	55.37	54.29	3.15	0.19		0.27	0.04
28 d	40.89	39.12	41.99	42.75	40.86	1.26	0.47		0.91	0.55
Pre-weaning	47.51	46.01	48.53	45.57	47.46	2.56	0.32	0.58	0.92	0.82
Post-weaning	54.44	45.51	47.16	54.68	53.13	2.02	0.13	0.63	0.25	0.02
Overall	51.66	45.71	47.71	51.04	50.86	1.46	0.29	0.81	0.40	0.22

HP, haptoglobin; IL-8, interleukin-8, T, treatment; D, day.

3.4. Oxidative Status

Table 4 shows the antioxidant capacity in serum. The T-AOC in L-Gln supplementation groups, during the post-weaning and whole experiment period, were significantly higher than that in the CON group, and it linearly and quadratically increased with L-Gln added to the diets. Furthermore, T-AOC in L-Gln supplementation groups showed a significant difference and tendency compared to that in the CON group, respectively, and it presented a linear increasing and quadratic increasing tendency with increasing L-Gln levels at 14 and 16 d of experiment, respectively. GSH-Px level in L-Gln supplementation groups during the pre-weaning, post-weaning, and whole experiment period were significantly higher than that in the CON group, and it showed a linear and quadratic increase with L-Gln added to the diets. Moreover, we also found that the GSH-Px level in L-Gln supplementation groups at 7, 14, and 16 as well as 18 d of experiment showed a significant increase and increasing tendency, respectively, and it also showed a linear and quadratic increase with L-Gln added to the diets. Although there was no significant difference for SOD between L-Gln supplementation groups and CON, the SOD level showed a quadratic increase with increasing L-Gln level added to the diets at 16 d of experiment. At 16 and 18 d of experiment, the CAT in L-Gln supplementation groups were significantly higher than that in the CON group, and increasing L-Gln level added in diets quadratically increased the CAT level. In contrast, MDA in L-Gln supplementation groups were significantly lower than that in the CON group, and it quadratically decreased with increasing Gln levels at 18 d of the experiment.

Table 4. Serum antioxidant indicators of calves among different groups (N = 9 per group).

Items	Treatment					SEM	*p*-Value			
	CON	1%Gln	2%Gln	3%Gln	4%Gln		T	T × D	Linear	Quadratic
T-AOC, U/mL										
0 d	23.3	22.2	23.3	26.8	24.6	2.39	0.70		0.77	0.78
7 d	23.2	28.0	28.6	26.1	26.7	2.91	0.32		0.07	0.80
14 d	20.9	27.5	26.1	23.9	26.7	1.79	0.09		0.19	0.09
16 d	18.6	24.2	26.7	26.4	25.8	1.84	0.03		0.01	0.16
18 d	21.4	25.3	24.7	28.7	26.5	2.75	0.48		0.33	0.20
21 d	22.2	28.5	26.3	25.8	26.2	2.13	0.42		0.27	0.15
28 d	22.0	24.5	25.2	25.1	24.6	1.94	0.79		0.29	0.56
Pre-weaning	22.5	25.9	26.0	25.6	26.0	2.04	0.18	0.57	0.11	0.47
Post-weaning	21.1	25.6	25.7	26.5	25.8	1.88	<0.01	0.62	<0.01	<0.01
Overall	21.7	25.7	25.8	26.1	25.9	0.66	<0.01	0.23	<0.01	<0.01
GSH-Px, U/mL										
0 d	60.3	60.9	61.1	63.0	61.8	5.03	1.00		0.86	0.84
7 d	87.1	116.2	112.0	110.2	109.4	6.20	0.04		0.02	0.03
14 d	59.5	87.7	85.7	83.8	85.0	6.50	0.05		0.02	0.04
16 d	47.4	60.3	61.1	52.6	53.4	3.26	0.04		0.02	0.04
18 d	51.4	64.4	67.7	63.0	58.1	8.28	0.07		0.03	0.02
21 d	55.8	66.7	71.6	67.4	64.5	7.48	0.12		0.14	0.08
28 d	60.3	73.1	76.2	69.4	69.4	12.06	0.93		0.43	0.80
Pre-weaning	69.0	88.3	86.3	85.7	85.4	4.64	<0.01	<0.01	<0.01	<0.01
Post-weaning	53.7	66.1	69.2	63.1	61.4	7.52	<0.01	0.82	<0.01	<0.01
Overall	60.3	75.6	76.5	72.8	71.7	2.87	<0.01	0.23	<0.01	0.05
SOD, U/mL										
0 d	73.2	74.6	79.1	75.9	77.0	2.57	0.60		0.22	0.81
7 d	85.1	84.7	83.1	82.0	80.1	1.92	0.77		0.83	0.68
14 d	85.4	86.4	87.0	90.0	86.6	2.56	0.78		0.44	0.63
16 d	89.0	93.8	91.6	93.0	93.1	1.64	0.29		0.34	0.05
18 d	85.2	90.3	91.9	92.4	90.6	3.92	0.73		0.24	0.53
21 d	88.1	91.3	91.6	93.5	90.9	2.05	0.51		0.16	0.34
28 d	80.3	82.1	81.5	83.9	82.3	4.75	0.99		0.79	0.71

Table 4. *Cont.*

Items	Treatment					SEM	p-Value			
	CON	1%Gln	2%Gln	3%Gln	4%Gln		T	T × D	Linear	Quadratic
Pre-weaning	81.2	81.9	83.1	82.6	81.2	2.04	0.95	0.91	0.84	0.54
Post-weaning	85.7	89.4	89.2	90.7	89.2	3.16	0.76	0.36	0.21	0.52
Overall	83.8	86.2	86.5	87.2	85.8	1.27	0.42	0.66	0.13	0.33
CAT, U/mL										
0 d	28.7	31.9	31.0	29.1	32.3	1.67	0.42		0.80	0.36
7 d	32.1	33.3	34.6	32.3	37.1	2.68	0.63		0.98	0.80
14 d	35.1	35.3	35.3	35.2	38.1	1.70	0.69		0.66	0.63
16 d	35.3	37.4	38.6	38.9	34.6	1.01	0.03		0.17	0.04
18 d	36.8	38.9	38.3	38.4	37.8	1.03	0.04		0.14	0.03
21 d	41.2	42.9	44.4	44.9	44.0	1.37	0.37		0.11	0.52
28 d	43.4	44.3	46.0	49.4	44.9	2.31	0.44		0.21	0.63
Pre-weaning	32.0	33.5	33.6	32.2	35.8	1.93	0.64	0.88	0.48	0.79
Post-weaning	39.2	40.9	41.8	42.9	40.3	0.81	0.11	0.72	0.17	0.12
Overall	36.1	37.7	38.3	38.3	38.4	0.71	0.19	0.92	0.04	0.53
MAD, nmol/mL										
0 d	2.97	2.95	3.03	3.00	2.85	0.13	0.88		0.52	0.59
7 d	3.17	3.27	3.19	3.26	3.20	0.05	0.49		0.56	0.11
14 d	3.18	3.19	3.22	3.20	3.20	0.04	0.97		0.51	0.98
16 d	3.51	3.37	3.29	3.43	3.51	0.09	0.32		0.66	0.06
18 d	3.41	3.28	3.24	3.32	3.27	0.03	0.04		0.57	0.04
21 d	3.15	3.20	3.18	3.14	3.17	0.03	0.67		0.84	0.60
28 d	3.12	3.14	3.16	3.15	3.14	0.02	0.65		0.16	0.75
Pre-weaning	3.11	3.14	3.15	3.15	3.08	0.04	0.66	0.93	0.58	0.65
Post-weaning	3.30	3.25	3.22	3.26	3.27	0.03	0.18	0.94	0.49	0.13
Overall	3.22	3.20	3.19	3.21	3.19	0.02	0.41	0.98	0.13	0.43

T-AOC, total oxidative capacity; GSH–Px, glutathione peroxidase; SOD, superoxide dismutase; CAT, catalase; MDA, malonaldehyde, T, treatment; D, day.

3.5. Immune Status

Table 5 shows the immune capacity among groups. IgG concentration in L-Gln supplementation groups, during the pre-weaning, post-weaning, and whole experiment period, were significantly higher than that in the CON group, and it showed a linear and quadratic increase with L-Gln supplementation in the diets. We also found that, except for at 0 d of experiment, L-Gln supplementation groups boosted significantly or had a significant trend on IgG concentration at other sampled times. For IgA concentration, it appeared to have a quadratic increasing tendency at 14 d of the experiment. The serum IgM in L-Gln supplementation groups was significantly higher than that in the CON group, and increasing L-Gln level added to the diets quadratically increased the serum IgM concentration at 7 and 14 d of experiment and during the pre-weaning and whole experiment period. Furthermore, the serum IgM concentration showed a linear increase and linear increasing tendency at 7 d of experiment and during the pre-weaning period as well as at 14 d of experiment, respectively.

Table 5. Serum immune indicators of calves among different groups (N = 9 per group).

Items	Treatment					SEM	*p*-Value			
	CON	1%Gln	2%Gln	3%Gln	4%Gln		T	T × D	Linear	Quadratic
IgG, g/L										
0 d	10.78	11.71	10.76	11.26	12.11	1.32	0.93		0.84	0.46
7 d	4.14	5.72	5.05	4.38	4.78	0.41	0.09		0.30	0.17
14 d	4.65	6.95	6.31	6.24	6.74	0.49	0.02		0.09	0.01
16 d	2.62	4.47	4.02	3.85	3.75	0.40	0.04		0.03	0.04
18 d	1.80	3.09	2.85	2.73	2.80	0.21	<0.01		0.01	<0.01
21 d	2.29	4.08	3.92	3.67	3.74	0.31	<0.01		<0.01	0.01
28 d	5.11	7.84	7.85	7.29	7.30	0.74	0.09		0.03	0.13
Pre-weaning	6.52	8.13	7.37	7.29	7.88	0.55	<0.01	0.58	<0.01	<0.01
Post-weaning	2.96	4.87	4.66	4.39	4.40	0.33	<0.01	0.31	<0.01	<0.01
Overall	4.48	6.27	5.82	5.63	5.89	0.30	<0.01	0.52	0.02	<0.01
IgA, g/L										
0 d	0.81	0.78	0.81	0.80	0.77	0.01	0.29		0.55	0.08
7 d	0.88	0.80	0.90	0.87	0.87	0.03	0.30		0.83	0.07
14 d	0.68	0.74	0.74	0.75	0.73	0.02	0.26		0.10	0.13
16 d	0.71	0.79	0.74	0.74	0.73	0.03	0.27		0.57	0.12
18 d	0.82	0.83	0.80	0.83	0.87	0.05	0.86		0.55	0.45
21 d	0.70	0.70	0.69	0.71	0.69	0.02	0.86		0.94	0.84
28 d	0.74	0.79	0.80	0.80	0.80	0.03	0.53		0.20	0.40
Pre-weaning	0.79	0.77	0.82	0.81	0.79	0.02	0.29	0.91	0.42	0.55
Post-weaning	0.74	0.78	0.76	0.77	0.77	0.03	0.79	0.88	0.34	0.42
Overall	0.76	0.78	0.78	0.79	0.78	0.01	0.74	0.91	0.31	0.49
IgM, g/L										
0 d	2.39	2.30	2.31	2.29	2.36	0.08	0.85		0.33	0.62
7 d	2.36	2.65	2.88	2.75	2.80	0.11	0.03		0.01	0.40
14 d	2.35	2.55	2.53	2.71	2.58	0.08	0.05		0.06	0.03
16 d	2.45	2.69	2.50	2.55	2.56	0.08	0.20		0.69	0.04
18 d	2.64	2.79	2.73	2.77	3.02	0.14	0.43		0.85	0.25
21 d	2.57	2.57	2.51	2.60	2.57	0.07	0.94		0.69	0.62
28 d	2.60	2.68	2.67	2.69	2.67	0.09	0.95		0.58	0.58
Pre-weaning	2.37	2.50	2.57	2.58	2.58	0.09	<0.01	0.56	<0.01	<0.01
Post-weaning	2.57	2.68	2.60	2.65	2.71	0.06	0.37	0.63	0.36	0.51
Overall	2.48	2.60	2.59	2.62	2.65	0.04	0.02	0.47	0.11	0.01

T, treatment; D, day.

3.6. Small Intestinal Morphology

As shown in Figure 2, we observed that increasing the L-Gln quadratically increased the villus height and V/C of the duodenum and jejunum. In contrast, the crypt depth of the duodenum and jejunum decreased quadratically with increasing L-Gln levels.

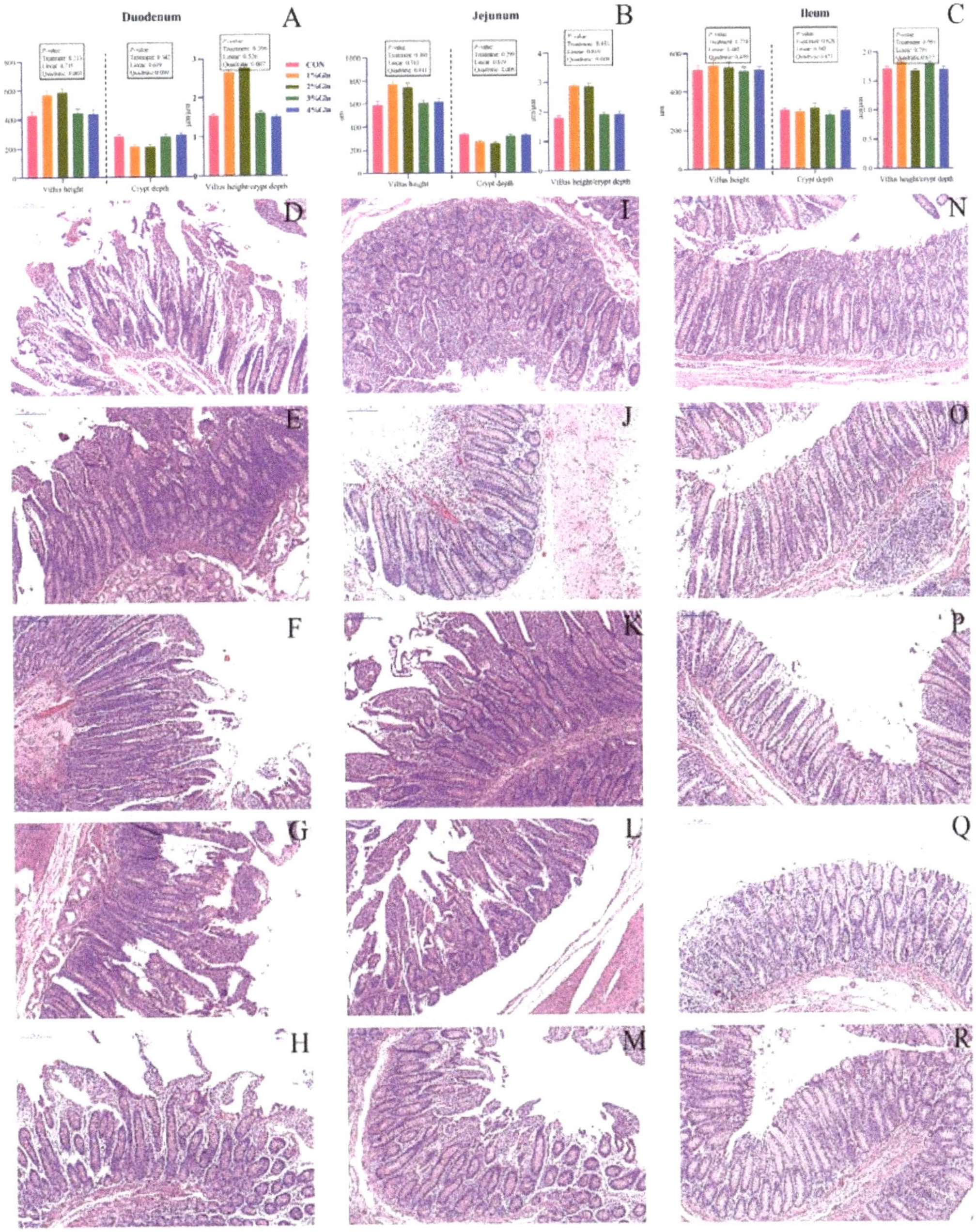

Figure 2. Histomorphology of the small intestine ((**A**): duodenum, (**B**): jejunum, and (**C**): ileum; mean ± SEM) among different groups (N = 6 per group). Light micrographs (80× magnification, blue bar represents 200 μm) of the duodenum (**D–H**), Jejunum (**I–M**), Ileum (**N–R**) of calves in CON (**D**,**I**,**N**) or 1%Gln (**E**,**J**,**O**) or 2%Gln (**F**,**K**,**P**) or 3%Gln (**G**,**L**,**Q**) or 4%Gln (**H**,**M**,**R**).

4. Discussion

L-Gln, as a functional amino acid, is widely used as a feed additive in pigs [16,17] and chickens [24] to combat stress injury. However, research on glutamine in young ruminants is limited. Considerable development in gastrointestinal physiology occurs during the first few weeks of life, allowing calves to quickly adapt to the liquid–solid dietary transition. During the transition, the calf often experiences weaning stress, which poses great challenges to its health and adult productivity. To our knowledge, this study is the first to examine the effects of rumen-protected L-Gln supplementation in the diet on calves during weaning period. The findings revealed that low L-Gln supplementation (1 and 2% of DMI) in the diet better relieved the negative effects of weaning on calves via enhancing the immune and antioxidant capacity as well as improving intestinal morphology.

Growth performance is a crucial indicator of evaluating weaning stress status; calves with severe weaning stress have lower ADG and feed efficiency [3]. Although the ADG and feed efficiency of weaned calves were lower than pre-weaning in this study, we found that the addition of L-Gln to the diets improved the growth performance of weaned calves, and low L-Gln level supplementation (1 and 2% of DMI) had better effects. Similar to our result, Ma et al. reported that medium L-Gln level (160 g/d) supplementation could promote the compensatory growth of growth-retarded yaks compared to the high level (180 g/d) [25]. Another recent study also found that compared with no supplementation, the ADG of weaned calves at 35 d of age was increased by adding L-Gln with 2% of DMI to diet, and weaned completely 3 d earlier [20]. However, Hu et al. found that different doses of glutamine administered intravenously did not improve the growth performance of calves during the post-weaning period, possibly because intravenous administration of L-Gln first entered the liver and kidney for metabolic utilisation, whereas L-Gln entering the small intestine did not reach the threshold of effective concentration [21]. In addition, we found that L-Gln supplementation also increased the DMI of calves during the pre-weaning and total experiment period, but had no effect on feed efficiency. Da Silva et al. found a similar result when they added L-Gln in milk replacers for calves [19]. These results also indicate that L-Gln has potential as a functional food attractant for calves in the future.

Cortisol and HP are key indicators of reacting stress degree [26,27]. IL-8 is a biomarker of weaning stress [28]. In contrast to DMI and feed efficiency, cortisol, HP, and IL-8 of weaned calves were lower than pre-weaning in this study. Furthermore, they showed a quadratic decrease with increasing L-Gln level added in the diet. In particular, 1 and 6 d after weaning the calves, the lower-level L-Gln (1 and 2 % of DMI) groups also had better results, which further confirmed that the lower levels of L-Gln can alleviate the adverse effects of weaning in calves. Interestingly, L-Gln supplementation had no effect on serum L-Gln concentration in this study, which is similar to previous studies about piglets [29,30]. These responses indicate that L-Gln added to the diet will be metabolized by the digestive tract and other organs before entering the bloodstream. Prior to weaning, L-Gln supplementation increased serum insulin concentration and linearly increased with supplemental level, whereas serum glucose concentration was not affected. Wu reported that although L-Gln was essential for the integrity and development of the gut, 90% of it is not utilized efficiently in normal conditions [31]. Combined with the SUN results before weaning, we thought that part of L-Gln, which could not be utilized by the small intestine, might be converted into glucose in blood, and the body maintains glucose homeostasis by raising insulin levels. After weaning, L-Gln supplementation linearly increased SUN concentration. The reason for the results that L-Gln might be conversed to the glutamate and ammonium providing the energy for small intestine by deamination in the small intestine's mitochondria. The ammonium will eventually contribute to hepatic ureagenesis and therefore to the increase the SUN concentrations. Interestingly, we still found that the lower-level L-Gln (1% and 2% of DMI) had higher SUN concentrations at 1 d after weaning. The result also suggested that the lower-level L-Gln could be utilized efficiently by the small intestine after weaning in this study, which might also be why higher-level

L-Gln (3% and 4% of DMI) had no better effect than low-level L-Gln (1% and 2% of DMI) supplementation groups on alleviating the negative effects of sudden weaning.

The negative effect of sudden weaning is usually accompanied by oxidative stress; therefore, the antioxidant capacity of calves can reflect their ability to resist weaning stress. Normally, there is a redox balance between the production of reactive oxygen species in the body's intracellular environment and the ability of biological antioxidant systems to neutralise these active intermediates [32,33]. Oxidative stress disrupts this balance and causes tissue damage [34]. In turn, the body produces antioxidant enzymes to reduce ROS; for example, GSH-Px is an important antioxidant in the body, and glutamate, the metabolite of glutamine in cells, is one of its main components [35,36]; SOD and CAT can clear free radicals and hydrogen peroxide in cells, respectively [34]; MDA is the oxidation end-product of hydrogen peroxide and can also indirectly reflect the degree of tissue peroxide damage [37]. Therefore, the greater the levels of SOD, CAT, and GSH-Px and the lower the levels of MDA, the stronger the antioxidant capacity of the body. Combined with the results of this study, these results suggest that dietary L-Gln supplementation can improve resistance to oxidative stress in weaned calves, and lower-level groups have better capacity because of higher GSH-Px activity. Similarly, Wu et al. found that glutamate supplementation reduced oxidative stress by inhibiting the mTOR signalling pathway [38]. In fish, glutamate supplementation can increase the activity of GSH-PX and SOD to reduce intestinal oxidative damage [39]. Therefore, we think L-Gln with 1 and 2% of DMI enhance the antioxidant capacity of weaned calves by breaking it down into glutamate, which is consistent with higher SUN concentration at 1 d after weaning. Immunoglobulin, which is present in the serum, is a glycoprotein produced by the proliferation and differentiation of B cells into plasma cells after antigen stimulation. The determination of serum IgG, IgA, and IgM levels can reflect the body's humoral immunity status. In this study, serum IgM and IgG levels in L-Gln supplementation groups were elevated after weaning. Therefore, we believe that L-Gln supplementation can enhance the immune response of calves, and lower-level L-Gln supplementation can maximise the immune response in this study. IgG, the most abundant immunoglobulin in serum, may be more sensitive to L-Gln stimulation, so it is still significantly elevated after weaning. Hu et al. found that intravenous injection of L-Gln (16 g/d) significantly increased the proportion of $CD4^+$ T cells in calves after weaning [21]. $CD4^+$ T cells play an important role in the immune system [40]. Thus, intravenous injection of L-Gln (16 g/d) can improve the immunity of calves. Although there is different effect on weaned calves between the parenteral administration and in-feed administration, these results indicate that L-Gln can enhance the immune capacity of weaned calves. Similarly, a recent study found that feeding rumen-protected L-Gln to postpartum cows increased the serum total protein concentration and decreased the somatic cell count of milk [41]. In addition, intravenous injection of L-Gln into postpartum cows also increased the proportion of monocytes and $CD8^+$ T cells [42]. In conclusion, low-level L-Gln supplementation can better improve the antioxidant and immune capacities of weaned calves. Interestingly, calves of L-Gln supplementation groups in this study also had better immune and antioxidant capacities before weaning. These results further suggest that L-Gln has potential as a novel antioxidant and immunomodulator for calf production.

Weaning stress causes damage to the small intestine, mainly reflected in the villus becoming shorter and crypt depth becoming higher, and then affects the digestion and absorption of nutrients and body health [43]. In this study, calves fed with 1% and 2% L-Gln had longer villus height, shorter crypt depth, and larger V/C in the duodenum and jejunum. Van Keulen et al. also found that 9 L/d milk allowance with 1% DM of milk L-Gln supplementation significantly increased villus length in the duodenum and jejunum but did not affect crypt depth and V/C [18]. We believe that the difference in the results may be due to the limitations of van Keulen's study. The addition of L-Gln to 1% DMI of milk may not meet the needs of calves for L-Gln, which is similar to the finding that feeding 1% DMI of milk replacer containing plant proteins L-Gln does not affect the histological morphology of the small intestine [44]. In the present study, we added L-Gln according to the percentage of DMI in the calves. Therefore, compared with the two studies [18,44], we had higher a L-Gln allowance in calves and found that L-Gln added at 2% of total DMI may also be the threshold for small intestine development in calves. Similar to van Keulen et al. [18], no effect of L-Gln on the villus structure of the ileum was found in our study, which may be the reason that L-Gln is mainly absorbed in the duodenum and jejunum.

Interestingly, high-level L-Gln (3% and 4% of DMI) supplementation groups can relieve the negative effects of weaning; however, there is no better effect than low-level L-Gln (1% and 2% of DMI) supplementation groups. Although similar results have been reported in piglets [29], chickens [24], and yaks [25], the reasons are unclear. Higher-level L-Gln supplementation will release large amounts of ammonium in the gut, and the ammonium can increase metabolic stress of the kidneys and liver, which may damage liver and kidney function; large amounts of ammonium lead to large-scale autophagy of small intestinal cells [45], which may affect nutrient absorption; moreover, it is known that intestinal microbiome plays the crucial role on the utilization of exogenous amino acids, especially epithelial microbiome [31,46,47]. The large amounts of ammonium will alter the pH of the epithelial and lumen environment of intestine, affecting microbial colonization in the small intestine and perhaps causing microbial niche shifts [48]. However, these potential mechanisms need to be verified and explored using molecular biological methods and high-throughput sequencing techniques in the future.

In addition, the rumen-protected L-Gln used in this study was evaluated in situ using Holstein bulls, and the manner was similar to that used in Kong et al. [49]. Holstein bulls have a similar energy allocation to that in Holstein calves, and bulls also consume energy for maintenance and growth rather than for lactation. Calves have a distinct rumen microbiota and morphology from adult cattle [50,51]. The rumen maturity of weaned calves is about 80% of that of adult cattle, and feed degradability in the rumen decreases as DMI rises. Although these factors may influence the true rumen pass rate of the rumen-protected L-Gln in calves, we believe that the deviation does not affect the results of supplementing different levels of L-Gln in this study. Molano et al. found that rumen-protected methionine had a good effect on calves by feeding calves with different methionine sources [52]; Kong et al. also found valid results by feeding calves with rumen-protected lysine [49]. In the research field of rumen-protected products, in vitro and in situ methods are commonly used to evaluate rumen-protected products' performance, with the in situ method being favoured. In fact, one of the field's acknowledged limitations is that the degradation rate of rumen-protected products evaluated in adult cattle differs from that of calves; however, calves cannot be used to evaluated for rumen-protected products. As a result, an advanced device that can simulate the rumen motility and environment of calves must be developed in the future.

5. Conclusions

The results of this study provide evidence that the addition of L-Gln in the diet improved the negative effects of sudden weaning in calves. Importantly, we found that lower-level L-Gln supplementation (1 and 2% of DMI) in diets had better effect in the growth performance, immune and antioxidant capacity, and morphology of the duodenum and jejunum of weaned calves compared with the higher-level L-Gln supplementation (3 and 4% of DMI) groups. These results provided effective reference for relieving weaning stress of calves.

Author Contributions: Conceptualization, Y.B. and S.L.; methodology, F.K.; software, S.W. and F.W.; validation, S.W.; resources, Z.C., H.Y. and Y.W.; writing—original draft preparation, S.W. and F.W.; writing—review and editing, S.W., F.W., Y.B. and S.L.; visualization, W.W.; supervision, Y.B. and S.L.; project administration, S.L.; funding acquisition, S.L. All authors have read and agreed to the published version of the manuscript.

Funding: This work was supported by the "China Agriculture Research System of MOF and MARA" (grant number CARS36), and the 2115 Talent Development Program of China Agricultural University.

Institutional Review Board Statement: This study was conducted in Bengbu Pasture, Modern Farming, Bengbu City, Anhui, China. The animal study protocol was approved by the Institutional Review Board of the China Agricultural University (protocol code AW81102202-1-3 and 28.4.2017 of approval), and the animal welfare and handling procedures were strictly followed by China Agricultural University experimental guidelines during the experiment.

Informed Consent Statement: Not applicable.

Data Availability Statement: Data is contained within the article.

Conflicts of Interest: The authors declare no conflict of interest.

References

1. Myers, S.E.; Faulkner, D.B.; Ireland, F.A.; Parrett, D.F. Comparison of three weaning ages on cow-calf performance and steer carcass traits. *J. Anim. Sci.* **1999**, *77*, 323–329. [CrossRef] [PubMed]
2. Diao, Q.; Zhang, R.; Fu, T. Review of Strategies to Promote Rumen Development in Calves. *Animals* **2019**, *9*, 490. [CrossRef] [PubMed]
3. Arthington, J.D.; Spears, J.W.; Miller, D.C. The effect of early weaning on feedlot performance and measures of stress in beef calves. *J. Anim. Sci.* **2005**, *83*, 933–939. [CrossRef]
4. Enríquez, D.; Hötzel, M.J.; Ungerfeld, R. Minimising the stress of weaning of beef calves: A review. *Acta Vet. Scand.* **2011**, *53*, 28. [CrossRef] [PubMed]
5. Pluske, J.R.; Hampson, D.J.; Williams, I.H. Factors influencing the structure and function of the small intestine in the weaned pig: A review. *Livest. Prod. Sci.* **1997**, *51*, 215–236. [CrossRef]
6. Wang, S.; Ma, T.; Zhao, G.; Zhang, N.; Tu, Y.; Li, F.; Cui, K.; Bi, Y.; Ding, H.; Diao, Q. Effect of Age and Weaning on Growth Performance, Rumen Fermentation, and Serum Parameters in Lambs Fed Starter with Limited Ewe-Lamb Interaction. *Animals* **2019**, *9*, 825. [CrossRef] [PubMed]
7. Wang, J.; Chen, L.; Li, P.; Li, X.; Zhou, H.; Wang, F.; Li, D.; Yin, Y.; Wu, G. Gene expression is altered in piglet small intestine by weaning and dietary glutamine supplementation. *J. Nutr.* **2008**, *138*, 1025–1032. [CrossRef]
8. Souba, W.W.; Klimberg, V.S.; Plumley, D.A.; Salloum, R.M.; Flynn, T.C.; Bland, K.I.; Copeland, E.M., 3rd. The role of glutamine in maintaining a healthy gut and supporting the metabolic response to injury and infection. *J. Surg. Res.* **1990**, *48*, 383–391. [CrossRef]
9. Cao, Y.; Feng, Z.; Hoos, A.; Klimberg, V.S. Glutamine enhances gut glutathione production. *JPEN J. Parenter. Enter. Nutr.* **1998**, *22*, 224–227. [CrossRef]
10. Wang, D.; Du, Y.; Huang, S.; You, Z.; Zheng, D.; Liu, Y. Combined supplementation of sodium humate and glutamine reduced diarrhea incidence of weaned calves by intestinal microbiota and metabolites changes. *J. Anim. Sci.* **2021**, *99*, skab305. [CrossRef]
11. Heyland, D.; Muscedere, J.; Wischmeyer, P.E.; Cook, D.; Jones, G.; Albert, M.; Elke, G.; Berger, M.M.; Day, A.G. A randomized trial of glutamine and antioxidants in critically ill patients. *N. Engl. J. Med.* **2013**, *368*, 1489–1497. [CrossRef] [PubMed]
12. Teixeira, A.D.O.; Nogueira, E.T.; Kutschenko, M.; Rostagno, H.S.; Lopes, D.C. Inclusion of glutamine associated with glutamic acid in the diet of piglets weaned at 21 days of age. *Rev. Bras. Saude E Prod. Anim.* **2014**, *15*, 881–896. [CrossRef]
13. Wang, B.; Wu, Z.; Ji, Y.; Sun, K.; Dai, Z.; Wu, G. L-Glutamine Enhances Tight Junction Integrity by Activating CaMK Kinase 2-AMP-Activated Protein Kinase Signaling in Intestinal Porcine Epithelial Cells. *J. Nutr.* **2016**, *146*, 501–508. [CrossRef] [PubMed]
14. Gatel, F.; Guion, P. Effects of monosodium l glutamate on diet palatability and piglet performance during the suckling and weaning periods. *Anim. Prod.* **1990**, *50*, 365–372. [CrossRef]
15. Reeds, P.J.; Burrin, D.G.; Stoll, B.; Jahoor, F.; Wykes, L.; Henry, J.; Frazer, M.E. Enteral glutamate is the preferential source for mucosal glutathione synthesis in fed piglets. *Am. J. Physiol.* **1997**, *273*, E408–E415. [CrossRef]
16. Johnson, I.R.; Ball, R.O.; Baracos, V.E.; Field, C.J. Glutamine supplementation influences immune development in the newly weaned piglet. *Dev. Comp. Immunol.* **2006**, *30*, 1191–1202. [CrossRef]
17. Ji, F.J.; Wang, L.X.; Yang, H.S.; Hu, A.; Yin, Y.L. Review: The roles and functions of glutamine on intestinal health and performance of weaning pigs. *Animal* **2019**, *13*, 2727–2735. [CrossRef]
18. van Keulen, P.; Khan, M.A.; Dijkstra, J.; Knol, F.; McCoard, S.A. Effect of arginine or glutamine supplementation and milk feeding allowance on small intestine development in calves. *J. Dairy Sci.* **2020**, *103*, 4754–4764. [CrossRef]
19. Da Silva, J.T.; Manzoni, T.; Rocha, N.B.; Santos, G.; Miqueo, E.; Slanzon, G.S.O.; Bittar, C.M.M. Evaluation of milk replacer supplemented with lysine and methionine in combination with glutamate and glutamine in dairy calves. *J. Appl. Anim. Res.* **2018**, *46*, 960–966. [CrossRef]
20. Wickramasinghe, H.; Kaya, C.A.; Baumgard, L.H.; Appuhamy, J. Early step-down weaning of dairy calves from a high milk volume with glutamine supplementation. *J. Dairy Sci.* **2022**, *105*, 1186–1198. [CrossRef]
21. Hu, Z.Y.; Su, H.W.; Li, S.L.; Cao, Z.J. Effect of parenteral administration of glutamine on autophagy of liver cell and immune responses in weaned calves. *J. Anim. Physiol. Anim. Nutr.* **2013**, *97*, 1007–1014. [CrossRef] [PubMed]
22. Xiao, Z.; Wang, S.; Suo, D.; Wang, R.; Huang, Y.; Su, X. Enzymatic probe sonication for quick extraction of total bisphenols from animal-derived foods: Applicability to occurrence and exposure assessment. *Environ. Pollut.* **2022**, *292*, 118457. [CrossRef] [PubMed]
23. NRC. *Nutrient Requirement of Dairy Cattle*; National Academic Press: Washington, DC, USA, 2001.

24. Bartell, S.M.; Batal, A.B. The effect of supplemental glutamine on growth performance, development of the gastrointestinal tract, and humoral immune response of broilers. *Poult. Sci.* **2007**, *86*, 1940–1947. [CrossRef] [PubMed]
25. Ma, J.; Shah, A.M.; Wang, Z.S.; Hu, R.; Zou, H.W.; Wang, X.Y.; Zhao, S.N.; Kong, X.Y. Dietary supplementation with glutamine improves gastrointestinal barrier function and promotes compensatory growth of growth-retarded yaks. *Animal* **2021**, *15*, 100108. [CrossRef] [PubMed]
26. Zavy, M.T.; Juniewicz, P.E.; Phillips, W.A.; VonTungeln, D.L. Effect of initial restraint, weaning, and transport stress on baseline and ACTH-stimulated cortisol responses in beef calves of different genotypes. *Am. J. Vet. Res.* **1992**, *53*, 551–557.
27. El Deeb, W.M.; El Bahr, S.M. Acute phase proteins and oxidative stress biomarkers in water buffalo calves subjected to transportation stress. *Comp. Clin. Pathol.* **2014**, *23*, 577–582. [CrossRef]
28. O'Loughlin, A.; McGee, M.; Doyle, S.; Earley, B. Biomarker responses to weaning stress in beef calves. *Res. Vet. Sci.* **2014**, *97*, 158–163. [CrossRef]
29. Lee, D.N.; Cheng, Y.H.; Wu, F.Y.; Sato, H.; Yen, H.T. Effect of Dietary Glutamine Supplement on Performance and Intestinal Morphology of Weaned Pigs. *Asian Australas. J. Anim. Sci.* **2003**, *16*, 1770–1776. [CrossRef]
30. Yoo, S.S.; Field, C.J.; McBurney, M.I. Glutamine supplementation maintains intramuscular glutamine concentrations and normalizes lymphocyte function in infected early weaned pigs. *J. Nutr.* **1997**, *127*, 2253–2259. [CrossRef]
31. Wu, G. Amino acids: Metabolism, functions, and nutrition. *Amino Acids* **2009**, *37*, 1–17. [CrossRef]
32. Venè, R.; Castellani, P.; Delfino, L.; Lucibello, M.; Ciriolo, M.R.; Rubartelli, A. The cystine/cysteine cycle and GSH are independent and crucial antioxidant systems in malignant melanoma cells and represent druggable targets. *Antioxid. Redox Signal.* **2011**, *15*, 2439–2453. [CrossRef] [PubMed]
33. Chaiswing, L.; Zhong, W.; Liang, Y.; Jones, D.P.; Oberley, T.D. Regulation of prostate cancer cell invasion by modulation of extra- and intracellular redox balance. *Free Radic. Biol. Med.* **2012**, *52*, 452–461. [CrossRef] [PubMed]
34. Fink, R.C.; Scandalios, J.G. Molecular evolution and structure—Function relationships of the superoxide dismutase gene families in angiosperms and their relationship to other eukaryotic and prokaryotic superoxide dismutases. *Arch. Biochem. Biophys.* **2002**, *399*, 19–36. [CrossRef] [PubMed]
35. Sakai, O.; Uchida, T.; Roggia, M.F.; Imai, H.; Ueta, T.; Amano, S. Role of Glutathione Peroxidase 4 in Glutamate-Induced Oxytosis in the Retina. *PLoS ONE* **2015**, *10*, e0130467. [CrossRef] [PubMed]
36. Jin, L.; Li, D.; Alesi, G.N.; Fan, J.; Kang, H.B.; Lu, Z.; Boggon, T.J.; Jin, P.; Yi, H.; Wright, E.R.; et al. Glutamate dehydrogenase 1 signals through antioxidant glutathione peroxidase 1 to regulate redox homeostasis and tumor growth. *Cancer Cell* **2015**, *27*, 257–270. [CrossRef] [PubMed]
37. Cordis, G.A.; Maulik, N.; Das, D.K. Detection of oxidative stress in heart by estimating the dinitrophenylhydrazine derivative of malonaldehyde. *J. Mol. Cell. Cardiol.* **1995**, *27*, 1645–1653. [CrossRef]
38. Wu, M.; Xiao, H.; Ren, W.; Yin, J.; Tan, B.; Liu, G.; Li, L.; Nyachoti, C.M.; Xiong, X.; Wu, G. Therapeutic effects of glutamic acid in piglets challenged with deoxynivalenol. *PLoS ONE* **2014**, *9*, e100591. [CrossRef]
39. Jiang, W.D.; Feng, L.; Qu, B.; Wu, P.; Kuang, S.Y.; Jiang, J.; Tang, L.; Tang, W.N.; Zhang, Y.A.; Zhou, X.Q.; et al. Changes in integrity of the gill during histidine deficiency or excess due to depression of cellular anti-oxidative ability, induction of apoptosis, inflammation and impair of cell-cell tight junctions related to Nrf2, TOR and NF-κB signaling in fish. *Fish Shellfish. Immunol.* **2016**, *56*, 111–122. [CrossRef]
40. Taylor, P.A.; Noelle, R.J.; Blazar, B.R. CD4(+)CD25(+) immune regulatory cells are required for induction of tolerance to alloantigen via costimulatory blockade. *J. Exp. Med.* **2001**, *193*, 1311–1318. [CrossRef]
41. Nemati, M.; Menatian, S.; Joz Ghasemi, S.; Hooshmandfar, R.; Taheri, M.; Saifi, T. Effect of protected-glutamine supplementation on performance, milk composition and some blood metabolites in fresh Holstein cows. *Iran. J. Vet. Res.* **2018**, *19*, 225–228.
42. Doepel, L.; Lessard, M.; Gagnon, N.; Lobley, G.E.; Bernier, J.F.; Dubreuil, P.; Lapierre, H. Effect of postruminal glutamine supplementation on immune response and milk production in dairy cows. *J. Dairy Sci.* **2006**, *89*, 3107–3121. [CrossRef]
43. Santaolalla, R.; Fukata, M.; Abreu, M.T. Innate immunity in the small intestine. *Curr. Opin. Gastroenterol.* **2011**, *27*, 125–131. [CrossRef] [PubMed]
44. Drackley, J.K.; Blome, R.M.; Bartlett, K.S.; Bailey, K.L. Supplementation of 1% L-glutamine to milk replacer does not overcome the growth depression in calves caused by soy protein concentrate. *J. Dairy Sci.* **2006**, *89*, 1688–1693. [CrossRef]
45. Hu, Z.Y.; Li, S.L.; Cao, Z.J. Short communication: Glutamine increases autophagy of liver cells in weaned calves. *J. Dairy Sci.* **2012**, *95*, 7336–7339. [CrossRef]
46. Perlman, D.; Martínez-Álvaro, M.; Moraïs, S.; Altshuler, I.; Hagen, L.H.; Jami, E.; Roehe, R.; Pope, P.B.; Mizrahi, I. Concepts and Consequences of a Core Gut Microbiota for Animal Growth and Development. *Annu. Rev. Anim. Biosci.* **2021**, *10*, 177–201. [CrossRef]
47. O'Hara, E.; Neves, A.L.A.; Song, Y.; Guan, L.L. The Role of the Gut Microbiome in Cattle Production and Health: Driver or Passenger? *Annu. Rev. Anim. Biosci.* **2020**, *8*, 199–220. [CrossRef]
48. Zhang, Y.; Choi, S.H.; Nogoy, K.M.; Liang, S. Review: The development of the gastrointestinal tract microbiota and intervention in neonatal ruminants. *Animal* **2021**, *15*, 100316. [CrossRef]
49. Kong, F.; Gao, Y.; Tang, M.; Fu, T.; Diao, Q.; Bi, Y.; Tu, Y. Effects of dietary rumen-protected Lys levels on rumen fermentation and bacterial community composition in Holstein heifers. *Appl. Microbiol. Biotechnol.* **2020**, *104*, 6623–6634. [CrossRef]

50. Furman, O.; Shenhav, L.; Sasson, G.; Kokou, F.; Honig, H.; Jacoby, S.; Hertz, T.; Cordero, O.X.; Halperin, E.; Mizrahi, I. Stochasticity constrained by deterministic effects of diet and age drive rumen microbiome assembly dynamics. *Nat. Commun.* **2020**, *11*, 1904. [CrossRef]
51. Abdelsattar, M.M.; Zhuang, Y.; Cui, K.; Bi, Y.; Haridy, M.; Zhang, N. Longitudinal investigations of anatomical and morphological development of the gastrointestinal tract in goats from colostrum to postweaning. *J. Dairy Sci.* **2022**. [CrossRef]
52. Molano, R.A.; Saito, A.; Luchini, D.N.; Van Amburgh, M.E. Effects of rumen-protected methionine or methionine analogs in starter on plasma metabolites, growth, and efficiency of Holstein calves from 14 to 91 d of age. *J. Dairy Sci.* **2020**, *103*, 10136–10151. [CrossRef] [PubMed]

antioxidants

MDPI

Article

Supplementation of Rumen-Protected Glucose Increased the Risk of Disturbance of Hepatic Metabolism in Early Postpartum Holstein Cows

ZhiYuan Ma [1,2,†], LuoYun Fang [3,†], Emilio Ungerfeld [4], XiaoPeng Li [1], ChuanShe Zhou [1], ZhiLiang Tan [1], LinShu Jiang [3,*] and XueFeng Han [1,*]

1 CAS Key Laboratory for Agro-Ecological Processes in Subtropical Region, National Engineering Laboratory for Pollution Control and Waste Utilization in Livestock and Poultry Production, South-Central Experimental Station of Animal Nutrition and Feed Science in Ministry of Agriculture, Institute of Subtropical Agriculture, The Chinese Academy of Sciences, Changsha 410125, China; mzy@lzu.edu.cn (Z.M.); lixiaopeng123lover@163.com (X.L.); zcs@isa.ac.cn (C.Z.); zltan@isa.ac.cn (Z.T.)

2 College of Pastoral Agriculture Science and Technology, Lanzhou University, Lanzhou 730000, China

3 Beijing Key Laboratory for Dairy Cow Nutrition, Beijing University of Agriculture, Beijing 102206, China; fangly@bac.edu.cn

4 Centro Regional de Investigación Carillanca, Instituto de Investigaciones Agropecuarias INIA, Vilcún 4880000, Chile; emilio.ungerfeld@inia.cl

* Correspondence: jls@bau.edu.cn (L.J.); xfhan@isa.ac.cn (X.H.); Tel.: +86-1081798101 (L.J.); +86-7314619702 (X.H.)

† These authors contributed equally to this work.

Abstract: The dual stress of reduced feed intake and increased milk yield in dairy cows early postpartum results in a negative energy balance. Rumen-protected glucose (RPG) has been reported to replenish energy, increase milk yield, and improve gut health. However, early postpartum cows often develop an insulin resistance, implying that RPG may not be well utilized and increased milk production may increase the liver's fat oxidization burden. This study aimed to investigate the effects of RPG on the hepatic oxidative/antioxidative status and protein profile. Starting 7 d before expected calving, six pairs of cows were supplemented with rumen-protected glucose (RPG, n = 6) or with an equal amount of rumen-protecting coating fat (CON, n = 6). Liver samples were obtained from 10 cows 14 d after calving (d 14). Concentration of malondialdehyde and activity of glutathione peroxidase were increased and the activities of catalase and superoxide dismutase tended to increase in the livers of the RPG cows compared to the CON cows. The revised quantitative insulin sensitivity check index (RQUICKI) was decreased by RPG, but triacylglycerol concentration in liver was increased by RPG supplementation. The overall profiles of hepatic proteins were similar between CON and RPG. A partial least square regression was conducted to identify the proteins associated with liver lipidosis, oxidative stress, and antioxidative capacity. The top twenty proteins, according to their variable importance value, were selected for metabolic pathway enrichment analysis. Eighteen enriched KEGG pathways were identified, including metabolism, the citrate cycle, propanoate metabolism, the peroxisome, and type II diabetes mellitus. Our study showed that RPG supplementation reduced insulin sensitivity but increased the liver triglyceride concentration and the oxidative stress in early postpartum cows. Liver proteins related to lipidosis, oxidative stress, and antioxidative capacity, were positively associated with the glutamine metabolism, citric acid cycle, peroxisome, and type II diabetes pathways, which may indicate an increased risk of liver metabolic disorders caused by RPG supplementation in early postpartum cows.

Keywords: rumen-protected glucose; liver; oxidative stress; proteomics

Citation: Ma, Z.; Fang, L.; Ungerfeld, E.; Li, X.; Zhou, C.; Tan, Z.; Jiang, L.; Han, X. Supplementation of Rumen-Protected Glucose Increased the Risk of Disturbance of Hepatic Metabolism in Early Postpartum Holstein Cows. *Antioxidants* **2022**, *11*, 469. https://doi.org/10.3390/antiox11030469

Academic Editor: Stanley Omaye

Received: 6 February 2022
Accepted: 25 February 2022
Published: 26 February 2022

1. Introduction

Early postpartum dairy cows experience a negative energy balance under the dual stress of insufficient energy intake due to anorexia and the increased energy demand for milk production [1,2]. The negative energy balance in the early postpartum dairy cow is associated with metabolic disorders such as ketosis and hepatic lipidosis [3–5].

Considerable research has been conducted seeking means to alleviate the negative energy balance: dietary supplementation with cereals [6], linoleic acid [7], rumen-protected methionine [8], and glucose [9]. In our previous studies, fat-coated, rumen-protected glucose (RPG) was supplemented to dairy cows during the transition period, resulting in an improved milk production [1] and gut health [10]. However, increased non-esterified fatty acids (NEFA) in plasma due to RPG supplementation, an indication of excessive lipolysis, was also observed [1]. Excessive lipolysis is a result of insufficient energy intake and mobilization of the body's energy reserves [11]. Fat mobilization from adipose tissue can exceed the oxidative capacity of the liver, resulting in metabolic disorders such as fatty liver disease, an increase in reactive oxygen species (ROS), a reduction of paraoxonase activity, and the onset of oxidative stress [12]. Rumen-protected glucose supplementation has been hypothesized to ameliorate the energy deficit but was instead shown to increase the negative energy balance of periparturient dairy cows [1]. We speculate that supplemented glucose in the form of RPG may stimulate the mammary gland to produce more milk, leading to a greater energy demand.

The liver is a hub organ in energy metabolism, with its activity being tightly controlled by hormones such as insulin [13]. Mammals in early postpartum tend to develop liver insulin resistance [14], which may be one of the mechanisms through which RPG failed to alleviate the negative energy balance of periparturient cows in a study by Li et al. (2019). It is possible that the supply of exogenous energy can either cause or aggravate liver metabolic disorders in transition cows, but that remains to be studied.

Recent advances in proteomic technologies have allowed researchers to characterize the proteomic profiles of the livers of dairy cows [15,16]. Tandem mass tags (TMT) technology, combined with UPLC-MS/MS, is a powerful technique for protein screening and is particularly well suited for small sample sizes [17]. In this study, TMT-based quantitative proteomic analyses were performed in the livers of early lactation Holstein cows to evaluate the effects of RPG supplementation. This study aimed to gain a better understanding of the effects of RPG on the hepatic oxidative/antioxidative status and protein profile.

2. Materials and Methods

2.1. Animals and Experimental Design

The animal trial was described in one of our previously published papers [1]. Out of a total of 22 cows with an expected calving date between May and July 2018, twelve 4–5 yr old Holstein cows (515 ± 42 kg body mass, 16.1 ± 3.7 kg milk/d, 2.5 ± 0.52 parity; mean ± SD) were selected for this animal trial. The twelve cows were partitioned into six pairs based on their milk production in the previous lactation cycle. One cow per pair was randomly assigned to either the control group (CON, $n = 6$) or to the rumen-protected glucose group (RPG, $n = 6$). The CON cows were fed the basal diet (Supplemental Table S1) plus 90 g/d of coating fat, and the RPG supplemented cows were fed the basal diet supplemented with 200 g of RPG, composed of 90 g glucose as the core with 90 g of coating fat and 20 g of water. One of the RPG supplemented cows developed acute cholecystitis unrelated to her treatment, so she was excluded from the trial along with her paired CON cow. All cows were individually fed the basal diet as a total mixed ration (TMR) twice daily at 07:30 and 14:30 h, allowing for a 10% refusal rate. Either the coating fat or RPG was provided to the corresponding cows from 7 d before calving to 14 d postpartum. The daily amount of coating fat and RPG were split into the two daily meals. A portion of the TMR mixed with the coating fat or RPG was fed first. The rest of the TMR without the coating fat or RPG was offered after the first portion of TMR containing either the coating fat or RPG was consumed. Water was freely available at all times.

2.2. Insulin Resistance Estimate

At 14 d postpartum, jugular venous blood was sampled before the morning feeding and after a 10 h overnight fasting. Concentrations of glucose, insulin, and NEFA in blood were measured as in our previous report [1]. A revised quantitative insulin sensitivity check index (RQUICKI) was calculated based on the reciprocal of the sum of the logarithm-transformed glucose, insulin, and NEFA concentrations [18]:

$$\text{RQUICKI} = \frac{1}{\log 10(G_b) + \log 10(I_b) + \log 10(NEFA_b)}$$

where G_b (mg/dL), I_b (μU/mL), and $NEFA_b$ (mM) are 10 h fasting concentrations of glucose, insulin, and NEFA in blood, respectively.

2.3. Liver Samples Collection

Five cows from each treatment were humanely euthanized with sedative xylazole at 14 d postpartum, and about 30 g of liver tissue was obtained from the middle of the right lobe. Bloodstains on the surface of the liver samples were rinsed with cold sterile saline. Samples were then divided into small pieces of about 2 g each and individually wrapped in sterile tin foil before being placed into sampling bags (Whirl-Pak™, Madison, WI, USA), which were quickly immersed in liquid nitrogen. The frozen liver samples were then transferred to a −80 °C refrigerator for long-term storage until analysis.

2.4. Liver Antioxidant Capacity and Triglyceride Measurement

Samples of liver tissue were ground with ice-cold saline (1/10, *v*/*v*). The resulting homogenates were then centrifuged at 1200× *g* at 4 °C for 10 min. The supernatants were then transferred to new Eppendorf tubes, and the protein concentration was determined through a Bradford assay. All oxidative stress and antioxidant indicators in liver samples were measured using commercially available kits provided by Beyotime (Shanghai, China). Malondialdehyde (MDA) concentration was determined by color reaction with thiobarbituric acid. Catalase (CAT) activity was determined by measuring the decrease in H_2O_2 concentration observed following incubation of the sample with an H_2O_2 standard solution. Total antioxidant capacity was determined using the ABTS method [19]. Total superoxide dismutase (SOD) activity was determined using the WST-8 method [20]. Activity of total glutathione peroxidase (GPx) was determined using the NADPH method [21]. Determination of liver triglyceride concentration was conducted according to the GPO-PAP method [22], using a kit provided by the Nanjing Jiancheng Biological Engineering Institute (Nanjing, China).

2.5. TMT-Based Quantitative Proteomics Analysis

2.5.1. Protein Pre-Treatment and TMT Labeling

Frozen liver samples were ground to powder in liquid nitrogen. Twenty milligrams of frozen liver powder were mixed with a 1.5 mL cold mix of tri-n-butyl phosphate/acetone/methanol (1:12:1, *v*/*v*/*v*) and left for 90 min at 4 °C to remove the lipids. Samples were then centrifuged at 10,000× *g* and 4 °C for 20 min. Pellets were air-dried and resuspended in a lysis buffer composed of 7 *M* urea, 1% (*v/v*) protease inhibitor cocktail, and 2 m*M* EDTA (all reagents were purchased from Solarbio Biotech, Beijing, China, except for the protease inhibitor cocktail, which was purchased by MERCK, Darmstadt, Germany). After centrifugation at 25,000× *g* and 4 °C for 20 min, the supernatants were transferred into new tubes. Proteins in the supernatants were reduced with 10 mM dithiothreitol (Thermo Scientific, San Jose, CA, USA) at 56 °C for 1 h, alkylated with 55 mM iodoacetamide (Thermo Scientific, San Jose, CA, USA) for 45 min in the dark, and precipitated with six times their volume of precooled acetone (Sinopharm, Shanghai, China) at −20 °C for 2 h. The protein pellets were then washed twice with ice-cold acetone. The pellet was dissolved with the lysis buffer. The Bradford method [23] was conducted to measure the protein concentration of the solution with a 2-D Quant kit (Amersham BioSciences Corp, Marlborough, MA, USA).

In order to generate peptides, protein pellets were digested overnight with trypsin (Solarbio, Beijing, China) with a trypsin:protein mass ratio of 1:50, followed by a second digestion for 4 h with a trypsin:protein mass ratio of 1:100. The resulting peptides were desalted using a Strata X C18 SPE column (Phenomenex, Los Angeles, CA, USA) according to the manufacturer's instructions. Peptides were then dissolved in 0.5 *M* tetraethylammonium bromide (Sigma-Aldrich, Saint Louis, MO, USA). Finally, the obtained peptides were labeled with the TMT kit (TMT 10 plex™ Isobaric Label Reagent Set, Thermo Scientific, San Jose, CA, USA).

2.5.2. HPLC Fractionation and LC-MS/MS Analysis

The labeled peptides were fractionated on a Gemini C18 Column (5 μm, 250 × 4.6 mm column; Phenomenex, Torrance, CA, USA) using 5% acetonitrile (solvent A; U.S. Pharmacopeia, Rockville, MD, USA) and 95% acetonitrile (solvent B), and then fractionated on a SHIMADZU-LC-20AB system monitored at 214 nm. The fractions were collected using 1–5% solvent B for 10 min, 5–35% solvent B for 40 min, 35–95% solvent B for 1 min, solvent B for 3 min and then 5% solvent B for 10 min at a flow rate of 1 mL/min. A total of 20 fractions were obtained and vacuum dried for further LC-MS/MS analysis.

Before loading, each sample was dissolved in mobile phase A (2% acetonitrile, 0.1% formic acid; *v*/*v*). The peptide mixtures were separated in a SHIMADZU-LC-20AB system (SHIMADZU Corporation, Saitama, Japan) with an Acclaim Pep Map RSLC C18 analytical column (75 μm × 25 cm, 2 μm particle size; Thermo Scientific, Waltham, MA, USA) at a flow rate of 300 nL/min. The liquid gradient was set as followed: 8 min of 5% mobile phase B (0.1% formic acid, 98% acetonitrile), 35 min of 8–35% B, 5 min of 35–60% mobile phase B, 2 min of 60–80% mobile phase B, 5 min of 80% mobile phase B, and 10 min of 5% mobile phase B. The peptides separated by the SHIMADZU-LC-20AB system were injected into a Q Exactive (Thermo Fisher Scientific, Waltham, MA, USA). The peptides were ionized at 1.6 kV and data-dependent acquisition was conducted by Q Exactive. The MS1 spectrum was set at 350–1600 m/z, and the scanning resolution was set at 70,000 dpi. The MS2 spectrum was set at 100 m/z, and the scanning resolution was set at 17,500 dpi. Dynamic exclusion was set at 15 s. The automatic gain control setting of MS1 and MS2 were set at 3E6 and 1E5, respectively. The twenty most intense ions were selected for fragmentation by entering a higher-energy collision dissociation pool. Subsequently, secondary mass spectrometry analysis was performed. The TMT-proteomic analysis was conducted by BGI-Shenzhen, Shenzhen, China.

2.5.3. Data Processing and TMT Quantification

Secondary mass spectrometry data were processed using *Mascot Search Engine* v 2.3.02 (Matrix Science, London, UK) and *Proteome Discoverer Software* v 2.1.0.81 (Thermo Scientific, USA) against *Bos taurus* sequences in Uniprot KB (*Bos taurus* database, 31,889 entries, downloaded 14 September 2019, http://www.uniprot.org/). Mascot and Proteome Discoverer were searched with a fragment ion mass tolerance of 0.050 Da and a parent ion tolerance of 20.0 ppm. IQuant (The Beijing Genomics Institute, Shenzhen, China, OR) was used to validate MS/MS-based peptide and protein identifications (Wen et al., 2014). The false discovery rate (FDR) thresholds for protein, peptide, and modification sites were set at 1%. For protein quantification, unique peptides with a report ion intensity of >10,000 and FDR of <0.01 were used, and the signal between different samples was normalized by the sum of the total intensity of each reported ion channel. Total protein abundances were normalized to one million by taking the median of all quantified proteins in a sample.

2.5.4. Protein and Gene Ontology Identification

We used the Mascot Search Engine 2.3.02 (Matrix Science, London, UK) classification system to identify significantly enriched Gene Ontology (GO) terms and pathways.

2.5.5. Identification of Proteins Related to the Phenotype of Interest

A partial least square regression (PLSR) analysis with sparsity-inducing penalized regression was fitted. The PLSR was performed by using the *R* package *caret* v 6.0 [24], with the formula set as train(NEFA + MDA + GPx + triacylglycerol ~., method = 'pls'). Specific engines for variable importance on a model-by-model basis were performed by using the vamp() function in *caret* v 6.0 [24].

2.5.6. KEGG Pathway Enrichment

The set of the top 20 variable importance proteins was enriched using KABOS 3.0 [25]. The background gene list was set as *Bos taurus*, the pathway database was set to the KEGG pathway, and a hypergeometric test/Fisher's exact test was conducted. All defined KEGG pathways and their correlations were presented in a landscape, and the clusters of KEGG pathways were produced online (http://kobas.cbi.pku.edu.cn/, 22 September 2021) using the default settings.

2.6. Statistical Analyses

A pairwise *t*-test was performed for the phenotypic data by using *t.test()* in *R v 4.0* [26], with the parameter of *paired = TRUE*. A *p*-value of $\leq$0.05 was set as the criterion for significance, and $0.05 < p \leq 0.1$ was defined as a tendency.

3. Results

The concentration of MDA was higher in the livers of RPG cows when compared to CON cows ($p = 0.02$; Table 1). The GPx activity was increased ($p = 0.03$) and the activities of CAT and SOD had a tendency to increase ($p \leq 0.07$) in the livers of RPG supplemented cows. TAC was not influenced by RPG supplementation ($p = 0.35$), but triacylglycerol content in the liver was increased by RPG supplementation ($p = 0.02$). The RQUICKI insulin resistance index was decreased by RPG supplementation ($p = 0.018$, Figure 1).

Table 1. Effects of rumen-protected glucose (RPG) on the variables of oxidative stress, antioxidant capacity, and triglyceride content in the liver of early postpartum cows ($n_{\text{pairs}} = 5$).

Item [1]	Treatment [2]		SEM	*p*-Value
	CON	RPG		
Oxidative stress				
MDA, µmol/mg protein	0.68	1.09	0.116	0.02
Antioxidant				
TAC, µmol/mg protein	243	217	24.2	0.35
SOD, U/mg protein	1.24	2.02	0.315	0.07
CAT, U/mg protein	25.1	25.7	0.18	0.06
GPx, U/mg protein	27.6	52	8.04	0.03
Triacylglycerol, µmol/g	540	958	115.8	0.02

[1] MDA—Malondialdehyde; TAC—total antioxidant capacity; SOD—Superoxide dismutase; CAT—Catalase; GPx—Glutathione peroxidase. [2] CON—cows supplemented with coating fat; RPG—cows supplemented with rumen-protected glucose.

A total of 4890 proteins were identified by GO in the liver samples (Supplemental Tables S2 and S3). Of these, most proteins were involved in synthesis of cellular components (3991/4890), molecular functions (3829/4890), and biological processes (3787/4890) (Figure S1). Principal coordinate plots did not cluster by treatment ($p = 0.46$, Figure 2).

We filtered 35, 24, and 17 bovine proteins from the KEGG pathways of *Bos tarus* [27], and 71.4%, 58.3%, and 58.8% of them were identified by proteomics to be involved in glycolysis/glycogenesis, citric acid cycle, and fatty acid β oxidation, respectively (Figure 3). Most of the detected proteins were uninfluenced by RPG supplementation ($p \geq 0.15$, Figure 3, Table S3), except for Q3ZBY4, involved in the production of fructose 1, 6-diphosphate from glyceraldehyde 3-phosphate, which was down-regulated by RPG supplementation ($p = 0.03$).

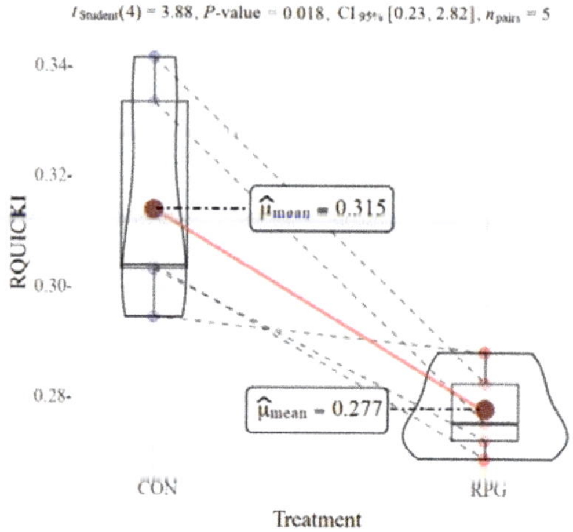

Figure 1. Effects of rumen-protected glucose (RPG) on RQUICKI, an insulin resistance index estimate according to the concentrations of NEFA, glucose, and insulin in plasma. Paired cows are connected with dashed lines.

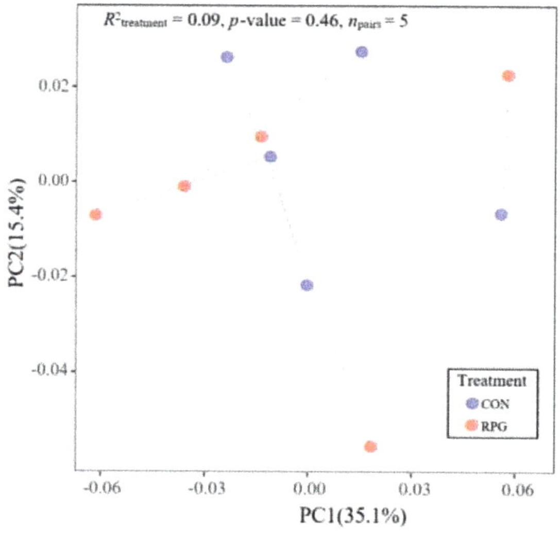

Figure 2. Principal coordinate analysis based on the Bray–Curtis dissimilarity matrix of liver protein profiles. Paired cows are connected with dashed lines.

The top twenty proteins most associated with oxidative stress, fatty acid metabolism, and concentration of NEFA in plasma, and of MDA, GPx, and triacylglycerol in the liver, were F6Q751, Q9NOV4, E1BJF9, F1N2L9, A5PKH3, A5D9G3, Q3MHX5, P41976, E1BAS6, P02070, A1A4L7, F1N0S6, D4QBB3, G5E5T5, Q2ABB1, E1B92, F1MFN2, E1BAI7, Q2YDG3, and Q8SPU8 (Figure 4A and Supplemental Table S4). These proteins, identified through partial least squares regression, were involved in 37 KEGG pathways (Supplemental Table S5), which were grouped into six clusters (C1-C6; Figure 4B and Supplemental Table S5). Eighteen KEGG pathways were identified as enriched (Figure 4C, $p < 0.05$): glutathione metabolism, drug metabolism by cytochrome P450, metabolism of xenobiotics by cytochrome P450, chemical carcinogenesis, and platinum drug resistance in C1; citrate cycle and propanoate metabolism in C2; African trypanosomiasis pathway in C3; peroxisome in C4; type II diabetes mellitus in C5; and tyrosine metabolism in C6.

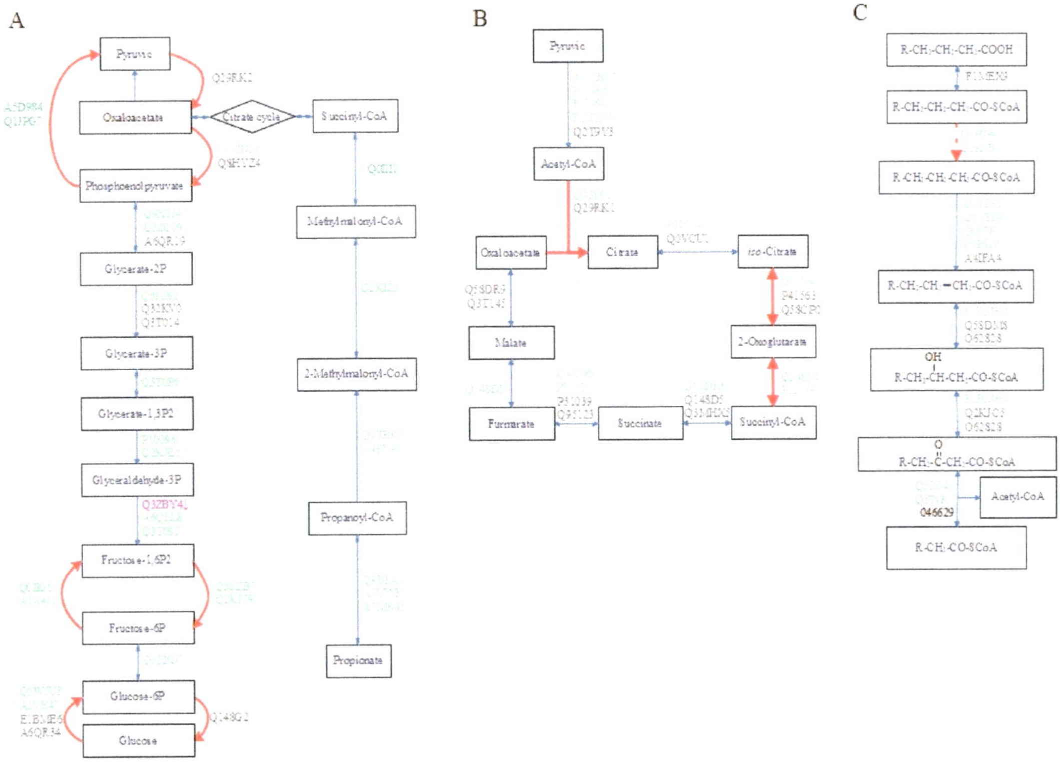

Figure 3. Hepatic metabolism pathway of glucose, citric acid cycle and fatty acid metabolism as influenced by rumen-protected protein (RPG) supplementation. (**A**) Glycolysis/glycogenesis. (**B**) Citrate cycle. (**C**) β-oxidation of fatty acid. The red arrows show the rate-limiting steps in each metabolic pathway. For each protein code, the gray color indicates no detection, green indicates no effect caused by RPG supplementation, and purple indicates a decrease in the RPG treatment group.

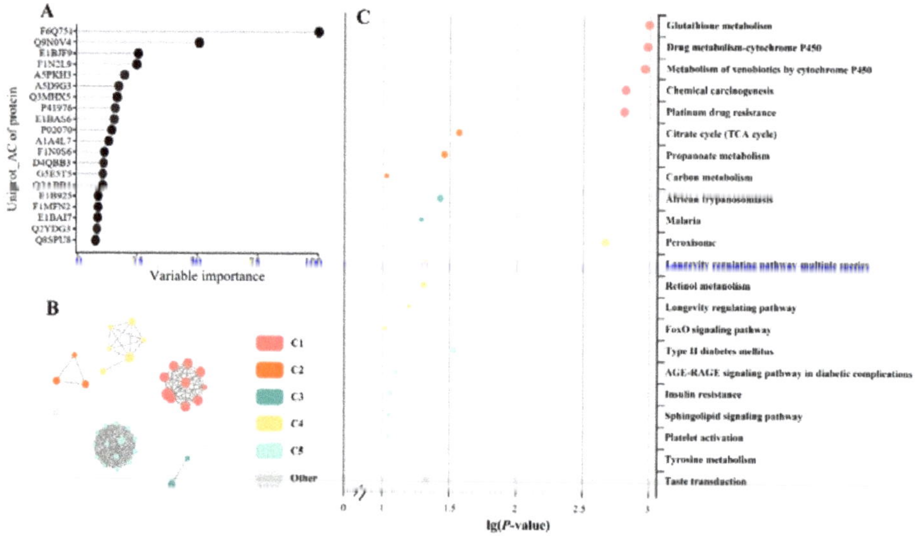

Figure 4. Liver protein profiles. (**A**) Top 20 most important proteins as found by a partial least squares regression associated with oxidative stress, fatty acid metabolism, and the concentration of NEFA in plasma, and of MDA, GPx, and triacylglycerol in the liver. (**B**) Network of enriched KEGG pathways. The nodes represent KEGG pathways, and the node colors represent different clusters; the node size represents 6 levels of enriched p-value (from smallest to largest): [0.05, 1], [0.01, 0.05], [0.001, 0.0), [0.0001, 0.00), [10^{-10}, 0.0001), (0, 10^{-10}]; and correlated pathways are linked with lines. The pathways in each of the clusters are listed in Table S5. (**C**) Bubbles of KEGG pathways enriched in the RPG treatment. The color and size of the bubbles are the same as the color and size in the circular network. If there are more than 5 enriched KEGG pathways per cluster, the top 5 with the highest enrichment ratio are displayed.

4. Discussion

Milk yield of our cows was considerably lower than the average for developed countries, but likely representative of the developing world [28]. This may explain why some of our results were inconsistent with those obtained with high-yielding cows. We observed increased concentration of circulating insulin and glucose [1], while Sauls-Hiesterman et al. [29] reported that no changes was caused by an even higher level of RPG supplementation (1500 g/d). Loncke et al. [30] reported a mean liver outflow of glucose of 0.995 (0.467–1.405) mmol kg BM^{-1} h^{-1}, which for 515 kg cows (closer with the cows in our study), is equivalent to 2214 (1039–3126) g/d. In that regard, the 90 g/d RPG supplementation in our experiment would represent only 4.1% of the average gluconeogenesis flow, i.e., rather small. However, the lactation animals in the meta-analysis by Loncke et al. [30] were consuming on average 26.4 g DM/kg BM, which would be equivalent to 13.6 kg DM/d. Because cows in this study consumed approximately 8 kg DM/d, if their glucose outflow was proportional to their DMI (and without considering differences in diet composition), their expected liver glucose outflow would have been 1302 instead of 2214 g/d. Then, after correcting for this, RPG supplementation in the present study would represent about 6.9% of gluconeogenesis; that may partially explain why low-producing cows had a milk production response to a relatively small amount of RPG supplementation.

Milk production increases rapidly after calving, and as a consequence, the mammary gland has high demands for energy and glucose [31]. Although RPG supplementation increased the circulating glucose concentration [1], proteomic analysis showed little change in the liver glucose metabolism, with one of the enzymes involved in gluconeogenesis even experiencing a decrease in relative abundance. The concentration of circulating insulin drops significantly after calving, allowing the mammary gland, rather than other peripheral tissues, to preferentially take up and utilize glucose [32]. Meanwhile, glucose utilization by peripheral tissues other than the mammary gland is reduced due to reduced insulin sensitivity [33]. The most accurate way to measure the insulin resistance is through hyperinsulinemic-euglycemic clamp studies [34]. However, the hyperinsulinemic-euglycemic clamp test is time-consuming and may stress the postpartum cows. Some simple and cheap surrogate indices have been developed in human medicine to assess insulin sensitivity in patients with diabetes. We conducted a simple test called the RQUICKI index [18], which is based on a single blood sample, to predict the insulin resistance of cows. Rumen-protected glucose was originally designed to relieve the negative energy balance betweem energy and glucose. We observed that RPG supplementation further reduced insulin sensitivity in cows, which partly explains the increased lipid mobilization, as reflected by a higher amount of circulating NEFA, caused by RPG. Most [35–37], although, not all [9], previous studies showed the result that cows had lower insulin sensitivity when infused with glucose in blood, in agreement with this study.

Fat mobilization in the post-peripartum period is a consequence of a negative energy balance and insulin resistance [38]. The preferential supply of milk to the offspring increases the burden on the maternal energy metabolism. The release of NEFA to the blood may exceed the oxidative capacity of the liver [39]. Under these conditions, triglycerides can accumulate in the liver, causing the metabolic disease known as fatty liver disease [40]. In this study, RPG supplementation increased triglyceride concentration in the liver, indicating that supplementation with RPG increased the risk of disturbances in the hepatic metabolism in early postpartum dairy cows. In contrast, a previous study with early postpartum ewes showed that intravenous glucose infusion reduced the mobilization of lipid reserves [41]. However, the energy deficiency in early postpartum ewes may not be as severe as in dairy cows, as indicated by similar plasma glycerol and milk yield between ewes infused with glucose and their control counterparts [41]. The lack of changes observed in the liver proteins suggest that adipose tissue can increase the liver's ability to mobilize fat, but it has no ability to oxidize it. Other lipid marker in liver cells such as Annexin V/PI and Oil Red Staining could be useful to confirm lipid accumulation [42].

A high level of lipolysis in body adipose tissue may lead to increased oxidative stress in the liver. ROS-induced oxidative damage to lipids causes MDA accumulation, which is an index of lipid peroxidation. Oxidative stress caused by excess lipolysis is common in postpartum cows [40]. In this study, although ROS were not determined, the cause of the increased MDA with RPG supplementation could be explained by an increased oxidative stress. In the early postpartum dairy cow, the glucose absorbed in the intestinal tract is mainly used to synthesize lactose and glucogenic amino acids in milk, increasing the burden on the liver to oxidize fatty acids. Hepatic oxidative stress induced by glucose infusion has been reported in rats [43,44], but we are not aware of studies investigating the effects of glucose infusion on oxidative stress and antioxidant status in the liver of early postpartum cows. Mammals are equipped with an integrated antioxidant system that removes the harmful effects of ROS [45]. Hepatic oxidative stress state is not always reflected by the antioxidant markers in blood, as indicated by the inconsistency of superoxide dismutase activity between plasma and liver [46]. This inconsistency in superoxide dismutase activity further emphasizes the need to use liver tissue to study the antioxidant capacity. The superoxide is a primary ROS, and SOD is considered to be the first enzymatic antioxidant converting the superoxide to hydrogen peroxide [47]. Hydrogen peroxide is further degraded to water by CAT and GPx [45]. In this study, both SOD and CAT tended to increase and GPx increased in RPG supplemented cows, which indicates that RPG supplementation increased the oxidative stress and antioxidant activity. Oxidative stress is a consequence of an increased generation of ROS and/or the reduced physiological activity of antioxidants. Supplementation with RPG led to an increase of both oxidative stress and the activities of antioxidant enzymes, suggesting that RPG did not negatively affect the liver's antioxidant response and that the aggravation of oxidative stress in the liver mainly resulted from the production of ROS exceeding the functional capacity of antioxidants. Previous evidence showed that increased oxidative stress exacerbates insulin resistance by impairing the secretion of insulin by β-cells [48,49]; however, RPG supplementation to the same cows used in this study elicited a positive response in circulating insulin concentration before feeding [1]. In a previous study, supplementation with RPG did not decrease circulating insulin concentration in early lactation dairy cows, in agreement with our results [9]. Increased insulin concentration suggests that the oxidative stress induced by RPG supplementation was not severe enough as to affect β-cell function.

The analysis of proteomic profiles in tissues has been proposed as an approach to identify metabolic alterations for medical and nutritional treatments, and markers of clinical status [15,16]. The amount of protein we identified was comparable to a previous report with early postpartum Holstein dairy cows [50]. As far as we know, this is the first study reporting the effects of exogenous glucose on liver protein profiles in dairy cows. The TMT-based untargeted proteomics analysis could not classify animals based on the proteomics alteration induced by RPG supplementation. Likewise, PCoA plots did not separate by treatment, indicating that major proteins in the liver were not affected by RPG supplementation. A minor proportion (68/2741) of proteins in the livers of cows suffering from a negative energy balance were reported to be different from those of cows not suffering from a negative energy balance [51]. Proteins differing between the cows with a negative energy balance and cows not undergoing a negative energy balance were involved in inflammatory response, mitochondrial dysfunction, and fatty acid uptake [51]. Replenishing energy with RPG was expected to shift hepatic protein profiles to be more similar to cows not suffering negative energy balance. Based on the hepatic metabolism induced by RPG, we selected a group of proteins that may contribute to increased lipolysis, ROS, and antioxidants by RPG supplementation. A protein may participate in more than one metabolic pathway; therefore, multiple KEGG metabolic pathways were found to be enriched [25]. These pathways were grouped into six major clusters based on the network features of the proteins involved. The glutamine pathway associated with antioxidant stress appeared in the C1 metabolic pathway cluster, and the most notable proteins associated with our response variables were F6Q751 (Glutathione transferase) and Q9N0V4 (Glutathione S-transferase Mu 1), which are members of the glutathione transferase (GST) superfamily. The GST family represents one of the most abundant and important series of detoxification enzymes in the liver [52]. The main function of GSTs is to conjugate electrophilic compounds with glutathione (GSH), thereby making these compounds less active and enabling their excretion [53]. In this way, GSTs contribute to the metabolism of drugs, pesticides, and other xenobiotics [54]. GPx reduces hydrogen peroxide and organic hydroperoxides to mitigate oxidative stress; therefore, the enriched metabolic pathway of glutamine could help prevent deleterious cellular events that might develop in the livers of early postpartum dairy cows supplemented with RPG. The citrate cycle pathway, associated with antioxidant stress, appeared in the C2 metabolic pathway cluster, and the most notable protein was succinate-CoA ligase. According to the characteristics of substrate formation, succinate-CoA ligase can be divided into two categories: ATP and GTP biosynthesis [55]. The former is mainly found in the brain and heart, while the latter is mainly found in the liver and kidneys [55]. Succinate-CoA ligase catalyzes the hydrolysis of succinyl-CoA thioester bond to produce GTP and succinate, which enters the citrate cycle. Inhibition of succinate-CoA ligase in rat liver mitochondria was shown to cause metabolic disorders such as severe lactic acidosis and fatty liver disease [55]. The peroxisome and type II diabetes mellitus pathways were selected in pathway clusters C4 and C5, respectively. The main protein that caused peroxisome pathway enrichment was P41976, or Mn-superoxide dismutase. TSODs are enzymes that catalytically convert superoxide radicals to oxygen and hydrogen peroxide. The active center of SODs binds to a metal atom, which can be Cu, Fe, Mn, or Ni [56]. Mn-SODs are well conserved throughout evolution and across kingdoms. The highest contents of Mn-SOD are found in the liver, followed by the kidney and heart [46]. Under severe oxidative stress such as aging [57] or high intake of alcohol [58], the expression of Mn-SOD in the liver increases dramatically, indicating its importance in resisting oxidative stress. A0JNH7, or protein kinase C, was the main protein enriched in the type 2 diabetes pathway. The protein kinase C family is composed of lipid-dependent kinases with wide-ranging roles in modulating insulin function [59]. Increased protein kinase C activity is linked to reduced insulin receptor autophosphorylation in the livers of starved rats [60]. Thus, an increase in its abundance with RPG supplementation suggests that the livers of RPG cows are in a state of energy deficiency.

5. Conclusions

Our study showed that RPG supplementation reduced insulin sensitivity and increased the triacylglycerol contents and oxidative stress in the livers of early postpartum cows. Proteins related to oxidative stress, lipolysis, and antioxidant function that were differentially expressed between the control and RPG treatments were classified as being involved in metabolic pathways of the glutamine pathway, citrate cycle, peroxisome, and type II diabetes mellitus, which could indicate an increased risk of metabolic disorders in the livers of early postpartum dairy cows supplemented with RPG. In this study, the risk of metabolic disorders seemed to be augmented by RPG supplementation to relatively low-yielding cows; further research with high-yielding dairy cows, which are more exposed to liver metabolic disorders, is recommended.

Supplementary Materials: The following supporting information can be downloaded at: https://www.mdpi.com/article/10.3390/antiox11030469/s1, Table S1. Ingredients and chemical composition of basal diet; Table S2. Protein annotation; Table S3. Protein profiles; Table S4. Variable importance. Table S5. Pathway enrichment analysis; Figure S1: Gene Ontology (GO) terms of total proteins for categories of cellular component.

Author Contributions: Conceptualization, X.H., L.J. and Z.T.; methodology, X.L. and C.Z.; investigation, Z.M. and L.F.; data curation, Z.M. and X.H.; writing—original draft preparation, Z.M. and L.F.; writing—review and editing, Z.M. and E.U.; project administration, L.J. and X.H. All authors have read and agreed to the published version of the manuscript.

Funding: This work was supported by the National Key Research and Development Program of China (Grant No. 2016YFD0501206) and by the Open Project of Beijing Key Laboratory of Dairy Cow Nutrition, Beijing University of Agriculture (Grant No. BJKLDN2018-1).

Institutional Review Board Statement: The experiment was approved by the Animal Care Committee of the Institute of Subtropical Agriculture, Chinese Academy of Sciences, Changsha, China (Authorization No. ISA000259).

Informed Consent Statement: Not applicable.

Data Availability Statement: The proteomic data is contained within the article For more details, please check https://ngdc.cncb.ac.cn/omix: accession no. OMIX916; 9 February 2022.

Conflicts of Interest: The authors declare no conflict of interest. The funders had no role in the design of the study, in the collection, analyses, interpretation of data, in the writing of the manuscript, or in the decision to publish the results.

References

1. Li, X.P.; Tan, Z.L.; Jiao, J.Z.; Long, D.L.; Zhou, C.S.; Yi, K.L.; Liu, C.H.; Kang, J.H.; Wang, M.; Duan, F.H.; et al. Supplementation with fat-coated rumen-protected glucose during the transition period enhances milk production and influences blood biochemical parameters of liver function and inflammation in dairy cows. *Anim. Feed Sci. Tech.* **2019**, *252*, 92–102. [CrossRef]
2. Ceciliani, F.; Lecchi, C.; Urh, C.; Sauerwein, H. Proteomics and metabolomics characterizing the pathophysiology of adaptive reactions to the metabolic challenges during the transition from late pregnancy to early lactation in dairy cows. *J. Proteom.* **2018**, *178*, 92–106. [CrossRef] [PubMed]
3. Grummer, R.R.; Mashek, D.G.; Hayirli, A. Dry matter intake and energy balance in the transition period. *Vet. Clin. N. Am. Food Anim. Pract.* **2004**, *20*, 447–470. [CrossRef]
4. Odens, L.J.; Burgos, R.; Innocenti, M.; VanBaale, M.J.; Baumgard, L.H. Effects of varying doses of supplemental conjugated linoleic acid on production and energetic variables during the transition period. *J. Dairy Sci.* **2007**, *90*, 293–305. [CrossRef]

5. Selim, S.; Salin, S.; Taponen, J.; Vanhatalo, A.; Kokkonen, T.; Elo, K.T. Prepartal dietary energy alters transcriptional adaptations of the liver and subcutaneous adipose tissue of dairy cows during the transition period. *Physiol. Genom.* **2014**, *46*, 328–337. [CrossRef] [PubMed]
6. Aguilar-Pérez, C.; Ku-Vera, J.; Centurión-Castro, F.; Garnsworthy, P.C. Energy balance, milk production and reproduction in grazing crossbred cows in the tropics with and without cereal supplementation. *Livest. Sci.* **2009**, *122*, 227–233. [CrossRef]
7. García, A.; Cardoso, F.C.; Campos, R.; Thedy, D.X.; González, F. Metabolic evaluation of dairy cows submitted to three different strategies to decrease the effects of negative energy balance in early postpartum. *Pesqui. Veterinária Bras.* **2011**, *31*, 11–17. [CrossRef]
8. Rulquin, H.; Delaby, L. Effects of the energy balance of dairy cows on lactational responses to rumen-protected methionine. *J. Dairy Sci.* **1997**, *80*, 2513–2522. [CrossRef]
9. Wang, Y.P.; Cai, M.; Hua, D.K.; Zhang, F.; Jiang, L.S.; Zhao, Y.G.; Wang, H.; Nan, X.M.; Xiong, B.H. Metabolomics reveals effects of rumen-protected glucose on metabolism of dairy cows in early lactation. *Anim. Feed Sci. Tech.* **2020**, *269*, 114620. [CrossRef]
10. Zhang, X.; Wu, J.; Han, X.; Tan, Z.; Jiao, J. Effects of rumen-protected glucose on ileal microbiota and genes involved in ileal epithelial metabolism and immune homeostasis in transition dairy cows. *Anim. Feed Sci. Tech.* **2019**, *254*, 114199. [CrossRef]
11. Morris, D.G.; Waters, S.M.; McCarthy, S.D.; Patton, J.; Earley, B.; Fitzpatrick, R.; Murphy, J.J.; Diskin, M.G.; Kenny, D.A.; Brass, A.; et al. Pleiotropic effects of negative energy balance in the postpartum dairy cow on splenic gene expression: Repercussions for innate and adaptive immunity. *Physiol. Genom.* **2009**, *39*, 28–37. [CrossRef] [PubMed]
12. Turk, R.; Juretic, D.; Geres, D.; Svetina, A.; Turk, N.; Flegar-Mestric, Z. Influence of oxidative stress and metabolic adaptation on PON1 activity and MDA level in transition dairy cows. *Anim. Reprod. Sci.* **2008**, *108*, 98–106. [CrossRef] [PubMed]
13. Rui, L. Energy metabolism in the liver. *Compr. Physiol.* **2014**, *4*, 177–197. [CrossRef] [PubMed]
14. Lebovitz, H.E. Insulin resistance: Definition and consequences. *Exp. Clin. Endocrinol. Diabetes* **2001**, *109*, S135–S148. [CrossRef] [PubMed]
15. Bendixen, E.; Danielsen, M.; Hollung, K.; Gianazza, E.; Miller, I. Farm animal proteomics—A review. *J. Proteom.* **2011**, *74*, 282–293. [CrossRef] [PubMed]
16. Conrad, D.H.; Goyette, J.; Thomas, P.S. Proteomics as a method for early detection of cancer: A review of proteomics, exhaled breath condensate, and lung cancer screening. *J. Gen. Intern. Med.* **2008**, *23*, 78–84. [CrossRef]
17. Norton, D.; Crow, B.; Bishop, M.; Kovalcik, K.; George, J.; Bralley, J.A. High performance liquid chromatography-tandem mass spectrometry (HPLC/MS/MS) assay for chiral separation of lactic acid enantiomers in urine using a teicoplanin based stationary phase. *J. Chromatogr. B* **2007**, *850*, 190–198. [CrossRef]
18. Perseghin, G.; Caumo, A.; Caloni, M.; Testolin, G.; Luzi, L. Incorporation of the fasting plasma FFA concentration into QUICKI improves its association with insulin sensitivity in nonobese individuals. *J. Clin. Endocrinol. Metab.* **2001**, *86*, 4776–4781. [CrossRef]
19. Miller, N.J.; Rice-Evans, C.; Davies, M.J.; Gopinathan, V.; Milner, A. A novel method for measuring antioxidant capacity and its application to monitoring the antioxidant status in premature neonates. *Clin. Sci.* **1993**, *84*, 407–412. [CrossRef]
20. Misra, H.P.; Fridovich, I. The role of superoxide anion in the autoxidation of epinephrine and simple assay for superoxide dismutase. *J. Biol. Chem.* **1972**, *244*, 6049–6055. [CrossRef]
21. Roberts, C.K.; Barnard, R.J.; Sindhu, R.K.; Jurczak, M.; Ehdaie, A.; Vaziri, N.D. Oxidative stress and dysregulation of NAD(P)H oxidase and antioxidant enzymes in diet-induced metabolic syndrome. *Metab.-Clin. Exp.* **2006**, *55*, 928–934. [CrossRef] [PubMed]
22. Albers, J.J.; Wahl, P.W.; Cabana, V.G.; Hazzard, W.R.; Hoover, J.J. Quantitation of apolipoprotein A-I of human plasma high density lipoprotein. *Metab.-Clin. Exp.* **1976**, *25*, 633–644. [CrossRef]
23. Bradford, M.M. Bradford protein assay (Determination of protein concentrations). *Anal. Biochem.* **1976**, *72*, 248. [CrossRef]
24. Kuhn, M. Caret: Classification and regression training. *Astrophys. Source Code Libr.* **2015**, *10*, 4426. [CrossRef]
25. Bu, D.; Luo, H.; Huo, P.; Wang, Z.; Zhang, S.; He, Z.; Wu, Y.; Zhao, L.; Liu, J.; Guo, J.; et al. KOBAS-i: Intelligent prioritization and exploratory visualization of biological functions for gene enrichment analysis. *Nucleic Acids Res.* **2021**, *49*, W317–W325. [CrossRef]
26. R Core Team. R: A Language and Environment for Statistical Computing. Available online: https://www.R-project.org/ (accessed on 24 January 2020).
27. KEGG. Available online: https://www.kegg.jp/pathway/bta01100 (accessed on 1 January 2022).
28. FAO. FAOStat: Food and Agriculture Data. Available online: http://www.fao.org/faostat/en/#data/QC (accessed on 12 January 2021).
29. Sauls-Hiesterman, J.A.; Banuelos, S.; Atanasov, B.; Bradford, B.J.; Stevenson, J.S. Physiologic responses to feeding rumen-protected glucose to lactating dairy cows. *Anim. Reprod. Sci.* **2020**, *216*, 106346. [CrossRef]
30. Loncke, C.; Nozière, P.; Vernet, J.; Lapierre, H.; Bahloul, L.; Al-Jammas, M.; Sauvant, D.; Ortigues-Marty, I. Net hepatic release of glucose from precursor supply in ruminants: A meta-analysis. *Animal* **2020**, *14*, 1422–1437. [CrossRef]
31. Greenbaum, A.L.; Salam, A. Regulation of mammary gland metabolism: Pathways of glucose utilization, metabolite profile and hormone response of a modified mammary gland cell preparation. *Eur. J. Biochem.* **1978**, *87*, 505–516. [CrossRef]
32. Ingvartsen, K.L.; Andersen, J.B. Integration of metabolism and intake regulation: A review focusing on periparturient animals. *J. Dairy Sci.* **2000**, *83*, 1573–1597. [CrossRef]
33. Weber, M.; Locher, L.; Huber, K.; Kenez, A.; Rehage, J.; Tienken, R.; Meyer, U.; Daenicke, S.; Sauerwein, H.; Mielenz, M. Longitudinal changes in adipose tissue of dairy cows from late pregnancy to lactation. Part 1: The adipokines apelin and resistin and their relationship to receptors linked with lipolysis. *J. Dairy Sci.* **2016**, *99*, 1549–1559. [CrossRef]

34. Kim, J.K. Hyperinsulinemic-euglycemic clamp to assess insulin sensitivity in vivo. *Methods Mol. Biol.* **2009**, *560*, 221–238. [CrossRef] [PubMed]
35. Shingu, H.; Hodate, K.; Kushibiki, S.; Ueda, Y.; Watanabe, A.; Shinoda, M.; Matsumoto, M. Breed differences in growth hormone and insulin secretion between lactating Japanese Black cows (beef type) and Holstein cows (dairy type). *Comp. Biochem. Physiol. Part C Toxicol. Pharmacol.* **2002**, *132*, 493–504. [CrossRef]
36. Prodanovic, R.; Kirovski, D.; Vujanac, I.; Djuric, M.; Koricanac, G.; Vranjes-Djuric, S.; Ignjatovic, M.; Samanc, H. Insulin responses to acute glucose infusions in Busa and Holstein-Friesian cattle breed during the peripartum period: Comparative study. *Acta Vet.* **2013**, *63*, 373–384. [CrossRef]
37. Hammon, H.M.; Bellmann, O.; Voigt, J.; Schneider, F.; Kuehn, C. Glucose-dependent insulin response and milk production in heifers within a segregating resource family population. *J. Dairy Sci.* **2007**, *90*, 3247–3254. [CrossRef] [PubMed]
38. Komaragiri, M.; Casper, D.P.; Erdman, R.A. Factors affecting body tissue mobilization in early lactation dairy cows. 2. Effect of dietary fat on mobilization of body fat and protein. *J. Dairy Sci.* **1998**, *81*, 169–175. [CrossRef]
39. Grummer, R.R. Etiology of lipid-related metabolic disorders in periparturient dairy cows. *J. Dairy Sci.* **1993**, *76*, 3882. [CrossRef]
40. Emery, R.S.; Liesman, J.S.; Herdt, T.H. Metabolism of long chain fatty acids by ruminant liver. *J. Nutr.* **1992**, *122*, 832. [CrossRef]
41. Rutter, L.M.; Manns, J.G. Changes in metabolic and reproductive characteristics associated with lactation and glucose infusion in the postpartum ewe. *J. Anim. Sci.* **1986**, *63*, 538. [CrossRef]
42. Wang, Y.; Liu, B.; Wu, P.; Chu, Y.; Gui, S.; Zheng, Y.; Chen, X. Dietary selenium alleviated mouse liver oxidative stress and NAFLD Induced by obesity by regulating the KEAP1/NRF2 Pathway. *Antioxidants* **2022**, *11*, 349. [CrossRef]
43. Ling, P.R.; Mueller, C.; Smith, R.J.; Bistrian, B.R. Hyperglycemia induced by glucose infusion causes hepatic oxidative stress and systemic inflammation, but not STAT3 or MAP kinase activation in liver in rats. *Metab. Clin. Exp.* **2003**, *52*, 868–874. [CrossRef]
44. Ling, P.R.; Smith, R.J.; Bistrian, B.R. Acute effects of hyperglycemia and hyperinsulinemia on hepatic oxidative stress and the systemic inflammatory response in rats. *Crit. Care Med.* **2007**, *35*, 555–560. [CrossRef]
45. Birben, E.; Sahiner, U.M.; Sackesen, C.; Erzurum, S.; Kalayci, O. Oxidative stress and antioxidant defense. *World Allergy Organ. J.* **2012**, *5*, 9–19. [CrossRef]
46. Marklund, S.L. Extracellular superoxide dismutase and other superoxide dismutase isoenzymes in tissues from nine mammalian species. *Biochem. J.* **1984**, *222*, 649–655. [CrossRef] [PubMed]
47. Halliwell, B.; Chirico, S. Lipid peroxidation: Its mechanism, measurement, and significance. *Am. J. Clin. Nutr.* **1993**, *57*, 724S–725S. [CrossRef] [PubMed]
48. Evans, J.L.; Goldfine, I.D.; Maddux, B.A.; Grodsky, G.M. Are oxidative stress activated signaling pathways mediators of insulin resistance and β-cell dysfunction? *Diabetes* **2003**, *52*, 1–8. [CrossRef] [PubMed]
49. Shah, S.; Iqbal, M.; Karam, J.; Salifu, M.; Mcfarlane, S.I. Oxidative stress, glucose metabolism, and the prevention of type 2 diabetes: Pathophysiological insights. *Antioxid Redox Signal.* **2007**, *9*, 911–929. [CrossRef] [PubMed]
50. Kuhla, B.; Ingvartsen, K.L. Proteomics and the characterization of fatty liver metabolism in early lactation dairy cows. In *Proteomics in Domestic Animals: From Farm to Systems Biology*; Springer: Berlin/Heidelberg, Germany, 2018; pp. 219–231. [CrossRef]
51. Swartz, T.H.; Moallem, U.; Kamer, H.; Kra, G.; Levin, Y.; Mamedova, L.K.; Bradford, B.J.; Zachut, M. Characterization of the liver proteome in dairy cows experiencing negative energy balance at early lactation. *J. Proteom.* **2021**, *246*, 104308. [CrossRef]
52. Ryu, C.S.; Choi, Y.J.; Nam, H.S.; Jeon, J.S.; Jung, T.; Park, J.E.; Choi, S.J.; Lee, K.; Lee, M.Y.; Kim, S.K. Short-term regulation of the hepatic activities of cytochrome P450 and glutathione S-transferase by nose-only cigarette smoke exposure in mice. *Food Chem. Toxicol.* **2019**, *125*, 182–189. [CrossRef]
53. Hayes, J.D.; Pulford, D.J. The glutathione s-transferase supergene family: Regulation of GST and the contribution of the isoenzymes to cancer chemoprotection and drug resistance part I. *Crit. Rev. Biochem. Mol. Biol.* **1995**, *30*, 521–600. [CrossRef]
54. Oakley, A.J. Glutathione transferases: New functions. *Curr. Opin. Struct. Biol.* **2005**, *15*, 716–723. [CrossRef]
55. Stumpf, D.A.; Mcafee, J.; Parks, J.K.; Eguren, L. Propionate inhibition of succinate:CoA ligase (GDP) and the citric acid cycle in mitochondria. *Pediatric Res.* **1980**, *14*, 1127–1131. [CrossRef] [PubMed]
56. Abreu, I.A.; Cabelli, D.E. Superoxide dismutases—A review of the metal-associated mechanistic variations. *Biochim. Biophys. Acta (BBA)—Proteins Proteom.* **2010**, *1804*, 263–274. [CrossRef]
57. Yen, T.C.; KING, K.; Lee, H.C.; Yeh, S.H.; Wei, Y.H. Age-dependent increase of mitochondrial DNA deletions together with lipid peroxides and superoxide dismutase in human liver mitochondria. *Free Radic. Bio. Med.* **1994**, *16*, 207–214. [CrossRef]
58. Koch, O.R.; Deleo, M.E.; Borrello, S.; Palombini, G.; Galeotti, T. Ethanol treatment up regulates the expression of mitochondrial manganese superoxide dismutase in rat liver. *Biochem. Biophys. Res. Commun.* **1994**, *201*, 1356–1365. [CrossRef] [PubMed]
59. Schmitz-Peiffer, C.; Biden, T.J. Protein kinase C function in muscle, liver, and beta-cells and its therapeutic implications for type 2 diabetes. *Diabetes* **2008**, *57*, 1774–1783. [CrossRef] [PubMed]
60. Karasik, A.; Rothenberg, P.L.; Yamada, K.; White, M.F.; Kahn, C.R. Increased protein kinase C activity is linked to reduced insulin receptor autophosphorylation in liver of starved rats. *J. Biol. Chem.* **1990**, *265*, 10226–10231. [CrossRef]

antioxidants

Article

Increased Ingestion of Hydroxy-Methionine by Both Sows and Piglets Improves the Ability of the Progeny to Counteract LPS-Induced Hepatic and Splenic Injury with Potential Regulation of TLR4 and NOD Signaling

Meng Liu [1], Ying Zhang [1], Ke-Xin Cao [1], Ren-Gui Yang [2], Bao-Yang Xu [1,*], Wan-Po Zhang [3], Dolores I. Batonon-Alavo [4], Shu-Jun Zhang [5] and Lv-Hui Sun [1,*]

1 Hubei Hongshan Laboratory, College of Animal Science and Technology, Huazhong Agricultural University, Wuhan 430070, China; liumeng0821@webmail.hzau.edu.cn (M.L.); zhangysunny@163.com (Y.Z.); caokexin131@hotmail.com (K.-X.C.)
2 Tang Ren Shen Group Co., Ltd., Zhuzhou 412007, China; yrg5068@trsgroup.cn
3 College of Veterinary Medicine, Huazhong Agricultural University, Wuhan 430070, China; zwp@mail.hzau.edu.cn
4 Adisseo France S.A.S., 10, Place du Général de Gaulle, 92160 Antony, France; dolores.batonon-alavo@adisseo.com
5 Key Laboratory of Breeding and Reproduction of Ministry of Education, Huazhong Agricultural University, Wuhan 430070, China; sjxiaozhang@mail.hzau.edu.cn
* Correspondence: baoyangxu@mail.hzau.edu.cn (B.-Y.X.); lvhuisun@mail.hzau.edu.cn (L.-H.S.)

Citation: Liu, M.; Zhang, Y.; Cao, K.-X.; Yang, R.-G.; Xu, B.-Y.; Zhang, W.-P.; Batonon-Alavo, D.I.; Zhang, S.-J.; Sun, L.-H. Increased Ingestion of Hydroxy-Methionine by Both Sows and Piglets Improves the Ability of the Progeny to Counteract LPS-Induced Hepatic and Splenic Injury with Potential Regulation of TLR4 and NOD Signaling. *Antioxidants* **2022**, *11*, 321. https://doi.org/10.3390/antiox11020321

Academic Editors: Min Xue, Junmin Zhang, Zhenyu Du, Jie Wang, Wei Si and Stanley Omay

Received: 6 January 2022
Accepted: 4 February 2022
Published: 6 February 2022

Publisher's Note: MDPI stays neutral with regard to jurisdictional claims in published maps and institutional affiliations.

Abstract: Methionine, as an essential amino acid, play roles in antioxidant defense and the regulation of immune responses. This study was designed to determine the effects and mechanisms of increased consumption of methionine by sows and piglets on the capacity of the progeny to counteract lipopolysaccharide (LPS) challenge-induced injury in the liver and spleen of piglets. Primiparous sows (n = 10/diet) and their progeny were fed a diet that was adequate in sulfur amino acids (CON) or CON + 25% total sulfur amino acids as methionine from gestation day 85 to postnatal day 35. A total of ten male piglets were selected from each treatment and divided into 2 groups (n = 5/treatment) for a 2 × 2 factorial design [diets (CON, Methionine) and challenge (saline or LPS)] at 35 d old. After 24 h challenge, the piglets were euthanized to collect the liver and spleen for the histopathology, redox status, and gene expression analysis. The histopathological results showed that LPS challenge induced liver and spleen injury, while dietary methionine supplementation alleviated these damages that were induced by the LPS challenge. Furthermore, the LPS challenge also decreased the activities of GPX, SOD, and CAT and upregulated the mRNA and(or) protein expression of TLR4, MyD88, TRAF6, NOD1, NOD2, NF-kB, TNF-α, IL-8, p53, BCL2, and COX2 in the liver and (or) spleen. The alterations of GPX and SOD activities and the former nine genes were prevented or alleviated by the methionine supplementation. In conclusion, the maternal and neonatal dietary supplementation of methionine improved the ability of piglets to resist LPS challenge-induced liver and spleen injury, potentially through the increased antioxidant capacity and inhibition of TLR4 and NOD signaling pathway.

Keywords: sows; piglets; methionine; lipopolysaccharide; tissue damage

1. Introduction

Methionine is an essential amino acid for mammals and birds, and thus its dietary uptake is indispensable for animal maintenance, growth, and development [1]. It is the second or third most limiting amino acid for pigs that are fed corn-soybean meal diets [2,3]; it is used for protein synthesis and is involved in the methylation reactions of DNA and choline metabolism [4,5]. Additionally, methionine acts as the precursor for glutathione and taurine and plays a critical role in antioxidant defense and regulates immune responses in various animal species [6–8].

Weaning, when carried out at a very young age, is a stressful period psychologically and physiologically for piglets, which are susceptible to gastrointestinal diseases that are caused by pathogens, including pathogenic *Escherichia coli* (*E. coli*) [9]. The bacterial endotoxin lipopolysaccharide (LPS) can induce inflammation and oxidative stress, which causes damage to different organs and leads to adverse effects on animal productivity [10]. Our previous study showed that an increased consumption of methionine by both maternal and neonatal pigs can improve the capability of the progeny to cope with the LPS-induced negative effects [8]. However, the underlining mechanisms in this regard remain unclear.

Transmembrane toll-like receptors (TLRs) and cytoplasmic nucleotide-binding oligomerization domain proteins (NODs) play critical roles in modulating the innate and adaptive immune responses [11,12]. Toll-like receptor 4 (TLR4), the receptor of LPS, is a major player and triggers the activation of different intracellular signaling cascades, such as the activation of nuclear factor-κB (NF-κB) and the production of reactive oxygen species [13]. NOD1 and NOD2, as specialized NODs among the NOD family, which can connect with the peptidoglycan and LPS, and trigger the signal transduction pathway [14,15]. Therefore, TLRs or NODs can initiate a downstream signaling event that leads to the activation of NF-kB, which then stimulates the expression of inflammatory genes including interleukin (IL)-1β, IL-6, IL-8, and tumor necrosis factor-α (TNF-α). Consequently, the overproduction of proinflammatory cytokines can induce host tissue injury [13–15]. The main difference between TLRs and NODs is that TLRs are transmembrane sensors, and NODs are intracellular sensors [11]. Among the TLR family, TLR4 is the best studied member which responds primarily to LPS and induces an inflammatory response [12]. Among the NOD family, NOD1 and NOD2 are the best-characterized members, which can connect with the LPS and peptidoglycan, and trigger a signal transduction pathway [16].

Therefore, we hypothesize that increased the consumption of methionine by both sows and piglets might improve the capacity of the piglets to counteract LPS-induced injury to organs with the regulation of TLR4 and NOD signaling. The current study was conducted to investigate whether the increased consumption of methionine by both sows and progeny could alleviate the LPS-induced liver and spleen injury via the regulation of the TLR4 and NOD signaling in weaning pigs.

2. Materials and Methods

2.1. Animals, Treaztments, and Sample Collection

The animal protocol (HZAUSW-2018-022) of this study was approved by the Institutional Animal Care and Use Committee of Huazhong Agricultural University, China. In total, 20 primiparous sows (Landrace × Yorkshire) were divided into two groups (n = 10 sows/treatment) on day 85 of gestation based on their body weight and backfat thickness. A schematic of this experimental procedure to investigate the role of hydroxy-methionine (OH-Met) supplementation in piglets is shown in Supplemental Figure S1. Sows from the control group were fed a corn/soybean-control diet (CON), which was formulated to meet the nutritional requirements of sows (NRC, 2012, Table 1) [8,17]. Another group of sows were fed the control diet that was supplemented with OH-Met (Rhodimet AT88, Adisseo, France) at 25% above the total sulfur amino acids that were present in the CON. The detailed housing and feeding procedures for the sows were described in our previous study [8]. The feeding trial for the sows was lasted until the weaning of the piglets.

The piglets were weaned at the lactation day 21, and piglets from the same sow were kept in a pen. The piglets that were produced from the CON group of sows were fed a control diet which met the piglet's nutrients recommendations (CON; NRC, 2012, Table 1) [8,17]. The piglets from the OH-Met group of sows were fed the CON that was supplemented with OH-Met at 25% above the total sulfur amino acids that were present in the CON [8]. The piglets were allowed free access to the feed and water. The feeding trial for the piglets was lasted 14 days. On the day 35, 20 male piglets from the two groups (10 piglets/group) were selected according to their average body weight. They were divided into 4 groups (n = 5 piglets/group) for a 2-by-2 factorial design trial that included the dietary treatments (CON and OH-Met) and immunological challenge [saline vs. LPS (100 μg/kg BW, *E. coli* 0111: B4, Sigma)] by intraperitoneal injection [8]. After 24 h post-challenge, the piglets were humanely euthanized by intravenous injection of sodium pentobarbital (40 mg/kg body weight) to collect the liver and spleen for histologic examination as previously described [18]. Meanwhile, the liver and spleen tissues were washed with ice-cold isotonic saline, snap-frozen in liquid nitrogen, and stored at −80 °C until use [19].

Table 1. Ingredients and nutrients composition [1].

Ingredients (%)	Sows		Piglets
	Gestation	Lactation	Post-Weaning
Corn	61.77	65.74	17.60
Expanded corn	-	-	15.0
Wheat flour	-	-	10.0
Wheat bran	15	-	-
Soybean meal	14	28	-
Expanded soybeans	-	-	8.0
Fermented soybean meal	-	-	5.0
Corn gluten feed	2.0	-	-
Fish meal	-	-	4.0
Whey powder	-	-	12.0
Soybean oil	3.5	2.5	-
Sugar	-	-	8.0
Glucose	-	-	6.0
Emulsified fat powder	-	-	5.0
Plasma protein	-	-	5.0
$CaCO_3$	1.00	0.60	0.50
$CaHPO_4$	1.20	1.70	1.50
Salt	0.30	0.30	0.30
DL-Met	0.07	0.06	0.30
L-Lys	0.16	0.10	0.50
L-Thr	-	-	0.30
Vitamin premix [2]	0.50	0.50	0.50
Mineral premix [3]	0.50	0.50	0.50
Crude protein (%)	14.6	17.6	21.0
Digestible energy (MJ/kg)	13.7	14.2	14.2
Total Lys (%)	0.75	0.98	1.45
Total Met (%)	0.29	0.32	0.48
Total Met + Cys (%)	0.52	0.60	0.82
D Lys (%)	0.65	0.85	1.30
SID Met (%)	0.26	0.28	0.45
SID Met + Cys (%)	0.45	0.52	0.72
Calcium (%)	0.69	0.69	0.65
Total phosphorus (%)	0.60	0.63	0.64

[1] The OH-Met treatment diets during gestation, lactation, and days 21–35 were prepared by adding 1.477, 1.705, or 2.330 kg OH-Met (88%), respectively, to 1000 kg of the control diet at the expense of corn, to obtain TSSA levels in OH-Met treatments for gestation, lactation, and day 21–35 are 125% of the CON treatments, respectively. [2] Vitamin premix provided per kg of diet: retinyl acetate, 10,000 IU; dl-α-tocopheryl acetate, 50 IU; cholecalciferol 2500 IU; menadione, 5.0 mg; thiamin, 2.0 mg; pantothenic acid, 12.0 mg; riboflavin, 5.0 mg; pyridoxine, 10.0 mg; niacin, 30.0 mg; d-biotin, 0.2 mg; cyanocobalamin, 0.05 mg; folic acid, 1.5 mg; choline chloride 1500 mg. [3] Mineral premix provided per kg of diet: $FeSO_4 \cdot 7H_2O$, 498 mg; $ZnSO_4 \cdot 7H_2O$, 440 mg; $CuSO_4 \cdot 5H_2O$, 78.7 mg; Na_2SeO_3, 0.66mg; $MnSO_4 \cdot 5H_2O$, 110 mg; KI, 0.4 mg.

2.2. Histopathological and Redox Status Analysis

The liver and spleen tissues were examined microscopically after being fixed in 10% neutral buffered formalin, embedded in paraffin, sectioned at 5 μm, and stained with hematoxylin and eosin [18]. The total antioxidant capacity (T-AOC) and the activities of glutathione peroxidase (GPX), superoxide dismutase (SOD), and catalase (CAT), along with concentrations of malondialdehyde (MDA) were measured by specific assay kits (A003,

A005, A001, A007–1, and A087–1–2) that were purchased from the Nanjing Jiancheng Bioengineering Institute of China. The protein concentrations were measured by the bicinchoninic acid assay.

2.3. Real-Time q-PCR and Western-Blot Analyses

Real-time q-PCR analyses of the pertaining samples were conducted as previously described [20]. The primer pairs that were designed using Primer Express 3.0 (Applied Biosystems) are shown in Table 2. The expression of the target genes relative to the housekeeping gene (β-actin) was analyzed by the 2^{-ddCT} method. The relative mRNA expression level of each target gene was normalized to the control group (as 1). Western blot analyses of the pertaining samples were performed as previously described [21]. The primary antibodies that were used for each gene product are presented in Supplemental Table S1. The protein concentrations were measured by the bicinchoninic acid assay.

Table 2. List of primers that were used for q-PCR analysis [1].

Gene	Accession	Forward (5′-3′)	Reverse (5′-3′)
TLR4	NM_001113039.2	TGCTTTCTCCGGGTCACTTC	TTAGGAACCACCTGCACGC
MyD88	NM_001099923.1	GGCCCAGCATTGAAGAGGA	GACATCCAAGGGATGCTGCTA
TRAF6	NM_001105286.1	TTGGCTGCCATGAAAAGATGC	CTGAGCAACAGCCAGAGGAA
NOD1	NM_001114277.1	CAACCAAATCGGCGACGAAG	GCCGTTGAATGCAAGACTCAG
NOD2	NM_001105295.1	CTGTGAGCAGCTGCAGAAGT	TGGTTGTTTCCCAGCCTCAAT
NF-κB	NM_001048232.1	AGTACCCTGAGGCTATAACTCGC	TCCGCAATGGAGGAGAAGTC
COX2	NC_000845.1	ATGATCTACCCGCCTCACAC	AAAAGCAGCTCTGGGTCAAA
TNF-α	NM_214022.1	GGCCCAAGGACTCAGATCAT	CTGTCCCTCGGCTTTGACAT
IL-6	NM_214399.1	CCCTGAGGCAAAAGGGAAAGAA	CTCAGGTGCCCCAGCTACAT
IL-1β	NM_214055.1	CCCAATTCAGGGACCCTACC	TTTTGGGTGCAGCACTTCAT
IL-8	NM_213867.1	CTTCCAAACTGGCTGTTGCC	GTTGTTGTTGCTTCTCAGTTCTCT
p53	NM_213824.3	TGTAACCTGCACGTACTCCC	TCGGCCCGTAAATTCCCTTC
BCL2	XM_021099593.1	AGGATAACGGAGGCTGGGATG	CACTTATGGCCCAGATAGGCA
β-actin	XM_003124280.5	CTACACCGCTACCAGTTCGC	AGGGTCAGGATGCCTCTCTT

[1] TLR4, toll-like receptor 4; MyD88, myeloid differentiation factor 88; TRAF6, TNF-α receptor-associated factor 6; NOD, nucleotide binding oligomerization domain containing; NF-κB, nuclear transcription factor kappaB; COX2, cyclooxygenase 2; TNF-α, tumor necrosis factor-α; IL, interleukin; p53, tumor protein p53; BCL2, B-cell lymphoma 2.

2.4. Statistical Analysis

Statistical analysis was performed using SPSS (version 13, Chicago, IL, USA). The data are presented as the means ± SE. The data were analyzed by a two-way ANOVA with a significance level of $p < 0.05$. A Tukey test was used to do post hoc comparisons of the means if there was a significant effect.

3. Results

3.1. Liver and Spleen Histopathology

The histopathology of the liver and spleen are presented in Figure 1. Specifically, no obvious histopathological changes were observed in the liver and spleen of piglets in the control and OH-Met groups (Figure 1A,C,E,G). However, the histopathological results showed that LPS induced liver injury including cell swelling, narrowing of liver sinusoids, and an increase of neutrophil infiltration (Figure 1B). Meanwhile, LPS induced spleen damages such as congestion, moderate lymphocytosis, and neutrophilia infiltration in the splenic red pulp (Figure 1F). Notably, the dietary supplementation of OH-Met alleviated these LPS-induced damages in the liver (Figure 1D) and spleen (Figure 1H).

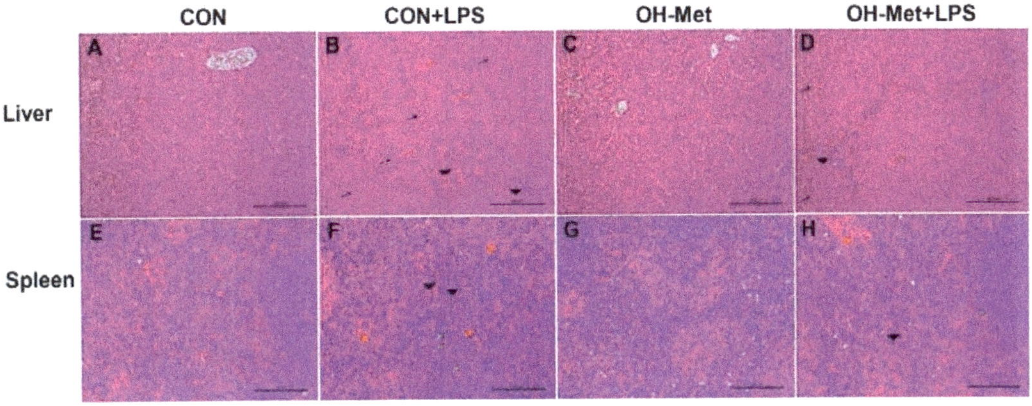

Figure 1. The effects of OH-Met supplementation on liver and spleen morphology after 24 h LPS-challenge in weaned piglets. The representative photomicrographs of liver and spleen sections that were stained with hematoxylin and eosin; photomicrographs are shown at 100× magnification (liver) and 200× magnification (spleen) magnification. Scale bars = 22.4 μm. The red arrow indicates narrowing of liver sinusoids; The black arrow indicates swelling; The black arrowhead indicates neutrophils infiltration; The red arrowhead indicates congestion; The green arrow indicates moderate lymphocytosis; CON, piglets receiving a control diet and injected with saline; CON+LPS, piglets receiving the control diet and challenged with LPS; OH-Met, piglets receiving a diet that was supplemented OH-Met at 25% above the total SAA present in the CON diet and injected with saline; OH-Met+LPS, piglets receiving a diet that was supplemented OH-Met at 25% above the total SAA present in the CON diet and challenged with LPS.

3.2. Antioxidant Parameters in Liver and Spleen

The antioxidant variables in the liver and spleen were significantly affected by the LPS challenge, dietary OH-Met supplementation, or their interaction (Figure 2). In the liver (Figure 2F–J), the LPS challenge did not affect ($p \geq 0.05$) the activity of T-AOC and MDA concentration in both diets and decreased the activities of GPX and SOD only in the CON diet. It led to a 45.4% or 36.2% decrease ($p < 0.05$) in the CAT activity in both the diets with or without OH-Met. Notably, dietary OH-Met supplementation increased ($p < 0.05$) the hepatic GPX activity (30.5%) in the LPS-challenge groups. In the spleen (Figure 2F–J), LPS led to a 18.8% decrease in CAT activity ($p < 0.05$) in the dietary supplementation with OH-Met groups, while it only led to 8.8% decrease in the SOD activity ($p < 0.05$) in the diets without OH-Met supplementation. Notably, dietary OH-Met supplementation increased ($p < 0.05$) the splenic SOD activity (19.9%) in the LPS groups. The dietary OH-Met supplementation decreased ($p < 0.05$) the MDA concentration compared to the other three treatments that were injected or not with LPS.

3.3. Expression of TLR4 and NODs Signaling in Liver and Spleen

Among the 12 assayed genes of TLR4 and NODs signaling, 12 genes in the liver and nine genes in the spleen were affected the LPS challenge, OH-Met supplementation, or their interaction (Figure 3). Specifically, the LPS challenge increased ($p < 0.05$) the mRNA levels of TLR4, myeloid differentiation factor 88 (*MyD88*), TNF-α receptor-associated factor 6 (TRAF6), NOD1, NOD2, NF-κB, cyclooxygenase 2 (COX2), TNF-α, IL-8, and tumor protein p53 (p53) in the liver of the CON groups (Figure 3A). Interestingly, the changes of 10 genes including TLR4, MyD88, TRAF6, NOD1, NOD2, NF-κB, COX2, TNF-α, IL-8, and p53 by the LPS challenge were prevented or mitigated ($p < 0.05$) by the OH-Met supplement (Figure 3A). In the spleen, LPS challenge increased ($p < 0.05$) the mRNA levels of MyD88 and decreased ($p < 0.05$) the mRNA levels of B-cell lymphoma 2 (BCL2) in the spleen of the CON and OH-Met groups (Figure 3B). Notably, dietary OH-Met supplementation alleviated ($p < 0.05$) the LPS-induced changes on the BCL2 expression in the spleen (Figure 3B). Meanwhile, the dietary OH-Met supplementation reduced ($p < 0.05$) TLR4 and IL-8 expression, as well as increased ($p < 0.05$) NF-κB expression (Figure 3B).

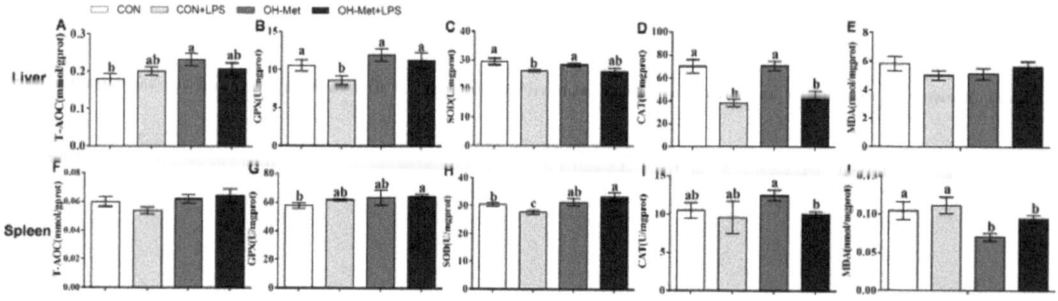

Figure 2. The effects of OH-Met supplementation on antioxidant indexes of the (**A–E**) liver and (**F–J**) spleen after 24 h LPS challenge in weaned piglets. The values are the means ± SEs, $n = 5$. Labeled means in a row without a common letter differ, $p < 0.05$. T-AOC, total antioxidant capacity; GPX, glutathione peroxidase; SOD, superoxide dismutase; CAT, catalase; MDA, malondialdehyde; CON, piglets receiving a control diet and injected with saline; CON+LPS, piglets receiving the control diet and challenged with LPS; OH-Met, piglets receiving a diet that was supplemented OH-Met at 25% above the total SAA present in the CON diet and injected with saline; OH-Met+LPS, piglets receiving a diet that was supplemented OH-Met at 25% above the total SAA present in the CON diet and challenged with LPS.

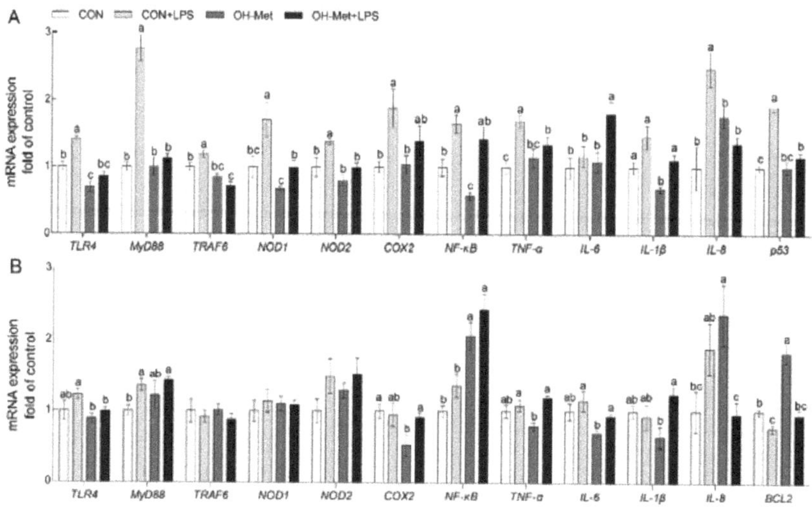

Figure 3. The effects of OH-Met supplementation on TLR4 and NODs signal-related genes of the (**A**) liver and (**B**) spleen after 24 h LPS challenge in weaned piglets. The values are the means ± SEs, $n = 5$. The labeled means without a common letter differ, $p < 0.05$. CON, piglets receiving a control diet and injected with saline; CON+LPS, piglets receiving the control diet and challenged with LPS; OH-Met, piglets receiving a diet that was supplemented OH-Met at 25% above the total SAA present in the CON diet and injected with saline; OH-Met+LPS, piglets receiving a diet that was supplemented OH-Met at 25% above the total SAA present in the CON diet and challenged with LPS.

3.4. Production of Selected TLR4 and NODs Signaling Proteins in Liver and Spleen

In the liver (Figure 4A), Western blot results showed that the LPS challenge increased ($p < 0.05$) the protein levels of MyD88, NF-κB, and p53 in the CON groups. Notably, the changes of MyD88, NF-κB, and p53 proteins by the LPS challenge were prevented or mitigated by OH-Met supplementation. Interestingly, the dietary OH-Met supplementation reduced ($p < 0.05$) the hepatic TLR4 protein in the LPS challenge groups. In the spleen (Figure 4B), Western blot results showed that LPS challenge increased ($p < 0.05$) the protein levels of TRAF6 and NF-κB but reduced ($p < 0.05$) the protein levels of BCL2 in the CON groups. Notably, the changes of these proteins by the LPS challenge were inhibited by OH-Met supplementation. Additionally, dietary OH-Met supplementation also reduced ($p < 0.05$) splenic TLR4 protein in both the saline and the LPS challenge groups.

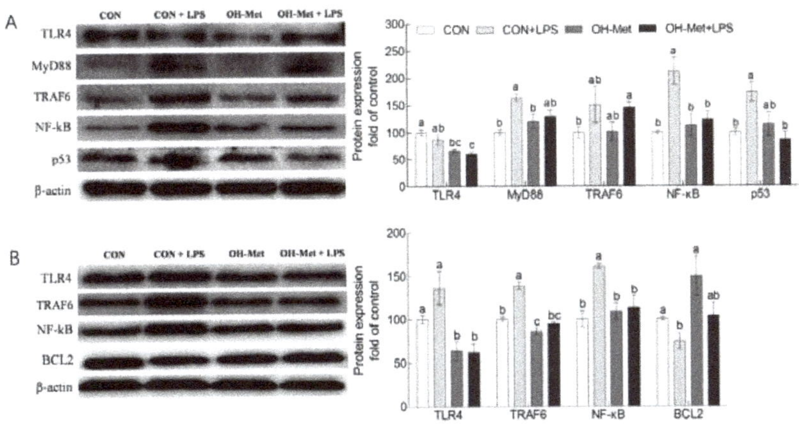

Figure 4. The effects of OH-Met supplementation onTLR4 and NODs signal-related protein production of the (**A**) liver and (**B**) spleen after 24 h LPS challenge in weaned piglets. The values are the means ± SEs, n = 3–4. The labeled means without a common letter differ, $p < 0.05$. CON, piglets receiving a control diet and injected with saline; CON+LPS, piglets receiving the control diet and challenged with LPS; OH Met, piglets receiving a diet that was supplemented OH-Met at 25% above the total SAA present in the CON diet and injected with saline; OH-Met+LPS, piglets receiving a diet that was supplemented OH-Met at 25% above the total SAA present in the CON diet and challenged with LPS.

4. Discussion

The interesting finding in this study was that the maternal and neonatal dietary methionine supplementation during the late gestation, lactation, and postweaning periods can mitigate the LPS challenge-induced damages in the liver and spleen of piglets. In this study, LPS challenge induced hepatic and splenic injury, as evidenced by swelling, sinusoidal narrowing, increased inflammatory cells, neutrophils in sinusoids and lobules of liver, as well as bruising, moderate lymphocytosis, and neutrophilia in the red pulp of spleen. These outcomes are consistent with previous studies, which provided evidence that damages in the liver and spleen were induced by LPS challenge [22–24]. Interestingly, the histopathological changes in the liver and spleen that were induced by the LPS challenge were moderated by dietary-supplemented OH-Met in this study. These findings are in line with our previous study which showed that methionine supplementation alleviated LPS challenge-induced negative changes on the plasma biochemistry biomarkers that were related to the liver function and inflammation of piglets [8]. Similar findings were also reported in cows and showed that methionine supplementation exerted protective effects against LPS challenge-induced negative effects in bovine mammary epithelial and polymorphonuclear cells [25,26].

Consistent with previous studies [27,28], the piglets that were challenged by LPS in the present study experienced oxidative stress, as indicated by the reduction of antioxidant capacity parameters (GPX, SOD, and T-AOC) and an increase of MDA concentration in the liver and/or spleen. Strikingly, the LPS challenge-induced imbalance in the redox status in the liver and the spleen were alleviated by the OH-Met supplementation in the current study. These outcomes were consistent with previous reports that methionine plays a particularly important role in providing Cys for GSH synthesis to improve antioxidant capacity [8,29,30]. Taken together, these results agreed with previous studies, which showed that dietary supplementation of methionine beyond sulfur amino acids growth requirements can reduce the LPS challenge-induced oxidative stress in pigs, poultry, and fish [8,31,32].

It is well documented that both the integral membrane TLRs and the cytosolic NODs are the receptors of LPS and play pivotal roles in the host defense against LPS-induced challenges [33,34]. In agreement with previous studies [33–35], the LPS challenge induced the mRNA and(or) protein production of TLRs (TLR4, MyD88, and TRAF6) and NODs (NOD1, NOD2) in the liver and (or) spleen of the piglets. Then, as the downstream signaling of TLRs and NODs, the inflammatory reaction-related genes including NF-kB, COX2, TNF-α, IL-6, IL-1β, and IL-8, were upregulated at the mRNA and (or) protein levels in the liver and (or) spleen of the piglets in the current study. Meanwhile, the LPS challenge also upregulated the apoptotic gene (p53) but downregulated anti-apoptotic gene (BCL2) at the mRNA and (or) protein levels in the liver and (or) spleen of the piglets. Consequently, the overproduction of pro-inflammatory cytokines and apoptosis-related signaling that was induced by the LPS challenge may be attributed to the damages in the liver and spleen of piglets that were observed in the present study [13–15,36]. Taken together, these outcomes are consistent with those that have been reported in previous studies [37,38], which showed that LPS challenge can induce tissue damage via the activation of TLRs and NODs signaling and further induce excessive inflammation and apoptosis. Strikingly, the increased consumption of sulfur amino acids, such as OH-Met, by sows and piglets prevented and (or) alleviated the changes of most of the genes in the liver and (or) spleen that were induced by the LPS challenge. These findings revealed that the increased consumption of methionine by sows and piglets alleviated the LPS challenge-induced damages in the liver and the spleen that were associated with the potential regulation of the TRLs and NODs signaling. In agreement with previous studies, methionine can downregulate TLR4 and (or) NODs signaling in osteoclast precursors and thus decrease bone loss during osteoporosis [39] in bovine mammary epithelial cells and improve the immune and antioxidant status [40]. Similar findings were also obtained for other amino acids in previous studies which showed that dietary supplementation of aspartate, glycine, and glutamate attenuate the LPS challenge-induced negative effects via TLR4 and (or) NODs signaling in tissues of piglets [41–45].

Nevertheless, several seemingly conflicting or inconsistent scenarios were observed in the current study. For example, the protein production of TLR4 and TRAF6 in the liver and NF-κB in the spleen between the saline and LPS challenge in CON diet groups, did not correlate well with their mRNA abundance. This discrepancy may be explained by a complex feedback or post-transcriptional mechanism regulating protein synthesis [38].

In summary, this study has illustrated that the maternal and neonatal dietary supplementation of OH-Met at 25% above the total sulfur amino acid requirements exert beneficial effects in alleviating the LPS challenged-induced damages in the liver and spleen of piglets. Moreover, the protective mechanism of OH-Met against LPS challenge-induced adverse effects may be associated with (1) an enhancement of the animal's antioxidant capacities or (2) an inhibition of TLR4 and NOD signaling pathway. This finding indicated that the recommendations from NRC (2012) for sulfur amino acids for sows during gestation and lactation might require an update. Attention should also be paid to the provision of sulfur amino acids for weaned piglets that were often challenged by oxidation and inflammation.

Supplementary Materials: The following are available online at https://www.mdpi.com/article/10.3390/antiox11020321/s1. Table S1: Name, type, dilution, and source of the primary antibodies; Figure S1: Schematic of the experimental procedure of the animal trial; Figure S2: Validation of the specificity of antibodies against and the correct bands of TRL4, MyD88, TRAF6, NF-κB, p53, and BCL2.

Author Contributions: L.-H.S., B.-Y.X., D.I.B.-A. and S.-J.Z. designed the research; M.L., Y.Z., K.-X.C., R.-G.Y., B.-Y.X. and W.-P.Z. conducted the experiments and analyzed the data; M.L., B.-Y.X. and L.-H.S. wrote the manuscript. All authors have read and agreed to the published version of the manuscript.

Funding: This project was supported in part by the National Key Research and Development Program of China (2021YFD1300405) and a research grant (16MEI536) by Adisseo France S.A.S., 10, Place du Général de Gaulle, 92160 Antony, France.

Institutional Review Board Statement: The animal protocol of this study was approved by the Institutional Animal Care and Use Committee of Huazhong Agricultural University, China (HZAUSW-2018-022).

Informed Consent Statement: Not applicable.

Data Availability Statement: Data are contained within the article.

Acknowledgments: The authors acknowledge the financial support of the National Key Research and Development Program of China (2021YFD1300405) and a research grant (16MEI536) by Adisseo France S.A.S., 10, Place du Général de Gaulle, 92160 Antony, France.

Conflicts of Interest: M.L., Y.Z., K.-X.C., R.-G.Y., B.-Y.X., W.-P.Z., S.-J.Z., and L.-H.S. declare they have no conflict of interest; D.I.B.-A. is a member of Adisseo France S.A.S.

References

1. Burley, H.K. Enrichment of Methionine from Naturally Concentrated Feedstuffs for Use in Organic Poultry Diets. Ph.D. Thesis, The Pennsylvania State University, State College, PA, USA, 2012.
2. Schingoethe, D.J.; Ahrar, M. Protein solubility, amino acid composition, and biological value of regular and heat-treated soybean and sunflower meals. *J. Dairy Sci.* **1979**, *62*, 925–931. [CrossRef]
3. Conde-Aguilera, J.A.; Floc'h, N.L.; Huërou-Luron, I.L.; Mercier, Y.; Tesseraud, S.; Lefaucheur, L.; Milgen, J.V. Splanchnic tissues respond differently when piglets are offered a diet 30% deficient in total sulfur amino acid for 10 days. *Eur. J. Nutr.* **2016**, *55*, 2209–2219. [CrossRef] [PubMed]
4. Baker, D.H. Comparative species utilization and toxicity of sulfur amino acids. *J. Nutr.* **2006**, *136*, 1670–1675. [CrossRef]
5. Giguére, A.; Girard, C.L.; Matte, J.J. Methionine, folic acid and vitamin B12 in growing-finishing pigs: Impact on growth performance and meat quality. *Arch. Anim. Nutr.* **2008**, *62*, 193–206. [CrossRef]
6. Zhao, L.; Zhang, N.Y.; Pan, Y.X.; Zhu, L.Y.; Batonon-Alavo, D.I.; Ma, L.B.; Khalil, M.M.; Qi, D.S.; Sun, L.H. Efficacy of 2-hydroxy-4-Methylthio-butanoic acid compared to DL-Methionine on growth performance, carcass traits, feather growth, and redox status of Cherry Valley ducks. *Poult. Sci.* **2018**, *97*, 3166–3175. [CrossRef]
7. Fang, Z.; Yao, K.; Zhang, X.; Zhao, S.; Sun, Z.; Tian, G.; Yu, B.; Lin, Y.; Zhu, B.Q.; Jia, G.; et al. Nutrition and health relevant regulation of intestinal sulfur amino acid metabolism. *Amino Acids* **2010**, *39*, 633–640. [CrossRef]
8. Zhang, Y.; Xu, B.Y.; Zhao, L.; Zhu, L.Y.; Batonon-Alavo, D.; Jachacz, J.; Qi, D.S.; Zhang, S.J.; Ma, L.B.; Sun, L.H. Increased consumption of sulfur amino acids by both sows and piglets enhances the ability of the progeny to adverse effects induced by lipopolysaccharide. *Animals* **2019**, *9*, 1048. [CrossRef] [PubMed]
9. Goswami, P.S.; Friendship, R.M.; Gyles, C.L.; Poppe, C.; Boerlin, P. Preliminary investigations of the distribution of Escherichia coli O149 in sows, piglets, and their environment. *Can. J. Vet. Res.* **2011**, *75*, 57–60. [PubMed]
10. Hou, X.; Zhang, J.; Ahmad, H.; Zhang, H.; Xu, Z.; Wang, T. Evaluation of antioxidant activities of ampelopsin and its protective effect in lipopolysaccharide-induced oxidative stress piglets. *PLoS ONE* **2014**, *9*, e108314. [CrossRef]
11. Fukata, M.; Vamadevan, A.S.; Abreu, M.T. Toll-like receptors (TLRs) and Nod-like receptors (NLRs) in inflammatory disorders. *Semin. Immunol.* **2009**, *21*, 242–253. [CrossRef]
12. Takeuchi, O.; Akira, S. Pattern recognition receptors and inflammation. *Cell* **2010**, *140*, 805–820. [CrossRef] [PubMed]
13. Sabroe, I.; Parker, L.C.; Dower, S.K.; Whyte, M.K.B. The role of TLR activation in inflammation. *J. Pathol.* **2008**, *214*, 126–135. [CrossRef]
14. Lappas, M. NOD1 and NOD2 regulate proinflammatory and prolabor mediators in human fetal membranes and myometrium via nuclear factor-kappa B. *Biol. Reprod.* **2013**, *89*, 14. [CrossRef] [PubMed]
15. Takada, H.; Uehara, A. Enhancement of TLR-mediated innate immune responses by peptidoglycans through NOD signaling. *Curr. Pharm. Des.* **2006**, *12*, 4163–4172. [CrossRef]

16. Moreira, L.O.; Zamboni, D.S. NOD1 and NOD2 signaling in infection and inflammation. *Front. Immunol.* **2012**, *3*, 328. [CrossRef] [PubMed]
17. National Research Council (NRC). *Nutrient Requirements of Swine*, 11th ed.; National Academy Press: Washington, DC, USA, 2012.
18. Sun, L.H.; Zhang, N.Y.; Zhu, M.K.; Zhao, L.; Zhou, J.C.; Qi, D.S. Prevention of aflatoxin B1 hepatoxicity by dietary selenium is associated with inhibition of cytochrome P450 isozymes and up-regulation of 6 selenoprotein genes in chick liver. *J. Nutr.* **2016**, *146*, 655–661. [CrossRef]
19. Zhao, L.; Sun, L.H.; Huang, J.Q.; Briens, M.; Qi, D.S.; Xu, S.W.; Lei, G.X. A novel organic selenium compound exerts unique regulation of selenium speciation, selenogenome, and selenoproteins in broiler chicks. *J. Nutr.* **2017**, *147*, 789–797. [CrossRef]
20. Zhou, J.C.; Zhao, H.; Li, J.G.; Xia, X.J.; Wang, K.N.; Zhang, Y.J.; Liu, Y.; Zhao, Y.; Lei, G.X. Selenoprotein gene expression in thyroid and pituitary of young pigs is not affected by dietary selenium deficiency or excess. *J. Nutr.* **2009**, *139*, 1061–1066. [CrossRef]
21. Huang, J.Q.; Ren, F.Z.; Jiang, Y.Y.; Xiao, C.; Lei, X.G. Selenoproteins protect against avian nutritional muscular dystrophy by metabolizing peroxides and regulating redox/apoptotic signaling. *Free Radic. Biol. Med.* **2015**, *83*, 129–138. [CrossRef] [PubMed]
22. Masaki, T.; Chiba, S.; Tatsukawa, H.; Yasuda, T.; Noguchi, H.; Seike, M.; Yoshimatsu, M. Adiponectin protects LPS-induced liver injury through modulation of TNF-α in KK-Ay obese mice. *Hepatology* **2004**, *40*, 177–184. [CrossRef] [PubMed]
23. Véronique, B. Effects of Proinflammatory Agents on Oxygen Species Production by Bovine Mammary Epithelial and Immune Cells. Master's Thesis, McGill University, Montreal, QC, Canada, 2000.
24. Schmöcker, C.; Weylandt, K.H.; Kahlke, L.; Wang, J.; Lobeck, H.; Tiegs, G.; Berg, T.; Kang, J.X. Omega-3 fatty acids alleviate chemically induced acute hepatitis by suppression of cytokines. *Hepatology* **2007**, *45*, 864–869. [CrossRef] [PubMed]
25. Vailati-Riboni, M.; Xu, T.; Qadir, B.; Buckrout, R.; Parys, C.; Loor, J.J. In vitro methionine supplementation during lipopolysaccharide stimulation modulates immune metabolic gene network expression in isolated polymorphonuclear cells from lactating Holstein cows. *J. Dairy Sci.* **2019**, *102*, 8343–8351. [CrossRef]
26. Dai, H.; Coleman, D.N.; Hu, L.; Martinez-Cortés, I.; Wang, M.; Parys, C.; Shen, X.; Loor, J.J. Methionine and arginine supplementation alter inflammatory and oxidative stress responses during lipopolysaccharide challenge in bovine mammary epithelial cells in vitro. *J. Dairy Sci.* **2020**, *103*, 676–689. [CrossRef]
27. Li, Q.; Liu, Y.L.; Che, Z.Q.; Zhu, H.L.; Meng, G.Q.; Hou, Y.Y. Dietary L-arginine supplementation alleviates liver injury caused by Escherichia coli LPS in weaned pigs. *Innate Immun.* **2012**, *18*, 804–814. [CrossRef]
28. Li, Y.; Zhao, X.L.; Jiang, X.M.; Chen, L.; Hong, L.; Zhuo, Y.; Lin, Y.; Fang, Z.F.; Che, L.Q.; Feng, B.; et al. Effects of dietary supplementation with exogenous catalase on growth performance, oxidative stress, and hepatic apoptosis in weaned piglets challenged with lipopolysaccharide. *J. Anim. Sci.* **2020**, *98*, skaa067. [CrossRef] [PubMed]
29. Swennen, Q.; Geraert, P.A.; Mercier, Y.; Everaert, N.; Stinckens, A.; Willemsen, H.; Li, Y.; Decuypere, E.; Buyse, J. Effects of dietary protein content and 2-hydroxy-4-methylthiobutanoic acid or DL-methionine supplementation on performance and oxidative status of broiler chickens. *Br. J. Nutr.* **2011**, *106*, 1845–1854. [CrossRef]
30. Toborek, M.; Kopieczna-Grzebieniak, E.; Dr´ozdz, M.; Wieczorek, M. Increased lipid peroxidation as a mechanism of methionine-induced atherosclerosis in rabbits. *Atherosclerosis* **1995**, *115*, 217–224. [CrossRef]
31. Wang, Z.G.; Pan, X.J.; Zhang, W.Q.; Peng, Z.Q.; Zhao, R.Q.; Zhou, G.H. Methionine and selenium yeast supplementation of the maternal diets affects antioxidant activity of breeding eggs. *Poult. Sci.* **2010**, *89*, 931–937. [CrossRef]
32. Noor, Z.; Noor, M.; Khan, S.A.; Younas, W.; Ualiyeva, D.; Hassan, Z.; Yousafzai, A.M. Dietary supplementations of methionine improve growth performances, innate immunity, digestive enzymes, and antioxidant activities of rohu (Labeo rohita). *Fish Physiol. Biochem.* **2021**, *47*, 451–464. [CrossRef] [PubMed]
33. Sun, L.; Pham, T.T.; Cornell, T.T.; McDonough, K.L.; McHugh, W.M.; Blatt, N.B.; Dahmer, M.K.; Shanley, T.P. Myeloid-specific gene deletion of protein phosphatase 2A magnifies MyD88- and TRIF-dependent inflammation following endotoxin challenge. *J. Immunol.* **2017**, *198*, 404–416. [CrossRef]
34. Qiao, J.Y.; Sun, Z.Y.; Liang, D.M.; Li, H.H. Lactobacillus salivarius alleviates inflammation via NF-κB signaling in ETEC K88-induced IPEC-J2 cells. *J. Anim. Sci. Biotechnol.* **2020**, *11*, 76. [CrossRef]
35. Bulgari, O.; Dong, X.W.; Roca, A.L.; Carli, A.M.; Loor, J.J. Innate immune responses induced by lipopolysaccharide and lipoteichoic acid in primary goat mammary epithelial cells. *J. Anim. Sci. Biotechnol.* **2017**, *8*, 28. [CrossRef] [PubMed]
36. Walter, K.R.; Lin, X.; Jacobi, S.K.; Käser, T.; Esposito, D.; Odle, J. Dietary arachidonate in milk replacer triggers dual benefits of PGE2 signaling in LPS-challenged piglet alveolar macrophages. *J. Anim. Sci. Biotechnol.* **2019**, *10*, 13. [CrossRef]
37. Chen, L.; Li, S.; Zheng, J.; Li, W.T.; Jiang, X.M.; Zhao, X.L.; Li, J.; Che, L.Q.; Lin, Y.; Xu, S.Y.; et al. Effects of dietary Clostridium butyricum supplementation on growth performance, intestinal development, and immune response of weaned piglets challenged with lipopolysaccharide. *J. Anim. Sci. Biotechnol.* **2018**, *9*, 62. [CrossRef] [PubMed]
38. Wang, K.L.; Chen, G.Y.; Cao, G.T.; Xu, Y.L.; Wang, Y.X.; Yang, C.M. Effects of Clostridium butyricum and Enterococcus faecalis on growth performance, intestinal structure, and inflammation in lipopolysaccharide-challenged weaned piglets. *J. Anim. Sci.* **2019**, *97*, 4140–4151. [CrossRef] [PubMed]
39. Vijayan, V.; Khandelwal, M.; Manglani, K.; Gupta, S.; Surolia, A. Methionine down-regulates TLR4/MyD88/NF-κB signalling in osteoclast precursors to reduce bone loss during osteoporosis. *Br. J. Pharmacol.* **2014**, *171*, 107–121. [CrossRef] [PubMed]
40. Dai, H.; Coleman, D.N.; Lopes, M.G.; Hu, L.; Martinez-Cortés, I.; Parys, C.; Shen, X.; Loor, J.J. Alterations in immune and antioxidant gene networks by gamma-d-glutamyl-meso-diaminopimelic acid in bovine mammary epithelial cells are attenuated by in vitro supply of methionine and arginine. *J. Dairy Sci.* **2021**, *104*, 776–785. [CrossRef]

41. Leng, W.B.; Liu, Y.L.; Shi, H.F.; Li, S.; Zhu, H.L.; Pi, D.A.; Hou, Y.Q. Aspartate alleviates liver injury and regulates mRNA expressions of TLR4 and NOD signaling-related genes in weaned pigs after lipopolysaccharide challenge. *J. Nutr. Biochem.* **2014**, *25*, 592–599. [CrossRef]
42. Xu, X.; Wang, X.Y.; Wu, H.T.; Zhu, H.L.; Liu, C.C.; Hou, Y.Q.; Dai, B.; Liu, X.T.; Liu, Y.L. Glycine relieves intestinal injury by maintaining mTOR signaling and suppressing AMPK, TLR4, and NOD signaling in weaned piglets after lipopolysaccharide challenge. *Int. J. Mol. Sci.* **2018**, *19*, 1980. [CrossRef]
43. Kang, P.; Wang, X.Y.; Wu, H.T.; Zhu, H.L.; Hou, Y.Q.; Wang, L.M.; Liu, Y.L. Glutamate alleviates muscle protein loss by modulating TLR4, NODs, Akt/FOXO and mTOR signaling pathways in LPS-challenged piglets. *PLoS ONE* **2017**, *12*, e0182246. [CrossRef]
44. Liu, Y.L.; Wang, X.Y.; Hou, Y.Q.; Yin, Y.L.; Qiu, Y.S.; Wu, G.Y.; Hu, C.A. Roles of amino acids in preventing and treating intestinal diseases: Recent studies with pig models. *Amino Acids* **2017**, *49*, 1277–1291. [CrossRef] [PubMed]
45. Zhang, Y.C.; Mu, T.Q.; Jia, H.; Yang, Y.; Wu, Z.L. Protective effects of glycine against lipopolysaccharide-induced intestinal apoptosis and inflammation. *Amino Acids* **2021**, 1–12. [CrossRef] [PubMed]

antioxidants

MDPI

Article

Comparison of Growth Performance, Immunity, Antioxidant Capacity, and Liver Transcriptome of Calves between Whole Milk and Plant Protein-Based Milk Replacer under the Same Energy and Protein Levels

Shuo Wang [1,2,†], Fengming Hu [1,2,†], Qiyu Diao [1,2], Shuang Li [1], Yan Tu [1,2,*] and Yanliang Bi [1,2,*]

1 Key Laboratory of Feed Biotechnology of the Ministry of Agriculture and Rural Affairs, Institute of Feed Research, Chinese Academy of Agricultural Sciences, Beijing 100081, China; 82101185153@caas.cn (S.W.); hufengming@caas.cn (F.H.); diaoqiyu@caas.cn (Q.D.); 82101202361@caas.cn (S.L.)
2 Beijing Key Laboratory for Dairy Cow Nutrition, Institute of Feed Research, Chinese Academy of Agricultural Sciences, Beijing 100081, China
* Correspondence: tuyan@caas.cn (Y.T.); biyanliang@caas.cn (Y.B.)
† These authors contributed equally to this work.

Abstract: High-cost milk proteins necessitate cheaper, effective milk replacer alternatives, such as plant proteins. To examine plant protein-based milk replacer's impact on growth performance, serum immune and antioxidant indicators, and liver transcriptome profiles in suckling calves. We assigned 28 newborn Holstein calves (41.60 ± 3.67 kg of body weight at birth) to milk (M) or milk replacer (MR) and starter diets pre-weaning (0–70 d of age) but with the same starter diet post-weaning (71–98 d of age). During the pre-weaning period, compared with the M group, MR group had significantly lower body weight, withers height, heart girth, average daily gain, feed efficiency, serum immunoglobulin (Ig) M concentration, superoxide dismutase concentration, and total antioxidant capacity; whereas they had significantly higher serum aspartate aminotransferase concentration. During the post-weaning period, MR group presented significantly higher average daily gain, alanine transaminase, aspartate aminotransferase, and malonaldehyde concentrations; whereas they had significantly lower serum IgA and IgM concentrations than the M group. Transcriptome analysis revealed 1, 120 and 293 differentially expressed genes (DEGs; MR vs. M group) in the calves from pre- and post-weaning periods, respectively. The DEGs related to xenobiotic and lipid metabolism and those related to energy metabolism, immune function, and mineral metabolism were up- and downregulated, respectively, during the pre-weaning period; during the post-weaning period, the DEGs related to osteoclast differentiation and metabolic pathways showed difference. In this study, compared with M group, MR group had the same growth performance during the overall experimental period; however, MR affected the hepatic metabolism, immune, and antioxidant function of calves. These observations can facilitate future studies on milk replacers.

Keywords: calf; liver; milk replacer; plant protein; transcriptome

Citation: Wang, S.; Hu, F.; Diao, Q.; Li, S.; Tu, Y.; Bi, Y. Comparison of Growth Performance, Immunity, Antioxidant Capacity, and Liver Transcriptome of Calves between Whole Milk and Plant Protein-Based Milk Replacer under the Same Energy and Protein Levels. *Antioxidants* **2022**, *11*, 270. https://doi.org/10.3390/antiox11020270

Academic Editor: Stanley Omaye

Received: 6 January 2022
Accepted: 27 January 2022
Published: 29 January 2022

1. Introduction

Successful calf production, especially dairy calf management, is crucial for the profitability and sustainability of the dairy industry [1,2]. Due to the underdeveloped rumen, suckling calves cannot meet their entire nutritional requirements through digesting solid feed; therefore, their nutrition primarily relies on milk and milk replacers [3]. Rearing calves using milk replacers has become a common practice in dairy farming worldwide. In the United States, more than 85% of calves are fed milk replacers before weaning [1]. The use of milk replacers, besides increasing the quantity of fresh milk available for human consumption, improves the health of the calves by avoiding disease transmission from the dam to the calf, reduces the cost of feeding, and provides farmers the opportunity to manipulate the nutrition of the calves for different production needs [4–6]. Therefore, an appropriate choice of milk replacer is essential for meeting the long-term targets of dairy farmers.

Protein is one of the main nutrients in the milk replacer, and the source as well as the concentration of the protein directly affects the growth and development of calves. In the 1980s, milk protein sources, including skim milk powder, casein, and whey protein, were the main protein sources in the milk replacers due to their balanced nutrition, fast digestion and absorption, and high utilization rate [7]. However, owing to the large-scale international trade of milk powder and the constant rise in the prices of casein and whey proteins, the use of plant proteins, especially the soy protein, in milk replacers is being studied extensively [8–10]. Compared with milk protein, plant proteins are high in crude proteins (CP) and functional amino acids [11]. However, compared to milk protein, the plant proteins have lower digestibility and solubility, poor amino acid profiles, and they contain anti-nutritional factors (ANF), such as protease inhibitors and phytic acid in soy protein, and non-starch polysaccharides in wheat protein [7,11,12]. Many studies have shown that the outcome of feeding calves with milk replacers containing plant protein in the pre-weaning period is generally unsatisfactory; however, it can increase the rate of rumen development and lower the weaning age through higher starter feed, increasing the average daily age (ADG) once the calves are weaned [5,8,13–15]. Meanwhile, wheat and rice proteins have been studied as potential replacements for milk protein in the formulation of milk replacers [16]. Wheat proteins, which contain glutamic acid and glutamine, are involved in the synthesis and metabolism of nucleic acids and proteins in intestinal cells; they also play a major role in intestinal mucosa cell regeneration and the maintenance of intestinal fitness [17]. Meanwhile, rice proteins, which contain lysine, methionine, and low ANF, possess antioxidative properties and regulates lipid metabolism [18,19]. To date, many studies have focused on the substitution of milk proteins with single plant proteins. Since the different kinds of plant proteins have different advantageous characteristics, we think that a combination of various plant proteins is necessary for the successful implementation of plant proteins in milk replacers.

As the metabolic powerhouse of the body, the liver is involved in the metabolism of the milk replacer and milk. Transcriptomics is a major tool in studying animal nutrition and health, and it is widely employed in biological system research [20,21]. Transcriptomics provides insights into the biological functions of genes that are upregulated or downregulated in experimental animals in response to the test materials [20,22]. Recently, with the development of RNA-Seq technology, transcriptomics has promoted genome research of biological systems, enabling the generation of biological information about experimental targets [22,23]. However, there have not been any studies on the gene expression in the livers of calves fed on milk replacers. Therefore, understanding the differentially expressed genes (DEGs) between the livers of milk-fed and milk replacer-fed calves might be helpful for future research on the effects of milk replacers on calves.

In this study, we hypothesized that the calves fed with a combination of various plant protein-based milk replacers have similar growth performance compared to whole milk-fed calves under the same energy and protein levels, which probably should be the case due to a better adaptability of their liver metabolism to milk replacer. Based on these hypotheses, this study aimed to examine the mechanism by which plant protein-based milk replacers influence the growth of the calves and identify possible hub-genes and molecular pathways involved in the metabolism of plant protein-based milk replacers using RNA-Seq technologies to characterize the liver transcriptome of the calves and provide the latest insights into the research and development of milk replacers.

2. Materials and Methods

2.1. Animal Ethics

This research was conducted in Chabei Pasture, Modern Farming, Zhangjiakou City, Hebei, China. The Animal Ethics Committee of the Chinese Academy of Agricultural

Sciences (No. AEC-CAAS-20180224) approved the experimental procedures, and animal welfare and handling procedures were strictly followed during the experiment.

2.2. Animal, Study Design, and Diets

Twenty-eight newborn, female, and healthy Holstein calves (41.60 ± 3.67 kg of body weight at birth) were selected for this study. All calves were kept in individual calf pens ($3 \times 1.5 \times 1.6$-m^3; length × width × height) and the bedding for calves was straw. The pens were cleaned and changed the straw before 1600 h every day to ensure the health and hygiene of the calves. Calves were drenched a total of 6 L of colostrum, with 4 L drenched within 1 h after birth and 2 L drenched 5 h after the first feeding. Before being drenched, colostrum quality was qualified (IgG > 55 g/L; total number of bacteria < 50,000 CFU/mL). During the pre-weaning period (2–70 d), the calves were fed liquid feed (whole milk or milk replacer) with 39 °C twice a day (08:00 and 18:00 h) using a bucket. Along with the liquid feed, the pelleted starter feed and clean water (39 °C) were provided ad libitum from 14 days of age. Subsequently, during the post-weaning period (71–98 d) the calves were fed only the pelleted starter feed.

These 28 calves were assigned to either the M or the MR group, according to their body weight at birth (mean body at birth of M vs. MR = 41.9 vs. 41.4 kg); each group consisted of 14 calves. Whole milk was obtained from milk tanks and pasteurized; the milk replacer was reconstituted as an emulsion (12.5%, *w*/*v*) using cooled (50–60 °C) boiled water (drinking water from pasture). After feeding with colostrum, the calves in the M group were fed on 8 L whole milk per day from 2 to 63 d, whereas the calves in the MR group were fed on 8 L whole milk per day from 2 to 7 d and then transitioned from whole milk to the plant protein-based milk replacer from 8 to 13 d (transition period). During the transition period, the ratio of milk replacer to whole milk was gradually increased from 1:2 to 2:1 (*v*:*v*). From 14 to 63 d, the calves in the MR group were fed on 8 L milk replacer per day. All the calves started weaning on 64 d, and the weaning period ended on 70 d. During the weaning period, the feeding amount of milk or milk replacer (according to their respective groups) was decreased 1 L per day. The milk replacer contained 50% plant proteins, including wheat, rice, and soybean protein. During the experimental period (lasting for 98 d), all the calves were reared in the same individual pens and fed the same starter feed. We had two rows of pens (12 individual pens per row) in this study, and all calves were randomly assigned to the pens. To calculate the daily matter intake (DMI) of each calf, the starter feed was changed at 08:00 h each day. The amounts of starter feed and milk or milk replacer that was remaining and the amount that was added were recorded. Moreover, the milk was sampled before the morning feeding every 15 d throughout the study period to measure its chemical composition according to the method described by Kong et al. [24]. The ingredients and the chemical compositions of the milk and milk replacer, and the starter feed are shown in Table 1 and Table S1, respectively.

Table 1. Ingredient and chemical composition of milk and milk replace.

Items	Treatment	
	Milk Powder [1]	**Milk Replacer** [2]
Chemical composition (% of DM except for Dry matter; mean ± SD)		
Dry matter	96.76 ± 0.61	96.32
Gross energy MJ/kg	24.81 ± 0.41	24.62
Crude Protein	27.01 ± 0.36	26.93
Ether extract	30.67 ± 0.53	16.28
Ash	5.58 ± 0.05	5.53
Calcium	1.06 ± 0.03	1.02
Phosphorus	0.75 ± 0.03	0.77

[1] Milk powder was obtained by drying the fresh whole milk, which was sampled every 15 days. [2] The ingredient compositions are 5% wheat protein powder, 4% rice protein powder, 18% whole soy powder, 6% concentrate whey protein, 22% whole milk power, 20% high protein whey powder, 12% fat powder, 0.2% vitamin complex, 4% Trace elements and mineral complexes, and 8.8% soluble carrier, and the lysine, Threonine, Methionine, essential amino acid, nonessential amino acid of milk replacer are 1.43%, 1.24%, 0.91%, 9.91%, and 15.59%, respectively (% of DM). Per kg milk replacer (DM basis) contains 15,000 IU vitamin A, 5000 IU vitamin D, 50 mg vitamin E, 6.5 mg vitamin B1, 6.5 mg vitamin B2, 6.5 mg vitamin B6, 0.07 mg vitamin B12, 20 mg vitamin B5, 13 mg vitamin B3, 0.1 mg vitamin H, 10 ppm Cu, 100 ppm Fe, 40 ppm Mn, 40 ppm Zn, 0.5 ppm I, 0.3 ppm Se, 0.1 ppm Co.

2.3. Determination of Growth Performance and Serum Indicators

Body weights of the calves were recorded at birth and every 14 d thereafter before the morning feeding; additionally, other body measurements, including heart girth, withers height, and body length of the calves, were recorded before the morning feeding on 1, 70, and 98 d of age according to the method described by Kargar and Kanani [25]. ADG of the calves was calculated every 14 d. Feed efficiency was calculated as ADG/Total DMI (liquid DMI + starter feed DMI; kg/d). On 35, 42, 49, 84, 91, and 98 d of age, six healthy calves were selected from each group, and 10 mL of their blood samples was collected from the jugular vein before the morning feeding. Blood samples were centrifuged at 3000× g for 15 min (Tiangen OSE-MP25, Beijing, China) to obtain the serum, which was then stored in 1.5 mL centrifuge tubes at −20 °C for further analyses. The concentration of serum immunoglobulin (Ig) G, IgA, and IgM were determined using ELISA kits (F4042-A, F3995-A, and F6685-A, respectively). Aspartate aminotransferase (AST), Alanine transaminase (ALT), and alkaline phosphatase (ALP) concentrations were determined using an automatic biochemical analyzer (kehua-zy KHB-1280, Shanghai, China). The total antioxidant capacity (T-AOC), superoxide dismutase (SOD), glutathione peroxidase (GSH-PX), malondialdehyde (MDA), and catalase (CAT) concentrations in the serum were determined using commercial kits (Nanjing Jian Cheng Bioengineering Institute, Nanjing, China).

2.4. Liver Biopsy

Liver samples (the liver was sampled from the same calves from whom blood samples had been obtained) were collected from the calves (four calves per group) for liver biopsy using a handcrafted, stainless steel, biopsy gun according to the method described by Kong et al. [26]; samples were collected from the same eight calves (four calves per group) at 2 weeks before [liver samples from the M group before weaning (MBW) and liver samples from the MR group before weaning (RBW)] and after weaning [liver samples from the M group after weaning (MAW) and liver samples from the MR group after weaning (RAW)]. An approximately 15 cm^2 skin area above each calf's liver was shaved and disinfected with 75% ethyl alcohol. Thereafter, 10 mL of toluene thiazide hydrochloride (20 mg/mL) was injected subcutaneously and through the intercostal muscles at the sampling site. The sampling site was anesthetized, and a 1-cm incision was made using a sterile blade. Approximately 0.5 g of the liver tissue was collected into cryopreservation tubes at each sampling time using a handcrafted, stainless steel, biopsy gun for liver biopsy, and the liver tissue was stored at −80 °C for transcriptome assay.

2.5. Total RNA Extraction and Sequencing

The total RNA from the ground liver samples (total number of samples = 16) was extracted using TRIzol (Invitrogen, Carlsbad, CA, USA). The purity of RNA was determined using a Nano Drop®ND-1000 spectrophotometer (Implen, Santa Clara, CA, USA). The quality of the extracted RNA was assessed using an Agilent 2100 bioanalyzer (Agilent Technologies, Santa Clara, CA, USA). The RIN value of all samples was >8; 2 μg of RNA extracted from each sample was sent to Allwegene Technology Inc. (Beijing, China) for library building and sequencing. Briefly, mRNA was enriched using magnetic beads with Oligo(dT) and was broken into short fragments with fragmentation buffer. These fragments were reverse transcribed to cDNA for PCR amplification and cDNA library construction. After successful library construction, Illumina Hiseq 4000 PE150 (Illumina, San Diego, CA, USA) was used for sequencing.

2.6. Validation of RNA-Seq Data by Quantitative Real-Time PCR (qRT-PCR)

To verify the accuracy of RNA-Seq, we selected eight DEGs, including *CCND1, PCK1, ACACA, GSTM3, STAT3, MMP9, CYP4A22*, and *NAT10*. The IQ5 qRT-PCR detection system (Bio-Rad; Hercules, CA, USA) was used for the analyses of all the samples. The primers required for the *q*RT-PCR analyses of these eight DEGs were designed using NCBI primer-BLAST, and they were synthesized by Sangon Bioengineering (Shanghai, China) Co. Ltd. β-Actin was used as an internal control to normalize the expression data [27]. All the primers are listed in Table S2.

2.7. Statistical Analyses

Statistical analyses were conducted using GraphPad Prism 8 (GraphPad Software Inc., CA, USA) and R software (v 4.0.3). The statistical power for the serum samples of this study was >0.8 using the "Pwr" package (https://cran.r-project.org/web/packages/pwr/, accessed on 11 November 2021). Data were checked for normality using the shapiro.test function of the R software before analyses, and the data fitting non-normality were log transformed. Data on body weight, withers height, heart girth, and body length at 0 d of age were analyzed using *t*-test. Furthermore, at 70 and 98 d of age they were analyzed using AVCOVA as described by Wang et al [28]. The relevant linear model was used as follows:

$$Y_{ijk} = \mu + T_i + \alpha\left(R_j - \overline{R}\right) + e_{ijk}$$

where Y_{ijk} is the dependent variable; μ is the average experimental value; T_i is the fixed effect of treatments i (i = M or MR group); $\alpha\left(R_j - \overline{R}\right)$ designates the covariate variable of initial body weight, where the α is the regression coefficient relating initial body weight to the variable measured, R_j is the initial body weight for the jth calf (j = 1, 2, 3, … … , 27, and 28), and $\overline{R}$ is the overall mean of the initial body weight; e_{ijk} is the error term. If there was a co-effect in the progress of analyzing, we used the effects function in the 'effects' package (https://CRAN.R-project.org/web/packages/effects/, accessed on 11 November 2021) to remove covariate factors and to correct the average value of these indicators.

Data on intake, ADG, and feed efficiency were analyzed using mixed model in R. The relevant model used was as follows:

$$Y_{ijklm} = \mu + T_{ik} + D_{jk} + P_k + TD_{ijk} + TP_{ik} + Calf_l + e_{ijklm}$$

where Y_{ijklm} is the dependent variable; μ is the average experimental value; T_{ik} is the fixed effect of treatments i (i = M or MR group) during the period k (k = pre-weaning or post-weaning); D_{jk} designates the repeated effect of the period k (k = pre-weaning or post-weaning); P_k is the effect of period k (k = pre-weaning or post-weaning); TD_{ijk} is the interaction effect of treatment and day during the period k (k = pre-weaning or post-weaning); TP_{ik} is the interaction effect of treatment and period; $Calf_l$ is the random effect of lth calf (l = 1, 2, 3, … … , 27, and 28); e_{ijklm} is the error term.

The data on serum indicators were analyzed using another mixed model in R. The relevant model used was as follows:

$$Y_{ijklm} = \mu + T_{ik} + D_{jk} + TD_{ijk} + Calf_l + e_{ijklm}$$

where Y_{ijklm} is the dependent variable; μ is the average experimental value; T_{ik} is the fixed effect of treatments i (i = M or MR group) during the period k (k = pre-weaning or post-weaning); D_{jk} designates the repeated effect during the period k (k = pre-weaning or post-weaning); TD_{ijk} is the interaction effect of treatment and day during the period k (k = pre-weaning or post-weaning); $Calf_l$ is the random effect of lth calf (l = 1, 2, 3, ... , 27, and 28); e_{ijklm} is the error term.

The raw data of RNA-seq were processed with a Perl script. The clean data were obtained by removing reads containing adapters, reads containing poly-N, and reads of low quality from the raw data. The clean reads were then mapped to the bovine reference genome (http://bovinegenome.org, accessed on 2 November 2021) using the TopHat2 software (v 2.1.0). Gene expression levels in each library were normalized to fragments per kilobase of exon model per million mapped reads (FPKM). Differential gene expression analysis of the two groups (MR vs. M) was performed using the DESeq R package (1.10.1). Genes with p-value ≤ 0.05 and |log2(fold-change)| > 1.3 found by DESeq were assigned as DEGs (Kong et al., 2017; Papah et al., 2018). MCODE with standard parameters (node score cutoff, 0.2; K-Core, 2; maximum depth from seed, 100) from Cytoscape version 3.7.2 [29] was used for modular analysis. Gene Ontology (GO) and Kyoto Encyclopedia of Genes and Genomes (KEGG) for DEGs were analyzed using the CluoGo in Cytosacpe. STRING 11 (http://string-db.org/, accessed on 5 November 2021) was used to predict protein-protein interactions (PPI) of DEGs. A PPI network was drawn using Cytoscape and the cytoHubba application was used to identify the hub-genes [30,31].

The 'corrplot' package of R software was used to analyze the relationship between the DEGs and the apparent indicators that included feed efficiency and serum indicators (based on Spearman's coefficient). All the data were reported as means, and the differences with a $p < 0.05$ were considered statistically significant.

3. Results

3.1. Intake and Growth Performance

The growth performance of the calves is shown in Table 2. There were no interactions among feeding group and day-age of the calves. The body weight, withers height, body length, and heart girth of the calves were not significantly different at birth (0 d) and 98 d of age. At 70 d of age, the calves in the MR group presented significantly lower body weight, withers height, and heart girth than those in the M group; however, body length was not significantly different between the two groups. During the pre-weaning period, the DMI of the liquid feed, ADG, and feed efficiency in the calves of the MR group were significantly lower compared to the calves of the M group; however, the DMI of the starter feed in the calves of the MR group were significantly higher than those in the M group. During the post-weaning period, the DMI of the starter feed and ADG in the MR group presented significantly higher levels than those in the M group; however, their feed efficiency was significantly lower relative to the M group. During the total experiment period (0–98 d), the calves of the MR group presented significantly higher overall DMI, yet showed a lower feed efficiency compared to the calves of the M group. However, the overall ADG was not significantly different between the two groups.

Table 2. Growth performance of calves fed milk or milk replacer. (N = 14 per group).

Items [4]	Treatment (T) [1]		SEM	p-Value				
	MR	M		T	Days (D) [2]	T × D	P [3]	T × P
Pre-weaning (0–70 d)								
DMI of starter feed, kg/d	0.23	0.16	0.02	0.009	<0.001	0.601		
DMI of liquid feed, kg/d	1.00	1.05	0.01	0.043	<0.001	0.622		
Total DMI [4], kg/d	1.23	1.22	0.02	0.513	<0.001	0.742		
ADG, kg/d	0.83	0.92	0.02	0.007	<0.001	0.461		
Feed efficiency [5]	0.67	0.76	0.03	<0.001	<0.001	0.731		
Post-weaning (70–98 d)								
DMI of starter feed, kg/d	2.79	2.28	0.10	<0.001	<0.001	0.231		
ADG, kg/d	1.09	0.97	0.03	0.024	<0.001	0.814		
Feed efficiency	0.39	0.43	0.01	0.006	<0.001	0.731		
Overall (0–98 d)								
Total DMI, kg/d	1.68	1.51	0.03	0.002			<0.001	0.004
ADG, kg/d	0.91	0.94	0.02	0.334			<0.001	0.193
Feed efficiency	0.45	0.54	0.01	<0.001			<0.001	0.241
Body weight, kg								
Initial (d 0)	41.4	41.9	1.65	0.781				
Weaning (d 70)	98.2	107.4	2.83	0.004				
Final (d 98)	128.1	132.7	3.68	0.225				
Skeletal growth								
Withers height, cm								
Initial (d 0)	77.7	76.0	0.93	0.072				
Weaning (d 70)	94.6	97.2	1.13	0.026				
Final (d 98)	98.7	101.3	1.30	0.055				
Heart girth, cm								
Initial (d 0)	77.3	76.1	0.98	0.233				
Weaning (d 70)	107.1	109.9	1.10	0.018				
Final (d 98)	116.9	118.2	1.35	0.341				
Body length, cm								
Initial (d 0)	69.8	69.4	1.06	0.702				
Weaning (d 70)	97.3	99.4	1.04	0.131				
Final (d 98)	105.5	106.5	1.73	0.584				

[1] MR = milk replacer; M = milk. [2] For all variables, days of age were used as 14-d period. [3] P = calf phase (pre-weaning vs. post-weaning period). [4] Total DMI = total daily matter intake (starter feed intake + liquid feed intake); [5] Feed efficiency = ADG/Total DMI; Standard error of the mean (SEM).

3.2. Serum Indicators

The serum indicators are shown in Table 3. Based upon our observations, there were no interactions between the feeding group (M or MR) and the day-age of the calves that would affect the concentrations of serum variables. The serum concentrations of ALP, IgG, GSH-PX, and CAT were not significantly affected in the calves of the MR group. However, during the pre-weaning period, the calves of the MR group showed significantly higher AST concentrations and significantly lower IgM, SOD, and T-AOC concentrations compared to the calves of the M group. Similarly in the post-weaning period, the calves of the MR group showed significantly higher ALT, AST, and MDA concentrations and lower IgA and IgM concentrations compared to the calves of the M group.

Table 3. Serum variables as influenced by feeding milk versus milk replacer to calves. (*N* = 6 per group).

Items	Treatment (T) [1]		SEM	*p*-Value		
	MR	M		T	Days (D) [2]	T × D
ALT, U/L						
pre-weaning (35–49 d)	7.5	7.1	0.99	0.524	0.979	0.621
post-weaning (84–98 d)	19.9	15.2	1.05	0.035	0.837	0.179
AST, U/L						
pre-weaning (35–49 d)	121.6	91.5	2.88	0.028	0.778	0.632
post-weaning (84–98 d)	183.2	148.7	4.23	0.036	0.001	0.712
ALP, U/L						
pre-weaning (35–49 d)	159.6	149.4	7.93	0.205	0.730	0.453
post-weaning (84–98 d)	188.1	186.1	11.49	0.693	0.334	0.818
IgG, g/L						
pre-weaning (35–49 d)	9.9	10.2	0.50	0.942	0.706	0.579
post-weaning (84–98 d)	12.7	12.3	0.44	0.994	0.206	0.765
IgA, g/L						
pre-weaning (35–49 d)	1.3	1.6	0.04	0.571	0.002	0.442
post-weaning (84–98 d)	1.1	1.2	0.02	0.028	0.004	0.668
IgM, g/L						
pre-weaning (35–49 d)	2.1	2.8	0.06	0.012	0.725	0.971
post-weaning (84–98 d)	2.1	2.7	0.06	0.001	0.819	0.715
SOD, U/mL						
pre-weaning (35–49 d)	85.3	99.1	3.06	0.045	0.679	0.391
post-weaning (84–98 d)	128.3	125.8	2.26	0.278	0.138	0.442
MDA, nmol/mL						
pre-weaning (35–49 d)	7.2	5.9	0.23	0.035	0.673	0.229
post-weaning (84–98 d)	4.7	4.2	0.11	0.015	0.714	0.814
GSH-PX, U/mL						
pre-weaning (35–49 d)	824	837	26.48	0.595	0.115	0.415
post-weaning (84–98 d)	1173	1202	30.46	0.698	<0.001	0.458
T-AOC, U/mL						
pre-weaning (35–49 d)	6.2	7.1	0.16	0.005	0.541	0.818
post-weaning (84–98 d)	9.8	10.3	0.42	0.103	0.328	0.643
CAT, U/mL						
pre-weaning (35–49 d)	8.6	8.5	0.33	0.596	0.333	0.992
post-weaning (84–98 d)	11.4	12.1	0.48	0.286	0.215	0.710

[1] MR = milk replacer; M = milk. [2] For all variables, days of age were used as 8-d period. Standard error of the mean (SEM), Serum immunoglobulin (Ig), aspartate aminotransferase (AST), glutamic-pyruvic transaminase (ALT), alkaline phosphatase (ALP), total antioxidant capacity (T-AOC), superoxide dismutase (SOD), glutathione peroxidase (GSH-PX), malondialdehyde (MDA), catalase (CAT).

3.3. Mapping Summary Statistics

For each library, RNA-Seq produced more than 24,731,334 raw reads (Table S3). The sequencing generated 18,673,574–23,908,557 clean reads for each sample. The GC content in the libraries ranged from 51.49% to 52.41%. The 16 samples had at least 98.22% reads with ≥Q20, and 95.03% reads with ≥Q30. The majority of reads in each library were mapped to the bovine reference genome, and the average mapping rates were 92.58%, 93.59%, 93.83%, and 93.54% for the MAW, MBW, RAW, and RBW groups, respectively. Simultaneously, the MAW, MBW, RAW, and RBW groups had an average of 90.94%, 91.77%, 92.11%, and 91.72% reads mapped to the bovine genome, respectively.

3.4. Differentially Expressed Genes (DEGs)

The principal component analysis (PCA) of all genes from the liver samples showed separations between MBW and RBW or MAW and RAW based on the feeding group (M or MR) and the time of sample collection (pre-weaning or post-weaning) (Figure 1A,B). Compared to the gene expression in the MBW group, 1120 DEGs were found in the liver tissue of the RBW group, including 649 upregulated and 471 downregulated DEGs (Figure 1C).

Similarly, 293 DEGs were identified in the liver samples of the calves in the RAW and MAW groups, among which 160 were upregulated and 133 were downregulated (Figure 1D). The heat plot was generated to visualize the distribution of the DEGs (Figure S1).

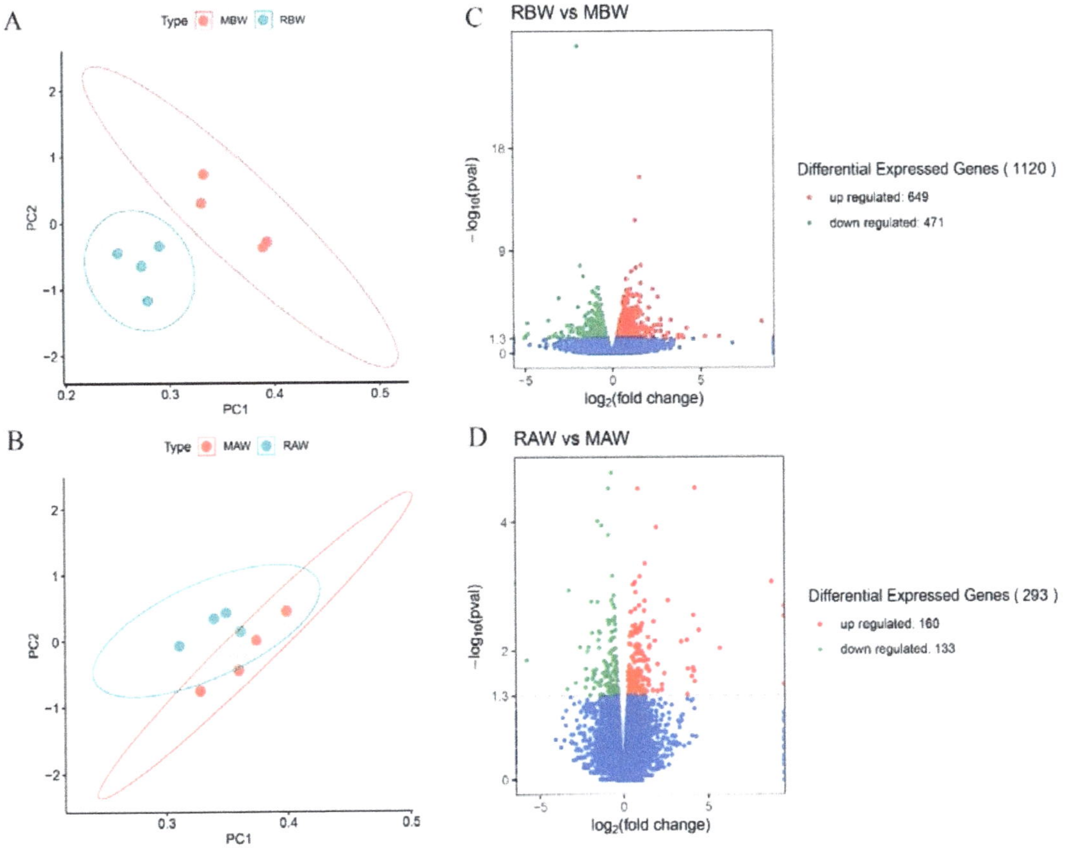

Figure 1. Gene expression between milk and milk replacer groups. (**A**) The principal component analysis (PCA) of all genes in liver between MBW and RBW. (**B**) The principal component analysis (PCA) of all genes in liver between MAW and RAW. (**C**) The expression of differentially expressed genes between two groups during pre-weaning period. (**D**) The expression of differentially expressed genes between two groups during post-weaning period. The red dots indicate upregulated differentially expressed genes and the green dots indicate downregulated differentially expressed genes. The blue dots reveal that there is no difference in the expression of genes between the two groups. MBW (liver samples from the milk group before weaning); RBW (liver samples from the milk replacer group before weaning); MAW (liver samples from the milk group after weaning); RAW (liver samples from the milk replacer after weaning).

3.5. Enrichment Analysis of the DEGs

In the pre-weaning period, we found 25 clusters in 1120 DEGs using the MCODE between two groups (Supplementary File S2), of which the Top 4 clusters (based on network score, Figure 2A,C,E,G), including *ACACA*, *FASN*, *CPT1A*, *PCK1*, *SLC2A4*, *NAT10*, *ACTB*, *CCND1*, *GSTA5*, *ADH5*, and *ADH6*, among others, mainly enriched some functions including fatty acid metabolism, AMPK signaling pathway, xenobiotic metabolism, immunity and so on (Figure 2B,D,F,H). Then we analyzed the up- and down-regulated DEGs, respectively. Interestingly, the upregulated DEGs were primarily enriched in xenobiotic and lipid metabolism, including the metabolism of xenobiotics by cytochrome P450, chemical carcinogenesis, and the peroxisome proliferator-activated receptor (PPAR) signaling pathway, among others (Table S4). The downregulated DEGs were mainly enriched in energy metabolism, immune function, and mineral metabolism pathways, including mineral absorption, apoptosis, fatty acid biosynthesis, oxidoreductase activity, oxidation-reduction process, and NADH dehydrogenase activity, among others (Tables S4 and S5).

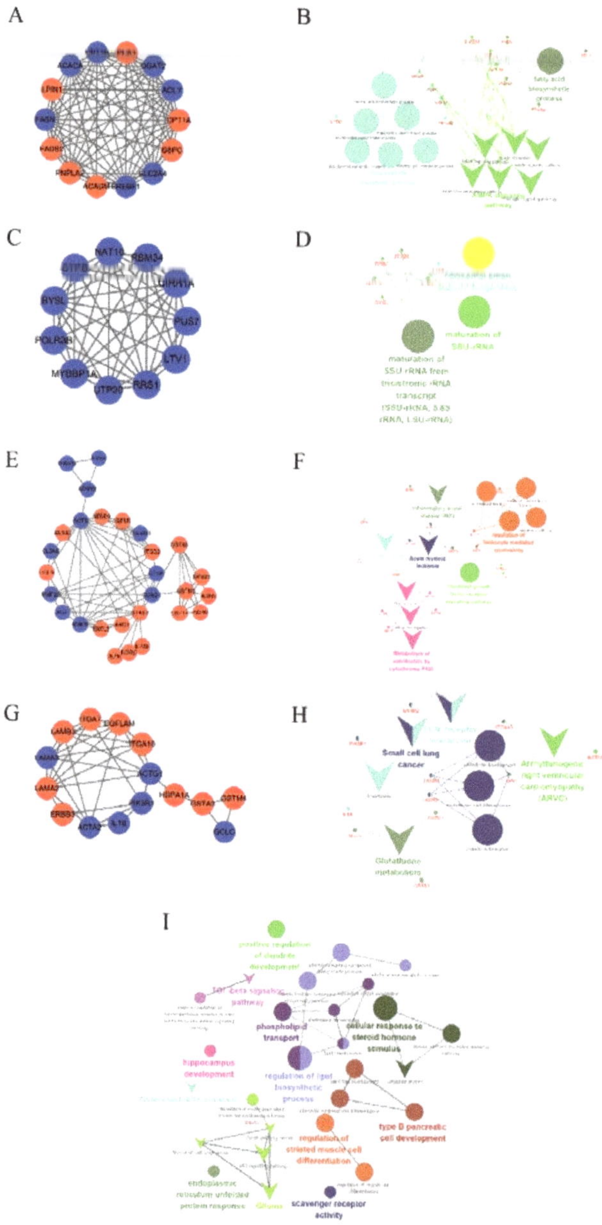

Figure 2. Enrichment analysis of differentially expressed genes (DEGs). (**A**,**B**) The gene interactive network

and the enrichment result of Top 1 Module between RBW and MBW. (**C**,**D**) The gene interactive network and the enrichment result of Top 2 Module between RBW and MBW. (**E**,**F**) The gene interactive network and the enrichment result of Top 3 Module between RBW and MBW. (**G**,**H**) The gene interactive network and the enrichment result of Top 4 Module between RBW and MBW. (**I**) The enrichment result of all DEGs between RAW and MAW. Red represents the upregulated genes and blue represents the downregulated genes in the gene interactive network. Different colors represent different enrichment groups (based on the Cohen's kappa score) in enrichment analysis. The triangle and circle graph represent the KEGG and GO results, respectively, in the enrichment analysis. MBW (liver samples from the milk group before weaning); RBW (liver samples from the milk replacer group before weaning). MAW (liver samples from the milk group after weaning); RAW (liver samples from the milk replacer group after weaning).

During the post-weaning period, osteoclast differentiation, muscle cell differentiation, and lipid metabolism appeared enrichment (Figure 2I). Similarly, we also analyzed the up- and down-regulated DEGs, respectively, and found that the metabolic pathway, Osteoclast differentiation, cell cycle, and cell projections appeared to have been enriched (Table S6; Figure S2).

3.6. Protein-Protein Interaction Analysis of DEGs Using STRING

We selected the top 200 DEGs (based on FC and p-value) to construct an interaction network (Figure S3). The genes in the center of the interaction network included *CCND1*, *CAT*, *ADH5*, *ADH6*, *CS*, *CCL2*, *PCK1*, *CYP26A1*, and *ACACA*. Additionally, we identified the top 10 hub genes (based on degree) among the top 200 DEGs, including *CAT*, *CCND1*, *UGT2B15*, *PCK1*, *ADH5*, *GSTA5*, *GSTM3*, *ADH6*, *GCLC*, and *CS*, using the cytoHubba application (Figure 3A).

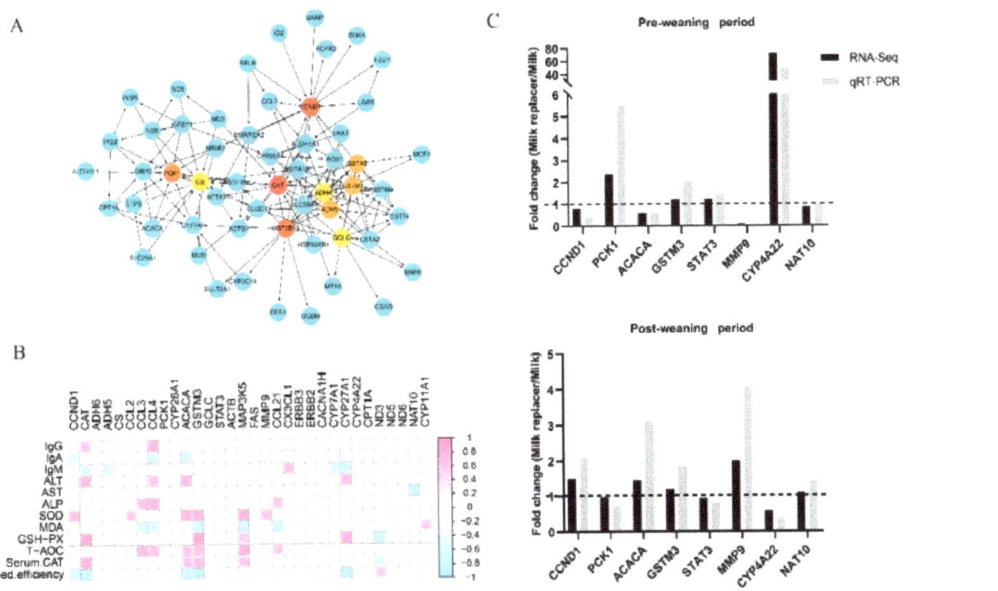

Figure 3. Relationship between hub-DEGs and phenotypes, and verification of transcriptome accuracy. (**A**) The top 10 hub-genes by cytoHubba in the top 200 differentially expressed genes between two groups. Blue genes are the first-stage genes of the hub-genes; (**B**) Correlation analysis between the apparent indicators and differently expressed genes. The pink color represents a positive correlation, and the blue color represents a negative correlation. The red to blue scale bar running from 1.0 to −1.0 represents r = 1.0 to −1.0. The blank represents $p > 0.05$. Serum immunoglobulin (Ig), glucose

(GLU), triacylglycerol (TG), aspartate aminotransferase (AST), glutamic-pyruvic transaminase (ALT), alkaline phosphatase (ALP), total protein (TP), albumin (ALB), globulin (GLB), total antioxidant capacity (T-AOC), superoxide dismutase (SOD), glutathione peroxidase (GSH-PX), malondialdehyde (MDA), serum catalase (Serum CAT); (**C**) Validation of eight differentially expressed genes using quantitative real-time PCR (qRT-PCR). During pre- and post-weaning period, genes were validated. The mRNA level of each gene was normalized to that of β-actin (n = 4). The ratio of expression in livers of milk replacer calves to milk (Milk replacer/Milk) via RNA-Seq and RT-PCR is shown (n = 1).

3.7. Correlation Analysis between Hub-Genes among DEGs and Apparent Indicators

To identify the correlation between gene expression and the apparent indicators, we further analyzed Spearman's rank correlation between the selected some key DEGs and the serum variables and the feed efficiency of the calves (Figure 3B). Interestingly, we observed that *CCND1*, *CAT*, *CCL3*, *CCL4*, *ACACA*, *GSTM3*, *MAP3K5*, *CYP27A1*, and *ND3* had correlations with the apparent indicators.

3.8. Validation of RNA-Seq Data by qRT-PCR

Furthermore, we validated the eight DEGs obtained from the pre- and post-weaning period samples using qRT-PCR technology. As seen in Figure 3C, the qRT-PCR results showed that the DEGs had the same expression pattern as observed in the RNA-Seq, indicating that our transcriptome data were highly reliable.

4. Discussion

Plant protein in calf milk replacer improves the quantity of animal protein available for human consumption, however plant protein-based milk replacer has a negative impact on calves. To our knowledge, this study is the first to systematically compare calves' response between whole milk and plant protein-based milk replacer. The findings revealed that milk replacer had an influence on calves' liver function, immunity, and antioxidant capacity, however had no effect on growth performance throughout the study.

The amount and type of the liquid feed are most likely to be the main factors that affect the solid feed intake of calves during the pre-weaning period. Previous studies have also shown that both restricted and unrestricted intake of milk as well as milk replacer can affect the solid feed intake of calves [15,32]. Zhang et al. found that the calves fed on milk replacers containing soy protein in the pre-weaning period had greater solid intake compared to the calves fed on whole milk [15]. Similarly, Ghorbani et al. reported that soymilk used as a novel milk replacer can stimulate early intake of starter feed in calves [5]. Likewise, in this study, lower liquid feed intake and higher solid feed intake were observed in the calves of the MR group compared to the calves of the M group in the pre-weaning period. Furthermore, we found lower ADG in the MR group, indicating that the calves may have better bioavailability of protein (casein, etc.) and energy (milk fat, lactose) from whole milk in comparison to the milk replacer. It is widely accepted that the plant proteins and the carbohydrates (glucose, etc.) in milk replacer cannot completely replace the milk proteins and lactose, respectively, of whole milk. In addition, there may be some unknown growth factors in whole milk that are responsible for the better ADG of the calves in the M group [33]. In this study, we discovered that the MR group had higher ADG after weaning than the R group. It is generally known that as calves digest solid feed, propionic acid and butyric acid are produced, which stimulate rumen development [34,35]. As a result, more solid feed consumption in the MR group will stimulate rumen development in advance before weaning, resulting in more energy from solid feed being available for growth after weaning. In addition, we also found that MR group could affect withers height and heart girth of calves. The calf stage is an essential period for bone growth and overall development, and calcium, phosphorus, and other minerals are also important for the development of bone [36]. In the present study, the mineral metabolism pathway was downregulated in the calves of the MR group; in particular, three genes (namely *ERBB2*, *ERBB3*, and *CACNA1H*) involved in calcium metabolism were downregulated in the calves fed on the milk replacer (Figure S4). Huang et al. reported that a plant protein-based milk replacer reduces calcium absorption [16]. In addition, Lee et al. reported that calves fed on milk replacers had a low rate of bone growth, similar with the observations of the present study [13]. Therefore, the outcome of this study indicates that exogenous mineral supplements, especially calcium, has a low bioavailability in the milk replacer; this suggests that we should pay attention to mineral element supplementation in future research related to the use of milk replacers for calves.

The liver is the largest metabolic organ of the body, and the health status of the liver is an indirect indicator of the health status of calves. Serum ALT and AST concentrations comprise a common indicator of hepatic health status. In the present study, during the pre-weaning period, the serum AST concentration in calves fed on the milk replacer was significantly higher than that of calves fed on whole milk. Therefore, we inferred that the milk replacer might have negatively affected the livers of the calves. Interestingly, TNF signaling and apoptosis pathway were significantly down-regulated, which also suggested MR might damage liver cell. Furthermore, we also found MR could influence immunity and antioxidant capacity in this study, which is similar to a previous study [16]. Seegraber et al. showed that replacing whey protein with plant proteins causes intestinal damage in suckling calves [37]. Moreover, the enrichment analysis of the liver DEGs during the pre-weaning period showed that the milk replacer regulated foreign material metabolism. Furtherly, we found that the key gene in foreign body metabolism, *GSTM3*, correlated with the serum antioxidant indicators. The GST genes are involved in the regulation of cellular glutathione. Moreover, studies have shown that FoxO signaling pathway was up-regulated, and the expression of *CAT* in the pathway involved in oxidative stress resistance and DNA repair [38], and it correlated with antioxidant indicators. As glutathione has been shown to be able to withstand oxidative stress in certain studies [39,40], the body boosts the expression of glutathione-related genes in response to oxidative stress damage. Milk replacer containing plant protein has been reported to contain ANFs, and these ANFs have adverse effects on the growth performance as well as the health of the calves [41,42]. Therefore, the observed reduction in the concentration of oxidation resistance indicators in the liver of the calves fed on the milk replacer can be attributed to the ANFs triggering oxidative stress. Similarly, the transcriptome analysis showed that *CCL2*, *CCL3*, *CCL21*, and *CX3CL1* were downregulated in the liver of the calves in the MR group. These DEGs are involved in the body's immune response and have been previously detected in cattle [43]. Furthermore, these DEGs are involved in some pathways and BPs, including the TNF signaling pathway, chemokine activity, G protein-coupled receptor binding, and immune response, indicating that the milk replacer could reduce the immunity of the calves. Interestingly, we found that *CAT*, *CCL4*, and *CX3CL1* are also correlated with serum immunity indicators, suggesting that the decreased immunity of calves is related to oxidative stress. In short, based on the observations of the present study, we inferred that milk replacer might injure liver cells in calves, leading to oxidative stress and impaired immunity. However, there were no improvements in the concentrations of the serum indicators in the calves of the MR group during the post-weaning period, and the expressions of the related genes remained the same during the pre- and post-weaning periods, indicating that the liver function was not recovered.

Feed efficiency, which measures how efficiently animals use feed to produce meat, eggs, and milk, has traditionally been a key indication in livestock production. In this study, the calves of the MR group showed a lower feed efficiency compared to the calves of the M group, which is evident not only in milk replacer's poor bioavailability and liver damage, but also in the expression pattern of genes associated to feed efficiency. The CYP genes play a significant role in feed efficiency, especially in functions related to liver metabolism [44–46]. Salleh et al. reported that *CYP11A1* upregulation results in declined efficiency of feed utilization in high residual feed intake dairy cattle. In the present study, *CYP27A1* has the same expression pattern as *CYP11A1* were upregulated and correlation with feed efficiency in the calves of the MR group [45]. In addition, the retinol metabolism pathway was significantly upregulated in the calves of the MR group compared to the calves of the M group. De Almeida Santana et al. had reported that retinol metabolism influences the feed conversion rate in cattle [47]. PPAR is an important fat metabolism pathway in the body. Interestingly, we found that bile acid, PPAR signaling, and fatty acid metabolism and degradation pathways were upregulated in the calves of the MR group; in contrast, the fatty acid biosynthesis pathway was downregulated, indicating that fat metabolism in calves is affected by feeding on milk replacers in the pre-weaning period. In the present study, we also observed that the upregulation of *STAT3* expression regulated the cell cycle and apoptosis via the JAK/STAT signaling pathway (Figure S5). *STAT3* stimulates fat deposition in cattle [48]. Therefore, the milk replacer used in the present study might have increased fat deposition in the liver, thereby negatively affecting the hepatic function. The protein encoded by *ACACA* plays an important role in converting acetyl CoA into fatty acids. Previous studies have shown that it is highly correlated with milk production, feed utilization, and fat deposition [49,50]. *FASN* is a rate-limiting gene in fatty acid synthesis, and it has a similar function to *ACACA*. *CPT1A* is important in regulating the oxidation of fatty acids in cells and transferring fatty acids from the outer membrane of the mitochondrion to the inner membrane [51]. Here, the GO cluster analysis showed that some BPs related to cellular respiration were downregulated. Oxidative phosphorylation occurs in the mitochondria and involves five complexes, which are essential for proton coupling and electron transfer [52,53]. In the present study, the expressions of *ND3*, *ND5*, *ND6*, and *NAT10* as well as the processes and pathways involved in cellular respiration were lower in the calves of the MR group as compared to the calves of the M group during the pre-weaning period. NADH dehydrogenase, which belongs to the NADH-CoQ dehydrogenase (complex I) family, is an enzyme that catalyzes the electron transfer from NADH to co-enzyme Q in the mitochondrial lining [54]. NADH is a marker in the mitochondrial chain producing energy. In the present study, downregulation of the gene regulating NADH dehydrogenase increased the NADH level in the mitochondria, thus affecting mitochondrial metabolism. The effects of Ca^{2+} on mitochondrial activities have been widely studied. Calcium plays an important role in regulating mitochondrial functions and ATP synthesis at different levels in organelles [55]. The Ca^{2+} concentration in the mitochondrial matrix has considerable effects on ROS generation, cytochrome C, and apoptosis [55]. Furthermore, *CS*, which plays an important role in regulating the citrate synthase activity in cells, was downregulated in the calves of the MR group. These results indicate that feeding on milk replacers can cause mitochondrial dysfunction in the calves. PPAR is closely related to the CYP genes. McCabe et al. reported that the expression of *CYP11A1* is upregulated during negative energy balance in cows, indicating that it plays an important role in the regulation of lipid and cholesterol synthesis in the liver [56]. Moreover, *CYP7A1* and *CYP27A1* are important regulators of cholesterol metabolism, bile acid biosynthesis, and steroid hormone pathways. Steroid hormone biosynthesis is an important metabolic pathway of negative energy balance in dairy cows, and this pathway was significantly enriched (RBW vs. MBW) in the present study. Pang et al. reported that energy-restricted feeding could lead to testis injury in sheep by affecting the process of cell apoptosis [57]. Therefore, a lack of energy might be harmful to the hepatic function in the calves of the MR group. We believe that the low energy of calves was linked to the content and source of fat present in the milk replacer. Although some studies have reported that higher fat content in milk replacers would influence the mammary gland development of heifers [58,59], Hu et al, have been reported there were no negative effect on growth performance of calves using palm oil and coconut oil instead of milk fat in milk replacer [60]. Therefore, the interaction between the source and concentration of fat for milk replacer may be the key factor affecting the performance of calves, which warrants further research.

Compensatory growth is a catch-up mechanism in animals, whereby an animal that had previously been affected by nutrient deficiency experiences accelerated growth when provided with feed that meets its nutritional requirements [61]. The greater ADG in the

MR group in this study might be due to compensatory growth, which could be linked to the activation of the osteoclast growth-related pathway and gene expression of *CCND1*, *MMP9*, etc., in the post-weaning period. In addition, this compensatory growth may be attributed to the normalization of the mineral content and the energy metabolism in the calves in the post-weaning period. In domestic ruminants, rationally using compensated growth can improve the feed efficiency and meat quality of beef cattle and sheep [62–64]. However, as a result of the different concerns, many studies have found that compensated growth in human childhood could increase the risk of death and illness in adults [65–67]. Unfortunately, the long-term influence of compensatory growth in young ruminants is still unclear. It is well known that dairy cows have a long-life cycle, which is highly beneficial for humans; thus, we must pay adequate attention to the influence of compensatory growth on the milk yielding potential of dairy cows in the later stages of their lives. Within a reasonable range of ADG, lower ADG of suckling calves implies lower milk yield of these individuals later in their life [2]. According to the current research results, milk replacers containing plant protein can reduce the ADG of suckling calves, thus making it a controversial topic in the field of plant protein-based milk replacers. The shortage of protein resources is a global problem, especially in China, where high-quality animal proteins such as meat, eggs, and milk are in high demand and so, commercialized at a high price. Meanwhile, milk powder accounts for the largest share in world trade [9]. Therefore, in China, the consumption of milk (milk source protein) by suckling calves must be minimized to meet the high demands of milk for human consumption. At the same time, plant protein milk substitutes can increase starter feed intake and lower the weaning age of the calves, and this is incredibly beneficial for the rumen development of the calves and the effective use of roughage in the post-weaning period [5,8,13–15]. Nevertheless, it is still unclear whether plant protein-based milk replacer has any effect on the milk yielding potentials of these calves, later in life; this is definitely worthy of further experimental research. Therefore, it is necessary to develop milk replacers containing appropriate plant protein sources under the pretext of the shortage of high-quality protein resources in the world, especially in China.

5. Conclusions

In the present study, we comprehensively examined the effects of a plant protein-based milk replacer on the growth performance, immunity, antioxidant capacity, and liver transcriptome in calves. The findings revealed that milk replacer had a negative effect on calves' liver function, immunity, and antioxidant capacity, however had no effect on growth performance throughout the study. The study provides insights into the metabolic mechanisms in the calves fed on a plant protein-based milk replacer, and it can serve as a basis for future studies on milk replacers.

Supplementary Materials: The following supporting information can be downloaded at: https://www.mdpi.com/article/10.3390/antiox11020270/s1. Supplementary File S1: Table S1. The composition and nutritional value of starter (dry matter basis). Table S2. Gene primers used in the study for quantitative RT-PCR. Table S3. Summary of the average statistics of the sequence quality and alignment information between two groups. Table S4. KEGG enrichment analysis of differentially expressed genes between RBW and MBW. Table S5. GO analysis (top 20) of downregulated genes of the liver of calves between RBW and MBW group. Table S6. KEGG enrichment analysis of differentially expressed genes between RAW and MAW. Figure S1. Heat-map showing the gene expression data of the significantly differentially expressed genes. Figure S2. GO analysis DEGs of the liver of calves between RAW and MAW group. Figure S3. STRING analysis shows that differentially expressed genes are involved in TOP 200 and predicted protein-protein interactions between RBW and MBW. Figure S4. Sketch map of calcium signaling pathway between RBW and MBW group. Figure S5. Sketch map of JAK-STAT signaling pathway RBW and MBW group. Supplementary File S2: MCODE results between RBW and MBW. Supplementary File S3: The detail information of DEGs between RBW and MBW. Supplementary File S4: The detail information of DEGs between RAW and MAW.

Author Contributions: Y.B. and Y.T. contributed to the design of the study and project management. S.W. and F.H. conducted the animal experiments, transcriptome sequencing and analysis, and wrote the manuscript. Q.D. revised the manuscript. S.L. collected biological materials. All authors have read and agreed to the published version of the manuscript.

Funding: This work was supported by the National Natural Science Foundation of China (grant number 31702136); the Fundamental Research Funds for Central Nonprofit Scientific Institution (grant number 1610382021004); the Subject Construction Funding of The Agricultural Science and Technology Innovation Program (grant number CAAS-ASTIP-2017-FRI-04); and the Earmarked Fund for Beijing Dairy Industry Innovation Consortium of Agriculture Research System (grant number BAIC06).

Institutional Review Board Statement: The Animal Ethics Committee of the Chinese Academy of Agricultural Sciences (No. AEC-CAAS-20180224) approved the experimental procedures, and animal welfare and handling procedures were strictly followed during the experiment.

Informed Consent Statement: Not applicable.

Data Availability Statement: The datasets used and analyzed during the current study are available from the corresponding author (Y.B.) on reasonable request.

Conflicts of Interest: The authors declare no conflict of interest.

References

1. Khan, M.A.; Bach, A.; Weary, D.M.; von Keyserlingk, M.A.G. Invited review: Transitioning from milk to solid feed in dairy heifers. *J. Dairy Sci.* **2016**, *99*, 885–902. [CrossRef] [PubMed]
2. Soberon, F.; Raffrenato, E.; Everett, R.W.; Van Amburgh, M.E. Preweaning milk replacer intake and effects on long-term productivity of dairy calves. *J. Dairy Sci.* **2012**, *95*, 783–793. [CrossRef] [PubMed]
3. Diao, Q.; Zhang, R.; Fu, T. Review of Strategies to Promote Rumen Development in Calves. *Animals* **2019**, *9*, 490. [CrossRef] [PubMed]
4. Castro, J.J.; Hwang, G.H.; Saito, A.; Vermeire, D.A.; Drackley, J.K. Assessment of the effect of methionine supplementation and inclusion of hydrolyzed wheat protein in milk protein-based milk replacers on the performance of intensively fed Holstein calves. *J. Dairy Sci.* **2016**, *99*, 6324–6333. [CrossRef]
5. Ghorbani, G.R.; Kowsar, R.; Alikhani, M.; Nikkhah, A. Soymilk as a novel milk replacer to stimulate early calf starter intake and reduce weaning age and costs. *J. Dairy Sci.* **2007**, *90*, 5692–5697. [CrossRef]
6. Soberon, M.A.; Liu, R.H.; Cherney, D.J. Short communication: Antioxidant activity of calf milk replacers. *J. Dairy Sci.* **2012**, *95*, 2703–2706. [CrossRef]
7. Petit, H.V.; Ivan, M.; Brisson, G.J. Digestibility measured by fecal and ileal collection in preruminant calves fed a clotting or a nonclotting milk replacer. *J. Dairy Sci.* **1989**, *72*, 123–128. [CrossRef]
8. Ansia, I.; Drackley, J.K. Graduate Student Literature Review: The past and future of soy protein in calf nutrition. *J. Dairy Sci.* **2020**, *103*, 7625–7638. [CrossRef]
9. Lagrange, V.; Whitsett, D.; Burris, C. Global market for dairy proteins. *J. Food Sci.* **2015**, *80* (Suppl. 1), A16–A22. [CrossRef]
10. Raeth, M.; Chester-Jones, H.; Ziegler, D.; Ziegler, B.; Schimek, D.; Cook, D.L.; Golombeski, G.; Grove, A.V. Pre- and postweaning performance and health of dairy calves fed milk replacers with differing protein sources. *Prof. Anim. Sci.* **2016**, *32*, 833–841. [CrossRef]
11. Wadhave, A.A.; Jadhav, A.I.; Arsul, V.A. Plant Proteins Applications: A Review. *World J. Pharm. Pharm. Sci.* **2014**, *3*, 702–712.
12. Kertz, A.F.; Hill, T.M.; Quigley, J.D., 3rd; Heinrichs, A.J.; Linn, J.G.; Drackley, J.K. A 100-Year Review: Calf nutrition and management. *J. Dairy Sci.* **2017**, *100*, 10151–10172. [CrossRef] [PubMed]
13. Lee, H.J.; Khan, M.A.; Lee, W.S.; Yang, S.H.; Kim, S.B.; Ki, K.S.; Kim, H.S.; Ha, J.K.; Choi, Y.J. Influence of equalizing the gross composition of milk replacer to that of whole milk on the performance of Holstein calves. *J. Anim. Sci.* **2009**, *87*, 1129–1137. [CrossRef] [PubMed]
14. Welboren, A.C.; Hatew, B.; López-Campos, O.; Cant, J.P.; Leal, L.N.; Martín-Tereso, J.; Steele, M.A. Effects of energy source in milk replacer on glucose metabolism of neonatal dairy calves. *J. Dairy Sci.* **2021**, *104*, 5009–5020. [CrossRef] [PubMed]
15. Zhang, R.; Zhang, W.B.; Bi, Y.L.; Tu, Y.; Beckers, Y.; Du, H.C.; Diao, Q.Y. Early Feeding Regime of Waste Milk, Milk, and Milk Replacer for Calves Has Different Effects on Rumen Fermentation and the Bacterial Community. *Animals* **2019**, *9*, 443. [CrossRef] [PubMed]
16. Huang, K.; Tu, Y.; Si, B.; Xu, G.; Guo, J.; Guo, F.; Yang, C.; Diao, Q. Effects of protein sources for milk replacers on growth performance and serum biochemical indexes of suckling calves. *Anim. Nutr.* **2015**, *1*, 349–355. [CrossRef] [PubMed]
17. Silano, M.; De Vincenzi, M. Bioactive antinutritional peptides derived from cereal prolamins: A review. *Nahrung* **1999**, *43*, 175–184. [CrossRef]
18. Komatsu, S.; Kajiwara, H.; Hirano, H. A rice protein library: A data-file of rice proteins separated by two-dimensional electrophoresis. *Appl. Genet.* **1993**, *86*, 935–942. [CrossRef] [PubMed]

19. Chen, Y.J.; Chen, Y.Y.; Wu, C.T.; Yu, C.C.; Liao, H.F. Prolamin, a rice protein, augments anti-leukaemia immune response. *J. Cereal Sci.* **2010**, *51*, 189–197. [CrossRef]
20. Kadarmideen, H.N. Genomics to systems biology in animal and veterinary sciences: Progress, lessons and opportunities. *Livest. Sci.* **2014**, *166*, 232–248. [CrossRef]
21. Suravajhala, P.; Kogelman, L.J.; Kadarmideen, H.N. Multi-omic data integration and analysis using systems genomics approaches: Methods and applications in animal production, health and welfare. *Genet. Sel. Evol.* **2016**, *48*, 38. [CrossRef] [PubMed]
22. Zhang, Z.H.; Jhaveri, D.J.; Marshall, V.M.; Bauer, D.C.; Edson, J.; Narayanan, R.K.; Robinson, G.J.; Lundberg, A.E.; Bartlett, P.F.; Wray, N.R.; et al. A comparative study of techniques for differential expression analysis on RNA-Seq data. *PLoS ONE* **2014**, *9*, e103207. [CrossRef] [PubMed]
23. Wang, Z.; Gerstein, M.; Snyder, M. RNA-Seq: A revolutionary tool for transcriptomics. *Nat. Rev. Genet.* **2009**, *10*, 57–63. [CrossRef] [PubMed]
24. Kong, F.; Li, Y.; Diao, Q.; Bi, Y.; Tu, Y. The crucial role of lysine in the hepatic metabolism of growing Holstein dairy heifers as revealed by LC-MS-based untargeted metabolomics. *Anim. Nutr.* **2021**, *7*, 1152–1161. [CrossRef]
25. Kargar, S.; Kanani, M. Reconstituted versus dry alfalfa hay in starter feed diets of Holstein dairy calves: Effects on growth performance, nutrient digestibility, and metabolic indications of rumen development. *J. Dairy Sci.* **2019**, *102*, 4051–4060. [CrossRef]
26. Kong, F.; Bi, Y.; Wang, B.; Cui, K.; Li, Y.; Fu, T.; Diao, Q.; Tu, Y. Integrating RNA-sequencing and untargeted LC-MS metabolomics to evaluate the effect of lysine deficiency on hepatic functions in Holstein calves. *Amino Acids* **2020**, *52*, 781–792. [CrossRef]
27. Flaga, J.; Korytkowski, Ł.; Górka, P.; Kowalski, Z.M. Age-related changes in mRNA expression of selected surface receptors in lymphocytes of dairy calves. *Pol. J. Vet. Sci.* **2018**, *21*, 213–216. [CrossRef]
28. Wang, S.; Diao, Q.Y.; Hu, F.M.; Bi, Y.L.; Piao, M.Y.; Jiang, L.S.; Sun, F.; Li, H.; Tu, Y. Development of ruminating behavior in Holstein calves between birth and 30 days of age. *J. Dairy Sci.* **2022**, *105*, 572–584. [CrossRef]
29. Shannon, P.; Markiel, A.; Ozier, O.; Baliga, N.S.; Wang, J.T.; Ramage, D.; Amin, N.; Schwikowski, B.; Ideker, T. Cytoscape: A software environment for integrated models of biomolecular interaction networks. *Genome Res.* **2003**, *13*, 2498–2504. [CrossRef]
30. Brohée, S.; van Helden, J. Evaluation of clustering algorithms for protein-protein interaction networks. *BMC Bioinform.* **2006**, *7*, 488. [CrossRef]
31. Chin, C.H.; Chen, S.H.; Wu, H.H.; Ho, C.W.; Ko, M.T.; Lin, C.Y. cytoHubba: Identifying hub objects and sub-networks from complex interactome. *BMC Syst. Biol.* **2014**, *8* (Suppl. 4), S11. [CrossRef] [PubMed]
32. Jasper, J.; Weary, D.M. Effects of ad libitum milk intake on dairy calves. *J. Dairy Sci.* **2002**, *85*, 3054–3058. [CrossRef]
33. Blum, J.W.; Baumrucker, C.R. Colostral and milk insulin-like growth factors and related substances: Mammary gland and neonatal (intestinal and systemic) targets. *Domest. Anim. Endocrinol.* **2002**, *23*, 101–110. [CrossRef]
34. Lin, L.; Xie, F.; Sun, D.; Liu, J.; Zhu, W.; Mao, S. Ruminal microbiome-host crosstalk stimulates the development of the ruminal epithelium in a lamb model. *Microbiome* **2019**, *7*, 83. [CrossRef] [PubMed]
35. Malmuthuge, N.; Liang, G.; Guan, L.L. Regulation of rumen development in neonatal ruminants through microbial metagenomes and host transcriptomes. *Genome Biol.* **2019**, *20*, 172. [CrossRef]
36. Gerber, H.P.; Ferrara, N. Angiogenesis and Bone Growth. *Trends Cardiovasc. Med.* **2000**, *10*, 223–228. [CrossRef]
37. Seegraber, F.J.; Morrill, J.L. Effect of protein source in calf milk replacers on morphology and absorptive ability of small intestine. *J. Dairy Sci.* **1986**, *69*, 460–469. [CrossRef]
38. Freedman, S.F.; Anderson, P.J.; Epstein, D.L. Superoxide dismutase and catalase of calf trabecular meshwork. *Investig. Ophthalmol. Vis. Sci.* **1985**, *26*, 1330–1335.
39. Gaetani, G.D.; Parker, J.C.; Kirkman, H.N. Intracellular restraint: A new basis for the limitation in response to oxidative stress in human erythrocytes containing low-activity variants of glucose-6-phosphate dehydrogenase. *Proc. Natl. Acad. Sci. USA* **1974**, *71*, 3584–3587. [CrossRef]
40. Schulz, J.B.; Lindenau, J.; Seyfried, J.; Dichgans, J. Glutathione, oxidative stress and neurodegeneration. *Eur. J. Biochem.* **2000**, *267*, 4904–4911. [CrossRef]
41. Montagne, L.; Toullec, R.; Savidge, T.; Lallès, J.P. Morphology and enzyme activities of the small intestine are modulated by dietary protein source in the preruminant calf. *Reprod. Nutr. Dev.* **1999**, *39*, 455–466. [CrossRef] [PubMed]
42. Castro, I.; Cerbón, M.A.; Pasapera, A.M.; Gutiérrez-Sagal, R.; Garcia, G.A.; Orozco, C.; Camacho-Arroyo, I.; Anzaldua, R.; Pérez-Palacios, G. Molecular mechanisms of the antihormonal and antiimplantation effects of norethisterone and its A-ring reduced metabolites. *Mol. Reprod. Dev.* **1995**, *40*, 157–163. [CrossRef] [PubMed]
43. Wilson, G.J.; Tuffs, S.W.; Wee, B.A.; Seo, K.S.; Park, N.; Connelley, T.; Guinane, C.M.; Morrison, W.I.; Fitzgerald, J.R. Bovine Staphylococcus aureus Superantigens Stimulate the Entire T Cell Repertoire of Cattle. *Infect. Immun.* **2018**, *86*, e00505-18. [CrossRef] [PubMed]
44. Grasfeder, L.L.; Gaillard, S.; Hammes, S.R.; Ilkayeva, O.; Newgard, C.B.; Hochberg, R.B.; Dwyer, M.A.; Chang, C.Y.; McDonnell, D.P. Fasting-induced hepatic production of DHEA is regulated by PGC-1alpha, ERRalpha, and HNF4alpha. *Mol. Endocrinol.* **2009**, *23*, 1171–1182. [CrossRef] [PubMed]
45. Salleh, M.S.; Mazzoni, G.; Höglund, J.K.; Olijhoek, D.W.; Lund, P.; Løvendahl, P.; Kadarmideen, H.N. RNA-Seq transcriptomics and pathway analyses reveal potential regulatory genes and molecular mechanisms in high- and low-residual feed intake in Nordic dairy cattle. *BMC Genom.* **2017**, *18*, 258. [CrossRef] [PubMed]

46. Tizioto, P.C.; Coutinho, L.L.; Decker, J.E.; Schnabel, R.D.; Rosa, K.O.; Oliveira, P.S.; Souza, M.M.; Mourão, G.B.; Tullio, R.R.; Chaves, A.S.; et al. Global liver gene expression differences in Nelore steers with divergent residual feed intake phenotypes. *BMC Genom.* **2015**, *16*, 242. [CrossRef]
47. Santana, M.H.d.A.; Junior, G.A.; Cesar, A.S.; Freua, M.C.; Gomes, R.d.C.; da Luz, E.S.S.; Leme, P.R.; Fukumasu, H.; Carvalho, M.E.; Ventura, R.V.; et al. Copy number variations and genome-wide associations reveal putative genes and metabolic pathways involved with the feed conversion ratio in beef cattle. *J. Appl. Genet.* **2016**, *57*, 495–504. [CrossRef] [PubMed]
48. Song, N.; Gui, L.S.; Xu, H.C.; Wu, S.; Zan, L.S. Identification of single nucleotide polymorphisms of the signal transducer and activator of transcription 3 gene (STAT3) associated with body measurement and carcass quality traits in beef cattle. *Genet. Mol. Res.* **2015**, *14*, 11242–11249. [CrossRef] [PubMed]
49. da Costa, A.S.; Pires, V.M.; Fontes, C.M.; Prates, J.A.M. Expression of genes controlling fat deposition in two genetically diverse beef cattle breeds fed high or low silage diets. *BMC Vet. Res.* **2013**, *9*, 118. [CrossRef]
50. Sumner-Thomson, J.M.; Vierck, J.L.; McNamara, J.P. Differential expression of genes in adipose tissue of first-lactation dairy cattle. *J. Dairy Sci.* **2011**, *94*, 361–369. [CrossRef]
51. Lee, K.; Kerner, J.; Hoppel, C.L. Mitochondrial carnitine palmitoyltransferase 1a (CPT1a) is part of an outer membrane fatty acid transfer complex. *J. Biol. Chem.* **2011**, *286*, 25655–25662. [CrossRef] [PubMed]
52. McKenzie, M.; Lazarou, M.; Thorburn, D.R.; Ryan, M.T. Analysis of mitochondrial subunit assembly into respiratory chain complexes using Blue Native polyacrylamide gel electrophoresis. *Anal. Biochem.* **2007**, *364*, 128–137. [CrossRef] [PubMed]
53. Menezes, M.J.; Riley, L.G.; Christodoulou, J. Mitochondrial respiratory chain disorders in childhood: Insights into diagnosis and management in the new era of genomic medicine. *Biochim. Biophys. Acta* **2014**, *1840*, 1368–1379. [CrossRef] [PubMed]
54. Steffen, W.; Gemperli, A.C.; Cvetesic, N.; Steuber, J. Organelle specific expression of subunit ND5 of human complex I (NADH dehydrogenase) alters cation homeostasis in Saccharomyces cerevisiae. *FEMS Yeast Res.* **2010**, *10*, 648–659. [CrossRef]
55. Fan, M.; Zhang, J.; Tsai, C.W.; Orlando, B.J.; Rodriguez, M.; Xu, Y.; Liao, M.; Tsai, M.F.; Feng, L. Structure and mechanism of the mitochondrial Ca(2+) uniporter holocomplex. *Nature* **2020**, *582*, 129–133. [CrossRef]
56. McCabe, M.; Waters, S.; Morris, D.; Kenny, D.; Lynn, D.; Creevey, C. RNA-seq analysis of differential gene expression in liver from lactating dairy cows divergent in negative energy balance. *BMC Genom.* **2012**, *13*, 193. [CrossRef]
57. Pang, J.; Li, F.; Feng, X.; Yang, H.; Han, L.; Fan, Y.; Nie, H.; Wang, Z.; Wang, F.; Zhang, Y. Influences of different dietary energy level on sheep testicular development associated with AMPK/ULK1/autophagy pathway. *Theriogenology* **2018**, *108*, 362–370. [CrossRef]
58. Bartlett, K.S.; McKeith, F.K.; VandeHaar, M.J.; Dahl, G.E.; Drackley, J.K. Growth and body composition of dairy calves fed milk replacers containing different amounts of protein at two feeding rates. *J. Anim. Sci.* **2006**, *84*, 1454–1467. [CrossRef]
59. Tikofsky, J.N.; Van Amburgh, M.E.; Ross, D.A. Effect of varying carbohydrate and fat content of milk replacer on body composition of Holstein bull calves. *J. Anim. Sci.* **2001**, *79*, 2260–2267. [CrossRef]
60. Hu, F.M.; Dong, L.F.; Bi, Y.L.; Ma, J.N.; Wang, B.; Diao, Q.Y.; Tu, Y. Effects of Different Fatty Acid Sources in Milk Replacer on Growth Performance, Digestion and Metabolism of Sucking Calves. *Chin. J. Anim. Nutr.* **2018**, *30*, 1736–1747. [CrossRef]
61. Coutinho, S.R.; With, E.; Rehfeld, J.F.; Kulseng, B.; Truby, H.; Martins, C. The impact of rate of weight loss on body composition and compensatory mechanisms during weight reduction: A randomized control trial. *Clin. Nutr.* **2018**, *37*, 1154–1162. [CrossRef] [PubMed]
62. Cui, K.; Wang, B.; Ma, T.; Si, B.W.; Zhang, N.F.; Tu, Y.; Diao, Q.Y. Effects of dietary protein restriction followed by realimentation on growth performance and liver transcriptome alterations of lamb. *Sci. Rep.* **2018**, *8*, 15185. [CrossRef] [PubMed]
63. Hansen, S.; Therkildsen, M.; Byrne, D.V. Effects of a compensatory growth strategy on sensory and physical properties of meat from young bulls. *Meat Sci.* **2006**, *74*, 628–643. [CrossRef] [PubMed]
64. Keady, S.M.; Waters, S.M.; Hamill, R.M.; Dunne, P.G.; Keane, M.G.; Richardson, R.I.; Kenny, D.A.; Moloney, A.P. Compensatory growth in crossbred Aberdeen Angus and Belgian Blue steers: Effects on the colour, shear force and sensory characteristics of longissimus muscle. *Meat Sci.* **2017**, *125*, 128–136. [CrossRef]
65. Eriksson, J.G.; Forsén, T.; Tuomilehto, J.; Winter, P.D.; Osmond, C.; Barker, D.J. Catch-up growth in childhood and death from coronary heart disease: Longitudinal study. *BMJ* **1999**, *318*, 427–431. [CrossRef] [PubMed]
66. Ibáñez, L.; Lopez-Bermejo, A.; Diaz, M.; de Zegher, F. Catch-up growth in girls born small for gestational age precedes childhood progression to high adiposity. *Fertil. Steril.* **2011**, *96*, 220–223. [CrossRef] [PubMed]
67. Silveira, P.P.; Pokhvisneva, I.; Gaudreau, H.; Rifkin-Graboi, A.; Broekman, B.F.P.; Steiner, M.; Levitan, R.; Parent, C.; Diorio, J.; Meaney, M.J. Birth weight and catch up growth are associated with childhood impulsivity in two independent cohorts. *Sci. Rep.* **2018**, *8*, 13705. [CrossRef]

antioxidants

MDPI

Article

Acremonium terricola Culture's Dose–Response Effects on Lactational Performance, Antioxidant Capacity, and Ruminal Characteristics in Holstein Dairy Cows

Fanlin Kong [1,†], Yijia Zhang [2,†], Shuo Wang [1], Zan Cao [3], Yanfang Liu [4], Zixiao Zhang [4], Wei Wang [1], Na Lu [4,*] and Shengli Li [1,*]

1 Beijing Engineering Technology Research Center of Raw Milk Quality and Safety Control, The State Key Laboratory of Animal Nutrition, Department of Animal Nutrition and Feed Science, College of Animal Science and Technology, China Agricultural University, No. 2 Yuanmingyuan West Road, Haidian District, Beijing 100094, China; fanlinkong@cau.edu.cn (F.K.); b20213040351@cau.edu.cn (S.W.); wei.wang@cau.edu.cn (W.W.)

2 Laboratory of Anatomy of Domestic Animals, Department of Basic Veterinary Medicine, College of Veterinary Medicine, China Agricultural University, No. 2 Yuanmingyuan West Road, Haidian District, Beijing 100094, China; BS20193050473@cau.edu.cn

3 Microbial Biological Engineering Company Limited, Fanhua Road Jingkai District, Hefei 230009, China; caozan1314@126.com

4 Beijing JingWa Agricultural Science and Technology Innovation Center, Mishan Road, Pinggu District, Beijing 101200, China; 13611235024@163.com (Y.L.); haoyangyi0928@cau.edu.cn (Z.Z.)

* Correspondence: luna2020@cau.edu.cn (N.L.); lishengli@cau.edu.cn (S.L.)

† These authors contributed equally to this work.

Citation: Kong, F.; Zhang, Y.; Wang, S.; Cao, Z.; Liu, Y.; Zhang, Z.; Wang, W.; Lu, N.; Li, S. *Acremonium terricola* Culture's Dose–Response Effects on Lactational Performance, Antioxidant Capacity, and Ruminal Characteristics in Holstein Dairy Cows. *Antioxidants* **2022**, *11*, 175. https://doi.org/10.3390/antiox11010175

Academic Editors: Min Xue, Junmin Zhang, Zhenyu Du, Jie Wang and Wei Si

Received: 25 December 2021
Accepted: 13 January 2022
Published: 17 January 2022

Publisher's Note: MDPI stays neutral with regard to jurisdictional claims in published maps and institutional affiliations.

Abstract: *Acremonium terricola* culture (ATC) has similar bioactive constituents to *Cordyceps* and is known for its nutrient and pharmacological value, indicating the potential of ATC as a new feed additive in dairy cow feeding. The primary aim of this experiment was to investigate the effects of increasing amounts of ATC in diets on milk performance, antioxidant capacity, and rumen fermentation, and the secondary aim was to evaluate the potential effects of high doses of ATC. A total of 60 multiparous Holstein cows (110 ± 21 days in milk; 2.53 ± 0.82 parity) were assigned into 15 blocks and randomly assigned to one of four groups: 0, 30, 60, or 300 g/d of ATC per cow for 97 days. Data were analyzed using repeated measures in the Mixed procedure. Dry-matter intake was not changed ($p > 0.05$), while energy-corrected milk and fat-corrected milk yields increased linearly and quadratically, and somatic cell count in milk decreased linearly and quadratically ($p < 0.05$). The lactation efficiency and the yields of milk fat and protein increased linearly ($p < 0.05$). On day 90, serum catalase level, total oxidative capacity, glutathione peroxidase, immunoglobulin A, and immunoglobulin M concentrations were significantly higher in the 60 and 300 g/d groups than in the 0 g/d group ($p < 0.05$). ATC addition showed linear effects on total volatile fatty acid (VFA), acetate, branched VFA concentrations, and rumen pH ($p < 0.05$). Supplementing 60 and 300 g/d ATC significantly affected the bacterial composition ($p < 0.05$). The relative abundance of *Christensenellaceae_R-7_group* and *Lachnospiraceae_NK3A20_group* were significantly increased by 60 g/d supplementation, and the relative abundance of *Erysipelotrichaceae_UCG_002*, *Acetitomaculum*, *Olsenella*, and *Syntrophococcus* were significantly increased by 300 g/d supplementation ($p < 0.05$). ATC was effective in enhancing rumen fermentation and reducing somatic cell count in milk, thereby improving milk yield. The optimized dose of ATC was 60 g/d for lactating cows, and there were no risks associated with high doses of ATC.

Keywords: *Acremonium terricola* culture; dairy cow; antioxidant capacity; milk performance; bacterial composition

1. Introduction

The health and biological function of livestock often have been prioritized in recent years [1]. Nutritional strategies have emerged, and they have been proposed as a key factor to improve the health status and welfare of animals, as well as to enhance productivity of livestock [2–4]. Feed additives, which are nutrients that when added to the feed can trigger the desired response of the animal's body on production parameters, have been widely used in animal nutrition for quite a long time [5–8]. The main compounds used for this purpose are fermented feed, probiotics, prebiotics, synbiotics, or various types of residues from plant production. *Cordyceps* has a long history, and has been used for the last 300 years [9]. It is a rare, naturally entomopathogenic fungus that is mostly collected on the Himalayan plateau in China. Although wild *Cordyceps* has many functional bioactive ingredients, including cordycepin, cordycepic acid, and cordyceps polysaccharide, it takes a long time to grow under strict conditions, cannot be cultivated artificially, and can only be harvested once [10]. In recent decades, many studies have shown that the major functional components in *Cordyceps* possess multiple resistances to viruses, inflammation, oxidants, tumors, diabetes, and thrombi [9,11–13] Thus, many investigators have been devoted to artificial cultivation to obtain prebiotics [14,15].

Acremonium terricola is a parasite isolated from *Cordyceps gunnii*; artificial solid fermentation is then used to obtain *Acremonium terricola* culture (ATC). ATC is a new feed additive with bioactive ingredients that are the same as those of *C. gunnii*, such as *D*-mannitol, galactomannan, cordycepin, and essential amino acids [16]. As reported in previous studies, ATC supplementation enhanced the growth performance, antioxidant capacity, and immune function of calves [17]. ATC supplementation increased the total tract apparent digestibility and milk performance in dairy cows, then decreased somatic cell counts in milk [18,19]. Dairy cows during the transition period are at risk of diseases, and undergo extensive tissue mobilization due to a negative energy balance [20–22]. Li et al. [23] found that 30 g/d ATC ameliorated the negative energy balance and improved milk production and antioxidant capacity. These studies indicated that ATC was an effective feed additive for dairy cows at 30 g/d, and the maximum addition amount (30 g/d) was more effective than other additional amounts. However, studies on the effects of the higher dose of ATC supplementation on milk production are lacking, which means that the appropriate level of ATC in the diet is still not clear. In this regard, galactomannans from ATC are water-soluble polysaccharide polymers, and several studies have reported that galactomannan plays a crucial role in manipulating gastrointestinal microbiota by providing nutrients for beneficial microbes [12,24,25]. Additionally, D-mannitol can also become the sole energy source for bacterial growth in the rumen [26]. It is known that highly dense and diverse microbial populations in the rumen cause many differences in feeding modes and physiology between ruminants and monogastric animals [27]. Thus, ATC supplementation in dairy cows may have the potential to further improve milk performance and rumen function.

The ingredients and bioactivities of ATC are the same as those of wild *Cordyceps*. Owing to the wild occurrence, natural *Cordyceps* is regarded as pharmacologically secure. However, dry mouth, nausea, abdominal distension, throat discomfort, headache, diarrhea, and allergic reactions have been reported in *Cordyceps* studies [28–30]. Tests of cordycepin in dogs involved gastrointestinal toxicity and bone marrow toxicity [31]. However, these toxicity tests were conducted using monogastric animals instead of ruminants. Thus, future in-depth evaluation of the security of ATC will benefit its mass production and use as a feed additive.

We hypothesized that ATC supplementation would improve the milk performance, antioxidant capacity, and immune function of midlactating dairy cows, and reshape their microbiota. Hence, this study investigated the effects of different amounts of ATC on feeding behavior, ruminal fermentation, bacterial composition, serum variables, and milk performance, as well as the possible adverse effects of high doses of ATC, on lactating dairy cows.

2. Materials and Methods

2.1. Animals and Treatments

On a commercial dairy farm (Jinyindao Dairy Farm; Sanyuan Food Co., Ltd., Beijing, China), 60 multiparous lactating dairy cows averaging 110 ± 21 days (mean ± SD) in milk, 28 ± 3 kg/d of milk, 2.53 ± 0.82 parity, and 630 ± 50 kg of body weight were selected and assigned into 15 blocks according to days in milk (DIM) and milk yield, and then randomly allocated to 1 of 4 groups: 0, 30, 60, or 300 g/d of ATC for each cow, added into a total mixed ration (TMR). The ATC was provided by Microbial Biological Engineering Ltd. (Hefei, China), and the functional composition of the ATC is listed in Supplementary Table S1. The recommended dose for ATC was 30 g/d [19,32]. In this experiment, we included double the recommended dose to determine whether ATC had a linear and quadratic effect on lactation performance, as well as 10 times the recommended dose to test the potential effects on dairy cow health. The same barn equipped with water bowls and automatic weighing feeding equipment (Insentec B.V., Marknesse, The Netherlands) were provided for these cows to obtain individual feed intake. All cows were fed and milked 3 times per day at 09:00, 16:00, and 22:00. The ATC was top-dressed onto a TMR; ingredients and nutrient contents of the TMR are listed in Supplementary Table S2. The diet was formatted using National Research Council (NRC, 2001) [33] to meet the nutrient requirements of lactating dairy cows. The experiment period was 97 days, including 7 days to allow for adaptation, during which ATC was supplemented gradually. According to the standard operating procedure on this commercial dairy farm, all cows had de-worming (ivermectin, Harbin Qianhe Animal Medicine Manufacturing Co., Ltd., Harbin, China) during the dry-off period, and the farm followed the epidemic prevention and control procedures. All experimental procedures were submitted to the experimental animal welfare and animal ethics committee of China Agricultural University and approved (CAU2021009-2).

2.2. Sampling and Analysis

Individual feed intake was recorded daily using feed-weighing equipment (RIC system, Insentec B.V., Marknesse, The Netherlands), and the samples of TMR and orts were collected on the same day (the first day of every week). The TMR, orts, and ATC were used to determine the nutrient composition, including dry matter (DM, 105 °C for 5 h), acid detergent fiber (ADF), neutral detergent fiber (NDF), and crude protein (CP) contents [34,35]. Daily dry-matter intake (DMI) was calculated using Equation (1):

$$\text{DMI} = \text{as fed TMR} \times \text{dry matter content} - \text{ort} \times \text{dry matter content} \quad (1)$$

The amino acid (AA) concentrations were analyzed with a Hitachi L-8900 automatic AA analyzer (HITACHI High-Tech Corporation Ltd., Shanghai, China). The concentrations of *D*-mannitol, galactomannan, 3′-deoxyadenosine, and ergosterol were determined using liquid chromatography [36].

Milk yield was recorded electronically, and 50 mL of representative milk from each cow was collected at 0, 30, 60, and 90 days using a diverter (BouMatic Company, Madison, WI, USA). The 50 mL milk sample included 20 mL of milk from the 09:00 milking, 15 mL from the 16:00 milking, and 15 mL from the 22:00 milking. Milk samples were stored at 4 °C for further analysis of milk composition (protein, fat, lactose, urea nitrogen, and somatic cell count) using a FOSS MilkoScanTM 7 (FOSS, Hillerod, Denmark).

Blood samples were collected from the tail vein of each cow into 5 mL Vacutainer K2EDTA tubes (OLABO Biotechnology Co., Ltd., Jinan, China) at 0, 45, and 90 days after morning feeding between 11:00 and 12:00. Blood serum was separated by centrifugation (3000× *g* for 10 min at 4 °C) after sampling immediately and stored at −20 °C. At the end of the experiment, all serum samples were analyzed using a GF-D200 Auto Analyzer (Caihong Analytical Instrument Co., Ltd., Gaomi, China) to determine urea nitrogen (SUN) and total amino acid (TAA) concentrations using standard commercial kits (BioSino BioTechnology and Science Inc., Beijing, China). The superoxide dismutase (SOD), catalase

(CAT), malondialdehyde (MDA), total oxidative capacity (TAC), and glutathione peroxidase (GSH-Px) levels were determined by colorimetry using commercial kits (Nanjing Jian Cheng Bioengineering Institute, Nanjing, China). Immunoglobulin A (IgA), immunoglobulin G (IgG), and immunoglobulin M (IgM) concentrations were determined with ELISA kits (Abcam, Cambridge, UK) using a Thermo Multiskan Ascent (Thermo Fisher Scientific, Shanghai, China). The intra- and interassay coefficients of variations were both <10%.

Ruminal fluid was collected using an oral stomach tube after 2 h morning feeding from six cows at 0 and 90 days in each group, which accounted for close to the average milk production. The tube was cleaned thoroughly with fresh warm water after each sampling. The first 150 mL of rumen fluid was useless, and subsequent rumen fluid was collected and filtered through four layers of cheesecloth to obtain the fluid fraction. The fluid fraction pH value was measured with a Sartorius PB-10 pH meter (Beijing Sartorius Instrument Systems Co., Ltd., Beijing, China). The rumen fluid was stored in a 20 mL centrifuge tube, then frozen at −20 °C for volatile fatty acid (VFA) analysis. The rest of the fluid was stored in a 2 mL centrifuge tube and frozen at −80 °C for 16S sequencing. VFA content was quantified using gas chromatography (Agilent 6890N, Agilent Technology, Inc., Beijing, China), according to Kong et al [37].

Total DNA was extracted from 1 mL rumen fluid samples (thawing for 10 min) using an E.Z.N.A.® Soil DNA Kit (Omega Bio-Tek, Norcross, GA, USA) according to the manufacturer's instructions. The DNA extract was analyzed on a 1% agarose gel, and DNA concentration and purity were determined using a NanoDrop 2000 UV–vis spectrophotometer (Thermo Scientific, Wilmington, DE, USA). The hypervariable region V3-V4 of the bacterial 16S rRNA gene was amplified with primer pairs 338F (5′-ACTCCTACGGGAGGCAGCAG-3′) and 806R (5′–GGACTACHVGGGTWTCTAAT-3′) using an ABI GeneAmp® 9700 PCR thermocycler (ABI, Foster City, CA, USA) [38]. The PCR amplification of the 16S rRNA gene was performed according to Kong et al [38]. The PCR product was extracted from a 2% agarose gel and purified using an AxyPrep DNA Gel Extraction Kit (Axygen Biosciences, Union City, CA, USA) according to the manufacturer's instructions, and quantified using a Quantus™ Fluorometer (Promega, Madison, WI, USA). Purified amplicons were pooled in equimolar amounts and paired-end sequenced on an Illumina MiSeq PE300 platform (Illumina, San Diego, CA, USA).

The raw 16S rRNA gene sequencing reads were demultiplexed, quality-filtered using Fastp (Version 0.20.1, Haplox, Shenzhen, China) [39], and merged using FLASH version 1.2.7 (The Center for Computational Biology at Johns Hopkins University, MD, USA) [40] with the criteria set according to Huang et al [41]. All samples were subsampled to equal size of 25,122 sequences for downstream analysis. Operational taxonomic units (OTUs) with a 97% similarity cutoff were clustered using UPARSE version 7.1 (Independent Investigator, CA, USA) [42], and chimeric sequences were identified and removed. The taxonomy of each OTU representative sequence was analyzed using the RDP Classifier version 2.2 (Center for Microbial Ecology, Michigan State University, MI, USA) [43] against the 16S rRNA database (e.g., Silva v138) using a confidence threshold of 0.7 [44]. The relative abundance of phylum or genus $\geq$ 1% was used to consider the predominant phylum or genus. Alpha diversity indices (observed richness, Chao1, ACE, Shannon, Simpson) and Good's coverage were obtained using the alpha rarefaction script in QIIME [45]. Principal coordinates analysis (PCoA) was conducted by Bray–Curtis matrices in R (http://www.rstudio.com, accessed on 3 May 2021). An analysis of similarities (ANOSIM) using Bray–Curtis distance matrices was performed to test the statistical differences among the observed microbial profiles.

2.3. Statistical Analysis

The effects on feeding behavior, milk performance, serum variables, and rumen fermentation parameters were analyzed using the Mixed procedure in Statistical Analysis System 9.4 (SAS Institute Inc., Cary, NC, USA). Before analyses, data were screened for normality using the Univariate procedure; the variables except for sequencing data met the assumptions for normality. A randomized block experimental design was used for the month, treatment, and interaction of treatment and month according to the following model:

$$Y_{ijkl} = \mu + A_i + T_j + AT_{ij} + B_k + \varepsilon_{ijkl} \quad (2)$$

where Y is the dependent variable, μ is the overall mean, A_i is the fixed effect of ATC supplementation, T_j is the repeated effect, AT_{ij} is the interaction effect of A and T, B_k is the block effect, and ε_{ijkl} is the random residual error. Cows were used as the experimental units. The linear and quadratic effects of the treatment on these variables were tested using orthogonal polynomials, accounting for unequal spacing of ATC supplementation levels. The results are presented as least-squares means and were separated using the PDIFF statement when the fixed effects were significant. Significance was declared at $p \leq 0.05$, and tendencies were declared at $0.05 \leq p \leq 0.10$.

The effects on alpha diversity indices and the relative abundance of genera were assessed using the Kruskal–Wallis H test, and Dunn's test was applied to conduct multiple comparisons. All *p*-values were corrected using a false discovery rate of 0.05, and a corrected *p*-value < 0.05 was considered significant. The data are presented as the mean ± standard error (SEM). Spearman's correlations were calculated between the relative abundances of genera. Significant correlations were defined as $-0.7 > r > 0.7$, and $p < 0.05$.

3. Results

3.1. Feeding Behavior

Table 1 shows the effects of ATC supplementation on feeding behavior. In the third month, 30 g/d ATC supplementation increased DMI ($p < 0.05$), whereas DMI/feeding frequency, average DMI, and average feeding frequency were not changed after ATC supplementation ($p > 0.05$). The different levels of ATC supplementation decreased feeding duration and feeding duration/feeding frequency, and increased DMI/feeding duration linearly ($p < 0.05$).

Table 1. Effect of different amounts of *Acremonium terricola* culture on feeding behavior in lactating dairy cows.

Items	ATC Supplementation, g/d per Head				SEM	*p*-Value			
	0	30	60	300		G	G × T	L	Q
DMI, kg/d									
Average	21.70	21.95	21.91	21.87	0.103	0.36	0.56	0.71	0.66
0–30 d	21.93	22.01	22.16	21.84	0.182	0.17			
30–60 d	21.38	21.52	21.59	21.68	0.170	0.21			
60–90 d	21.81 ^{b}	22.32 ^{a}	21.97 ^{ab}	22.10 ^{ab}	0.159	0.04			
Feeding frequency, bouts/d									
Average	20.33	20.71	21.65	21.15	0.540	0.27	0.32	0.35	0.90
0–30 d	19.85 ^{b}	21.28 ^{ab}	22.38 ^{a}	20.28 ^{ab}	0.884	0.04			
30–60 d	19.91	19.45	19.86	21.47	0.826	0.09			
60–90 d	21.24	21.41	22.70	21.70	0.769	0.22			
Feeding duration, h/d									
Average	4.40 ^{a}	3.67 ^{c}	3.57 ^{c}	3.97 ^{b}	0.038	<0.01	0.01	<0.01	<0.01
0–30 d	4.13 ^{a}	3.56 ^{b}	3.30 ^{c}	3.60 ^{b}	0.068	<0.01			
30–60 d	4.48 ^{a}	3.64 ^{c}	3.53 ^{c}	3.96 ^{b}	0.064	<0.01			
60–90 d	4.61 ^{a}	3.80 ^{c}	3.87 ^{c}	4.34 ^{b}	0.059	<0.01			

Table 1. *Cont.*

Items	ATC Supplementation, g/d per Head				SEM	*p*-Value			
	0	30	60	300		G	G × T	L	Q
DMI/Feeding frequency, kg/bouts									
Average	1.08	1.07	1.03	1.04	0.026	0.37	0.32	0.50	0.83
0–30 d	1.13	1.05	1.02	1.10	0.043	0.06			
30–60 d	1.08	1.12	1.09	1.01	0.040	0.06			
60–90 d	1.03	1.05	0.97	1.02	0.037	0.16			
Feeding duration/Feeding frequency, min/bouts									
Average	13.07 ^{a}	10.74 ^{b}	10.01 ^{c}	11.35 ^{b}	0.261	<0.01	0.45	<0.01	0.20
0–30 d	12.70 ^{a}	10.17 ^{bc}	9.13 ^{c}	10.90 ^{b}	0.464	<0.01			
30–60 d	13.52 ^{a}	11.32 ^{b}	10.68 ^{b}	11.12 ^{b}	0.434	<0.01			
60–90 d	12.99 ^{a}	10.73 ^{b}	10.24 ^{b}	12.04 ^{a}	0.404	<0.01			
DMI/Feeding duration, kg/h									
Average	4.94 ^{d}	6.01 ^{b}	6.18 ^{a}	5.55 ^{c}	0.066	<0.01	0.01	<0.01	0.01
0–30 d	5.37 ^{c}	6.20 ^{b}	6.73 ^{a}	6.08 ^{b}	0.107	<0.01			
30–60 d	4.75 ^{c}	5.93 ^{a}	6.12 ^{a}	5.48 ^{b}	0.100	<0.01			
60–90 d	4.71 ^{c}	5.89 ^{a}	5.70 ^{a}	5.10 ^{b}	0.093	<0.01			

ATC, *Acremonium terricola* culture; SEM, standard error; DMI, dry-matter intake. The effects included group effect (G) and the interaction effect of group and time (G × T), as well as linear (L) and quadratic effects (Q). Different lowercase letters in the same row show significant differences ($p < 0.05$). $n = 15$.

3.2. Milk Performance

As shown in Table 2, the performances including milk yield, fat-corrected milk yield (FCM), and energy-corrected milk yield (ECM) increased linearly and quadratically with ATC supplementation level ($p < 0.05$). There were no significant differences between the average milk yield and average ECM yield between the 60 g/d group and 300 g/d groups ($p > 0.05$), which were both higher than those of cows in the 30 g/d and 0 g/d groups ($p < 0.05$). The milk efficiency values both increased linearly with ATC supplementation level ($p < 0.05$), while there was no difference between the 60 g/d ATC treatment and 300 g/d ATC treatment groups ($p > 0.05$).

Table 2. Effects of different amounts of *Acremonium terricola* culture on milk performance in lactating dairy cows.

Items	ATC Supplementation, g/d per Head				SEM	*p*-Value			
	0	30	60	300		G	G × T	L	Q
Milk yield, kg/d									
Average	27.33 ^{c}	28.98 ^{b}	30.15 ^{a}	30.35 ^{a}	0.376	<0.01	0.98	<0.01	<0.01
0–30 d	28.56 ^{b}	29.75 ^{ab}	30.87 ^{a}	31.00 ^{a}	0.918	0.01			
30–60 d	27.53 ^{b}	29.28 ^{ab}	30.38 ^{a}	30.57 ^{a}	0.896	<0.01			
60–90 d	25.90 ^{b}	27.90 ^{a}	29.20 ^{a}	29.49 ^{a}	0.882	<0.01			
FCM yield, kg/d									
Average	32.73 ^{b}	34.95 ^{ab}	36.73 ^{a}	36.95 ^{a}	0.947	<0.01	1.00	<0.01	0.04
0–30 d	33.93	36.34	37.64	37.99	1.552	0.05			
30–60 d	33.30	34.98	37.22	36.95	1.451	0.06			
60–90 d	30.95 ^{b}	33.52 ^{ab}	35.32 ^{a}	35.92 ^{a}	1.350	0.02			

Table 2. *Cont.*

Items	ATC Supplementation, g/d per Head				SEM	p-Value			
	0	30	60	300		G	G × T	L	Q
ECM, kg/d									
Average	31.36 [c]	33.41 [b]	34.88 [a]	35.18 [a]	0.623	<0.01	0.98	<0.01	<0.01
0–30 d	33.26	34.89	35.96	36.13	1.107	0.05			
30–60 d	31.38 [b]	33.54 [ab]	35.46 [a]	35.30 [a]	1.035	0.01			
60–90 d	29.44 [b]	31.80 [ab]	33.24 [a]	34.10 [a]	0.963	<0.01			
Milk yield/DMI									
Average	1.26 [c]	1.32 [b]	1.38 [ab]	1.39 [a]	0.018	<0.01	0.99	<0.01	0.09
0–30 d	1.31 [b]	1.35 [ab]	1.40 [a]	1.42 [a]	0.021	0.01			
30–60 d	1.29 [b]	1.36 [ab]	1.41 [a]	1.41 [a]	0.020	0.01			
60–90 d	1.19 [b]	1.26 [ab]	1.33 [a]	1.34 [a]	0.017	<0.01			
FCM/DMI									
Average	1.51 [b]	1.60 [ab]	1.68 [a]	1.69 [a]	0.037	0.01	1.00	0.02	0.17
0–30 d	1.55	1.65	1.70	1.74	0.073	0.06			
30–60 d	1.56	1.63	1.73	1.70	0.068	0.08			
60–90 d	1.42	1.51	1.61	1.63	0.063	0.06			
ECM/DMI									
Average	1.45 [b]	1.53 [b]	1.60 [a]	1.61 [a]	0.032	<0.01	0.98	<0.01	0.07
0–30 d	1.52	1.59	1.63	1.66	0.053	0.05			
30–60 d	1.52 [b]	1.56 [ab]	1.65 [a]	1.66 [a]	0.050	0.01			
60–90 d	1.35 [b]	1.43 [ab]	1.52 [a]	1.55 [a]	0.046	0.01			

ATC, *Acremonium terricola* culture; SEM, standard error; ECM, energy-corrected milk; FCM, fat-corrected milk; DMI, dry-matter intake; Milk yield/DMI, kg of milk yield/kg of DMI; FCM/DMI, kg of FCM/kg of DMI; ECM/DMI, kg of ECM/kg of DMI; ECM = 0.3246 × milk yield + 13.86 × milk fat yield + 7.04 × milk protein yield; FCM = milk yield × 0.432 + milk fat yield × 16. The effects included group effect (G) and the interaction effect of group and time (G × T), as well as linear (L) and quadratic effects (Q). Different lowercase letters in the same row show significant differences ($p < 0.05$). $n = 15$.

Table 3 shows the effects of different amounts of ATC supplementation on milk composition. Milk protein, milk fat, milk lactose, and milk urea nitrogen (MUN) were not affected by ATC supplementation ($p > 0.05$). Milk protein yield, milk fat yield, and somatic cell count (SCC) were linearly influenced by ATC supplementation ($p < 0.05$). The milk protein yield in the first month was not increased after ATC supplementation ($p > 0.05$). The milk protein yield in the second and third months and the average yield were increased after 60 g/d and 300 g/d ATC supplementation ($p < 0.05$), but the milk protein yield and milk fat yield did not differ between the 60 g/d and 300 g/d ATC supplementation groups ($p > 0.05$). The SCC decreased in both a linear and quadratic manner after supplementation with ATC ($p < 0.05$), but there was no significant difference between the 60 g/d and 300 g/d ATC supplementation groups ($p > 0.05$).

Table 3. Effects of different amounts of *Acremonium terricola* culture on milk composition in lactating dairy cows.

Items	ATC Supplementation, g/d per Head				SEM	p-Value			
	0	30	60	300		G	G × T	L	Q
Milk protein, %									
Average	3.81	3.83	3.82	3.83	0.064	0.99	0.80	0.48	0.37
0–30 d	3.98	3.93	3.88	3.86	0.105	0.37			
30–60 d	3.71	3.81	3.88	3.83	0.098	0.22			
60–90 d	3.73	3.74	3.70	3.80	0.091	0.46			

Table 3. *Cont.*

Items	ATC Supplementation, g/d per Head				SEM	*p*-Value			
	0	30	60	300		G	G × T	L	Q
Milk protein yield, kg/d									
Average	1.04 [b]	1.11 [a]	1.15 [a]	1.16 [a]	0.026	<0.01	0.85	<0.01	<0.01
0–30 d	1.14	1.17	1.20	1.20	0.042	0.30			
30–60 d	1.02 [b]	1.12 [ab]	1.18 [a]	1.17 [a]	0.039	0.01			
60–90 d	0.97 [b]	1.04 [ab]	1.08 [a]	1.13 [a]	0.037	0.01			
Milk fat, %									
Average	4.73	4.77	4.83	4.84	0.152	0.94	1.00	0.80	0.99
0–30 d	4.68	4.87	4.85	4.91	0.270	0.51			
30–60 d	4.79	4.70	4.86	4.77	0.253	0.65			
60–90 d	4.71	4.74	4.79	4.85	0.235	0.69			
Milk fat yield, kg/d									
Average	1.29 [b]	1.38 [ab]	1.46 [a]	1.47 [a]	0.053	0.04	1.00	0.03	0.15
0–30 d	1.33	1.45	1.50	1.52	0.087	0.11			
30–60 d	1.32	1.38	1.48	1.46	0.081	0.16			
60–90 d	1.22	1.32	1.40	1.43	0.076	0.07			
Milk lactose, %									
Average	5.08	5.09	5.04	5.06	0.034	0.73	0.95	0.18	0.45
0–30 d	5.15	5.12	5.08	5.10	0.060	0.41			
30–60 d	4.99	5.05	5.02	5.03	0.056	0.48			
60–90 d	5.09	5.09	5.01	5.04	0.052	0.29			
Milk lactose yield, kg/d									
Average	1.39 [c]	1.48 [b]	1.52 [ab]	1.54 [a]	0.02	<0.01	<0.01	<0.01	0.04
0–30 d	1.47 [b]	1.52 [ab]	1.57 [a]	1.58 [a]	0.03	0.02			
30–60 d	1.38 [b]	1.48 [a]	1.52 [a]	1.54 [a]	0.03	<0.01			
60–90 d	1.32 [b]	1.42 [a]	1.46 [a]	1.49 [a]	0.03	<0.01			
Milk urea nitrogen, mg/dL									
Average	17.23	16.70	17.13	18.18	0.715	0.49	0.95	0.84	0.74
0–30 d	14.76	14.26	15.33	17.07	1.105	0.10			
30–60 d	16.32	15.29	15.70	16.10	1.270	0.56			
60–90 d	20.61	20.56	20.37	21.38	1.105	0.54			
SCC, ×1000/mL									
Average	68.06 [a]	52.23 [b]	47.90 [bc]	43.68 [c]	2.705	<0.01	0.42	<0.01	<0.01
0–30 d	67.23 [a]	58.60 [ab]	52.98 [bc]	41.85 [c]	4.809	<0.01			
30–60 d	70.73 [a]	44.73 [b]	44.64 [b]	45.25 [b]	4.496	<0.01			
60–90 d	66.23 [a]	53.35 [ab]	46.09 [b]	43.96 [b]	4.182	<0.01			

ATC, *Acremonium terricola* culture; SEM, standard error; SCC, somatic cell count. The effects included group effect (G) and the interaction effect of group and time (G × T), as well as linear (L) and quadratic effects (Q). Different lowercase letters in the same row show significant differences ($p < 0.05$). $n = 15$.

3.3. Serum Variables

Figure 1 shows the summary of the statistics for serum variables, including antioxidant capacity, immune function, and metabolites. On day 0, there was no significant difference on any serum variables among the four groups (Figure 1; $p > 0.05$). For antioxidant capacity, the CAT and TAC concentrations increased linearly after ATC supplementation (Figure 1A,C; $p < 0.05$). On day 90, the 60 g/d and 300 g/d ATC supplementation increased CAT, TAC, and GSH-Px concentrations (Figure 1A,C,E; $p < 0.05$). Only 300 g/d ATC supplementation increased SOD concentration on day 90 (Figure 1D; $p < 0.05$). For immune function, the ATC supplementation did not influence IgG concentration (Figure 1G; $p > 0.05$), but significantly increased IgA and IgM concentrations on day 90 (Figure 1F,H; $p < 0.05$). The IgA concentration increased linearly, and the IgM concentration increased quadratically after ATC supplementation (Figure 1F,H; $p < 0.05$). However, there were no differences among the 30 g/d, 60 g/d, and 300 g/d supplementation groups on day 45 and day 90 (Figure 1F,H; $p > 0.05$). For metabolites in the serum, the glucose concentration of the

300 g/d group was lower than in the 0 g/d group on day 45 (Figure 1I; $p < 0.05$). The SUN concentration was not affected by ATC supplementation (Figure 1J; $p > 0.05$). The TAA concentration increased linearly after ATC supplementation (Figure 1K; $p < 0.05$), and was higher in 300 g/d group on day 45 than in the 0, 30, and 60 g/d groups (Figure 1K; $p < 0.05$).

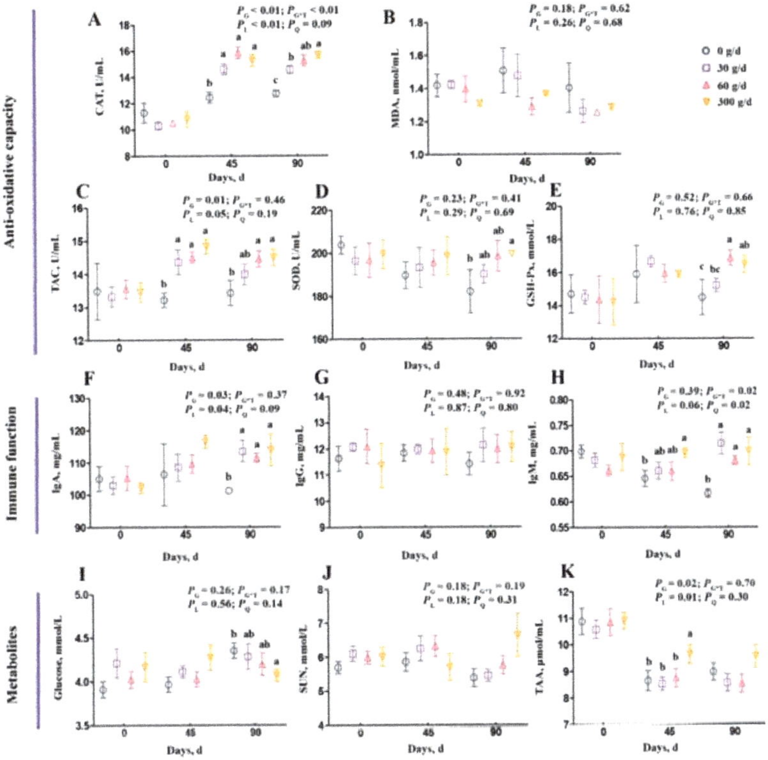

Figure 1. Effects of different amounts of *Acremonium terricola* culture on serum variables in lactating dairy cows. The serum variables were divided into three groups, including antioxidant capacity (**A–E**), immune function (**F–H**), and metabolites (**I–K**). Serum variables of dairy cows fed a diet with supplementation of ATC at 0, 30, 60, or 300 g/d were used. The effects included group effect and the interaction effect of group and time, as well as linear and quadratic effects. Different lowercase letters denote significant differences among treatments ($p < 0.05$). ATC, *Acremonium terricola* culture; CAT, catalase; MDA, malonaldehyde; TAC, total oxidative capacity; GSH–Px, glutathione peroxidase; SOD, superoxide dismutase; IgA, immunoglobulin A; IgG, immunoglobulin G; IgM, immunoglobulin M; SUN, serum urea nitrogen; TAA, total amino acids. Mean ± SEM. n = 15.

3.4. Rumen Fermentation Parameters

From the data in Figure 2, we observed a linear decrease in the rumen pH, and the pH of the 0 g/d group was higher than in other groups on day 90 (Figure 2A; $p < 0.05$). The supplementation did not affect ruminal NH_3-N, propionate, butyrate, or valerate concentrations (Figure 2B,E–G; $p > 0.05$). Treatment significantly affected the molar proportion of propionate with no linear or quadratic effects (Figure 2J), and the molar proportion of butyrate decreased after 60 g/d and 300 g/d supplementation at day 90 (Figure 2K). The total volatile fatty acid (TVFA), acetate, and branched volatile fatty acid (BVFA) concentrations increased linearly with ATC supplementation (Figure 2C,D,H; $p < 0.05$), but the molar proportion of acetate only significantly increased after 60 g/d and 300 g/d supplementation at day 90 compared with that of the 0 g/d group (Figure 2I; $p < 0.05$). The TVFA and acetate concentrations were higher in cows fed with 60 g/d and 300 g/d of ATC at day 45, and were higher in the 30, 60, and 300 g/d groups on day 90 than that of those fed with 0 g/d of ATC (Figure 2C,D; $p < 0.05$).

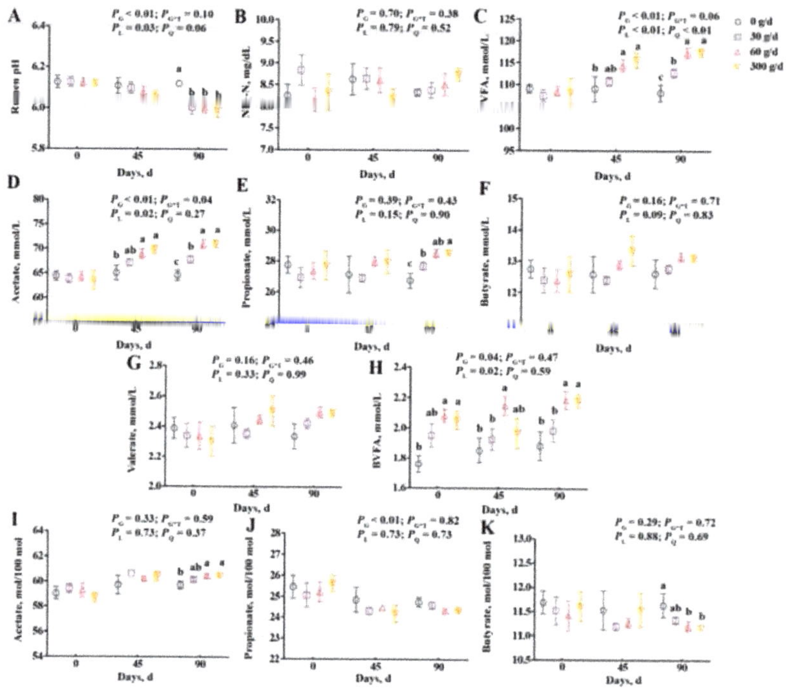

Figure 2. Effects of different amounts of *Acremonium terricola* culture on rumen pH (**A**), NH_3-N (**B**), and volatile fatty acids (**C–K**) in lactating dairy cows. The VFA are expressed as concentration (**C–H**) and molar proportion (**I–K**). The effects included group effect and the interaction effect of group and time, as well as linear and quadratic effects. Different lowercase letters denote significant differences among treatments ($p < 0.05$). ATC, *Acremonium terricola* culture; TVFA, total volatile fatty acid; BVFA, branched volatile fatty acid; Mean ± SEM. $n = 15$.

3.5. Diversity of Rumen Microbiota

Next, we assessed the diversity of rumen microbiota. The range for Good's coverage was 99.05–99.25%, indicating a good sequencing depth (Supplementary Table S3). The 16S rRNA sequencing showed a total of 1583 OTUs across all samples, with 97% similarity (Figure 3A–F). Rarefaction curves and Shannon curves showed a decreased number of new OTUs and Shannon index as the sequencing number increased (Figure 3G,H). A total of 1435, 1407, 1399, and 1198 OTUs were shared between day 0 and day 90 in the 0, 30, 60, and 300 g/d groups, respectively (Figure 3A–D). A total of 1365 and 1079 OTUs were shared among the four groups on day 0 and day 90, respectively. The alpha diversity analysis shows that there were no significant differences among these groups at day 0 (Figure 3I–M). At day 90, 300 g/d ATC supplementation significantly decreased Sobs, Chao1, ACE, and Shannon indices, and significantly increased the Simpson index when compared with those of other groups (Figure 3I–M; $p < 0.05$). There were significant differences between the 60 g/d group and the 0 g/d group with regards to alpha diversity indices, except for the Simpson index (Figure 3I–M; $p < 0.05$). The 30 g/d ATC supplementation only significantly decreased the Sobs index when compared with that of the 0 g/d group (Figure 3I; $p < 0.05$).

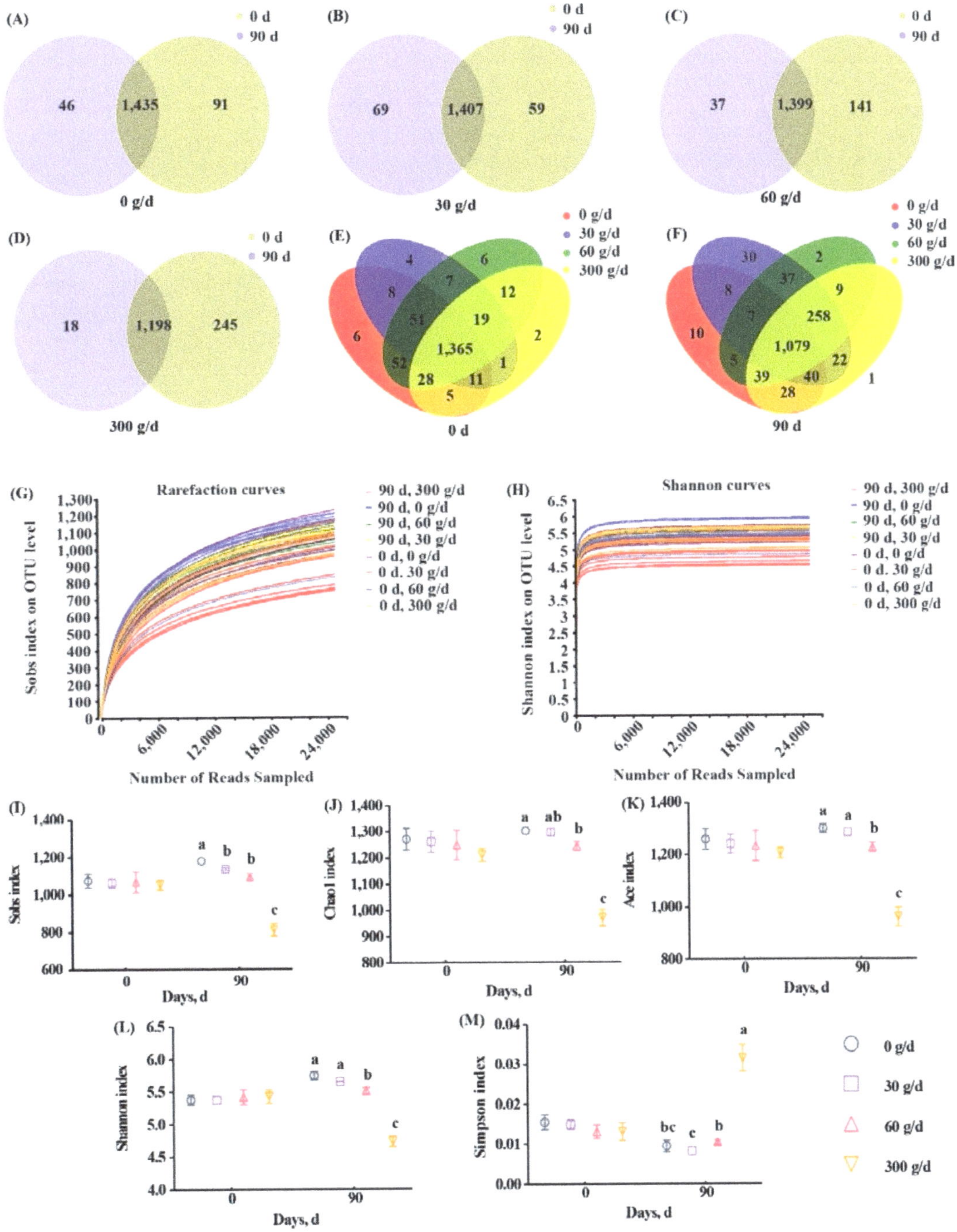

Figure 3. OTU number and alpha diversity responses of the ruminal microbiota to *Acremonium terricola* culture supplementation in lactating dairy cows. (**A–F**) Venn diagrams showing the number of

OTUs in the 0 g/d (**A**), 30 g/d (**B**), 60 g/d (**C**), and 300 g/d (**D**) groups at day 0 (**E**) and day 90 (**F**); (**G**) Shannon curves; (**H**) rarefaction curves; (**I–M**) alpha diversity. Different lowercase letters denote significant differences among treatments (Figure 3I–M, $p < 0.05$). OTU, operational taxonomic unit. Mean ± SEM. $n = 6$.

The PCoA was conducted based on the OTU level with ANOSIM analysis to test the statistical differences among the groups (Figure 4). No clear separation was found for rumen bacteria at day 0 ($p > 0.05$). The results showed distinct clustering according to the amount of ATC at day 90 ($p < 0.05$).

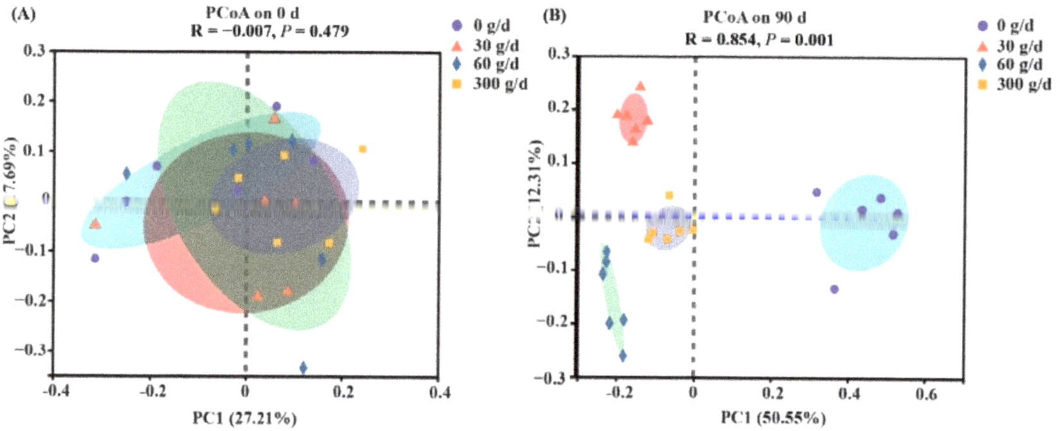

Figure 4. Bate diversity responses of the rumen microbiota to *Acremonium terricola* culture supplementation in lactating dairy cows on day 0 (**A**) and day 90 (**B**). Rumen microbiota of dairy cows fed a diet with supplementation of ATC at 0, 30, 60, or 300 g/d were used. Analysis of similarities (ANOSIM) using Bray–Curtis distance matrices was used to test the statistical differences. $n = 6$.

3.6. Microbial Profiles of Rumen Microbiota

Among the phyla detected in the rumen (Figure 5A), Firmicutes (52.65 ± 1.17), Bacteroidota (41.39 ± 1.24), Actinobacteriota (2.41 ± 0.40), Proteobacteria (1.20 ± 0.20), and Patescibacteria (1.11 ± 0.10) were predominant, followed by Spirochaetota (0.57 ± 0.05). The predominant genera (Figure 4B) were *Prevotella* (25.91 ± 1.94), *NK4A214_group* (6.18 ± 0.58), *Succiniclasticum* (6.04 ± 0.84), *Lachnospiraceae_NK3A20_group* (5.59 ± 0.52), *Ruminococcus* (3.83 ± 0.39), *Erysipelotrichaceae_UCG-002* (2.69 ± 0.95), *Christensenellaceae_R-7_group* (2.65 ± 0.28), *Acetitomaculum* (2.58 ± 0.32), *Rikenellaceae_RC9_gut_group* (2.44 ± 0.27), *Syntrophococcus* (1.92 ± 0.61), *Shuttleworthia* (1.73 ± 0.47), *Ruminococcus_gauvreauii_group* (1.28 ± 0.19), *Prevotellaceae_UCG-001* (1.26 ± 0.08), *Olsenella* (1.25 ± 0.37), and *Prevotellaceae_UCG-003* (1.02 ± 0.10). Notably, 14.87% of reads recovered from the rumen could not be confidently assigned at the genus level.

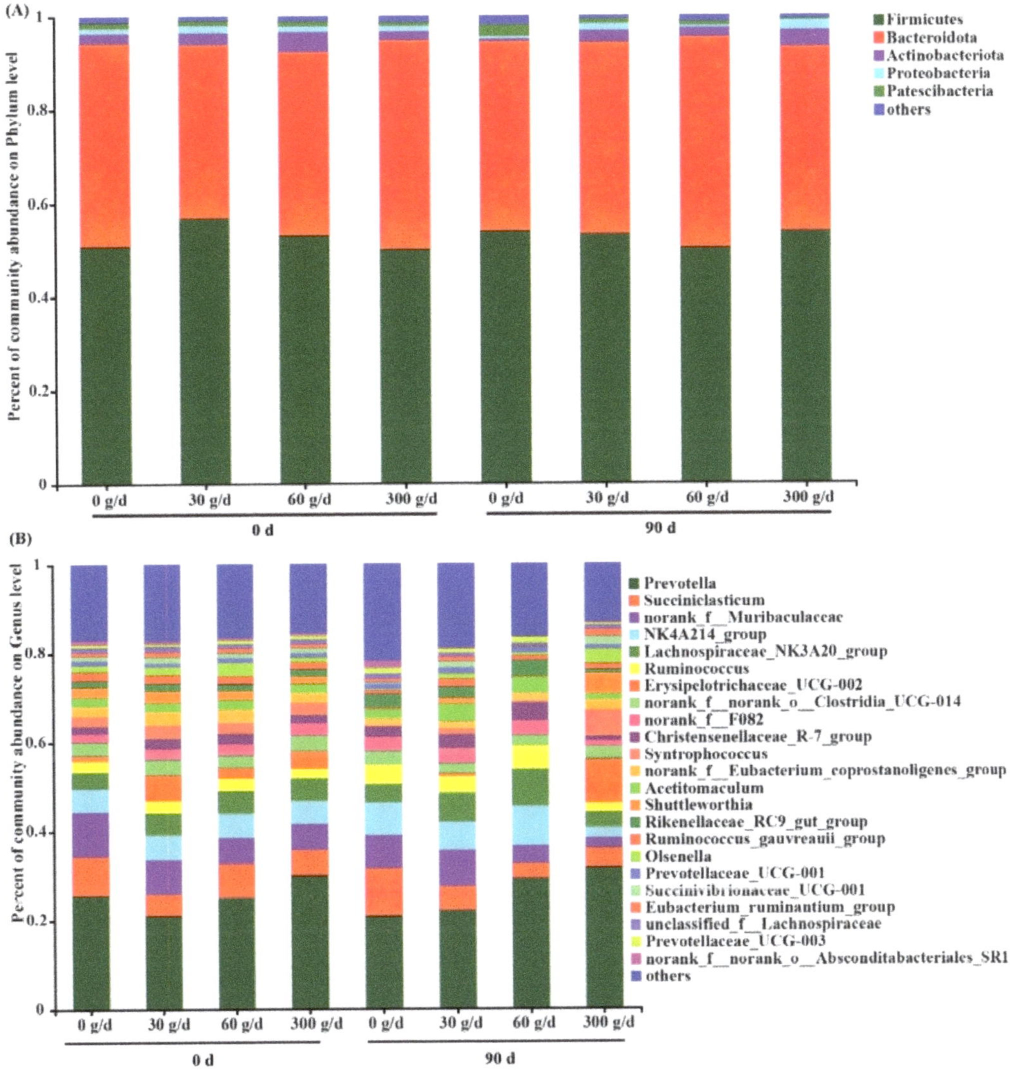

Figure 5. Composition of the rumen microbiota at the phylum level (**A**) and genus level (**B**) before (day 0) and after (day 90) of *Acremonium terricola* culture supplementation in lactating dairy cows. The relative abundance of taxa $\leq$ 0.01% belonged to others. n = 6.

3.7. Changes in the Rumen Bacterial Composition with Acremonium terricola Culture Supplementation

To explore the interactions within genera, we used network analysis based on strong and significant correlations of core genera between the 0 g/d group and ATC-treatment groups. The network, including both 0 g/d and 30 g/d groups (Figure 6A), consisted of eight nodes (core genera) and seven edges (relations). *Acetitomaculum* and *Lachnospiraceae_NK3A20_group* had positive correlations, and *Syntrophococcus* and *Ruminococcus_gauvreauii_group* also had positive correlations (Figure 6A). Similar to the network mentioned above, the network analysis from 0 g/d and 60 g/d showed that *Lachnospiraceae_NK3A20_group* had a positive correlation with *Acetitomaculum* and *Christensenellaceae_R–7_group* (Figure 6B).

This network consisted of 15 nodes and 19 edges, whereas 11 of the 15 nodes belonged to Firmicutes (Figure 6B). The last network of 0 g/d and 300 g/d showed 17 nodes and 33 edges. *Acetitomaculum* and *Christensenellaceae_R–7_group* were also found in the networks (Figure 6C). Furthermore, 10 of the 17 nodes belonged to Firmicutes (Figure 6C). The genera *Syntrophococcus*, *Shuttleworthia*, *Erysipelotrichaceae_UCG–003*, and *Olsenella* were part of the network (Figure 6C).

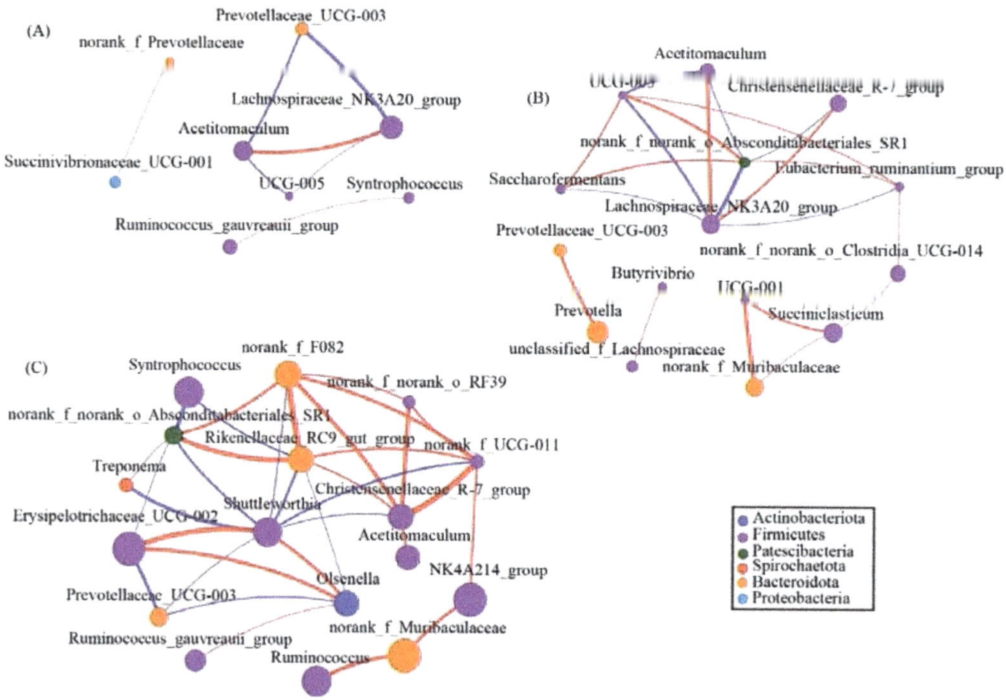

Figure 6. The network of co-occurring genera within the rumen microbiota supplied with different amounts of *Acremonium terricola* culture: (**A**) 0 and 30 g/d; (**B**) 0 and 60 g/d; (**C**) 0 and 300 g/d. The figure shows genera based on Spearman's correlation. The nodes represent the core genera, and the size of each node is proportional to the degree. The edges stand for strong (Spearman's correlation coefficient $r > 0.7$ or $r < -0.7$) and significant ($p < 0.05$) correlations between core genera. The nodes are colored based on phylum. Red and blue lines represent positive and negative correlations between two nodes, respectively.

We then conducted the Kruskal–Wallis H test to filter significantly different core genera (relative abundance $\geq$ 1%), as shown in Figure 7. The results showed that the relative abundance of *Lachnospiraceae_NK3A20_group* and *Christensenellaceae_R–7_group* increased after 60 g/d supplementation compared to that in the 0 g/d group ($p > 0.05$). The cows fed 300 g/d ATC had a lower relative abundance of *Lachnospiraceae_NK3A20_group*, *Christensenellaceae_R–7_group*, *Rilkenellaceae_RC9_gut_group*, and *Shuttleworthia* than that of cows fed 30 g/d and 60 g/d ($p < 0.05$), while there was a higher relative abundance of *Erysipelotrichaceae_UCG_002*, *Acetitomaculum*, and *Olsenella* than that of cows fed 0, 30, and 60 g/d ($p > 0.05$). The relative abundance of *Syntrophococcus* was also significantly increased in the 300 g/d group compared to that in the 0 g/d group ($p > 0.05$).

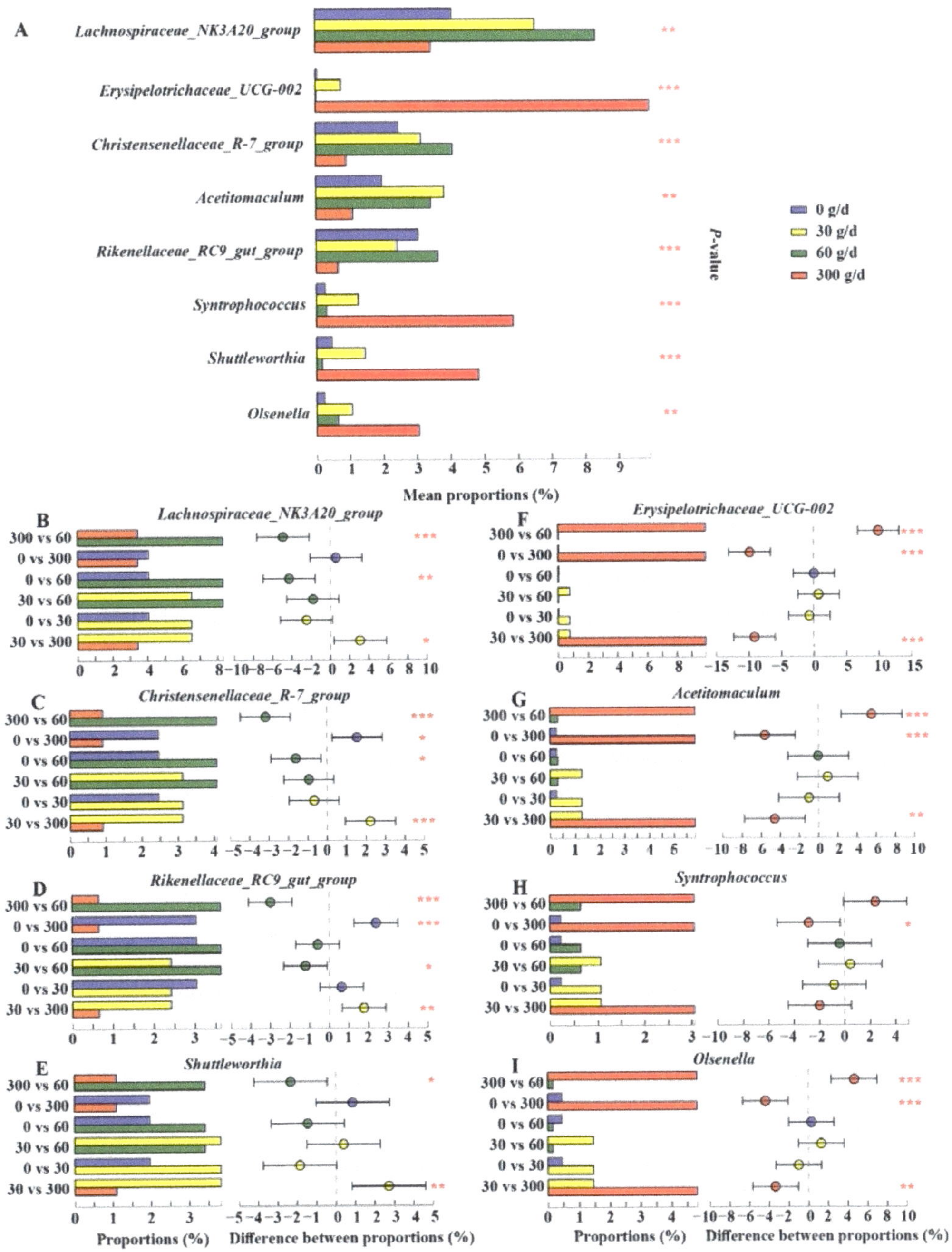

Figure 7. Effects of different levels of *Acremonium terricola* culture supplementation on the relative abundance of ruminal core genera (≥1%) in lactating dairy cows. The figure only shows the significantly different genera. The Kruskal–Wallis H test was used. *, $0.01 < p \leq 0.05$; **, $0.001 < p \leq 0.01$; ***, $p \leq 0.001$. $n = 6$.

4. Discussion

ATC is a new type of new feed additive with functional components similar to those of *Cordyceps*. The positive effects of ATC on livestock production and health have been reported in rats [16], calves [17], and dairy cows [19]. However, their pharmacological and biochemical actions, particularly in applications in ruminants, have not been clearly elucidated. In our study, DMI was not affected, regardless of the ATC dosage. Individual DMI is critical, because insufficient intake represents limited milk synthesis. The unaltered DMI agreed with the results of previous studies showing that ATC or *Cordyceps* spp. supplementation improved body weight and feed efficiency rather than DMI [17,19,46,47]. However, we noted differences in feeding behavior. Dairy cows treated with ATC had a shorter feeding duration. Although lying time and rumination time were not measured in this study, this result implied that the dairy cows might have a longer time to lie and ruminate after feeding. Increased rumination time and lying time have long been associated with increased milk performance, with the general belief that rumination increases the surface area of feed particles, making them more accessible to microbes [48], and increased lying time is associated with decreased lameness [49]. This is in agreement with the higher milk performance observed in this study.

Consistent with the report of Li et al. [32], we observed that supplemental ATC improved milk yield linearly without influencing DMI, resulting in improved milk efficiency. Although maximum yields of milk and milk efficiency were obtained at 300 g/d ATC, increasing the dosage from 60 g/d to 300 g/d had a limited positive effect on milk performance. Many factors may have contributed to this response. Lactose yield determines the amount of absorbed water in the alveoli, and thus, the volume of the produced milk [50]. Approximately 20% of the circulating blood glucose in dairy cows is converted into lactose [50]. Thus, it is not surprising that ATC supplementation increased milk lactose yield. Moreover, Costa et al. [51] demonstrated that SCC was negatively correlated with milk yield and milk lactose percentage. SCC is a well-established indicator of mammary gland inflammation, which is highly correlated with the presence of a mammary infection. The dairy cows used in our study had less than 150,000 cells/mL SCC among all treatments, which was considered healthy, and had low clinical mastitis incidence [52]. The SCC in milk decreased quadratically in cows fed different levels of supplemental ATC, with the highest decrease in cows fed ATC at 300 g/d. Many studies have demonstrated that cordycepin and galactomannan in *Cordyceps* or ATC functioned as antioxidants. Li et al. [16] indicated that ATC played anti-inflammatory and antioxidant roles by inhibiting the mitogen-activated protein kinase signaling pathway, and ATC increased the TAC and GSH-Px concentrations in dairy cows when fed up to 30 g/d [32]. Waewaree et al. [53] also indicated that weaned pigs fed a diet supplemented with the spent mushroom compost of *Cordyceps militaris* displayed greater immunoglobulin secretion, lower inflammation, and a lower pathogenic population. These data indicated that ATC had positive effects on immune and antioxidant functions, which was consistent with the serum variables and SCC changes in our study. We speculated that a possible explanation for our results was that ATC improved the immune function and antioxidant capacity of cows, thereby decreasing SCC and improving milk yield. Moreover, when cows were fed with 60 g/d and 300 g/d of ATC, ECM and FCM yields were greater at 60–90 days than those in 0 g/d and 30 g/d, which might have been due to the adaptation periods of the cows to the ATC.

According to the CP concentration in ATC in our study, 30, 60, and 300 g/d ATC supplementation supplied a total of 7.36, 14.72, and 73.59 g/d CP for cows, respectively. Barros et al. [54] found that a linear response to increasing dietary CP concentration increased MUN concentration and milk protein yield, which was inconsistent with the unchanged MUN and SUN concentrations in our study. This discrepancy is potentially explained by the benefits of balancing more adequate amounts of AA in the diet for N efficiency [55]. SUN and MUN concentrations can be used as indicators of microbial protein synthesis and N efficiency. Some N-NH_3 in the rumen, which is not used for microbial protein synthesis and absorbed from the rumen, would be used to produce urea in the liver, which consumes energy and reduces N utilization [56]. Thus, the unchanged N-NH_3 concentration in the rumen indicated that extra N intake from ATC supplementation contributed to microbial protein synthesis, rather than N-NH_3 production. Many studies have demonstrated that the supply of intestinally available limiting AA is pertinent to milk yield and milk protein yield by decreasing protein mobilization and increasing accretion [57,58]. According to the AA profile of ATC, TAA accounted for 17.27% of ATC (DM basis). Hence, under the current experimental conditions, ATC supplementation improved milk protein yield in a quadratic manner, possibly by reducing urea synthesis to save energy for protein synthesis and improve the balance of available AA in the mammary gland of lactating cows. Further studies on ruminal digestibility and intestinal availability of ingredients (cordycepin, D-mannitol, galactomannan, AA) of ATC will be useful in formulating an accurate ration.

Chen et al. [59] reviewed the toxicity of known metabolites identified from *Cordyceps* fungi and concluded that different compounds showed antimicrobial activity, and may have the potential to threaten the safety of the poultry industry. However, Ramos et al. [60] revealed that weaned pigs fed 1.5 g/kg of *Cordyceps militaris* spent mushroom compost showed reduced populations of pathogenic Gram-negative *E. coli* and increased populations of beneficial *Lactobacillus* spp. These different observations were partially attributed to the different species (over 1000) of *Cordyceps* fungi [59]. Hence, we evaluated the effects of ATC on the microbiota responses of dairy cows. No differences in bacterial diversity on day 0, and decreases in bacterial richness and evenness at day 90 after ATC treatment, were identified in this study, indicating that ATC supplementation depressed the microbiota. Furthermore, the common OTUs between day 0 and day 90 among these groups, as well as alpha diversity on day 90, changed with increasing levels of ATC supplementation, suggesting that ATC depressed the microbiota in a dose-dependent manner in dairy cows fed with increasing levels of ATC.

The pH and VFA concentrations are the indicators of rumen function and the ruminal environment. It has been reported that ATC increased CP digestibility of three types of roughage [19], and dairy cows fed ATC had an increased abundance of cellulolytic, proteolytic, and amylolytic bacteria in the rumen, which resulted in enhanced ruminal fermentation with increased VFA production and a decreased pH [32]. Unlike RT–PCR detection of selected bacterial species in a previous study [32], the changes in rumen microbiota in our study were analyzed using 16S gene sequencing. Although the microbiota was depressed, 30, 60, and 300 g/d ATC supplementation resulted in decreased pH and increased TVFA concentration. Reductions in the richness and diversity of microbiota in the rumen have been associated with higher production efficiency in dairy cattle [61]. According to the network of co-occurring genera and the Kruskal–Wallis H test, certain genera showed a strong and significant interaction with each other, and may contribute to increased milk performance. As the most abundant genera in the Firmicutes, *Lachnospiraceae_NK3A20_group* was characterized by cellulose-decomposing activity and starch hydrolysis, which could produce acetate and formate, among others [62]. The *Christensenellaceae_R-7_group*, belonging to the family Christensenellaceae, produces acetate and butyrate, with the ability to utilize arabinose, glucose, and mannose [63]. Galactomannan in ATC may support the growth of *Christensenellaceae_R-7_group*. *Acetitomaculum* is also an acetogenic genus that utilizes formate, glucose, and hydrogen to reduce methane production [64]. Finally, *Olsenella* ferments starch and glycogen substrates and produces lactic, acetic, and formic acids [65]. These findings corroborated our results, in which TVFA and acetate concentrations and molar proportions of acetate were increased. In truth, different taxa are associated with feed efficiency and methane emissions. Ramayo-Caldas et al. [66] identified a higher abundance of the Christensenellaceae and Lachnospiraceae families associated with methane production in dairy cows, and McLoughlin et al. [67] found that the genera *Olsenella* and *Acetitomaculum* exhibited negative associations with the feed conversion ratio and average daily gain, respectively. Our study was limited by the absence of methane emissions analysis, and further studies could provide more reliable estimates of the microbiota contribution to energy flow. Overall, although 30 g/d and

60 g/d ATC supplementation were linked to lower richness of microbiota, a simpler bacterial composition could result in increased dominance of specific functional components, which led to higher concentrations of products that were relevant to the dairy cows.

The extent of variability in OTU richness and diversity between groups was mostly associated with significant alterations in the microbiota [68,69]. The greatest changes in microbiota were obtained from the 0 g/d group to 300 g/d group, rather than the 30 g/d or 60 g/d groups. As described above, the abundance of *Lachnospiraceae_NK3A20_group*, *Christensenellaceae_R–7_group*, and *Acetitomaculum* was increased in the 30 g/d and 60 g/d groups, but decreased in the 300 g/d group, compared with that in the 0 g/d group, followed by that of *Rilkenellaceae_RC9_gut_group*. *Rilkenellaceae_RC9_gut_group* belongs to the family Rilkenellaceae, which are specialized for the tract environment, with anaerobic metabolisms, and the metabolic end-products (acetate, propionate, and succinic acid) are produced from glucose, lactose, and mannose [70]. Conversely, the decrease in the relative abundance of these genera may be attributed to high doses of cordycepin. A previous study found that just 30 g/d of ATC supplementation decreased the relative abundance of Fibrobacter succinogenes in the rumen of dairy cows, which are highly specialized in cellulose degradation [32]. This was similar to the findings of Huang et al. [71], who also showed cordycepin as an efficient killer of Mycobacterium tuberculosis and Mycobacterium bovis. Cordycepin showed antibacterial activity in these studies. Additionally, the interaction between protozoa and bacteria (influenced by ATC) also partly explained the depressed genera. The interaction of ciliates with other microbial groups in the rumen has been previously reported, documenting predation on other rumen microbes, such as bacteria, fungi, and protozoa as well [72]. Li et al. [32] reported that ATC supplementation improved the relative abundance of ciliate protozoa. Hence, we speculated that ATC reshaped the bacterial structure in the rumen of dairy cows by regulating ciliate protozoa. Moreover, we noted that inoculation of defaunated sheep with protozoa further promoted an increase in the abundance of *Syntrophococcus* [73], which agreed with our results.

The relative abundance of the three genera showed significant enrichment in the 300 g/d group; namely, *Erysipelotrichaceae_UCG_002*, *Syntrophococcus*, and *Shuttleworthia*, rather than in the 30 g/d or 60 g/d groups. *Erysipelotrichaceae_UCG_002* belongs to the Erysipelotrichaceae family and has been previously reported to be associated with VFA synthesis and energy generation. Hao et al. [74] indicated that *Erysipelotrichaceae_UCG_002*, *Syntrophococcus*, and *Shuttleworthia* were negatively correlated with rumen pH, and speculated that these genera were involved in fiber digestion. Acetate is the end product of fiber decomposition. Our results indicated that 300 g/d ATC supplementation changed the bacterial composition and enhanced fiber digestion differently compared with the 30 g/d or 60 g/d groups, and resulted in the same improvement of VFA production and modification of pH.

5. Conclusions

Feeding different levels of ATC to lactating dairy cows improved milk yield without affecting DMI, thus increasing milk protein yield. The improvement in milk yield was likely related to improved immune function and antioxidant capacity, which led to decreased SCC, or possibly was due to the improvement in rumen fermentation with a simpler bacterial composition, which favored more effective digestion to produce VFA. Under the current experimental conditions, the optimal dose of ATC supplementation was approximately 60 g/d. The 300 g/d high-dose ATC reshaped the microbiota differently without effects on milk performance when compared with 60 g/d ATC, suggesting that the microbiota responded differently to the individual active components in ATC at increased concentrations. Hence, experiments to test the effects of purified active components (e.g., cordycepin, *D*-mannitol, galactomannan) on microbiota are needed to provide more information on ATC as a feed additive.

Supplementary Materials: The following are available online at https://www.mdpi.com/article/10.3390/antiox11010175/s1, Table S1: Functional composition of *Acremonium terricola* culture; Table S2: Ingredients and nutrient composition of the experiment diet; Table S3: Sample information and sequencing statistics.

Author Contributions: Conceptualization, F.K.; methodology, S.W.; software, F.K. and Y.Z.; validation, Y.Z.; resources, Z.C., Y.L. and Z.Z.; writing—original draft preparation, F.K. and Y.Z.; writing—review and editing, F.K. and Y.Z.; visualization, F.K.; supervision, W.W., N.L. and S.L.; project administration, N.L. and S.L.; funding acquisition, N.L. and S.L. All authors have read and agreed to the published version of the manuscript.

Funding: This research was funded by the China Agriculture Research System of Ministry of Finance and Ministry of Agriculture and Rural Affairs (grant number: CARS36).

Institutional Review Board Statement: This study was approved by the experimental animal welfare and animal ethics committee of China Agricultural University (CAU2021009-2, approval on 15 May 2021).

Informed Consent Statement: Not applicable.

Data Availability Statement: The data presented in this study are available in this manuscript. The raw reads were deposited in the NCBI Sequence Read Archive (SRA) database (Accession Numbers: SRR16916794–SRR16916841).

Conflicts of Interest: The companies at which Z.C., Y.L., Z.Z., and N.L. are employed had no role in the design of the study; in the collection, analyses, or interpretation of data; in the writing of the manuscript; or in the decision to publish the results.

References

1. Fiore, E.; Arfuso, F.; Gianesella, M.; Vecchio, D.; Morgante, M.; Mazzotta, E.; Badon, T.; Rossi, P.; Bedin, S.; Piccione, G. Metabolic and hormonal adaptation in Bubalus bubalis around calving and early lactation. *PLoS ONE* **2018**, *13*, e0193803. [CrossRef]
2. Avondo, M.; Pagano, R.I.; Guastella, A.M.; Criscione, A.; Di Gloria, M.; Valenti, B.; Piccione, G.; Pennisi, P. Diet selection and milk production and composition in Girgentana goats with different alpha s1-casein genotype. *J. Dairy Res.* **2009**, *76*, 202–209. [CrossRef]
3. Armato, L.; Gianesella, M.; Morgante, M.; Fiore, E.; Rizzo, M.; Giudice, E.; Piccione, G. Rumen volatile fatty acids× dietary supplementation with live yeast and yeast cell wall in feedlot beef cattle. *Acta Agric. Scand. A—Anim. Sci.* **2016**, *66*, 119–124. [CrossRef]
4. Monteverde, V.; Congiu, F.; Vazzana, I.; Dara, S.; Di Pietro, S.; Piccione, G. Serum lipid profile modification related to polyunsaturated fatty acid supplementation in thoroughbred horses. *J. Appl. Anim. Res.* **2017**, *45*, 615–618. [CrossRef]
5. Đuričić, D.; Ljubić, B.B.; Vince, S.; Turk, R.; Valpotić, H.; Žaja, I.Ž.; Maćešić, N.; Benić, M.; Getz, I.; Samardžija, M. Effects of dietary clinoptilolite supplementation on β-hydroxybutirate serum level and milk fat to protein ratio during early lactation in holstein-friesian cows. *Microporous Mesoporous Mater.* **2020**, *292*, 109766. [CrossRef]
6. Folnožić, I.; Samardžija, M.; Đuričić, D.; Vince, S.; Perkov, S.; Jelušić, S.; Valpotić, H.; Ljubić, B.B.; Lojkić, M.; Gračner, D. Effects of in-feed clinoptilolite treatment on serum metabolic and antioxidative biomarkers and acute phase response in dairy cows during pregnancy and early lactation. *Res. Vet. Sci.* **2019**, *127*, 57–64. [CrossRef] [PubMed]
7. Folnožić, I.; Đuričić, D.; Žaja, I.Ž.; Vince, S.; Perkov, S.; Turk, R.; Valpotić, H.; Gračner, D.; Maćešić, N.; Lojkić, M. The influence of dietary clinoptilolite on blood serum mineral profile in dairy cows. *Vet. Arh.* **2019**, *89*, 447–462. [CrossRef]
8. Đuričić, D.; Sukalić, T.; Marković, F.; Kočila, P.; Žura Žaja, I.; Menčik, S.; Dobranić, T.; Benić, M.; Samardžija, M. Effects of dietary vibroactivated clinoptilolite supplementation on the intramammary microbiological findings in dairy cows. *Animals* **2020**, *10*, 202. [CrossRef] [PubMed]
9. Ashraf, S.A.; Elkhalifa, A.E.O.; Siddiqui, A.J.; Patel, M.; Awadelkareem, A.M.; Snoussi, M.; Ashraf, M.S.; Adnan, M.; Hadi, S. Cordycepin for Health and Wellbeing: A Potent Bioactive Metabolite of an Entomopathogenic Cordyceps Medicinal Fungus and Its Nutraceutical and Therapeutic Potential. *Molecules* **2020**, *25*, 2735. [CrossRef] [PubMed]
10. Chuang, W.Y.; Hsieh, Y.C.; Lee, T.-T. The Effects of Fungal Feed Additives in Animals: A Review. *Animals* **2020**, *10*, 805. [CrossRef]
11. Das, G.; Shin, H.-S.; Leyva-Gómez, G.; Prado-Audelo, M.L.D.; Cortes, H.; Singh, Y.D.; Panda, M.K.; Mishra, A.P.; Nigam, M.; Saklani, S.; et al. *Cordyceps* spp.: A Review on Its Immune-Stimulatory and Other Biological Potentials. *Front. Pharmacol.* **2021**, *11*, 602364. [CrossRef] [PubMed]
12. Li, R.; Tang, N.; Jia, X.; Xu, Y.; Cheng, Y. Antidiabetic activity of galactomannan from Chinese Sesbania cannabina and its correlation of regulating intestinal microbiota. *J. Funct. Foods* **2021**, *83*, 104530. [CrossRef]
13. Cho, H.J.; Cho, J.Y.; Rhee, M.H.; Park, H.J. Cordycepin (3′-deoxyadenosine) inhibits human platelet aggregation in a cyclic AMP- and cyclic GMP-dependent manner. *Eur. J. Pharmacol.* **2007**, *558*, 43–51. [CrossRef]

14. Kang, N.; Lee, H.-H.; Park, I.; Seo, Y.-S. Development of High Cordycepin-Producing Cordyceps militaris Strains. *Mycobiology* **2017**, *45*, 31–38. [CrossRef]
15. Raethong, N.; Wang, H.; Nielsen, J.; Vongsangnak, W. Optimizing cultivation of Cordyceps militaris for fast growth and cordycepin overproduction using rational design of synthetic media. *Comput. Struct. Biotechnol. J.* **2019**, *18*, 1–8. [CrossRef]
16. Li, Y.; Jiang, X.; Xu, H.; Lv, J.; Zhang, G.; Dou, X.; Zhang, Y.; Li, X. Acremonium terricola culture plays anti-inflammatory and antioxidant roles by modulating MAPK signaling pathways in rats with lipopolysaccharide-induced mastitis. *Food Nutr. Res.* **2020**, *64*. [CrossRef]
17. Li, Y.; Wang, Y.-Z.; Ding, X.; Zhang, Y.-g.; Xue, S.-c.; Lin, C.; Xu, W.-b.; Dou, X.-J.; Zhang, L.-Y. Effects of Acremonium terricola culture on growth performance, antioxidant status and immune functions in weaned calves. *Livest Sci.* **2016**, *193*, 66–70. [CrossRef]
18. Li, Y.; Sun, Y.-K.; Li, X.; Zhang, G.-N.; Xin, H.-S.; Xu, H.-J.; Zhang, L.-Y.; Li, X.-X.; Zhang, Y.-G. Effects of Acremonium terricola culture on performance, milk composition, rumen fermentation and immune functions in dairy cows. *Anim. Feed Sci. Technol.* **2018**, *240*, 40–51. [CrossRef]
19. Li, Y.; Wang, Y.-Z.; Zhang, G.-N.; Zhang, X.-Y.; Lin, C.; Li, X.-X.; Zhang, Y.-G. Effects of Acremonium terricola culture supplementation on apparent digestibility, rumen fermentation, and blood parameters in dairy cows. *Anim. Feed Sci. Technol.* **2017**, *230*, 13–22. [CrossRef]
20. Bradford, B.J.; Yuan, K.; Farney, J.K.; Mamedova, L.K.; Carpenter, A.J. Invited review: Inflammation during the transition to lactation: New adventures with an old flame. *J. Dairy Sci.* **2015**, *98*, 6631–6650. [CrossRef]
21. Dobos, A.; Fodor, I.; Kreizinger, Z.; Makrai, L.; Dénes, B.; Kiss, I.; Đuričić, D.; Kovačić, M.; Szeredi, L. Infertility in dairy cows–Possible bacterial and viral causes. *Vet. Stanica* **2022**, *53*, 35–43. [CrossRef]
22. Đuričić, D.; Vince, S.; Lojkić, M.; [illegible] Effects of dietary clinoptilolite on reproductive performance, serum progesterone and insulin-like growth factor-1 concentrations in dairy cows during pregnancy and lactation. *Pol. J. Vet. Sci.* **2020**, *23*, 69–75. [CrossRef] [PubMed]
23. Wang, Y.-Z.; Li, Y.; Xu, Q.-B.; Zhang, X.-Y.; Zhang, G.-N.; Lin, C.; Zhang, Y.-G. Effects of Acremonium terricola culture on production performance, antioxidant status, and blood biochemistry in transition dairy cows. *Anim. Feed Sci. Technol.* **2019**, *256*, 114261. [CrossRef]
24. Winarti, S.; Prasetyo, A. The Consumption of Galactomannan Effervescent Drinks made from Coconut Pulp Waste and Colonic Microbiota in Wistar Rats. *Int. J. Probiotics Prebiotics* **2020**, *15*, 52–56. [CrossRef]
25. Majeed, M.; Majeed, S.; Nagabhushanam, K.; Arumugam, S.; Natarajan, S.; Beede, K.; Ali, F. Galactomannan from Trigonella foenum-graecum L. seed: Prebiotic application and its fermentation by the probiotic Bacillus coagulans strain MTCC 5856. *Food Sci. Nutr.* **2018**, *6*, 666–673. [CrossRef] [PubMed]
26. Newbold, C.J.; McIntosh, F.M.; Wallace, R.J. Changes in the microbial population of a rumen-simulating fermenter in response to yeast culture. *Can. J. Anim. Sci.* **1998**, *78*, 241–244. [CrossRef]
27. Malmuthuge, N.; Guan, L.L. Understanding host-microbial interactions in rumen: Searching the best opportunity for microbiota manipulation. *J. Anim. Sci. Biotechnol.* **2017**, *8*, 8. [CrossRef]
28. Tuli, H.S.; Sandhu, S.S.; Sharma, A.K. Pharmacological and therapeutic potential of Cordyceps with special reference to Cordycepin. *3 Biotech* **2014**, *4*, 1–12. [CrossRef]
29. Shashidhar, M.G.; Giridhar, P.; Udaya Sankar, K.; Manohar, B. Bioactive principles from Cordyceps sinensis: A potent food supplement—A review. *J. Funct. Foods* **2013**, *5*, 1013–1030. [CrossRef]
30. Yu, X.; Mao, Y.; Shergis, J.L.; Coyle, M.E.; Wu, L.; Chen, Y.; Zhang, A.L.; Lin, L.; Xue, C.C.; Xu, Y. Effectiveness and Safety of Oral Cordyceps sinensis on Stable COPD of GOLD Stages 2-3: Systematic Review and Meta-Analysis. *Evid. Based Complement. Altern. Med.* **2019**, *2019*, 4903671. [CrossRef] [PubMed]
31. Rodman, L.E.; Farnell, D.R.; Coyne, J.M.; Allan, P.W.; Hill, D.L.; Duncan, K.L.K.; Tomaszewski, J.E.; Smith, A.C.; Page, J.G. Toxicity of Cordycepin in Combination with the Adenosine Deaminase Inhibitor 2′-Deoxycoformycin in Beagle Dogs. *Toxicol. Appl. Pharmacol.* **1997**, *147*, 39–45. [CrossRef] [PubMed]
32. Wang, C.; Zhao, F.; Liu, J.; Liu, H. Dipeptide (Methionyl-Methionine) Transport and Its Effect on β-Casein Synthesis in Bovine Mammary Epithelial Cells. *Cell. Physiol. Biochem.* **2018**, *49*, 479–488. [CrossRef]
33. National Research Council. *Nutrient Requirements of Dairy Cattle*; National Academy Press: Washington, DC, USA, 2001.
34. Association of Official Analytical Chemists. *Official Methods of Analysis*; AOAC International: Gaithersburg, MD, USA, 2000.
35. Van Soest, P.J.; Robertson, J.B.; Lewis, B.A. Methods for dietary fiber, neutral detergent fiber, and nonstarch polysaccharides in relation to animal nutrition. *J. Dairy Sci.* **1991**, *74*, 3583–3597. [CrossRef]
36. Chang, C.Y.; Lue, M.Y.; Pan, T.M. Determination of Adenosine, Cordycepin and Ergosterol Contents in Cultivated Antrodia camphorata by HPLC Method. *J. Food Drug Anal.* **2005**, *13*, 338–342. [CrossRef]
37. Kong, F.; Lu, N.; Liu, Y.; Zhang, S.; Li, S. Aspergillus oryzae and Aspergillus niger Co-Cultivation Extract Affects In Vitro Degradation, Fermentation Characteristics, and Bacterial Composition in a Diet-Specific Manner. *Animals* **2021**, *11*, 1248. [CrossRef] [PubMed]
38. Kong, F.; Gao, Y.; Tang, M.; Fu, T.; Diao, Q.; Bi, Y.; Tu, Y. Effects of dietary rumen–protected Lys levels on rumen fermentation and bacterial community composition in Holstein heifers. *Appl. Microbiol. Biotechnol.* **2020**, *104*, 6623–6634. [CrossRef] [PubMed]
39. Chen, S.; Zhou, Y.; Chen, Y.; Gu, J. fastp: An ultra-fast all-in-one FASTQ preprocessor. *Bioinformatics* **2018**, *34*, i884–i890. [CrossRef]

40. Magoč, T.; Salzberg, S.L. FLASH: Fast length adjustment of short reads to improve genome assemblies. *Bioinformatics* **2011**, *27*, 2957–2963. [CrossRef]
41. Huang, S.; Ji, S.; Wang, F.; Huang, J.; Alugongo, G.M.; Li, S. Correction to: Dynamic changes of the fecal bacterial community in dairy cows during early lactation. *AMB Express* **2021**, *11*, 40. [CrossRef]
42. Edgar, R.C. UPARSE: Highly accurate OTU sequences from microbial amplicon reads. *Nat. Methods* **2013**, *10*, 996–998. [CrossRef]
43. Wang, Q.; Garrity, G.M.; Tiedje, J.M.; Cole, J.R. Naive Bayesian classifier for rapid assignment of rRNA sequences into the new bacterial taxonomy. *Appl. Environ. Microbiol.* **2007**, *73*, 5261–5267. [CrossRef]
44. Pruesse, E.; Quast, C.; Knittel, K.; Fuchs, B.M.; Ludwig, W.; Peplies, J.; Glöckner, F.O. SILVA: A comprehensive online resource for quality checked and aligned ribosomal RNA sequence data compatible with ARB. *Nucleic Acids Res.* **2007**, *35*, 7188–7196. [CrossRef] [PubMed]
45. Caporaso, J.G.; Kuczynski, J.; Stombaugh, J.; Bittinger, K.; Bushman, F.D.; Costello, E.K.; Fierer, N.; Peña, A.G.; Goodrich, J.K.; Gordon, J.I.; et al. QIIME allows analysis of high-throughput community sequencing data. *Nat. Methods* **2010**, *7*, 335–336. [CrossRef] [PubMed]
46. Boontiam, W.; Wachirapakorn, C.; Wattanachai, S. Growth performance and hematological changes in growing pigs treated with Cordyceps militaris spent mushroom substrate. *Vet. World* **2019**, *13*, 768–773. [CrossRef]
47. Chanjula, P.; Cherdthong, A. Effects of spent mushroom Cordyceps militaris supplementation on apparent digestibility, rumen fermentation, and blood metabolite parameters of goats. *J. Anim. Sci.* **2018**, *96*, 1150–1158. [CrossRef]
48. Kaufman, E.I.; Asselstine, V.H.; LeBlanc, S.J.; Duffield, T.F.; DeVries, T.J. Association of rumination time and health status with milk yield and composition in early-lactation dairy cows. *J. Dairy Sci.* **2018**, *101*, 462–471. [CrossRef] [PubMed]
49. Galindo, F.; Broom, D.M. The relationships between social behaviour of dairy cows and the occurrence of lameness in three herds. *Res. Vet. Sci.* **2000**, *69*, 75–79. [CrossRef]
50. Costa, A.; Lopez-Villalobos, N.; Sneddon, N.W.; Shalloo, L.; Franzoi, M.; De Marchi, M.; Penasa, M. Invited review: Milk lactose—Current status and future challenges in dairy cattle. *J. Dairy Sci.* **2019**, *102*, 5883–5898. [CrossRef]
51. Costa, A.; Neglia, G.; Campanile, G.; De Marchi, M. Milk somatic cell count and its relationship with milk yield and quality traits in Italian water buffaloes. *J. Dairy Sci.* **2020**, *103*, 5485–5494. [CrossRef]
52. Rainard, P.; Foucras, G.; Boichard, D.; Rupp, R. Invited review: Low milk somatic cell count and susceptibility to mastitis. *J. Dairy Sci.* **2018**, *101*, 6703–6714. [CrossRef]
53. Boontiam, W.; Wachirapakorn, C.; Phaengphairee, P.; Wattanachai, S. Effect of Spent Mushroom (Cordyceps militaris) on Growth Performance, Immunity, and Intestinal Microflora in Weaning Pigs. *Animals* **2020**, *10*, 2360. [CrossRef] [PubMed]
54. Barros, T.; Quaassdorff, M.A.; Aguerre, M.J.; Colmenero, J.J.O.; Bertics, S.J.; Crump, P.M.; Wattiaux, M.A. Effects of dietary crude protein concentration on late-lactation dairy cow performance and indicators of nitrogen utilization. *J. Dairy Sci.* **2017**, *100*, 5434–5448. [CrossRef]
55. Schwab, C.G.; Broderick, G.A. A 100-Year Review: Protein and amino acid nutrition in dairy cows. *J. Dairy Sci.* **2017**, *100*, 10094–10112. [CrossRef]
56. Moharrery, A. Investigation of different levels of RDP in the rations of lactating cows and their effects on MUN, BUN and urinary N excretion. *Ital. J. Anim. Sci.* **2010**, *3*, 157–165. [CrossRef]
57. Morris, D.L.; Kononoff, P.J. Effects of rumen-protected lysine and histidine on milk production and energy and nitrogen utilization in diets containing hydrolyzed feather meal fed to lactating Jersey cows. *J. Dairy Sci.* **2020**, *103*, 7110–7123. [CrossRef] [PubMed]
58. Fehlberg, L.K.; Guadagnin, A.R.; Thomas, B.L.; Sugimoto, Y.; Shinzato, I.; Cardoso, F.C. Feeding rumen-protected lysine prepartum increases energy-corrected milk and milk component yields in Holstein cows during early lactation. *J. Dairy Sci.* **2020**, *103*, 11386–11400. [CrossRef]
59. Chen, B.; Sun, Y.; Luo, F.; Wang, C. Bioactive Metabolites and Potential Mycotoxins Produced by Cordyceps Fungi: A Review of Safety. *Toxins* **2020**, *12*, 410. [CrossRef] [PubMed]
60. Ramos, S.C.; Jeong, C.D.; Mamuad, L.L.; Kim, S.H.; Kang, S.H.; Kim, E.T.; Cho, Y.I.; Lee, S.S.; Lee, S.S. Diet Transition from High-Forage to High-Concentrate Alters Rumen Bacterial Community Composition, Epithelial Transcriptomes and Ruminal Fermentation Parameters in Dairy Cows. *Animals* **2021**, *11*, 838. [CrossRef]
61. Shabat, S.K.; Sasson, G.; Doron-Faigenboim, A.; Durman, T.; Yaacoby, S.; Berg Miller, M.E.; White, B.A.; Shterzer, N.; Mizrahi, I. Specific microbiome-dependent mechanisms underlie the energy harvest efficiency of ruminants. *ISME J.* **2016**, *10*, 2958–2972. [CrossRef] [PubMed]
62. Russell, J.B.; Rychlik, J.L. Factors that alter rumen microbial ecology. *Science* **2001**, *292*, 1119–1122. [CrossRef]
63. Waters, J.L.; Ley, R.E. The human gut bacteria Christensenellaceae are widespread, heritable, and associated with health. *BMC Biol.* **2019**, *17*, 83. [CrossRef] [PubMed]
64. Greening, R.C.; Leedle, J.A. Enrichment and isolation of Acetitomaculum ruminis, gen. nov., sp. nov.: Acetogenic bacteria from the bovine rumen. *Arch. Microbiol.* **1989**, *151*, 399–406. [CrossRef]
65. Göker, M.; Held, B.; Lucas, S.; Nolan, M.; Yasawong, M.; Glavina Del Rio, T.; Tice, H.; Cheng, J.F.; Bruce, D.; Detter, J.C.; et al. Complete genome sequence of Olsenella uli type strain (VPI D76D-27C). *Stand. Genom. Sci.* **2010**, *3*, 76–84. [CrossRef] [PubMed]
66. Ramayo-Caldas, Y.; Zingaretti, L.; Popova, M.; Estellé, J.; Bernard, A.; Pons, N.; Bellot, P.; Mach, N.; Rau, A.; Roume, H.; et al. Identification of rumen microbial biomarkers linked to methane emission in Holstein dairy cows. *J. Anim. Breed. Genet.* **2020**, *137*, 49–59. [CrossRef] [PubMed]

67. McLoughlin, S.; Spillane, C.; Claffey, N.; Smith, P.E.; O'Rourke, T.; Diskin, M.G.; Waters, S.M. Rumen Microbiome Composition Is Altered in Sheep Divergent in Feed Efficiency. *Front. Microbiol.* **2020**, *11*, 1981. [CrossRef]
68. Casanova-Martí, À.; Serrano, J.; Portune, K.J.; Sanz, Y.; Blay, M.T.; Terra, X.; Ardévol, A.; Pinent, M. Grape seed proanthocyanidins influence gut microbiota and enteroendocrine secretions in female rats. *Food Funct.* **2018**, *9*, 1672–1682. [CrossRef]
69. Dong, Y.; Cheng, H.; Liu, Y.; Xue, M.; Liang, H. Red yeast rice ameliorates high-fat diet-induced atherosclerosis in Apoe$^{-/-}$ mice in association with improved inflammation and altered gut microbiota composition. *Food Funct.* **2019**, *10*, 3880–3889. [CrossRef]
70. Graf, J. The Family Rikenellaceae. In *The Prokaryotes: Other Major Lineages of Bacteria and The Archaea*; Rosenberg, E., DeLong, E.F., Lory, S., Stackebrandt, E., Thompson, F., Eds.; Springer: Berlin/Heidelberg, Germany, 2014; pp. 857–859. [CrossRef]
71. Huang, F.; Li, W.; Xu, H.; Qin, H.; He, Z.G. Cordycepin kills Mycobacterium tuberculosis through hijacking the bacterial adenosine kinase. *PLoS ONE* **2019**, *14*, e0218449. [CrossRef]
72. Williams, A.G.; Coleman, G.S. The rumen protozoa. In *The Rumen Microbial Ecosystem*; Hobson, P.N., Stewart, C.S., Eds.; Springer: Dordrecht, The Netherlands, 1997; pp. 73–139.
73. De la Fuente, G.; Belanche, A.; Girwood, S.E.; Pinloche, E.; Wilkinson, T.; Newbold, C.J. Pros and cons of ion-torrent next generation sequencing versus terminal restriction fragment length polymorphism T-RFLP for studying the rumen bacterial community. *PLoS ONE* **2014**, *9*, e101435. [CrossRef]
74. Hao, Y.; Guo, C.; Gong, Y.; Sun, X.; Wang, W.; Wang, Y.; Yang, H.; Cao, Z.; Li, S. Rumen Fermentation, Digestive Enzyme Activity, and Bacteria Composition between Pre-Weaning and Post-Weaning Dairy Calves. *Animals* **2021**, *11*, 2527. [CrossRef]

antioxidants

MDPI

Article

Effects of Melatonin Supplementation during Pregnancy on Reproductive Performance, Maternal–Placental–Fetal Redox Status, and Placental Mitochondrial Function in a Sow Model

Xie Peng [1,†], Xuelin Cai [1,†], Jian Li [1,†], Yingyan Huang [1], Hao Liu [1], Jiaqi He [1], Zhengfeng Fang [1], Bin Feng [1], Jiayong Tang [1], Yan Lin [1], Xuemei Jiang [1], Liang Hu [2], Shengyu Xu [1], Yong Zhuo [1], Lianqiang Che [1] and De Wu [1,*]

1 Key Laboratory for Animal Disease Resistant Nutrition of the Ministry of Education, Animal Nutrition Institute, Sichuan Agricultural University, Chengdu 611130, China; pengxie@stu.sicau.edu.cn (X.P.); caixuelin@stu.sicau.edu.cn (X.C.); lijian@sicau.edu.cn (J.L.); huangyingyan@stu.sicau.edu.cn (Y.H.); liuhao@stu.sicau.edu.cn (H.L.); hejiaqi@stu.sicau.edu.cn (J.H.); fangzhengfeng@sicau.edu.cn (Z.F.); fengbin@sicau.edu.cn (B.F.); 13670@sicau.edu.cn (J.T.); linyan@sicau.edu.cn (Y.L.); jiangxuemei@sicau.edu.cn (X.J.); shengyuxu@sicau.edu.cn (S.X.); zhuoyong@sicau.edu.cn (Y.Z.); che.lianqiang@sicau.edu.cn (L.C.)

2 College of Food Science, Sichuan Agricultural University, Ya'an 625014, China; huliang@sicau.edu.cn

* Correspondence: wude@sicau.edu.cn

† These authors contributed equally to this work.

Citation: Peng, X.; Cai, X.; Li, J.; Huang, Y.; Liu, H.; He, J.; Fang, Z.; Feng, B.; Tang, J.; Lin, Y.; et al. Effects of Melatonin Supplementation during Pregnancy on Reproductive Performance, Maternal–Placental–Fetal Redox Status, and Placental Mitochondrial Function in a Sow Model. *Antioxidants* **2021**, *10*, 1867. https://doi.org/10.3390/antiox10121867

Academic Editor: Ana B Rodríguez Moratinos

Received: 8 September 2021
Accepted: 18 November 2021
Published: 24 November 2021

Abstract: Melatonin (MT) is a bio-antioxidant that has been widely used to prevent pregnancy complications, such as pre-eclampsia and IUGR during gestation. This experiment evaluated the impacts of dietary MT supplementation during pregnancy on reproductive performance, maternal–placental–fetal redox status, placental inflammatory response, and mitochondrial function, and sought a possible underlying mechanism in the placenta. Sixteen fifth parity sows were divided into two groups and fed each day of the gestation period either a control diet or a diet that was the same but for 36 mg of MT. The results showed that dietary supplementation with MT increased placental weight, while the percentage of piglets born with weight < 900 g decreased. Meanwhile, serum and placental MT levels, maternal–placental–fetal redox status, and placental inflammatory response were increased by MT. In addition, dietary MT markedly increased the mRNA levels of nutrient transporters and antioxidant-related genes involved in the Nrf2/ARE pathway in the placenta. Furthermore, dietary MT significantly increased ATP and NAD^+ levels, relative mtDNA content, and the protein expression of Sirt1 in the placenta. These results suggested that MT supplementation during gestation could improve maternal–placental–fetal redox status and reproductive performance by ameliorating placental antioxidant status, inflammatory response, and mitochondrial dysfunction.

Keywords: melatonin; oxidative stress; placenta; mitochondria; fetal growth; sow

1. Introduction

Rapid fetal growth during pregnancy leads to increased metabolic burdens on pregnant women or dams, causing elevated systemic oxidative stress [1,2]. Accumulating evidence suggests that maternal oxidative stress is associated with the occurrence of adverse pregnancy outcomes, such as preterm birth, preeclampsia, low birth weight, and fetal death [3,4]. The placenta is the only site for the transfer of nutrients to the fetus during gestation; thus, the placenta's health and function are closely associated with the development of a healthy fetus [5]. However, the placenta is extremely sensitive to oxidative stress due to its high metabolic activity and extensive cell division [6]. In the placenta, DNA damage, lipid peroxidation, and protein denaturation caused by reactive oxygen species (ROS) can alter placental function, leading to a reduced capacity for the transfer of oxygen and nutrients to the fetus [7]. Dietary antioxidants can enhance the antioxidant status of gestating mammals, which has been considered to be an effective strategy to prevent adverse pregnancy outcomes [8].

Melatonin (MT) is primarily synthesized and released by the pineal gland, and has antioxidant, anti-apoptotic, and anti-inflammatory effects [9,10]. As a robust antioxidant, MT can directly scavenge ROS and also stimulate antioxidant enzymes. In addition to the pineal gland, the placenta has been considered as the major extrapineal organ of MT synthesis during gestation [11]. MT can easily and quickly pass across the placental barrier and enter fetal circulation, and is considered to be vital for placental functions and fetal growth [12]. Maternal MT supplementation during gestation has arisen as a plausible way to improve reproductive performance in several animal models. A previous study reported that MT could protect mice against lipopolysaccharide-induced intrauterine fetal death and IUGR via its antioxidant and anti-inflammatory properties [13]. Additionally, maternal dietary MT supplementation from mid-to late-gestation has been linked to alterations in utero-placental hemodynamics and amino acid flux, negating the consequences of IUGR in ewes [14,15]. Although MT has been shown to improve fetal growth by increasing utero-placental blood flow and/or its antioxidant and anti-inflammatory effects, its underlying molecular mechanisms in placental growth and function have rarely been investigated.

With the progress of pregnancy, the placenta requires lots of energy to support rapid fetal growth, and mitochondria are critical as the primary sources of cellular energy [16]. Mitochondria are not only the main sites of ROS formation, but also a target of ROS attack, which may lead to changes in their function [17,18]. A previous report has found that maternal oxidative stress is closely related to placental mitochondrial dysfunction [19]. It has been suggested that placental mitochondrial dysfunction can affect subsequent fetal and placental growth [17,19]. Recently, Yang et al. reported that in aged oocytes, MT suppressed ROS production and reduced mitochondrial dysfunction [20]. However, there is limited research about the impacts of MT on placental mitochondrial function in pregnant mammals.

To our knowledge, no data are available currently regarding the effects of dietary MT supplementation during gestation on the reproductive performance and antioxidants status of sows, despite the fact that sows are increasingly used as animal models in biomedical researches on human pregnancy because of their similarity in terms of metabolic, inflammatory, gastrointestinal, and cardiovascular features [21]. Additionally, the placenta is a complex and transient organ that plays an important role in fetal development through its nutrients and hormone exchange functions between mother and fetus [22]. Placental dysfunction in human and sows has been implicated in disorders of maternal health and fetal growth [23,24]. Therefore, in this study, we hypothesized that dietary supplementation with MT in sow diets may improve reproductive performance by ameliorating maternal–placental–fetal redox status and placental mitochondrial dysfunction. The current study was carried out to verify the above hypotheses by evaluating the effects of MT on the reproductive performance, maternal–placental–fetal redox status, placental inflammatory response, and mitochondrial function.

2. Materials and Methods

The study was approved by the animal care and use committee of Sichuan Agricultural University (DKYB20131704).

2.1. Animals and Diet Design

A total of 16 Large White × Landrace fifth parity sows (3 years old) with similar backfat thickness were selected and inseminated with semen from the same Duroc boar. After artificial insemination, sows with their litters were randomly assigned to two treatment groups (n = 8 per group) and provided with a control diet (CON) or the same control diet containing 36 mg of MT (Sangon Biotech, Shangshai, China). Sows had similar starting weights between the two groups (CON: 225 ± 3.95 vs. MT: 216 ± 4.69 kg; p = 0.21). The dosage of MT was selected according to a previous study [25]. Feed was offered once daily at 14:00. Sows were fed 2.23 kg/d from mating until d 90 of gestation and 2.63 kg/d from d 91 of gestation until parturition. Sows were transferred to farrowing crates on d 107 of gestation.

MT was dissolved in absolute ethanol (12 mg/mL). The day prior to feeding, 3 mL of MT solution was absorbed onto 800 g of the CON diet in a plastic bag. The ethanol was allowed to evaporate overnight at room temperature and the individual plastic bags were sealed. The non-MT supplemented diet (800 g) was prepared in the same manner except that MT was not added to the ethanol. After the sows consumed the 800 g of modified feed, the remainder of the CON diet was given. In view of avoiding effects of lighting programs on the sows' physiology and MT secretion (MT secretion is inhibited by light and stimulated by darkness), a lighting schedule of 12 h light and 12 h dark (darkness from 20:00 to 08:00) was used for the whole experiment. The control diet was formulated according to National Research Council (2012) recommendations [26]. The dietary ingredients and nutritional levels are listed in the Supplementary Materials, Table S1. All sows were allowed to drink water ad libitum throughout this study.

The numbers of total piglets born, alive, stillborn, and mummified, were recorded, and their individual weights were obtained at parturition. The number of low BW piglets (piglets born alive with weight < 900 g) was recorded. In addition, 16 new-born piglets (1 piglet per litter with the BW closest to the average BW of the litter) were selected before ingesting colostrum, and the placentas of selected piglets were collected.

2.2. *Blood Sample Collection*

Maternal blood samples (5 mL) were obtained from the ear vein on days 90 and 110 of gestation and on farrowing day. In addition, maternal ear vein blood samples (3 mL) were collected at 14:00 (prior to feeding), 17:00, and 20:00 (lights off) on day 102 of gestation. Each selected piglet was anaesthetized with sodium pentobarbital (30 mg/kg BW), and blood samples (5 mL) were obtained from the jugular vein. All blood samples were centrifuged at 3000× *g* at 4 °C for 15 min to obtain the serum, and immediately frozen at −80 °C for subsequent analyses.

2.3. *Tissue Sample Collection*

Sows were monitored continuously throughout parturition, and each piglet was matched to the corresponding placenta using the umbilical tagging procedure, as previously reported [27]. During sow farrowing, each umbilical cord was tied with a short silk line which was attached to a numbered tag to match the birth order of the piglets, so that when the umbilical retracted into the birth canal, it could easily be identified [19]. After placental expulsion and weight recording, approximately 3 g of placental tissue (4 to 5 cm from the cord insertion point) was collected, then immediately placed in liquid nitrogen and stored at −80 °C for subsequent analyses.

2.4. *Analysis of Oxidative Stress Parameters in Serum and Placenta*

The contents of total antioxidant capacity (T-AOC; catalogue no. A015-2-1) and malondialdehyde (MDA; catalogue no. A003-1-2), and the activities of glutathione peroxidase (GSH-Px; catalogue no. A005-1-2), catalase (CAT; catalogue no. A007-1-1) and superoxide dismutase (SOD; catalogue no. A001-1-2) in serum and placenta were determined using assay kits (Jiancheng Bioengineering Institute, Nanjing, China). Before the assays, the placental tissues were homogenized in ice-cold saline solution (1:9, *w*/*v*), and centrifuged at 3000× *g* for 10 min at 4 °C. The supernatants were collected for the analysis. The T-AOC was measured using the colorimetric method described by Wan et al. [28] and detected the absorbance value at 520 nm with colored and stable chelates when combined with phenanthroline. One unit (U) of T-AOC was defined as per milligram of tissue protein or per milliliter of serum with an increasing absorbance of 0.01 in 1 min. MDA concentrations were determined using the thiobarbituric acid method [29], which is based on the reaction of MDA with thiobarbituric acid to form a pink chromogen that can be spectrophotometrically determined at 532 nm. The GSH-Px activity was determined according to the method of Zhang et al. [30] by quantifying the rate of hydrogen peroxide-induced oxidation of reduced glutathione (GSH) to oxidized glutathione (GSSG). A yellow product, with absorbance at 412 nm, was formed on reaction of GSH with 5,5′-dithiobis-(2-nitrobenzoic acid). The CAT activity was measured by the method described by Ozmen et al. [31]. The enzymatic reaction was terminated by the addition of ammonium molybdate, which generated a light-yellow composite that could be measured at 405 nm. The SOD activity was measured spectrophotometrically at 550 nm according to the method of Jia et al. [32], and 1 U of SOD was defined as the quantity of enzyme required to produce 50% inhibition of nitric ion production. There was less than 5% variation of intra-assay and inter-assay coefficients for each assay.

2.5. Hormonal and Biochemical Parameters Analysis

To validate the ELISA, all assays included positive quality controls (QCs) and assays were only accepted if R^2 was above 0.98, curve fit percentage recovery was within the 80–120% range, and intra-plate and inter-plate CV% had a threshold for acceptance below 20%. The serum concentrations of estradiol (E2; catalogue no. MM-048001) and progesterone (Prog; catalogue no. MM-120502) were determined using ELISA kits (Meimian Biotechnology, Nanjing, China). The minimal detection limit was 8 pmol/L for E2, 80 pmol/L for Prog. The intra- and inter-assay coefficients of variation were less than 10% and less than 12%, respectively. Concentrations of MT (catalogue no. RE54021) in the serum and placenta were measured with an ELISA kit (IBL, Hamburg, Germany). The sensitivity of this assay was 1.6 pg/mL. Both intra- and inter-assay coefficients of variation were less than 15%. Intra-and inter-assay CVs were less than 15. The serum concentrations of alanine aminotransferase (ALT; catalogue no. CH0101202), gamma-glutamyl transpeptidase (γ-GGT; catalogue no. CH0101204), and aspartate aminotransferase (AST; catalogue no. CH0101201) were measured with an automatic biochemical analyzer (Hitachi, Tokyo, Japan) according to corresponding commercial kits (Sichuan Maker Biotechnology Inc., Chengdu, China). ALT, AST, and γ-GGT were measured using an enzymatic rate method by ultraviolet and visible spectrophotometry. The minimal detection limit was 4 U/L for ALT, 3 U/L for AST, and 2 U/L for γ-GGT. There was less than 5% variation of intra-assay and inter-assay coefficients for each assay.

2.6. Measurement of Placental DNA, RNA and Protein

DNA, RNA, and protein were collected from snap-frozen placental samples (~0.1 g), using TRI Reagent RNA/DNA/Protein Isolation Reagent (Invitrogen Life Technologies, Carlsbad, CA, USA) and their concentrations were determined colorimetrically. DNA was analyzed fluorimetrically using the method of Prasad et al. [33]. RNA was determined by spectrophotometry using a modified Schmidt–Tannhauser method, as described by Munro and Fleck [34]. Protein concentration was analyzed according to the method of Lowry et al. [35] using reagents from Bio-Rad Laboratories (Hercules, CA, USA) and bovine serum albumin as the standard.

2.7. Measurement of ATP, NAD+, and NADH Levels in Placenta

Placental NAD^+ and NADH levels were measured by the NAD^+/NADH assay kit (catalogue no. S0175; Beyotime Biotechnology, Shanghai, China) according to the manufacturer's instructions. The absorbance was determined using a microplate spectrophotometer (Spectramax 190, Molecular Devices, Sunnyvale, CA, USA) at a wavelength of 450 nm. The NAD^+ and NADH levels were calculated according to the standard curve, and then, the ratio of NAD^+/NADH was calculated. Placental ATP levels were measured using the enhanced ATP assay kit (catalogue no. S0026; Beyotime Biotechnology, China) according to the manufacturer's instructions. ATP levels were calculated from relative light unit (RLU) values, which were measured using a GloMax 96 microplate luminometer (Promega, Stockholm, Sweden). Results were normalized to total protein concentration for inter-sample comparison.

2.8. Measurement of Mitochondrial Respiratory Chain Complex Activities

The activities of mitochondrial respiratory chain complexes I (catalogue no. FHTA-2-Y), II (catalogue no. FHTB-2-Y), and III (catalogue no. FHTC-2-Y) were measured with commercial kits (Suzhou Comin Biotechnology Co., Ltd., Suzhou, China). The activity of complex I was determined using the changing of NADH oxidation absorption at 340 nm. The activity of complex II was determined by calculating the alteration of the absorbance of 2,6-dichlorophenolindophenol at 605 nm. The activity of complex III was measured by calculating the alteration of the absorbance of cytochrome c at 550 nm.

2.9. Determination of Mitochondrial DNA (mtDNA) Content

The content of mtDNA relative to nuclear genomic DNA was determined by co-amplification of the mt D-loop and the nuclear-encoded β-actin using real-time PCR according to our previous study [36]. Total DNA of frozen placental tissue was extracted with DNAiso reagent (catalogue no. DP304; Tiangen Biotech, Beijing, China). The DNA samples were adjusted to a concentration of 100 ng/μL. The amounts of mt D-loop and β-actin gene were quantified by fluorescent probes. The primers and probe sequences are listed in the Supplementary Materials, Table S2. PCR amplification was carried out in a 20 μL total volume consisting of 1 μL DNA template (100 ng), 1 μL enhance solution, 1 μL probes, 8 μL TaqMan Universal Master Mix, 1 μL forward primer, 1 μL reverse primers, and 7 μL double-distilled H_2O. The fluorescence spectra were monitored with a Real-Time PCR Detection System (ABI 7900HT, Applied Biosystems, Foster City, CA, USA) as follows: 95 °C for 10 s, 50 cycles involving a combination of 95 °C for 5 s and 60 °C for 25 s, and 95 °C for 10 s. The $2^{-\Delta\Delta Ct}$ method was used to calculate the relative mtDNA content [37].

2.10. Measurement of Inflammatory Cytokine in Placenta

Approximately 0.1 g of placenta tissue samples were homogenized in 0.9% ice-cold physiological saline (1:9, *w*/*v*), then centrifuged at 3500× *g* at 4 °C for 15 min. The supernatant was collected to determine the concentrations of tumor necrosis factor α (TNF-α; catalogue no. MM-038301), interleukin 6 (IL-6; catalogue no. MM-041801), and IL-8 (catalogue no. MM-041701) with commercial ELISA kits (Meimian Biotechnology, Nanjing, China). The minimal detection limit was 10 pg/mL for TNF-α, 50 ng/mL for IL-6, and 15 ng/mL for IL-8. The intra-and inter-assay coefficients of variation were less than 10% and less than 12%, respectively. Results were normalized to total protein concentration for inter-sample comparison.

2.11. RNA Extraction and Gene Expression Analysis

Total RNA of frozen placental tissue was extracted with Trizol reagent (catalogue no. 9109; TaKaRa, Dalian, China). RNA integrity was checked by electrophoresis on a 1.0% agarose gel, and RNA quality and concentration were measured using a NanoDrop ND-2000 spectrophotometer (Thermo Fisher Scientific, Wilmington, DE, USA). Total RNA (1 μg) was reverse transcribed into complementary DNA using the PrimeScript RT Reagent Kit (catalogue no. RR047A; TaKaRa, Dalian, China). Real-time quantitative PCR was performed using SYBR Green (catalogue no. RR820A; TaKaRa, Dalian, China) with ABI-7900HT (Applied Biosystems, Foster City, CA, USA). The primers are listed in the Supplementary Materials, Table S3. The reaction mixture of 10 μL included 5 μL of SYBR Premix Ex Taq (2×), 0.4 μL of forward primer (10 μmol/L), 0.4 μL of reverse primer (10 μmol/L), 0.2 μL of ROX reference dye (50×), 1 μL of cDNA, and 3 μL of double-distilled water. The PCR procedure was as follows: pre-denaturating at 95 °C for 30 s, 40 cycles of denaturation at 95 °C for 5 s, annealing at 60 °C for 34 s, and a final extension at 72 °C for 6 min. β-actin was used as an internal control, and the data were calculated using the $2^{-\Delta\Delta Ct}$ method.

2.12. Western Blot Analysis

Frozen placental tissue samples were homogenized in liquid nitrogen and lysed in cell lysis buffer (Beyotime Biotechnology, Shanghai, China). Protein content was measured with the BCA kit (Beyotime Biotechnology, Shanghai, China). The Western blot analysis steps were conducted according to previously reported methods [38]. Primary antibodies against SIRT1(1:1000, 9475S, CST, Danvers, MA, USA) and GAPDH (1:1000, abs132004, Absin, Shanghai, China) were used in this study. Blots were analyzed with ImageJ software (NIH, Bethesda, MD, USA).

2.13. Statistical Analysis

An individual sow or piglet was considered as the experimental unit. The statistical analysis was performed using the SAS statistical software 9.2 (SAS Institute, Cary, NC, USA). The rate of low birthweight piglets (BW ≤ 900 g) was calculated with the chi-square test. The normal distribution of the other data in this study was calculated with the Shapiro–Wilk test, followed by Student's *t*-test. Data were presented as means ± SEM. Significant differences were set at $p \leq 0.05$, and a tendency was considered when $0.05 < p < 0.10$.

3. Results

3.1. Reproductive Performance

As shown in Table 1, no differences were found in the average litter size of the total of piglets born, live born, stillborn, and mummified piglets between the two groups ($p > 0.05$). Maternal MT supplementation reduced ($p < 0.05$) the percentage of piglets born alive with weight < 900 g compared with the CON. The average total litter weight of live-born and the average total placental weight for all live-born piglets were increased ($p < 0.05$) in the MT group (Figure 1a,c). Meanwhile, the average individual piglet weight of live-born piglets ($p = 0.09$) and the placental weight per live-born piglet ($p = 0.09$) tended to increase in the MT group (Figure 1b,d).

3.2. Hormonal and Biochemical Parameters in Serum and Placenta

As shown in Figure 2a, maternal MT supplementation significantly elevated serum MT concentrations at all time points tested on d 102 of gestation (G102), and the highest serum MT concentration occurred at 17:00 relative to the two other timepoints examined ($p < 0.01$). However, the serum concentrations of ALT, AST, and γ-GGT on G90 and farrowing day (Fd) did not differ between the two groups (Figure 2b–d). Besides, the serum concentrations of Prog and E2 on G90 and G110 ($p > 0.05$) were not influenced by MT supplementation (Figure 2e,f).

Table 1. Effects of maternal MT supplementation during gestation on the reproductive performance of sows.

	CON	MT	*p*-Value
Litter size, *n*	8	8	
Piglets, total *n*			
Total born	115	118	
Born alive	106	113	
Mummified piglets	3	3	
Stillborn piglets	6	2	
Piglets born alive with weight < 900 g	12	3	
Average litter size, *n*			
Total born	14.38 ± 0.82	14.75 ± 0.63	0.72
Born alive	13.25 ± 0.73	14.12 ± 0.40	0.31
Mummified piglets	0.38 ± 0.26	0.38 ± 0.18	1.00
Stillborn piglets	0.75 ± 0.37	0.25 ± 0.16	0.23
Rate of born alive piglets with weight < 900 g, %	11.32	2.65	0.01

CON = control, MT = melatonin. Rate of live-born piglets with weight < 900 g = 'the number of live-born piglets with weight < 900 g'/'total number of live-born piglets' × 100%. Data are expressed as means ± SEM, $n = 8$. Differences were considered significant at $p < 0.05$.

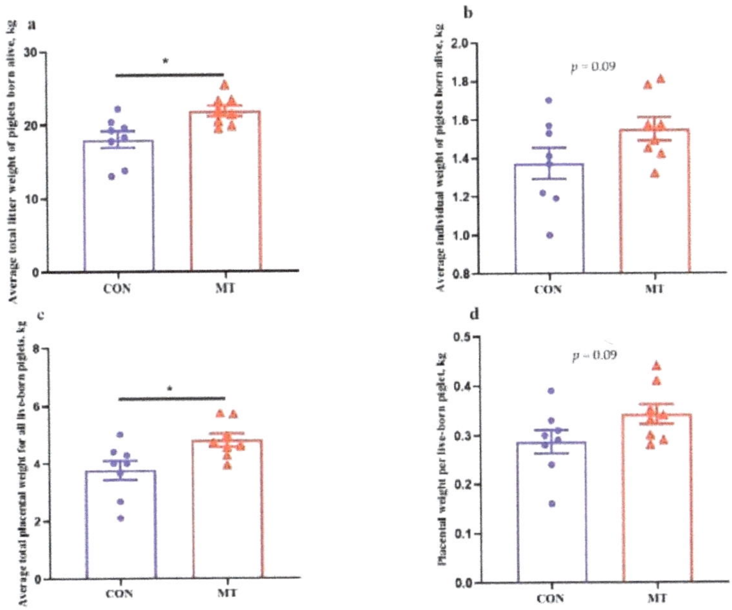

Figure 1. Effects of maternal MT supplementation during gestation on litter performance and placental weights. (**a**) Average total litter weight of piglets born alive. (**b**) Average individual weight of piglets born alive. (**c**) Average total placental weight for all live-born piglets. (**d**) Placental weight per live-born piglet. Data are presented as means ± SEM, $n = 8$. * $p < 0.05$. CON = control, MT = melatonin.

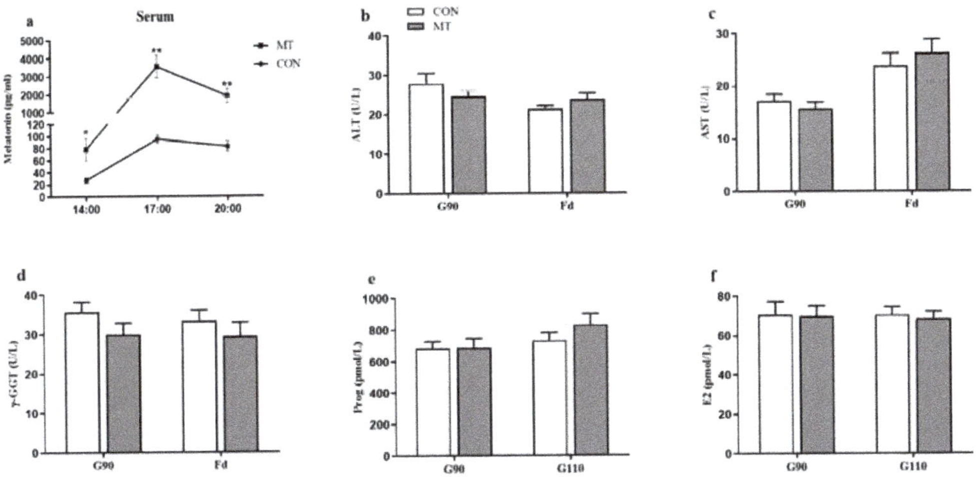

Figure 2. Effects of maternal MT supplementation during gestation on hormonal and biochemical parameters in the serum and placenta. (**a**) MT concentration in maternal serum on d 102 of gestation ($n = 7$/group). (**b**) ALT concentration, (**c**) AST concentration, (**d**) γ-GGT concentration, (**e**) Prog concentration, and (**f**) E2 concentration in maternal serum ($n = 8$/group). Data are presented as means ± SEM. * $p < 0.05$, ** $p < 0.01$. CON = control, MT = melatonin, G90 = d 90 of gestation, G110 = d 110 of gestation, Fd = farrowing day, ALT = alanine aminotransferase, AST = aspartate aminotransferase, γ-GGT = gamma-glutamyl transpeptidase, Prog = progesterone, E2 = estradiol-17β.

3.3. Antioxidant Capacity in Serum of Sows and New-Born Piglets

The content of MDA in the serum of sows at G90 and Fd and in NBP (new-born piglets) was decreased ($p < 0.05$) by MT supplementation (Figure 3a). The activity of GSH-Px and the content of T-AOC in serum of sows and NBP were not statistically different between the two groups ($p > 0.05$) (Figure 3b,c). CAT activity in the serum of sows at Fd and in NBP was increased ($p < 0.05$) in the MT group (Figure 3d). In addition, maternal MT supplementation increased ($p < 0.05$) SOD activity in the serum of sows at Fd, and tended to increase ($p = 0.07$) at G90 (Figure 3e).

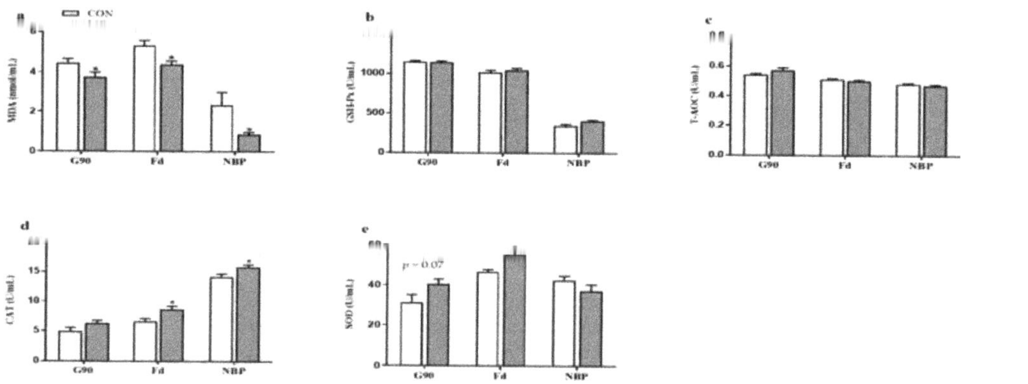

Figure 3. Effects of maternal MT supplementation during gestation on antioxidant status in serum of sows and new-born piglets. (**a**) MDA concentration, (**b**) GSH-Px activity, (**c**) T-AOC concentration, (**d**) CAT activity and (**e**) SOD activity in the serum of sows and new-born piglets. Data are presented as means ± SEM, $n = 8$. * $p < 0.05$. CON = control, MT = melatonin, G90 = d 90 of gestation, Fd = farrowing day, NBP = new-born piglets.

3.4. Antioxidant Status in the Placenta

As shown in Figure 4a,b, the placental MT concentration and the mRNA expression level of MT1 were increased by MT supplementation ($p < 0.01$). However, the content of MDA in the placenta was reduced ($p < 0.05$) by MT supplementation (Figure 4c). The GSH-Px, CAT, and SOD activities in the placenta were increased ($p < 0.05$) in the MT group (Figure 4d–g), while the content of T-AOC did not differ between the two groups (Figure 4e).

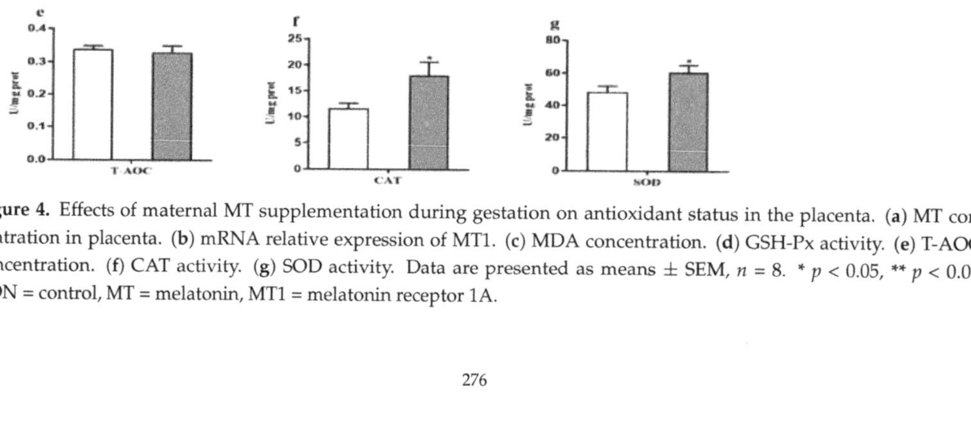

Figure 4. Effects of maternal MT supplementation during gestation on antioxidant status in the placenta. (**a**) MT concentration in placenta. (**b**) mRNA relative expression of MT1. (**c**) MDA concentration. (**d**) GSH-Px activity. (**e**) T-AOC concentration. (**f**) CAT activity. (**g**) SOD activity. Data are presented as means ± SEM, $n = 8$. * $p < 0.05$, ** $p < 0.01$. CON = control, MT = melatonin, MT1 = melatonin receptor 1A.

3.5. Placental DNA, RNA, and Protein Concentrations

As shown in Table 2, dietary MT significantly increased ($p < 0.01$) the placental protein/DNA ratio and tended to increase ($p = 0.07$) placental RNA/DNA ratio and protein concentration. However, the DNA and RNA concentrations in the placenta did not differ between the two groups.

Table 2. Effects of maternal MT supplementation during gestation on placental DNA, RNA, and protein concentrations.

	CON	MT	*p*-Value
DNA, μg/g	76.00 ± 6.21	64.25 ± 4.16	0.14
RNA, μg/g	77.42 ± 14.33	96.05 ± 9.57	0.30
RNA/DNA	0.98 ± 0.15	1.59 ± 0.27	0.07
Protein, mg/g	12.42 ± 2.24	17.15 ± 0.71	0.07
Protein/DNA, mg/μg	0.15 ± 0.02	0.28 ± 0.02	<0.01

CON = control, MT = melatonin. Data are expressed as means ± SEM, $n = 8$. Differences were considered significant at $p < 0.05$.

3.6. Placental ATP Levels and Mitochondrial Function

As shown in Figure 5, the placental ATP and NAD^+ levels (Figure 5a,c), the relative mtDNA content (Figure 5b), and complex I activity (Figure 5f) were increased ($p < 0.05$) by MT supplementation. Meanwhile, the protein expression level of SIRT1 in the placenta were also increased ($p < 0.05$) in the MT group (Figure 6). However, the NAD^+/NADH ratio, NADH level, and complex I and III activities in the placenta did not differ between the two groups.

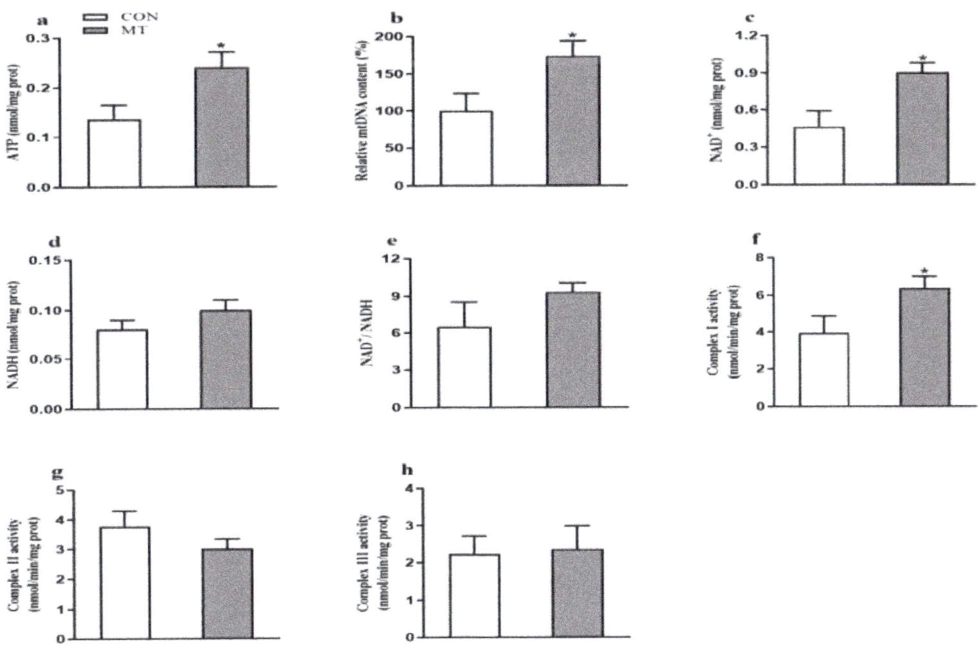

Figure 5. Effects of maternal MT supplementation during gestation on mitochondrial function in the placenta. (**a**) ATP levels. (**b**) Mitochondrial DNA (mtDNA) copy number. (**c**) NAD^+. (**d**) NADH. (**e**) The ratio of NAD^+/NADH. (**f**) Complex I activity. (**g**) Complex II activity. (**h**) Complex III activity. Data are presented as means ± SEM, $n = 8$. * $p < 0.05$. CON = control, MT = melatonin.

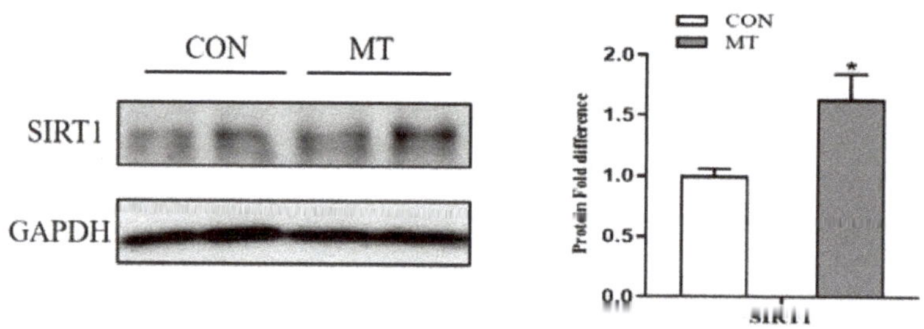

Figure 6. Effects of maternal MT supplementation during gestation on Sirt1 protein expression in placenta. Data are presented as means ± SEM, $n = 6$. * $p < 0.05$. CON = control, MT = melatonin.

3.7. *Placental Inflammatory Cytokine Concentrations*

As shown in Figure 7, dietary MT significantly decreased ($p < 0.01$) the concentration of placental TNF-α (Figure 7a) and tended to decrease ($p = 0.07$) the concentration of placental IL-8 (Figure 7c). However, the concentration of IL-6 in the placenta did not differ between the two groups (Figure 7b).

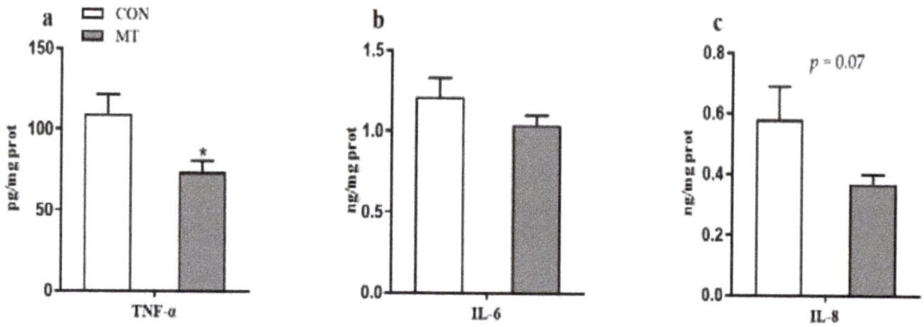

Figure 7. Effects of maternal MT supplementation during gestation on inflammatory cytokine in placenta. (**a**) TNF-α concentration. (**b**) IL-6 concentration. (**c**) IL-8 concentration. Data are presented as means ± SEM, $n = 8$. * $p < 0.05$. CON = control, MT = melatonin.

3.8. *Nrf2-Regulated Gene Expression in the Placenta*

The mRNA expression levels of *Nrf2*-regulated genes are presented in Figure 8a. Maternal MT supplementation increased ($p < 0.05$) the mRNA levels of *SOD*, *GPx1*, *Nrf2*, and *NQO1* in the placenta.

3.9. *Apoptosis and Proliferation-Related Gene Expression in the Placenta*

Maternal MT supplementation decreased ($p < 0.05$) the mRNA level of *Caspase-3*, while it tended to increase ($p = 0.09$) that of *Ki67* in the placenta (Figure 8b). There were no differences in mRNA levels of *Bax* and *Bcl2* between the two groups ($p > 0.05$).

3.10. *Nutrient Transporter Gene Expression in Placenta*

The mRNA expression levels of nutrient transporter genes, including *Glut3*, *SNAT2*, *SNAT3*, and *Pept1*, were increased by MT supplementation (Figure 8c). However, the mRNA levels of *Glut1* and *SNAT1* did not differ between the two groups.

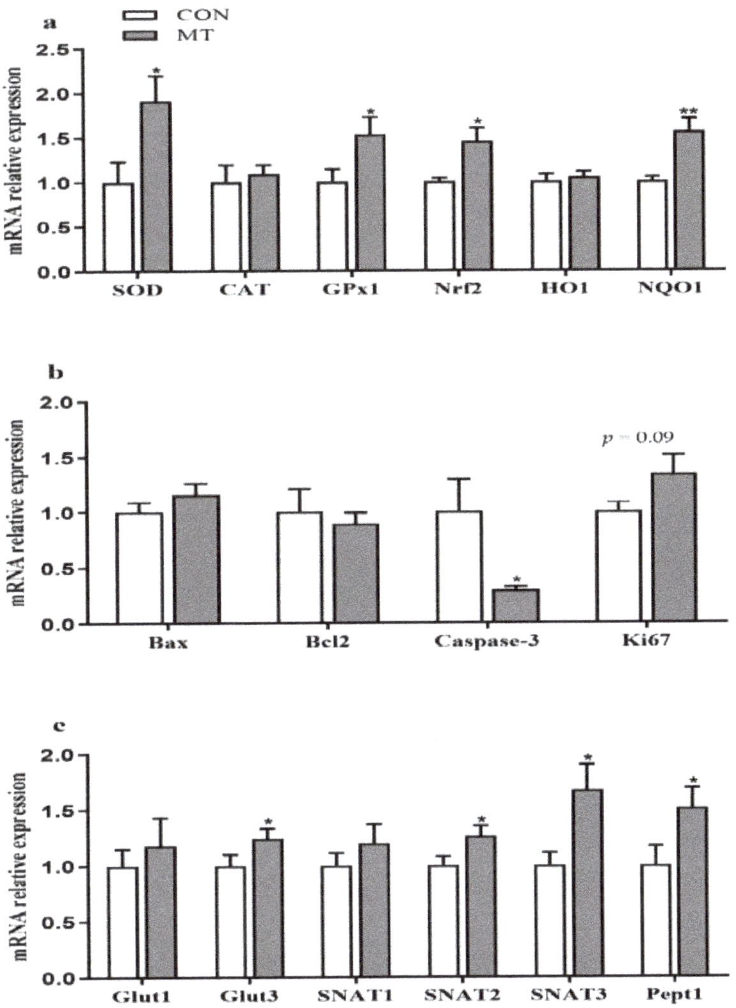

Figure 8. Effects of maternal MT supplementation during gestation on the relative expression levels of critical genes involved in the placental functions. (**a**) Nrf2-regulated gene expression. (**b**) Apoptosis and proliferation-related gene expression. (**c**) Nutrient transporter gene expression. Data are presented as means ± SEM, $n = 8$. * $p < 0.05$, ** $p < 0.01$. CON = control, MT = melatonin, Glut1 = Slc2a1, Glut3 = Slc2a3, SNAT1 = Slc38a1, SNAT2 = Slc38a2, SNAT3 = Slc38a4.

4. Discussion

MT is a safe molecule with low toxicity. It has been reported that MT is adequately safe to be administered during pregnancy, even in high doses (up to 200 mg/kg/day) [39]. In this study, dietary supplementation with MT significantly elevated serum melatonin concentrations at all time points tested, even prior to the onset of feeding. The patterns of circulating MT in pigs have been described as unchanging, irregular, or nocturnal [40–42]. The inconsistency in the MT profiles could be related to differences in assay methodology, geographical location of the study, lighting regimen, acclimation period, and method for administration of MT [42–45]. So far as we know, there are no published reports from sequential blood sampling in pregnant sows fed such a high dose of MT to compare with our study. In domestic gilts, oral application of 1 mg MT increased plasma MT concentrations within 30 min and that these remained high for at least 8 h [46]. In addition, oral application of MT (3 mg/d) at 15:00 h showed an elevated peak of melatonin ~6 h after lights off, and the overall patterns for MT in circulation appeared episodic at 3–6 h intervals while lights were on [42]. Furthermore, in this study, no statistically significant differences of MT supplementation on serum ALT, AST, and γ-GGT concentrations are an indication that the treatments have no obvious effect on liver function. Moreover, Prog and E2 are important regulators of reproduction, which play a crucial role in establishing and maintaining pregnancy [47,48]. In the current study, serum concentrations of Prog and E2 were not altered by MT supplementation. The findings of our study were not consistent with the previous results obtained in mice [49], which showed that intraperitoneal injection of MT (15 mg/kg) significantly decreased E2 concentration, with no obvious effects on Prog at day 6 of gestation. In addition, a previous study showed that MT dose- and time-dependently increased Prog production in the cultured luteal cells of pregnant sows [50]. Previous study also reported that suitable doses of MT (10^{-8}, 10^{-7}, and 10^{-6} M) could promote Prog secretion in cultured pig luteal cells, whereas a higher concentration of MT (10^{-5} M) exhibited no obvious difference between the groups [51]. Together, the discrepancies between studies suggest that the effect of melatonin on steroid hormone secretion could be highly complex, which might be explained by a number of factors, such as animal species, physiological conditions, as well as the dose and the duration of MT supplementation. However, the underlying mechanisms need further investigation. Based on the findings in the current study, it appears that the administration of 36 mg MT in pregnant sows showed no adverse maternal effects on the health status and secretion of reproductive hormones.

Melatonin works in a variety of ways as a circadian rhythm modulator, immunomodulator, direct free radical scavenger, and indirect antioxidant and cytoprotective agent in the maternal–placental–fetal unit, and it seems to be crucial for successful pregnancy [52,53]. Increasing evidence supports the idea that therapeutic use of melatonin during pregnancy may reduce materno-fetal complications and prevent neonatal diseases [12,54]. Oxidative stress, resulting from an antioxidant–prooxidant imbalance, has been implicated in the initiation or development of reproductive diseases (e.g., IUGR) affecting female reproductive processes [55]. Pregnancy is a state of high oxidative stress in humans and livestock, which is deleterious to placental development and fetal growth [56]. MDA is a primary marker of lipid peroxidation caused by ROS [57]. CAT, SOD, and GPX are important enzymes that constitute a first line antioxidant defense system to scavenge ROS [58]. In this study, maternal MT supplementation improved antioxidant status to a certain degree, and reduced MDA content in the serum of sows and new-born piglets. The new-born piglets were slaughtered before suckling colostrum in our study, which implied that the antioxidant defense capacity of new-born piglets may be enhanced in utero. During pregnancy, MT in maternal blood can easily pass across the placenta into fetal circulation and affect the fetus directly [52]. MT (10 mg/kg) administration to pregnant rats has been demonstrated to improve antioxidant activity and to protect against oxidative mitochondrial damage in the fetal rat brain [59,60]. In addition, pharmacological doses of melatonin (ranging from 0.1 to 4.0 mM) could reduce MDA content in rat brains in in vitro conditions [61]. Thus, the present results indicating that antioxidant activity in sows and new-born piglets was increased by dietary MT supplementation.

In general, maternal serum MT concentrations gradually increase during pregnancy, and this is mainly ascribed to placental production [11]. Moreover, high MT levels in the human placenta have been observed even during daytime in normal pregnancy, and lower placental MT levels were detected in pregnancies complicated by preeclampsia compared to normal pregnancies [52,62,63]. Our results showed that maternal MT supplementation increased placental MT concentrations and melatonin receptors, suggesting a beneficial effect of oral administration of MT in improving placental–fetal development. Supportively, MT has been reported to protect the villous trophoblast against hypoxia/reoxygenation-induced oxidative stress and proposed as a potential preventive option for IUGR [64]. In addition, our results showed that maternal MT supplementation tended to increase the average weight of live-born piglets, and markedly reduced the percentage of lower birth weight piglets (BW < 900 g). Similarly, a previous report demonstrated that MT supplementation during early-to mid-gestation can increase fetal weight at d 50 of gestation in gilts [25]. Additionally, consistent with the previous study that MT supplementation in undernourished pregnancy restored birth weight by increasing placental antioxidant enzymes [18], our data showed that GSH-Px, SOD, and CAT activities in the placenta were upregulated, and the content of MDA was decreased due to MT supplementation. Furthermore, a recent study showed that MT could attenuate intrauterine inflammation-induced placental oxidative stress via activating the Nrf2/ARE pathway [9]. Nrf2 performs a critical role in regulating antioxidant enzymes and phase II detoxification enzymes by the transcriptional activation of many genes containing ARE [65]. NQO1 is a detoxification enzyme that reduces NADPH oxidase activity and ROS production [66]. In this study, dietary supplementation with MT significantly increased the mRNA levels of antioxidant-related genes involved in the Nrf2/ARE pathway (*Nrf2*, *SOD*, *GPx1*, and *NQO1*). Therefore, these results may suggest that MT could promote fetal growth at least partly through its reduction of placental oxidative stress via activating the Nrf2/ARE pathway.

Cytokines play a vital role in immune status and inflammatory response [67]. Placental oxidative stress is often associated with increased production of pro-inflammatory cytokines, including IL-1β, IL-6, and IL-8 [68]. Excessive placental inflammation is associated with several pregnancy complications, such as IUGR and stillbirth [69]. A recent study showed that melatonin reverses the increase in IL-6 and TNF-α induced by hypoxia/reoxygenation in human primary villous trophoblasts [70]. In this study, maternal melatonin (36 mg/d) supplementation during pregnancy significantly decreased concentrations of placental pro-inflammatory cytokines, especially TNF-α, which was consistent with the results of a previous study in LPS-challenged mice [9]. TNF-α provokes various biological effects on placental and endometrial cell types, such as cell fusion, apoptosis, and hormone production [71]. A previous study has reported that TNF-α may inhibit the growth of trophoblast cells [72]. In addition, the increased TNF-α expression in the placenta was associated with impaired fetal development [73]. Thus, MT may improve placental and fetal growth through reducing placental inflammatory response.

During middle and late gestation, the placenta is a rapidly growing organ [74]. In this study, dietary MT supplementation increased placental weight per sow and tended to increase placental weight per fetus, indicating an increase in placental growth. The protein concentration, together with the ratio of RNA to DNA and protein to DNA in the placenta have been recognized as valuable biological parameters to determine placental growth and development [75]. DNA concentration was used as an index of hyperplasia, and the protein/DNA and RNA/DNA ratios were used as indices of hypertrophy and potential cellular protein synthetic activity, respectively [76]. In this study, maternal MT supplementation elevated the placental protein/DNA and RNA/DNA ratios, which is beneficial for the placental growth. In addition, the Ki67 protein is tightly linked to somatic cell proliferation. A rapid decrease of *Ki67* mRNA expression can be easily screened once the cell enters the non-proliferative state [77]. In the present study, maternal MT supplementation significantly decreased pro-apoptotic *Caspase-3* mRNA expression and had a tendency to increase *Ki67* mRNA expression, suggesting that placental cellular proliferation was enhanced. Supportively, data from ewes showed that maternal MT treatment had a tendency to increase placental cellular proliferation in cotyledonary tissue [78]. Previous study has found that pig placental weight is positively related to fetal weight [79]. Besides, placental weight is widely used as a parameter of placental functional capacity [80,81]. An important function of the placenta is to provide adequate oxygen and nutrients to the fetus to maintain fetal growth [5]. Previous studies have shown that MT treatment in pregnant ewes could improve oxygen supply to the fetus [82,83], and a potential mechanism may be associated with a decrease in oxidative stress and an increase in nitric oxide levels, leading to an increase in umbilical blood flow [82]. In this study, fetal blood gases were not measured; however, interestingly, the mRNA expression levels for a glucose transporter (*Glut3*), amino acid transporters (*SNAT2* and *SNAT3*), as well as peptide transporter 1 (*Pept1*) were significantly upregulated in the placentas of MT-supplemented sows. Upregulation of placental nutrient transporters can improve nutrient transfer to fetus, thus promoting fetal growth [84,85]. Similarly, a previous report in ewes has indicated that maternal MT supplementation improved fetal branched-chain amino acids uptake during maternal nutrient restriction, which could be applied to alleviate IUGR [14].

As a mitochondrial rich organ, the placenta requires a high level of ATP to support its growth and the active transport of nutrients. However, mitochondria are the main source of ATP and ROS formation, and also a target of ROS attack, which may lead to alterations in their structure and function [17]. Several studies have identified that mitochondrial dysfunction results from oxidative stress in the liver, intestine, and placenta [36,86]. In this study, the content of mtDNA and the antioxidant defense system in placenta were improved by dietary MT supplementation. Similarly, treatment of rotenone-induced impairment of porcine embryos with MT increased mtDNA content and decreased ROS generation [87]. It has been reported that abnormal mtDNA content can be indicative of mitochondrial dysfunction [88]. The NAD^+ reduction is closely associated with the dysregulation of mitochondria and energy homeostasis [89]. Our data showed that placental mitochondrial function was increased by MT supplementation, as evidenced by increased placental ATP, NAD^+ levels, and mtDNA content. In addition, mitochondrial complexes I and III are regarded as the major source of ROS generation [90]. Previous research has reported that elevated ROS generation would lead to the rapid loss of the activities of mitochondrial complexes [91]. Another study also showed a close link between enhanced ROS generation and reduced mitochondrial complex I activity in the hypoxic human placenta [92]. In this study, the mitochondrial complex I activity was increased by MT supplementation, suggesting a decrease in ROS formation in the placenta of MT-supplemented sows. Furthermore, the altered mtDNA amount is accompanied with changes in several transcriptional factors that participate in mitochondrial biogenesis [93]. Accumulating evidence indicates that SIRT1 activation reduces oxidative stress and stimulates mitochondrial biogenesis [94,95]. In our study, the protein abundance of SIRT1 was increased by dietary MT supplementation. Similarly, a previous study reported that MT could activate the SIRT1 pathway, thus promoting mitochondrial biogenesis and energy production [87]. Taken together, in this study, MT may play a protective role against oxidative stress-induced mitochondrial dysfunction and energy deficiency by improving mitochondrial biosynthesis.

5. Conclusions

In summary, our data indicated that dietary supplementation with MT in gestating sows could improve maternal–placental–fetal redox status and enhance placental growth and function, thereby improving pregnancy outcomes. The beneficial effects of MT might be closely related to ameliorating placental antioxidant status, inflammatory response, and mitochondrial dysfunction.

Supplementary Materials: The following are available online at https://www.mdpi.com/article/10.3390/antiox10121867/s1. Table S1: Ingredients and composition of the control diet. Table S2: Primer and probe sequences used for the determination of mtDNA content. Table S3: Primer sequences of the target and reference genes.

Author Contributions: D.W., L.C. and X.P. designed the study; X.C., J.L., Y.H., H.L., J.H. and X.P. conducted the research; Z.F., B.F., J.T., Y.L., X.J., L.H., S.X. and Y.Z. analyzed the data; X.P. wrote the manuscript; D.W. and X.P. revised the manuscript. All authors have read and agreed to the published version of the manuscript.

Funding: This study was supported by the National Natural Science Foundation of China (grants 31772616 and 31802078).

Institutional Review Board Statement: The study was approved by the animal care and use committee of Sichuan Agricultural University (DKYB20131704).

Informed Consent Statement: Not applicable.

Data Availability Statement: The data presented in this study are available in this manuscript or Supplementary Materials.

Acknowledgments: We would like to thank the staff at our laboratory for their ongoing assistance.

Conflicts of Interest: The authors declare that there are no conflict of interest.

References

1. Tan, C.; Ji, Y.; Zhao, X.; Xin, Z.; Li, J.; Huang, S.; Cui, Z.; Wen, L.; Liu, C.; Kim, S.W. Effects of dietary supplementation of nucleotides from late gestation to lactation on the performance and oxidative stress status of sows and their offspring. *Anim. Nutr.* **2021**, *7*, 111–118. [CrossRef]
2. Herrera, E.; Ortega-Senovilla, H. Maternal lipid metabolism during normal pregnancy and its implications to fetal development. *Clin. Lipidol.* **2010**, *5*, 899–911. [CrossRef]
3. Luo, Z.; Yao, J.; Xu, J. Reactive oxygen and nitrogen species ratio regulate porcine embryo development during pre-implantation period: A mini-review. *Anim. Nutr.* **2021**, *7*, 823–828. [CrossRef] [PubMed]
4. Pereira, A.C.; Martel, F. Oxidative stress in pregnancy and fertility pathologies. *Cell Biol. Toxicol.* **2014**, *30*, 301–312. [CrossRef] [PubMed]
5. Zhang, S.; Regnault, T.R.; Barker, P.L.; Botting, K.J.; McMillen, I.C.; McMillan, C.M.; Roberts, C.T.; Morrison, J.L. Placental adaptations in growth restriction. *Nutrients* **2015**, *7*, 360–389. [CrossRef] [PubMed]
6. Jauniaux, E.; Poston, L.; Burton, G.J. Placental-related diseases of pregnancy: Involvement of oxidative stress and implications in human evolution. *Hum. Reprod. Update* **2006**, *12*, 747–755. [CrossRef]
7. Fisher, J.J.; Bartho, L.A.; Perkins, A.V.; Holland, O.J. Placental mitochondria and reactive oxygen species in the physiology and pathophysiology of pregnancy. *Clin. Exp. Pharmacol. Physiol.* **2020**, *47*, 176–184. [CrossRef]
8. Meng, Q.; Guo, T.; Li, G.; Sun, S.; He, S.; Cheng, B.; Shi, B.; Shan, A. Dietary resveratrol improves antioxidant status of sows and piglets and regulates antioxidant gene expression in placenta by Keap1-Nrf2 pathway and Sirt1. *J. Anim. Sci. Biotechnol.* **2018**, *9*, 34. [CrossRef]
9. Lee, J.Y.; Li, S.; Shin, N.E.; Na, Q.; Dong, J.; Jia, B.; Jones-Beatty, K.; McLane, M.W.; Ozen, M.; Lei, J. Melatonin for prevention of placental malperfusion and fetal compromise associated with intrauterine inflammation-induced oxidative stress in a mouse model. *J. Pineal Res.* **2019**, *67*, e12591. [CrossRef]
10. Pang, Y.W.; Jiang, X.L.; Wang, Y.C.; Wang, Y.Y.; Hao, H.S.; Zhao, S.J.; Du, W.H.; Zhao, X.M.; Wang, L.; Zhu, H.B. Melatonin protects against paraquat-induced damage during in vitro maturation of bovine oocytes. *J. Pineal Res.* **2019**, *66*, e12532. [CrossRef]
11. Lanoix, D.; Beghdadi, H.; Lafond, J.; Vaillancourt, C. Human placental trophoblasts synthesize melatonin and express its receptors. *J. Pineal Res.* **2008**, *45*, 50–60. [CrossRef]
12. Voiculescu, S.; Zygouropoulos, N.; Zahiu, C.; Zagrean, A. Role of melatonin in embryo fetal development. *J. Med. Life* **2014**, *7*, 488. [PubMed]
13. Chen, Y.; Xu, D.; Wang, J.; Wang, H.; Wei, L.; Sun, M.; Wei, W. Melatonin protects against lipopolysaccharide-induced intra-uterine fetal death and growth retardation in mice. *J. Pineal Res.* **2006**, *40*, 40–47. [CrossRef]
14. Lemley, C.; Camacho, L.; Meyer, A.; Kapphahn, M.; Caton, J.; Vonnahme, K. Dietary melatonin supplementation alters uteroplacental amino acid flux during intrauterine growth restriction in ewes. *Animal* **2013**, *7*, 1500–1507. [CrossRef] [PubMed]
15. Lemley, C.O.; Meyer, A.M.; Camacho, L.E.; Neville, T.L.; Newman, D.J.; Caton, J.S.; Vonnahme, K.A. Melatonin supplementation alters uteroplacental hemodynamics and fetal development in an ovine model of intrauterine growth restriction. *Am. J. Physiol. Regul. Integr. Comp. Physiol.* **2012**, *302*, R454–R467. [CrossRef]
16. Le, Z.; Dong, S.; Zhang, R.; Cai, X.; Gao, A.; Xiao, R.; Yu, H. Placental mitochondrial biogenesis and function was slightly changed by gestational hypercholesterolemia in full-term pregnant women. *J. Dev. Orig. Health Dis.* **2018**, *9*, 395–400. [CrossRef] [PubMed]
17. Leduc, L.; Levy, E.; Bouity-Voubou, M.; Delvin, E. Fetal programming of atherosclerosis: Possible role of the mitochondria. *Eur. J. Obstet. Gynecol. Reprod. Biol.* **2010**, *149*, 127–130. [CrossRef] [PubMed]
18. Richter, H.G.; Hansell, J.A.; Raut, S.; Giussani, D.A. Melatonin improves placental efficiency and birth weight and increases the placental expression of antioxidant enzymes in undernourished pregnancy. *J. Pineal Res.* **2009**, *46*, 357–364. [CrossRef]
19. Hu, C.; Yang, Y.; Deng, M.; Yang, L.; Shu, G.; Jiang, Q.; Zhang, S.; Li, X.; Yin, Y.; Tan, C. Placentae for Low Birth Weight Piglets Are Vulnerable to Oxidative Stress, Mitochondrial Dysfunction, and Impaired Angiogenesis. *Oxid. Med. Cell. Longev.* **2020**, *2020*, 8715412. [CrossRef]
20. Yang, Q.; Dai, S.; Luo, X.; Zhu, J.; Li, F.; Liu, J.; Yao, G.; Sun, Y. Melatonin attenuates postovulatory oocyte dysfunction by regulating SIRT1 expression. *Reproduction* **2018**, *156*, 81–92. [CrossRef]

21. Barbaux, S.; Erwich, J.J.H.M.; Favaron, P.; Gil, S.; Gallot, D.; Golos, T.G.; Gonzalez-Bulnes, A.; Guibourdenche, J.; Heazell, A.; Jansson, T. IFPA meeting 2014 workshop report: Animal models to study pregnancy pathologies; new approaches to study human placental exposure to xenobiotics; biomarkers of pregnancy pathologies; placental genetics and epigenetics; the placenta and stillbirth and fetal growth restriction. *Placenta* **2015**, *36*, S5–S10.
22. Chavatte-Palmer, P.; Tarrade, A. Placentation in different mammalian species. *Ann. Endocrinol.* **2016**, *77*, 67–74. [CrossRef]
23. Yuan, T.-L.; Zhu, Y.-H.; Shi, M.; Li, T.-T.; Li, N.; Wu, G.-Y.; Bazer, F.W.; Zang, J.-J.; Wang, F.-L.; Wang, J.-J. Within-litter variation in birth weight: Impact of nutritional status in the sow. *J. Zhejiang Univ. Sci. B* **2015**, *16*, 417–435. [CrossRef]
24. Tetro, N.; Moushaev, S.; Rubinchik-Stern, M.; Eyal, S. The placental barrier: The gate and the fate in drug distribution. *Pharm. Res.* **2018**, *35*, 71.
25. Dearlove, B.; Kind, K.; Gatford, K.; van Wettere, W. Melatonin fed in early gestation increases fetal weight. *Anim. Prod. Sci.* **2017**, *57*, 2478. [CrossRef]
26. NRC. *Nutrient Requirements of Swine*, 11th ed.; National Academies Press: Washington, DC, USA, 2012.
27. Wilson, M.E.; Biensen, N.J.; Youngs, C.R.; Ford, S.P. Development of Meishan and Yorkshire littermate conceptuses in either a Meishan or Yorkshire uterine environment to day 90 of gestation and to term. *Biol. Reprod.* **1998**, *58*, 905–910. [CrossRef]
28. Wan, J.; Jiang, F.; Xu, Q.; Chen, D.; Yu, B.; Huang, Z.; Mao, X.; Yu, J.; He, J. New insights into the role of chitosan oligosaccharide in enhancing growth performance, antioxidant capacity, immunity and intestinal development of weaned pigs. *RSC Adv.* **2017**, *7*, 9669–9679. [CrossRef]
29. Livingstone, D.; Martinez, P.G.; Michel, X.; Narbonne, J.; O'hara, S.; Ribera, D.; Winston, G. Oxyradical production as a pollution-mediated mechanism of toxicity in the common mussel, *Mytilus edulis* L., and other molluscs. *Funct. Ecol.* **1990**, *4*, 415–424. [CrossRef]
30. Zhang, X.-D.; Zhu, Y.-F.; Cai, L.-S.; Wu, T.-X. Effects of fasting on the meat quality and antioxidant defenses of market-size farmed large yellow croaker (*Pseudosciaena crocea*). *Aquaculture* **2008**, *280*, 136–139. [CrossRef]
31. Özmen, B.; Özmen, D.; Erkin, E.; Güner, İ.; Habif, S.; Bayındır, O. Lens superoxide dismutase and catalase activities in diabetic cataract. *Clin. Biochem.* **2002**, *35*, 69–72. [CrossRef]
32. Jia, J.; Zhang, X.; Hu, Y.-S.; Wu, Y.; Wang, Q.-Z.; Li, N.-N.; Guo, Q.-C.; Dong, X.-C. Evaluation of in vivo antioxidant activities of Ganoderma lucidum polysaccharides in STZ-diabetic rats. *Food Chem.* **2009**, *115*, 32–36. [CrossRef]
33. Prasad, A.S.; DuMouchelle, E.; Koniuch, D.; Oberleas, D. A simple fluorometric method for the determination of RNA and DNA in tissues. *J. Lab. Clin. Med.* **1972**, *80*, 598–602.
34. Munro, H.; Fleck, A. Analysis of tissues and body fluids for nitrogenous constituents. *Mamm. Protein Metab.* **1969**, *3*, 423–525.
35. Lowry, O.H.; Rosebrough, N.J.; Farr, A.L.; Randall, R.J. Protein measurement with the Folin phenol reagent. *J. Biol. Chem.* **1951**, *193*, 265–275. [CrossRef]
36. Hu, L.; Peng, X.; Qin, L.; Wang, R.; Fang, Z.; Lin, Y.; Xu, S.; Feng, B.; Wu, D.; Che, L. Dietary nucleotides supplementation during the suckling period improves the antioxidative ability of neonates with intrauterine growth retardation when using a pig model. *RSC Adv.* **2018**, *8*, 16152–16160. [CrossRef]
37. Liu, J.; Zhang, Y.; Li, Y.; Yan, H.; Zhang, H. L-tryptophan enhances intestinal integrity in diquat-challenged piglets associated with improvement of redox status and mitochondrial function. *Animals* **2019**, *9*, 266. [CrossRef]
38. Hu, L.; Han, F.; Chen, L.; Peng, X.; Chen, D.; Wu, D.; Che, L.; Zhang, K. High nutrient intake during the early postnatal period accelerates skeletal muscle fiber growth and maturity in intrauterine growth-restricted pigs. *Genes Nutr.* **2018**, *13*, 23. [CrossRef] [PubMed]
39. Jahnke, G.; Marr, M.; Myers, C.; Wilson, R.; Travlos, G.; Price, C. Maternal and developmental toxicity evaluation of melatonin administered orally to pregnant Sprague-Dawley rats. *Toxicol. Sci.* **1999**, *50*, 271–279. [CrossRef]
40. Diekman, M.; Brandt, K.; Green, M.; Clapper, J.; Malayer, J. Lack of a nocturnal rise of serum melatonin in prepubertal gilts. *Domest. Anim. Endocrinol.* **1992**, *9*, 161–167. [CrossRef]
41. Green, M.; Clapper, J.; Andres, C.; Diekman, M. Serum concentrations of melatonin in prepubertal gilts exposed to either constant or stepwise biweekly alteration in scotophase. *Domest. Anim. Endocrinol.* **1996**, *13*, 307–323. [CrossRef]
42. Arend, L.S.; Knox, R.V.; Greiner, L.L.; Graham, A.B.; Connor, J.F. Effects of feeding melatonin during proestrus and early gestation to gilts and parity 1 sows to minimize effects of seasonal infertility. *J. Anim. Sci.* **2019**, *97*, 4635–4646. [CrossRef]
43. McConnell, S.; Ellendorff, F. Absence of nocturnal plasma melatonin surge under long and short artificial photoperiods in the domestic sow. *J. Pineal Res.* **1987**, *4*, 201–210. [CrossRef] [PubMed]
44. Diekman, M.A.; Arthington, J.A.; Clapper, J.A.; Green, M.L. Failure of melatonin implants to alter onset of puberty in gilts. *Anim. Reprod. Sci.* **1997**, *46*, 283–288. [CrossRef]
45. De Almeida, E.A.; Di Mascio, P.; Harumi, T.; Spence, D.W.; Moscovitch, A.; Hardeland, R.; Cardinali, D.P.; Brown, G.M.; Pandi-Perumal, S. Measurement of melatonin in body fluids: Standards, protocols and procedures. *Childs Nerv. Syst.* **2011**, *27*, 879–891. [CrossRef]
46. Paterson, A.; Maxwell, C.; Foldes, A. Seasonal inhibition of puberty in domestic gilts is overcome by melatonin administered orally, but not by implant. *Reproduction* **1992**, *94*, 97–105. [CrossRef] [PubMed]
47. Arck, P.; Hansen, P.J.; Mulac Jericevic, B.; Piccinni, M.P.; Szekeres-Bartho, J. Progesterone during pregnancy: Endocrine–immune cross talk in mammalian species and the role of stress. *Am. J. Reprod. Immunol.* **2007**, *58*, 268–279. [CrossRef]

48. Albrecht, E.D.; Aberdeen, G.W.; Pepe, G.J. The role of estrogen in the maintenance of primate pregnancy. *Am. J. Obstet. Gynecol.* **2000**, *182*, 432–438. [CrossRef]
49. He, C.; Wang, J.; Li, Y.; Zhu, K.; Xu, Z.; Song, Y.; Song, Y.; Liu, G. Melatonin-related genes expressed in the mouse uterus during early gestation promote embryo implantation. *J. Pineal Res.* **2015**, *58*, 300–309. [CrossRef]
50. Zhang, W.; Wang, Z.; Zhang, L.; Zhang, Z.; Chen, J.; Chen, W.; Tong, D. Melatonin stimulates the secretion of progesterone along with the expression of cholesterol side-chain cleavage enzyme (P450scc) and steroidogenic acute regulatory protein (StAR) in corpus luteum of pregnant sows. *Theriogenology* **2018**, *108*, 297–305. [CrossRef]
51. Wang, J.; Zhu, T.; Ma, X.; Wang, Y.; Liu, J.; Li, G.; Liu, Y.; Ji, P.; Zhang, Z.; Zhang, L. Melatonergic systems of AANAT, melatonin, and its receptor MT2 in the corpus luteum are essential for reproductive success in mammals. *Biol. Reprod.* **2021**, *104*, 430–444. [CrossRef]
52. Tamura, H.; Nakamura, Y.; Terron, M.P.; Flores, L.J.; Manchester, L.C.; Tan, D.-X.; Sugino, N.; Reiter, R.J. Melatonin and pregnancy in the human. *Reprod. Toxicol.* **2008**, *25*, 291–303. [CrossRef] [PubMed]
53. Aversa, S.; Pellegrino, S.; Barberi, I.; Reiter, R.J.; Gitto, E. Potential utility of melatonin as an antioxidant during pregnancy and in the perinatal period. *J. Matern. Fetal Neonatal Med.* **2012**, *25*, 207–221. [CrossRef] [PubMed]
54. Hsu, C.-N.; Huang, L.-T.; Tain, Y.-L. Perinatal use of melatonin for offspring health: Focus on cardiovascular and neurological diseases. *Int. J. Mol. Sci.* **2019**, *20*, 5681. [CrossRef] [PubMed]
55. Finkel, T. Oxidant signals and oxidative stress. *Curr. Opin. Cell Biol.* **2003**, *15*, 247–254. [CrossRef]
56. Al-Gubory, K.H.; Fowler, P.A.; Garrel, C. The roles of cellular reactive oxygen species, oxidative stress and antioxidants in pregnancy outcomes. *Int. J. Biochem. Cell Biol.* **2010**, *42*, 1634–1650. [CrossRef]
57. El-Sheikh, N.M.; Khalil, F.A. L-Arginine and L-glutamine as immunonutrients and modulating agents for oxidative stress and toxicity induced by sodium nitrite in rats. *Food Chem. Toxicol.* **2011**, *49*, 758–762. [CrossRef]
58. Valko, M.; Rhodes, C.; Moncol, J.; Izakovic, M.; Mazur, M. Free radicals, metals and antioxidants in oxidative stress-induced cancer. *Chem. Biol. Interact.* **2006**, *160*, 1–40. [CrossRef]
59. Okatani, Y.; Wakatsuki, A.; Kaneda, C. Melatonin increases activities of glutathione peroxidase and superoxide dismutase in fetal rat brain. *J. Pineal Res.* **2000**, *28*, 89–96. [CrossRef]
60. Wakatsuki, A.; Okatani, Y.; Shinohara, K.; Ikenoue, N.; Kaneda, C.; Fukaya, T. Melatonin protects fetal rat brain against oxidative mitochondrial damage. *J. Pineal Res.* **2001**, *30*, 22–28. [CrossRef] [PubMed]
61. Melchiorri, D.; Reiter, R.J.; Sewerynek, E.; Chen, L.D.; Nistic, G. Melatonin reduces kainate-induced lipid peroxidation in homogenates of different brain regions. *FASEB J.* **1995**, *9*, 1205–1210. [CrossRef]
62. Lanoix, D.; Guérin, P.; Vaillancourt, C. Placental melatonin production and melatonin receptor expression are altered in preeclampsia: New insights into the role of this hormone in pregnancy. *J. Pineal Res.* **2012**, *53*, 417–425. [CrossRef]
63. Vatish, M.; Steer, P.; Blanks, A.; Hon, M.; Thornton, S. Diurnal variation is lost in preterm deliveries before 28 weeks of gestation. *BJOG Int. J. Obstet. Gynaecol.* **2010**, *117*, 765–767. [CrossRef]
64. Lanoix, D.; Lacasse, A.A.; Reiter, R.J.; Vaillancourt, C. Melatonin: The watchdog of villous trophoblast homeostasis against hypoxia/reoxygenation-induced oxidative stress and apoptosis. *Mol. Cell Endocrinol.* **2013**, *381*, 35–45. [CrossRef]
65. Teixeira, T.M.; da Costa, D.C.; Resende, A.C.; Soulage, C.O.; Bezerra, F.F.; Daleprane, J.B. Activation of Nrf2-antioxidant signaling by 1,25-dihydroxycholecalciferol prevents leptin-induced oxidative stress and inflammation in human endothelial cells. *J. Nutr.* **2017**, *147*, 506–513. [CrossRef]
66. Dinkova-Kostova, A.T.; Talalay, P. NAD (P) H: Quinone acceptor oxidoreductase 1 (NQO1), a multifunctional antioxidant enzyme and exceptionally versatile cytoprotector. *Arch. Biochem. Biophys.* **2010**, *501*, 116–123. [CrossRef]
67. Praveena, P.E.; Periasamy, S.; Kumar, A.; Singh, N. Cytokine profiles, apoptosis and pathology of experimental Pasteurella multocida serotype A1 infection in mice. *Res. Vet. Sci.* **2010**, *89*, 332–339. [CrossRef]
68. Xie, C.; Wu, X.; Long, C.; Wang, Q.; Fan, Z.; Li, S.; Yin, Y. Chitosan oligosaccharide affects antioxidant defense capacity and placental amino acids transport of sows. *BMC Vet. Res.* **2016**, *12*, 243. [CrossRef] [PubMed]
69. Bartha, J.L.; Romero-Carmona, R.; Comino-Delgado, R. Inflammatory cytokines in intrauterine growth retardation. *Acta Obstet. Gynecol. Scand.* **2003**, *82*, 1099–1102. [CrossRef] [PubMed]
70. Sagrillo-Fagundes, L.; Assunção Salustiano, E.M.; Ruano, R.; Markus, R.P.; Vaillancourt, C. Melatonin modulates autophagy and inflammation protecting human placental trophoblast from hypoxia/reoxygenation. *J. Pineal Res.* **2018**, *65*, e12520. [CrossRef] [PubMed]
71. Haider, S.; Knöfler, M. Human tumour necrosis factor: Physiological and pathological roles in placenta and endometrium. *Placenta* **2009**, *30*, 111–123. [CrossRef]
72. Hunt, J.; Atherton, R.; Pace, J. Differential responses of rat trophoblast cells and embryonic fibroblasts to cytokines that regulate proliferation and class I MHC antigen expression. *J. Immunol.* **1990**, *145*, 184–189.
73. Holcberg, G.; Huleihel, M.; Sapir, O.; Katz, M.; Tsadkin, M.; Furman, B.; Mazor, M.; Myatt, L. Increased production of tumor necrosis factor-α TNF-α by IUGR human placentae. *Eur. J. Obstet. Gynecol. Reprod. Biol.* **2001**, *94*, 69–72. [CrossRef]
74. Krombeen, S.K.; Bridges, W.C.; Wilson, M.E.; Wilmoth, T.A. Factors contributing to the variation in placental efficiency on days 70, 90, and 110 of gestation in gilts. *J. Anim. Sci.* **2019**, *97*, 359–373. [CrossRef] [PubMed]
75. Zhang, H.; Sun, L.; Wang, Z.; Deng, M.; Nie, H.; Zhang, G.; Ma, T.; Wang, F. N-carbamylglutamate and L-arginine improved maternal and placental development in underfed ewes. *Reproduction* **2016**, *151*, 623–635. [CrossRef]

76. Scheaffer, A.; Caton, J.; Redmer, D.; Arnold, D.; Reynolds, L. Effect of dietary restriction, pregnancy, and fetal type on intestinal cellularity and vascularity in Columbia and Romanov ewes. *J. Anim. Sci.* **2004**, *82*, 3024–3033. [CrossRef] [PubMed]
77. Bullwinkel, J.; Baron-Lühr, B.; Lüdemann, A.; Wohlenberg, C.; Gerdes, J.; Scholzen, T. Ki-67 protein is associated with ribosomal RNA transcription in quiescent and proliferating cells. *J. Cell. Physiol.* **2006**, *206*, 624–635. [CrossRef] [PubMed]
78. Eifert, A.W.; Wilson, M.E.; Vonnahme, K.A.; Camacho, L.E.; Borowicz, P.P.; Redmer, D.A.; Romero, S.; Dorsam, S.; Haring, J.; Lemley, C.O. Effect of melatonin or maternal nutrient restriction on vascularity and cell proliferation in the ovine placenta. *Anim. Reprod. Sci.* **2015**, *153*, 13–21. [CrossRef]
79. Town, S.C.; Patterson, J.L.; Pereira, C.Z.; Gourley, G.; Foxcroft, G.R. Embryonic and fetal development in a commercial dam-line genotype. *Anim. Reprod. Sci.* **2005**, *85*, 301–316. [CrossRef]
80. Salafia, C.; Charles, A.; Maas, E. Placenta and fetal growth restriction. *Clin. Obstet. Gynecol.* **2006**, *49*, 236–256. [CrossRef]
81. Rumball, C.; Harding, J.; Oliver, M.; Bloomfield, F. Effects of twin pregnancy and periconceptional undernutrition on maternal metabolism, fetal growth and glucose–insulin axis function in ovine pregnancy. *J. Physiol.* **2008**, *586*, 1399–1411. [CrossRef]
82. Sales, F.; Peralta, O.A.; Narbona, E.; Mccoard, S.A.; Gonzalezbulnes, A.; Parraguez, V.H. Rapid Communication: Maternal melatonin implants improve fetal oxygen supply and body weight at term in sheep pregnancies. *J. Anim. Sci.* **2019**, *97*, 839–845. [CrossRef] [PubMed]
83. Tare, M.; Parkington, H.C.; Wallace, E.M.; Sutherland, A.E.; Lim, R.; Yawno, T.; Coleman, H.A.; Jenkin, G.; Miller, S.L. Maternal melatonin administration mitigates coronary stiffness and endothelial dysfunction, and improves heart resilience to insult in growth restricted lambs. *J. Physiol.* **2014**, *592*, 2695–2709. [CrossRef]
84. Kwan, S.T.; King, J.H.; Yan, J.; Wang, Z.; Jiang, X.; Hutzler, J.S.; Klein, H.R.; Brenna, J.T.; Roberson, M.S.; Caudill, M.A. Maternal choline supplementation modulates placental nutrient transport and metabolism in late gestation of mouse pregnancy. *J. Nutr.* **2017**, *147*, 2083–2092. [CrossRef]
85. Rosario, F.J.; Kanai, Y.; Powell, T.L.; Jansson, T. Mammalian target of rapamycin signalling modulates amino acid uptake by regulating transporter cell surface abundance in primary human trophoblast cells. *J. Physiol.* **2013**, *591*, 609–625. [CrossRef] [PubMed]
86. Tian, L.; Huang, J.; Wen, A.; Yan, P. Impaired Mitochondrial Function Results from Oxidative Stress in the Full-Term Placenta of Sows with Excessive Back-Fat. *Animals* **2020**, *10*, 360. [CrossRef]
87. Niu, Y.J.; Zhou, W.; Nie, Z.W.; Shin, K.T.; Cui, X.S. Melatonin enhances mitochondrial biogenesis and protects against rotenone-induced mitochondrial deficiency in early porcine embryos. *J. Pineal Res.* **2020**, *68*, e12627. [CrossRef] [PubMed]
88. Ott, M.; Gogvadze, V.; Orrenius, S.; Zhivotovsky, B. Mitochondria, oxidative stress and cell death. *Apoptosis* **2007**, *12*, 913–922. [CrossRef]
89. Canto, C.; Menzies, K.J.; Auwerx, J. NAD+ metabolism and the control of energy homeostasis: A balancing act between mitochondria and the nucleus. *Cell Metab.* **2015**, *22*, 31–53. [CrossRef]
90. Goguadze, N.; Zhuravliova, E.; Morin, D.; Mikeladze, D.; Maurice, T. Sigma-1 receptor agonists induce oxidative stress in mitochondria and enhance complex I activity in physiological condition but protect against pathological oxidative stress. *Neurotox. Res.* **2019**, *35*, 1–18. [CrossRef]
91. Gardner, P.R.; Nguyen, D.; White, C.W. Aconitase is a sensitive and critical target of oxygen poisoning in cultured mammalian cells and in rat lungs. *Proc. Nat. Acad. Sci. USA* **1994**, *91*, 12248–12252. [CrossRef]
92. Colleoni, F.; Padmanabhan, N.; Yung, H.-W.; Watson, E.D.; Cetin, I.; van Patot, M.C.T.; Burton, G.J.; Murray, A.J. Suppression of mitochondrial electron transport chain function in the hypoxic human placenta: A role for miRNA-210 and protein synthesis inhibition. *PLoS ONE* **2013**, *8*, e55194. [CrossRef] [PubMed]
93. Puigserver, P.; Spiegelman, B.M. Peroxisome proliferator-activated receptor-γ coactivator 1α (PGC-1α): Transcriptional coactivator and metabolic regulator. *Endocr. Rev.* **2003**, *24*, 78–90. [CrossRef] [PubMed]
94. Khader, A.; Yang, W.-L.; Kuncewitch, M.; Jacob, A.; Prince, J.M.; Asirvatham, J.R.; Nicastro, J.; Coppa, G.F.; Wang, P. Sirtuin 1 activation stimulates mitochondrial biogenesis and attenuates renal injury after ischemia-reperfusion. *Transplantation* **2014**, *98*, 148–156. [CrossRef] [PubMed]
95. Zhang, W.; Huang, Q.; Zeng, Z.; Wu, J.; Zhang, Y.; Chen, Z. Sirt1 inhibits oxidative stress in vascular endothelial cells. *Oxid. Med. Cell. Longev.* **2017**, *2017*, 7543973. [CrossRef] [PubMed]

antioxidants

MDPI

Article

The Effect of Oxidative Stress on the Chicken Ovary: Involvement of Microbiota and Melatonin Interventions

Jianping Wang [1,*,†], Ru Jia [1,†], Haojie Gong [1,†], Pietro Celi [2], Yong Zhuo [1], Xuemei Ding [1], Shiping Bai [1], Qiufeng Zeng [1], Huadong Yin [1], Shengyu Xu [1], Jingbo Liu [3], Xiangbing Mao [1] and Keying Zhang [1]

1 Animal Nutrition Institute, Sichuan Agricultural University, Chengdu 611130, China; rujia@sicau.edu.cn (R.J.); haojiegong@sicau.edu.cn (H.G.); zhuoyong@sicau.edu.cn (Y.Z.); dingxuemei@sicau.edu.cn (X.D.); shipingbai@sicau.edu.cn (S.B.); zqf@sicau.edu.cn (Q.Z.); yinhuadong@sicau.edu.cn (H.Y.); shengyuxu@sicau.edu.cn (S.X.); 13856@sicau.edu.cn (X.M.); zkeying@sicau.edu.cn (K.Z.)

2 Faculty of Veterinary and Agricultural Sciences, The University of Melbourne, Parkville 3010, Australia; pietro.celi@adisseo.com

3 School of Life Science and Engineering, Southwest University of Science and Technology, Mianyang 621010, China; liuswust@163.com

* Correspondence: wangjianping@sicau.edu.cn

† These authors shared first authorship.

Abstract: The poultry ovary is used as a classic model to study ovarian biology and ovarian cancer. Stress factors induced oxidative stress to cause follicle atresia, which may be a fundamental reason for the reduction in fertility in older laying hens or in aging women. In the present study, we set out to characterize the relationships between oxidative stress and ovarian function. Layers (62 weeks of age; BW = 1.42 ± 0.12 kg) were injected with tert-butyl hydroperoxide (tBHP) at 0 (CON) and 800 μmol/kg BW (oxidative stress group, OS) for 24 days and the role of melatonin (Mel) on tBHP-induced ovary oxidative stress was assessed through ovary culture in vitro. The OS (800 μmol/kg BW tert-butyl hydroperoxide) treatment decreased the reproduction performance and ovarian follicle numbers. OS decreased the expression of SIRT1 and increased the *P53* and *FoxO1* expression of the ovary. A decreased Firmicutes to Bacteroidetes ratio, enriched *Marinifilaceae* (family), *Odoribacter* (genus) and *Bacteroides_plebeius* (species) were observed in the cecum of the OS group. Using Mel in vitro enhanced the follicle numbers and decreased the ovary cell apoptosis induced by tBHP. In addition, it increased the expression of SIRT1 and decreased the P53 and FoxO1 expression. These findings indicated that oxidative stress could decrease the laying performance, ovarian function and influence gut microbiota and body metabolites in the layer model, while the melatonin exerts an amelioration the ovary oxidative stress through SIRT1-P53/FoxO1 pathway.

Keywords: follicle atresia; cecal microbiota; metabolomics; melatonin; SIRT1-P53/FoxO1 pathway; ovary stress biomarker

Citation: Wang, J.; Jia, R.; Gong, H.; Celi, P.; Zhuo, Y.; Ding, X.; Bai, S.; Zeng, Q.; Yin, H.; Xu, S.; et al. The Effect of Oxidative Stress on the Chicken Ovary: Involvement of Microbiota and Melatonin Interventions. *Antioxidants* **2021**, *10*, 1422. https://doi.org/10.3390/antiox10091422

Academic Editors: Min Xue, Junmin Zhang, Zhenyu Du, Jie Wang and Wei Si

Received: 15 July 2021
Accepted: 3 September 2021
Published: 6 September 2021

1. Introduction

In recent years, there has been a growing interest in the role of reactive oxygen species (ROS) and oxidative stress in female reproduction [1,2]. Oxidative stress refers to elevated intracellular levels of ROS derived from cellular metabolism or environmental stimuli that cause peroxidation of unsaturated lipids in cell membranes and oxidation of proteins and DNA, leading to further damage of the cell integrity and normal functions [3,4]. Oxidative stress has been proven to be linked to the internal mechanism for aging [5,6], many environmental stressors and chemical toxicants (gamma radiation, mycotoxins, heavy metals, pesticides, etc.), and health disorders [7–10]. Moreover, a large number of studies have shown that an excessive increase in ROS production will induce rapid primordial follicle loss and follicular atresia to lead to reproductive dysfunction [11–14]. However, the underlying pathological and molecular mechanisms in oxidative stress-induced fertility deterioration remain unexplored.

Recent studies in mammals have found that oxidative stress affected nutrient metabolism, altering the body's homeostasis and exerting detrimental effects on the gut microbiota and intestinal function [15,16]. In addition, it has been proved that gut microbial dysbiosis is closely related to inflammation, diseases, and other stress disorders [17,18]. However, until now, the underlying changes in microbiota and their relationship with reproductive function under oxidative stress conditions had not yet been elucidated.

Melatonin (Mel, N-acetyl-5-methoxytryptamine), is an indoleamine, can be mainly bio-synthesized in the pineal gland and the initial precursor of melatonin biosynthesis is tryptophan [19,20]. Mel is also synthesized in numerous peripheral organs, including the intestine, retina, skin and harderian gland [20,21]. It has been shown that Mel reduces oxidative stress by scavenging pro-oxidative molecules such as superoxide anions and detoxifying oxygen and nitrogen-based toxic reactant [21,22]. Melatonin has been shown to have direct effects on ovarian function and microbiota [6,18,19]. Melatonin is identified in human preovulatory follicular fluid, where its concentration is higher than that in peripheral serum [20]. Previous studies have been suggested that Mel may enhance follicle growth by increasing levels of antioxidant enzymes and reproductive hormones in laying hens [22,23]. These findings suggest that Mel is related to the reproductive process, but the underlying mechanism is unclear.

As an NAD+-dependent protein deacetylase, silent information regulator 1 (SIRT1) is involved in the deacetylation of histones and transcriptional factors regulating the cell cycle, and has been a principal modulator of metabolism and resistance to oxidative stress [24–26]. SIRT1 was found to be ubiquitously expressed in the ovaries of animals, and has also been involved in the regulation of ovarian aging, follicular development, and oocyte maturation [24,27]. Previous research has shown that SIRT1 is involved in the protective effects of melatonin [28]; however, the exact mechanism is not elucidated. Foxo1 (forkhead box O1), a member of the FOXO transcription factor family, is an important player in regulating cell fate and combating oxidative stress and a downstream target of SIRT1 [29,30]. Is the SIRT1-FoxO1 pathway involved in the antioxidative stress function of melatonin in the ovary?

The poultry ovary is a classic model for studying ovarian biology, follicular development and ovarian cancer. In this study, we hypothesized that oxidative stress would decrease ovarian function by changing body metabolism and gut microbiota, while Mel administration would prevent oxidative damage and maintain the ovarian function of laying hens. Thus, this study aimed to investigate the negative effect of oxidative stress on ovary function, gut microbiota and serum metabolome in a layer model. We also determined the modulating effect of melatonin on ovarian follicle atresia in order to identify the relationship between melatonin and oxidative stress-induced ovarian dysfunction in vitro.

2. Materials and Methods

2.1. Animals, Diets and Design

Thirty Lohmann laying hens (62 week of age; BW = 1.42 ± 0.12 kg) were randomly allocated into two experimental groups (*n* = 15). Layers were received an intraperitoneal injection of 5 mL of 0 (CON, phosphate-buffered saline) or 800 µmol/kg BW (oxidative stress group, OS; the dosage were determined according to Kučera et al. 2014) tert-butyl hydroperoxide (tBHP) at 9 a.m. on the 8th, 15th and 22nd day of the 24-day experiment. The experiment protocol is depicted in Figure 1(Aa). All hens were housed in an environmentally controlled room (45–60% relative humidity, 22 ± 2 °C temperature; lighting cycle, 16 h/day; 05:00 a.m. to 09:00 p.m. for light). Hens were supplied with water ad libitum and fed the same amount (100 g/kg) of complete feeding mixture in mash form (the diet nutrient composition is shown in Supplementary Materials Table S1).

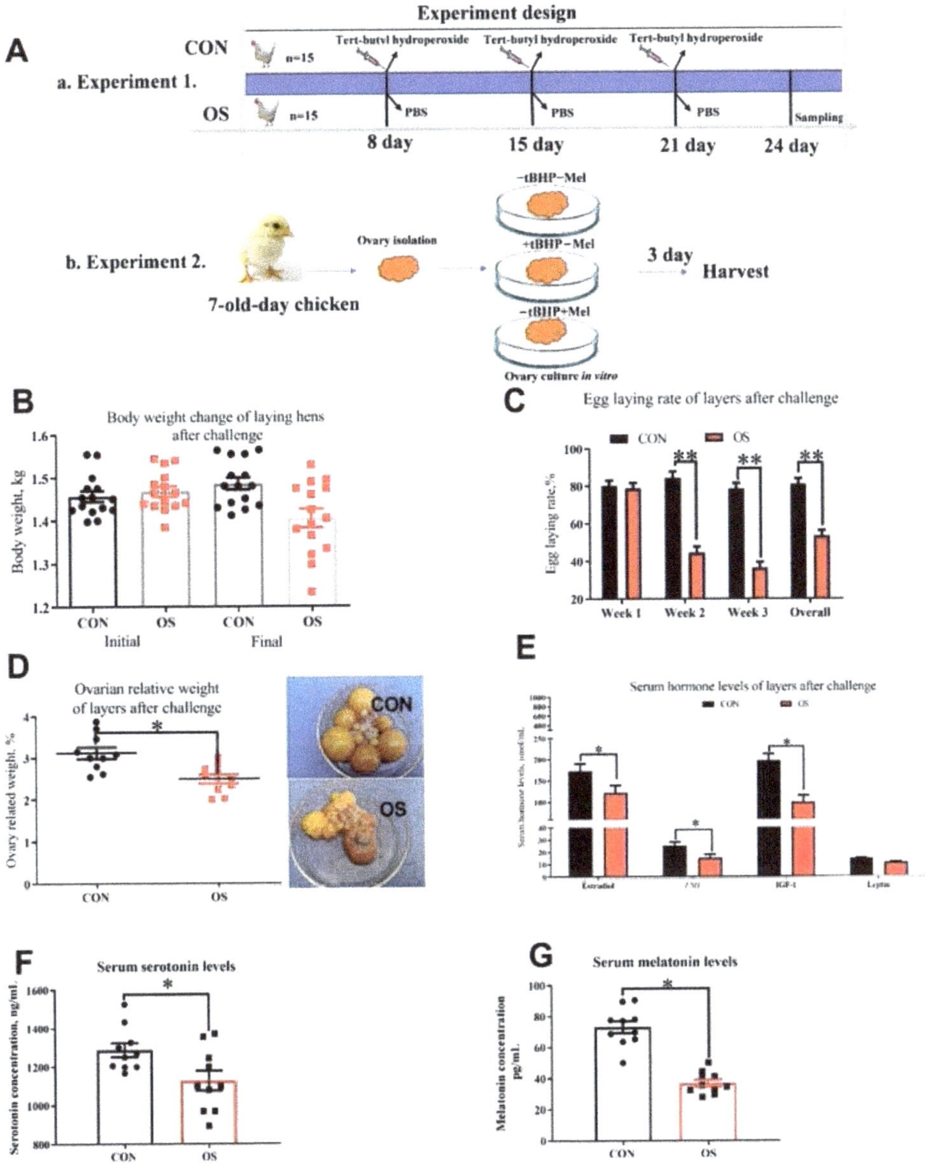

Figure 1. Oxidative stress (induced by tBHP) reduced egg-laying rate, ovary indices and serum hormone levels (Experiment 1). Data are means ± SEM represented by vertical bars or plot individual values ± SEM. (**A**) Schematic illustration of the experimental design. In Experiment 1, layers were fed the same basal diet for 24 days and with the tBHP (OS) or PBS (CON) injection at 9 AM of the d 8, 15, and 21. In Experiment 2, the ovaries of 7-day-old chickens are isolated for in vitro culture, and received different culture medium with 0 tBHP + 0 melatonin (−tBHP − Mel), 50 µmol/mL tBHP (+tBHP − Mel), and 50 µmol/mL tBHP + 100 ng/mL melatonin (+tBHP + Mel). (**B**) Body weight. (**C**) Egg-laying rate after challenge. (**D**) Ovarian relative weight. (**E**) Serum reproductive and metabolism-related hormone levels. (**F**) Serum serotonin levels, and (**G**) Serum melatonin levels. FSH = follicle-stimulating hormone, IGF-1 = insulin-like growth factor-1, CON = same dosage of PBS, OS = 800 µmol/kg BW of tBHP. Statistical significance was evaluated by *t*-test, * $p < 0.05$, ** $p < 0.01$.

2.2. Sample Collection and Measurements

Oviposition time of the laying hens was recorded, and blood samples of each hen were collected immediately following the first oviposition in the series according to the method described by Ren [31]. Serum samples were obtained from these blood samples by incubation at 4 °C for 30 min and subsequent centrifugation at 1500× *g* for 20 min. The same hens were then sacrificed with an overdose intravenous injection of sodium pentobarbital (300 mg/kg BW), the ovary was immediately harvested after washing with phosphate-buffered saline (PBS; pH = 7.2–7.4) and placed into 4% paraformaldehyde (pH = 7.2) fixation and paraffin for ovarian follicle counts (H&E) (described in 2.6) and cell apoptosis (TUNEL) analysis. The cecum contents were carefully collected, immediately placed in cryogenic vials, stored immediately at −20 °C in a portable freezer, delivered to the laboratory and stored at −80 °C until DNA extraction.

2.3. Culture of Layer Ovaries In Vitro

Ovaries organ from 7-day-old chickens were used to evaluate the effect of Mel on oxidative stress-induced follicle atresia. Ovary culture was performed in vitro as previously described [32,33]. Ovary were cultured in graded level of melatonin and tBHP (−tBHP − Mel = 0 Mel + 0 tBHP, +tBHP − Mel = 50 µmol/mL + 0 Mel + tBHP + Mel = 50 µmol/mL + 100 pg/mL) according to the previous study and also the melatonin levels in ovarian tissue. The in vitro experiment procedure is depicted in Figure 1(Ab).

2.4. Reproduction Performance and Blood Hormone Assay

Egg-laying rates were recorded every day and the body weight of the layer was recorded on the onset and end day of the experiment (n = 15). On day 24 of the animal trial, blood samples were collected (n = 10) from the wing after 12 h of fasting and the serum were separated by incubation at 4 °C for 30 min and subsequent centrifugation at 1500× *g* for 20 min. Commercial enzyme-linked immunosorbent assay (ELISA) kits were used to detect the serum concentration of estrogen (estradiol, ABIN5524298), follicle-stimulating hormone (FSH, ELISAGenie, CHFI00020, London, UK), insulin-like growth factor-1 (IGF-1, ELISAGenie, CHFI00088), leptin (ELISAGenie, CHEB0024), melatonin (IBL, #RE54021, Germany) and serotonin (Abcam, ab133053, Cambridge, UK).

2.5. Tissue Antioxidant Capacity

The activities of the following enzymes (n = 10) [superoxide dismutase (SOD), glutathione s-transferase (GST)], Cu-ZnSOD and MnSOD the levels of protein carbonyl (PC), total antioxidant capacity (T-AOC) and glutathione (GSH) and malondialdehyde (MDA) in ovarian tissue were determined by colorimetric enzymatic assays, and serum MDA concentration was measured by chemical colorimetric method, using an ELISA microplate reader (Tecan Co., Grodingen, Austria) and assay kits (T-SOD, A001-1-2; A004-1-1; Cu-ZnSOD, A001-4-1; GST, A004-1-1; PC, A087-1-2; T-AOC, A015-1-2; GSH, A005-1-2; MDA, A003-1), which were purchased from Nanjing Jiancheng Bioengineering Institute of China. All assays were conducted and interpreted according to the manufacturer's manual without any modification.

2.6. Histology and Follicle Counts

The quantification of ovarian follicles (n = 10) was performed as previously described with small modifications [29,30]. In brief, paraffin-embedded ovaries were serially sectioned at 5-µm thickness and stained with H&E for morphological observation. Every fifth section was counted for both ovaries in vivo experiment and in vitro cultured trial. Different classes of ovarian follicles were defined as previously described [22,34]. Those with oocytes surrounded by one layer of flattened five to eight somatic cells were defined as primordial follicles. Primary follicles consisted of an oocyte surrounded with one layer containing one enlarged cell or a whole layer of cuboidal pre-granulosa cells. Prehierarchical follicles contained more than one layer of granulosa cells, including small white follicles (SWFs, 2–4 mm), large white follicles (LWFs, 4–6 mm) and small yellow follicles (SYFs, 6–8 mm). Atretic follicles were defined as previously described [34,35]: early-aretic follicles were traversed by few or no blood vessels, and the granulosa cell layer had become either partially or completely detached from the basement membrane; the surface of the progressed-atretic follicles was opaque, the granulosa cell layer was totally disorganized within the follicle.

2.7. TUNEL Assay

At the end of experiment, the ovary tissue (n = 10) was quickly removed and placed immediately into methylaldehyde; then a TUNEL assay was performed, using an In Situ Cell Apoptosis Detection Kit I (POD), according to the manufacturer's protocol (Roche Group, Switzerland). Using BA200Digital (Mike Audi Industrial Group Co., Ltd., Xiamen, China) for image acquisition. The diaminobenzidine reacted with the labeled sample to generate an insoluble brown (light green in Experiment 2) signal, while blue-green to greenish tan signified nonapoptotic cells. Overall, 100 images were taken to measure cell apoptosis, and apoptosis rate was defined as the percentage of apoptotic cells (granulosa cells) in 100 granulosa cells counted. Observations were conducted in four different directional areas of each follicle. If the positive cells in the four areas of a follicle exceeded one-third, the follicle was defined as atretic. The percentage of atretic follicles (less than 2 mm in diameter) per section was calculated as the number of atretic follicles.

2.8. Ovary Function Related mRNA Expression by Real-Time PCR

The total RNA and real-time RT-PCR were carried out as described previously [9]. In brief, total RNA were extracted with Trizol followed by DNase1 treatment to remove genomic DNA. Gene expression of caspase 3, caspase 9, Bcl-2 (B-cell lymphoma-2), Bax, SIRT1 (silent information regulator 1), FoxO1 (forkhead box O1), and P53 was determined by quantitative real-time PCR in the ovary of layers by ABI 7900 Real-Time PCR system (ABI Biotechnology, Eldersburg, MD, USA). The primer information for all the genes is listed in Supplementary Materials Table S2. Each sample was assayed in triplicate and β–actin was used as the house-keeper genes. The $2^{-\Delta\Delta CT}$ method was used to calculate target gene expression, and mRNA expression in CON was used as baseline relative to treatment groups (i.e., fold-change).

2.9. Western Blotting Assay

Total protein expression in ovarian tissues (n = 10; Experiment 2) were detected by Western blotting as previously described [36,37]. Primary antibodies against FoxO1 (#2880; 1:1000 dilution; 78–82KD; Cell Signaling Technology), and P53 (#29453; 1:1000 dilution; 53KD; Novus), SIRT1 (#8469;1:1000 dilution; 110KD; Cell Signaling Technology) and β-actin (#4970S) were obtained and a pre-study was performed to confirm its specificity. The goat anti-rabbit IgG-HRP (1:10,000 dilution; Santa Cruz) was used as the secondary antibody. Meanwhile, a mouse monoclonal antibody against β-actin (1:5000 dilution; ImmunoWay, Plano, TX, USA) acted as a reference, and the goat anti-mouse IgG-HRP (1:10,000 dilution; Santa Cruz) was used as the secondary antibody. Bound antibodies were detected by the ECL Prime Western blotting detection reagent (GE-Healthcare) using ImageJ software (National Institutes of Health, Bethesda, MD, USA).

2.10. Gut Microbiota and Short-Chain Fatty Acids (SCFA) Analysis

The composition of the microbial community in the cecum digesta (n = 10) was assessed with high-throughput pyrosequencing as recently described [37]; the sequencing and bioinformatics analysis were performed by Novogene Bioinformatics Technology Co. (Tianjin, China). SCFA (n = 10) including acetate, propionate and butyrate in the cecum content, were also analyzed using Agilent 6890 gas chromatograph (Agilent Technologies, Santa Clara, CA, USA) following previous protocols [38].

2.11. Metabolic Profiling Analysis

The ovarian tissue sample (n = 4) from laying hens (2 ovarian tissue samples from the same treatment were pooled together to test the metabolic profiling, but 1 pooled sample from replicate 9 and 10 were contaminated so only 4 pooled sample were used here) were taken to analyze the effect of oxidative stress on ovarian metabolic and biochemical alternation in layers. The method was performed by Novogene Bioinformatics Technology Co. (Tianjin, China) according to the previous published (see the Supplementary Materials files). Differences were indicated when p-value was <0.05, VIP (Variable Importance in the Projection) > 1, and only fold changes >1.5 were considered.

2.12. Statistical Analysis

Data were analyzed by one-way analysis of variance (ANOVA) using GLM procedure of SAS 9.2 (SAS Institute, Cary, NC, USA) and GraphPad Prism (GraphPad Inc., La Jolla, CA, USA), and the difference between two treatments was performed by student t-test (Experiment 1) and Tukey's test (Experiment 2). The results are presented as mean ± SEM.

3. Results

3.1. Oxidative Stress Reduced Reproduction Performance, Ovary Indices and Serum Hormone Concentration

The hens in the OS group presented lower egg-laying rate throughout the trial compared to the CON group; moreover, the ovary indices were lower in hens challenged with tBHP (Figure 1C,D; $p < 0.05$). A decrease in the serum concentration of estradiol, FSH, IGF-1, serotonin and melatonin were observed in the OS group compared to the CON one (Figure 1E–G; $p < 0.05$). No differences in body weight were noted between the CON and OS groups (Figure 1B; $p > 0.05$).

3.2. tBHP Induced Oxidative Stress Decreased Ovarian Function

Compared to the CON group, layers challenged with tBHP had lower numbers of primordial, prehierarchical and total follicles (Figure 2A,B; $p < 0.05$). The number of atretic follicles and the ovary cell apoptosis rate were higher in the OS groups than in the CON ones (Figure 2A,B and Figure 3A,B; $p < 0.05$). The levels of antioxidant parameters (T-SOD, MnSOD, T-AOC) were decreased while concentration of oxidant products (PC, MDA) and GSH (intracellular antioxidant) was higher in the OS treatment (Figure 2E,F; $p < 0.05$), suggesting that the tBHP challenge induced oxidative stress in the ovary. The real-time PCR results indicated that the mRNA expression level related to ovarian follicular apoptosis factors (Bax, caspase 3, caspase 8, FoxO1, and P53) was greater, while the mRNA abundance of anti-apoptotic genes (Bcl-2 and SIRT1) was lower in the OS group compared to the CON group (Figure 2C,D; $p < 0.05$).

3.3. Oxidative Stress Reduced Cecal Short Chain Fatty Acids and Induced Gut Microbiota Dysbiosis

The concentration of the main short chain fatty acids (acetic, propionic and butyric) and the total SCFA in the cecum contents of layers treated with tBHP were lower compared to those in the CON group (Figure 4A; $p < 0.05$). As suggested by the decrease in the ACE, Chao1, and Simpson indexes relative to CON group, the tBHP challenge resulted in reduced microbial diversity (Figure 4B,C; $p < 0.05$). Structural changes in intestinal microbiota were assessed by PCoA based on the unweighted UniFrac metric, which indicated that layers in both groups had obvious clustering (Figure 4D).

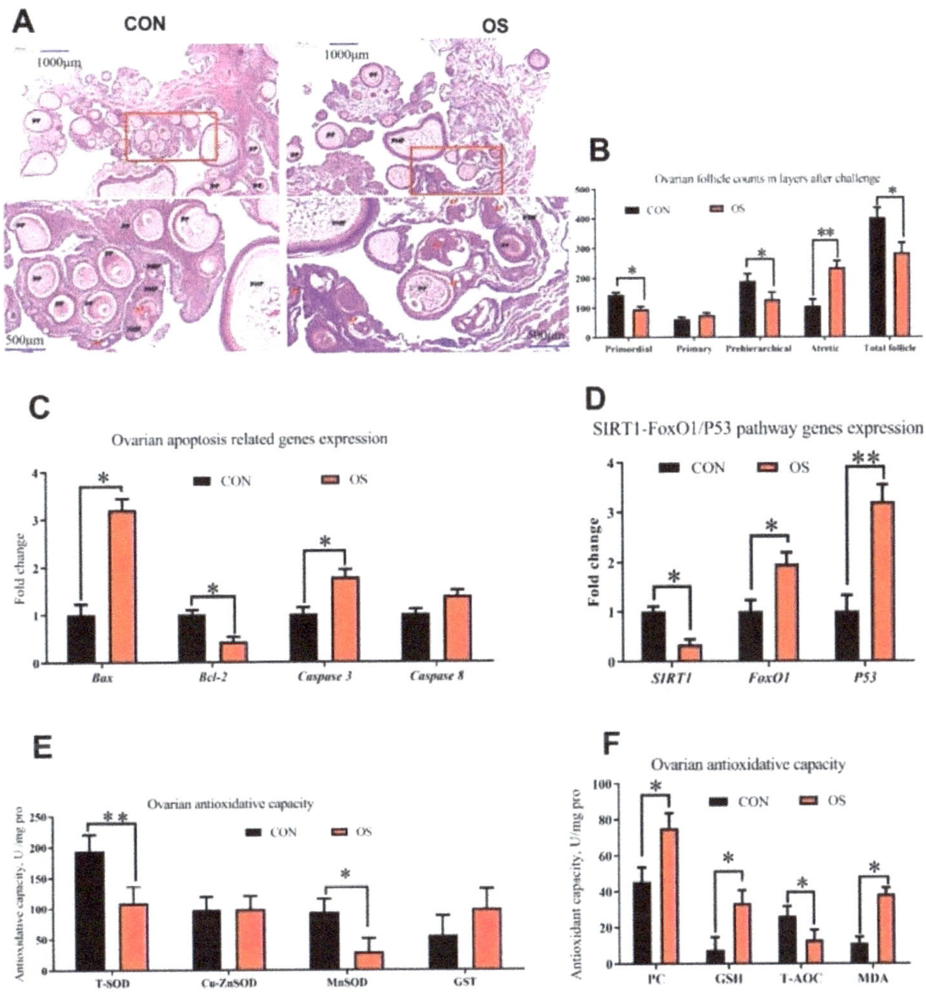

Figure 2. Oxidative stress (induced by tBHP) decreased ovary function (Experiment 1). (**A**,**B**) Ovarian histology of layers after 24 days treatment and follicle numbers at each developmental stage (PMF = primordial follicle; PF = primary follicle; PHF = prehierarchical follicle; AF = atretic follicle). (**C**,**D**) RT-PCR analysis for mRNA expression levels related to ovarian follicular apoptosis and antioxidant related gene expression. (**E**,**F**) Antioxidant capacity analysis for ovary with antioxidant enzyme activities and oxidation product. Data are means ± SEM represented by vertical bars or plot individual values ($n = 10$) ± SEM. CON = same dosage of PBS, OS = 800 μmol/kg BW of tBHP, Bcl-2 = B-cell lymphoma-2, SIRT1 = sirtuin 1, FoxO1 = forkhead box O1, T-SOD = total superoxide dismutase, GST = glutathione S transferase, PC = protein carbonyl, GSH = glutathione, T-AOC = total antioxidant capacity, MDA = malondialdehyde. Statistical significance was evaluated by *t*-test, * $p < 0.05$, ** $p < 0.01$.

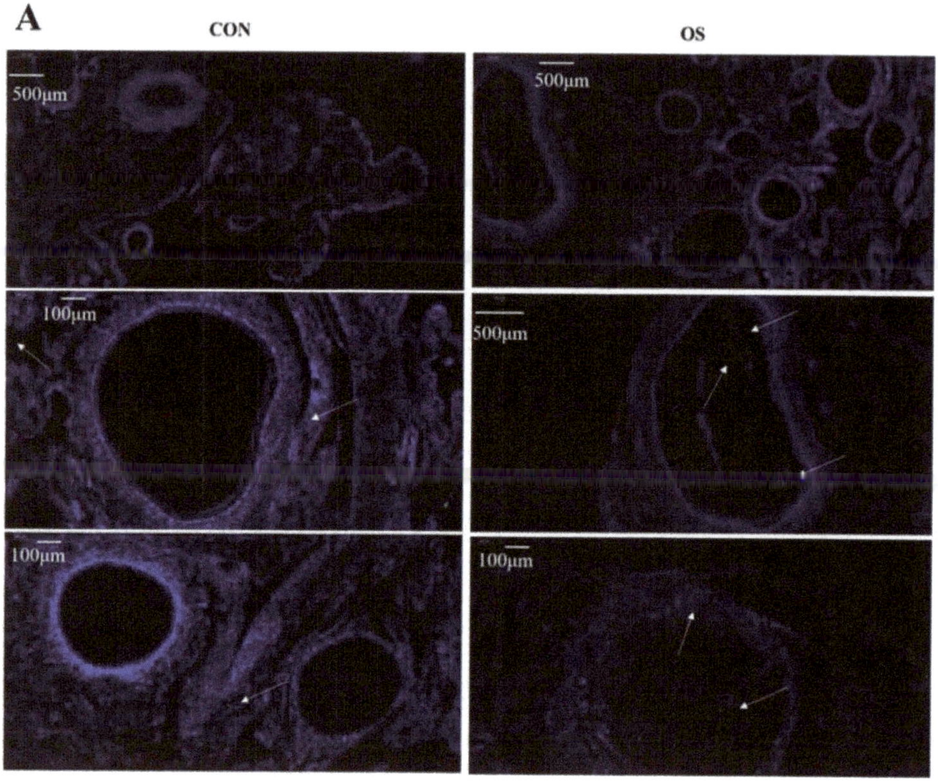

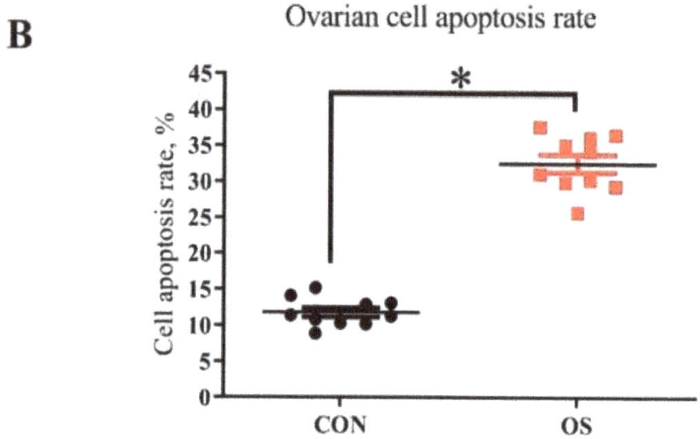

Figure 3. Oxidative stress (induced by tBHP) increased the cell apoptosis rate of ovary (Experiment 1). (**A**) TUNEL analysis for cell apoptosis in ovary; (**B**) The immunofluorescence results of TUNEL with the green color presents the positive cells. Data are plot individual values (n = 10). CON = same dosage of PBS, OS = 800 μmol/kg BW of tBHP. Statistical significance was evaluated by t-test, * $p < 0.05$.

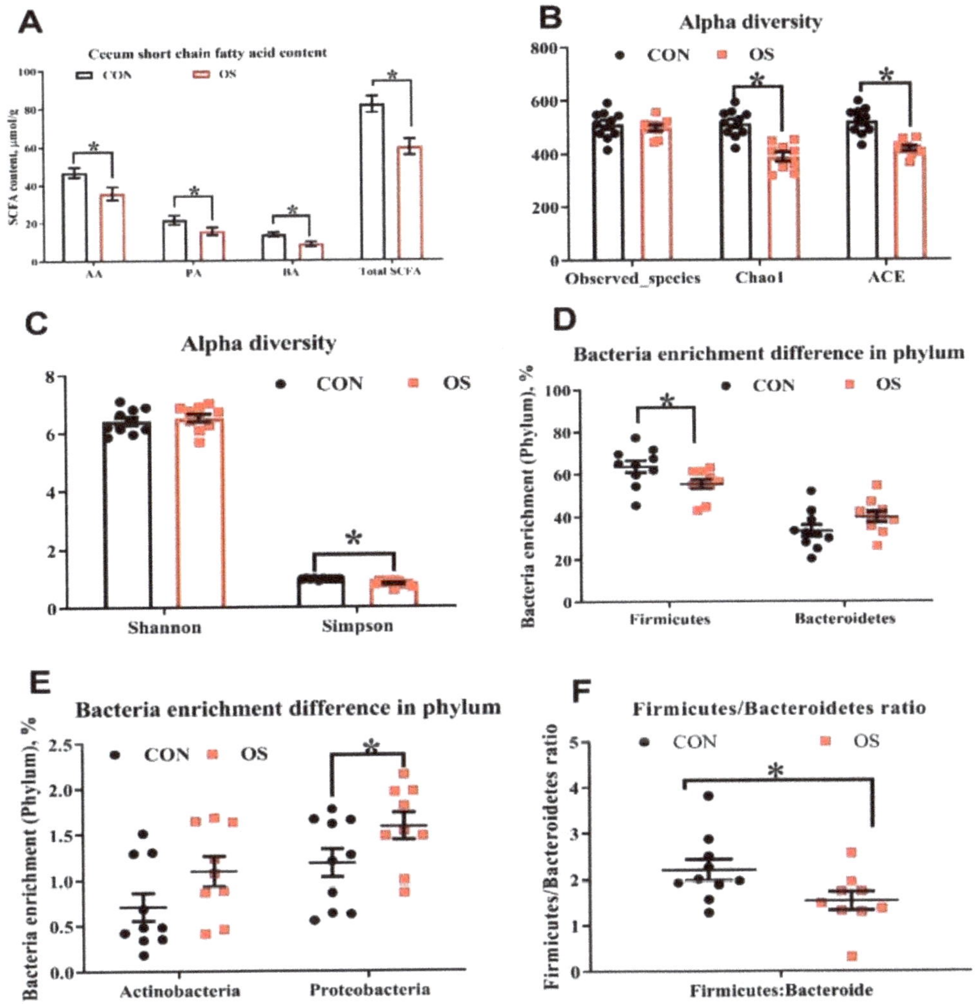

Figure 4. Oxidative stress (induced by tBHP) reduced cecal short chain fatty acid concentration (Experiment 1). Data are means ±SEM represented by vertical bars or plot individual values. (**A**) Short chain fatty acid concentration in cecal digesta. (**B**,**C**) Alpha diversity of cecum microbiota, with Observed species, Chao 1, ACE (**B**) and Shannon and Simpson index (**C**). (**D**–**F**). The difference $n = 10$. Statistical significance was evaluated by the Student's t-test, * $p < 0.05$.

As analyzed by Adonis and LEfSe (log10 LDA > 3), the tBHP challenge significantly reduced the relative abundance of Firmicutes (Phylum), *Lactobacillus* (genus), *Faecalibacterium* (genus), *unidentified_Prevotellaceae* (genus), *Prevotella_sp_KHD1* (Species), *Butyricicoccus* (Species), and *Clostridiales_bacterium_DJF-B152* (species) (Figure 4E,F; Figure 5A–E; Figure 6; $p < 0.05$), while it led to an increase in the enrichment of *Proteobacteria* (Phylum), *Marinifilaceae* (family), *Odoribacter* (genus) and *Clostridiales_bacterium_77_5d* (species). The *Firmicutes* to *Bacteroidetes* ratio was also lower in the OS group compared with the CON group (Figure 6; $p < 0.05$).

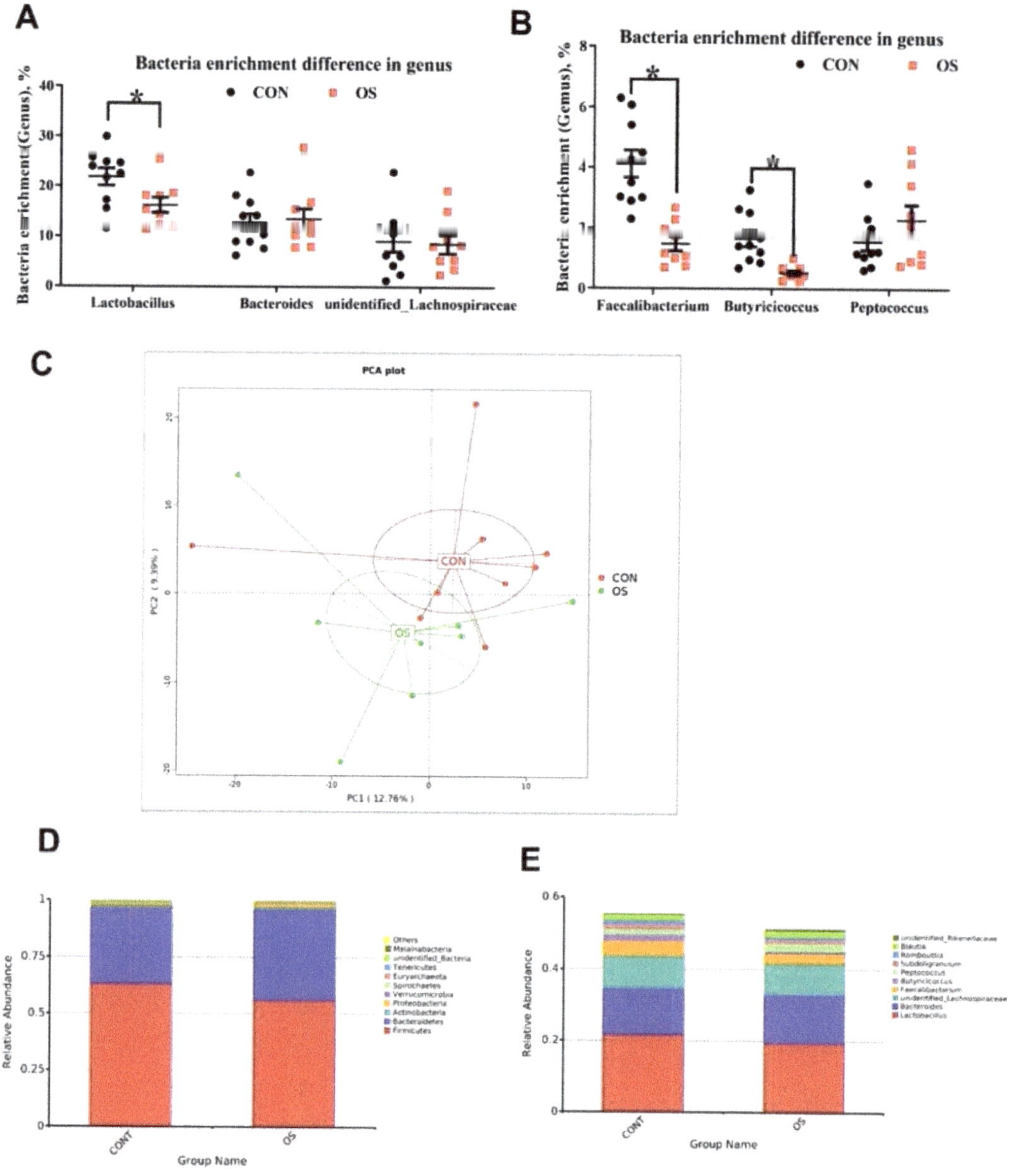

Figure 5. Oxidative stress (induced by tBHP) changed microbiota diversity (Experiment 1). Data are means ± SEM represented by vertical bars or plot individual values. (**A**,**B**) The bacteria abundance enrichment difference in the genus. (**C**) The principal coordinate analysis (PCoA) of the cecum microbiota based on unweighted UniFrac metrc. (**D**,**E**) The relative abundance of the top 10 phylum (**D**) and genus (**E**). The difference $n = 10$. Statistical significance was evaluated by the Student's t-test, * $p < 0.05$.

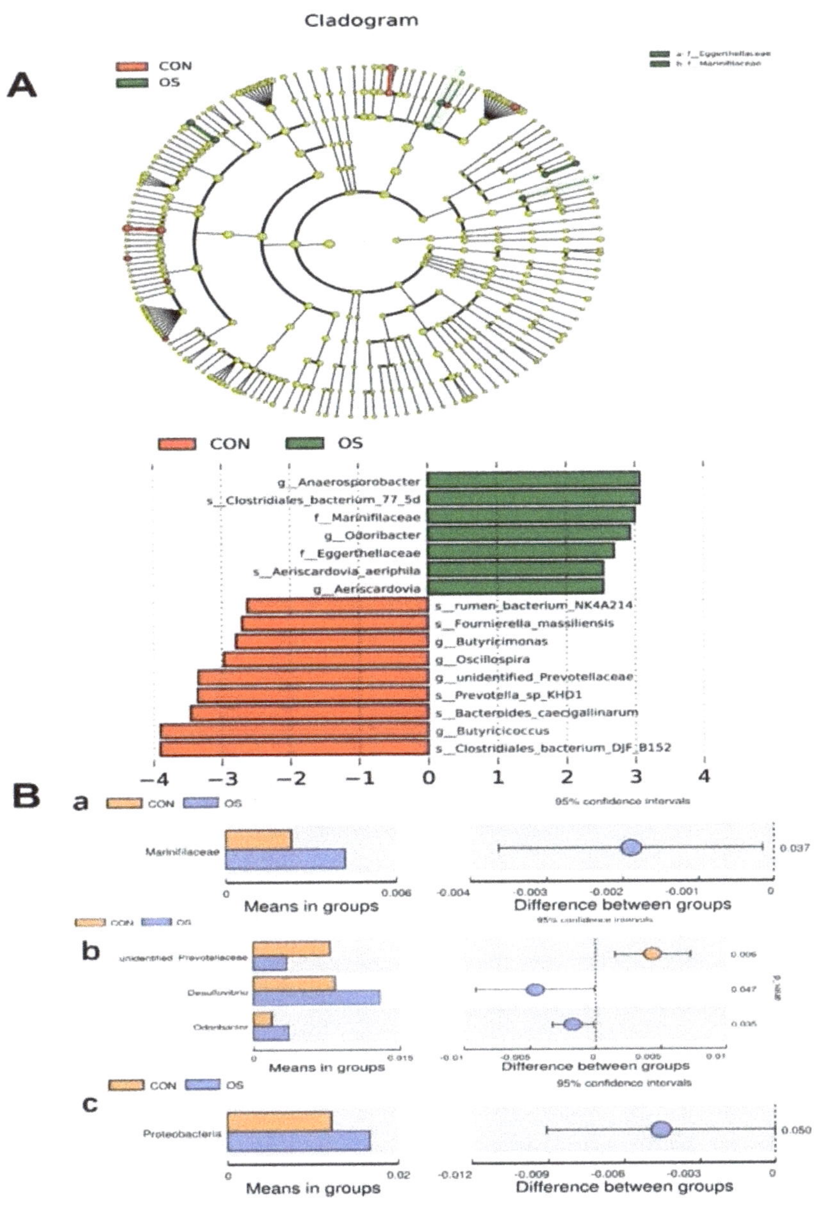

Figure 6. Oxidative stress (induced by tBHP) changed microbiota enrichment (Experiment 1). Data are means ±SEM represented by vertical bars or plot individual values. (**A**) Linear discrimination analysis coupled with effect size (LEfSe) identified most differentially abundant taxa in the cecum with LDA significant threshold > 3 were shown. (Red) CON enriched taxa; (Green) OS enriched taxa. (**B**) Analysis of different species between groups at different levels. (**a**) Family, (**b**) Genus, (**c**) Species. The difference n = 10. Statistical significance was evaluated by the Student's t-test.

3.4. Oxidative Stress Changed Amino Acid Biosynthesis and Ovarian Serotonin and Melatonin Concentration

In this study, a total of 481 metabolites were identified; the layer in the OS group showed a notable metabolite alteration as compared to the CON group, as showed by PLS-DA analysis (Figure 7A). We observed that 30 metabolites were altered between the OS and CON groups, of which 21 of them were downregulated, and the rest 9 were upregulated (Figure 7B,C; $p < 0.05$, VIP > 1). The significantly altered metabolites were distributed as amino acids, lipids, organic acids, nucleosides, sugar alcohol, aliphatic acyclic compounds, aromatic heteropolycyclic compounds, and aliphatic heteromonocyclic compounds. In order to identify possible pathways relevant to the development of oxidative stress, all of the attributed metabolites were subjected to a high-quality KEGG metabolic pathways database (Figure 7C). As shown in Figure 8D, the concentration of differential metabolites in the corresponding metabolic pathways of Tryptophan metabolism, Pyruvate metabolism, Gap junction and Citrate cycle (TCA cycle) ($p < 0.05$). The concentrations of serotonin and melatonin in the ovary of layers challenged with tBHP were increased (Figure 8A,B; $p < 0.05$), which was similar to the observed pattern of metabolomic profile (Figure 7B,C; $p < 0.05$, VIP > 1).

Correlations between metabolites and microbiota with significant differences between the two groups were obtained via Spearman's correlation analysis. As shown in Figure 8E, we observed that the bacteria genera, including *Oscillospira, unidentified_Prevotellaceae* and *Odoribacter* were most closely related to the changed metabolites in the ovary of OS group ($p < 0.05$). In particular, bacteria of the *Odoribacter* genera were negatively correlated to the concentration of S-Sulfo-L-Cysteine and positively correlated to the 5-Methoxyindole-3-acetic acid (indole and its derivatives), while bacteria of the Oscillospira genera were related mostly to the organic acids (anthranilic acid, (S)-2-Hydroxybutanoicacid) and nucleotide metabolism process (UMP, uridine5-diphosphate).

3.5. Melatonin Ameliorates tBHP Induced Oxidative Stress Ovarian Dysfunction In Vitro

Seven-day-old chicken ovaries were cultured and treated with tBHP and Mel for 3 days in vitro, and the histological analysis revealed that the addition of tBHP in the cell culture medium decreased the number of primordial follicles, primary and total follicles and increased the number of atretic follicles, cell apoptosis rate, and also the proapoptotic markers (Bax and caspase 3) in the ovary, whereas the addition of Mel was able to reverse the detrimental effect of tBHP (Figure 8A–D; $p < 0.05$). The concentration of protein oxidant products (PC) was higher in the ovary treated with tBHP, while it decreased after Mel administration (Figure 8E; $p < 0.05$). *SIRT1* gene expression and protein levels were decreased, while *FoxO1* and *P53* were increased in the OS group; the addition of Mel was able to reverse this effect (Figure 9F,G; $p < 0.05$).

A

B

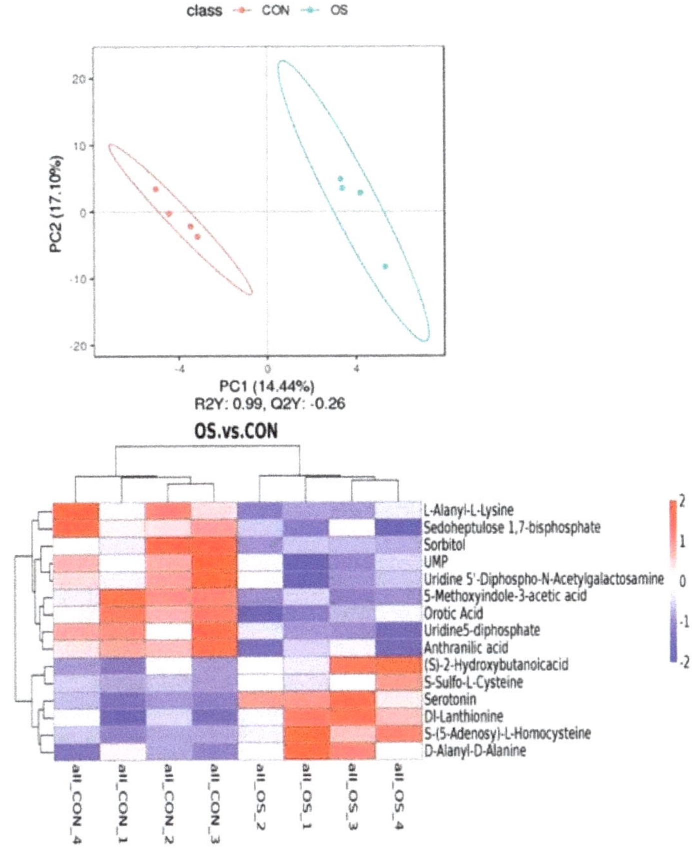

C

Ovarian different metabolites in OS group compared to controls

Name_des	VIP	FC	Up.Down
Uridine5-diphosphate	2.04	0.59	down
5-Methoxyindole-3-acetic acid	4.97	0.27	down
L-Alanyl-L-Lysine	3.90	0.36	down
Orotic Acid	1.65	0.66	down
Anthranilic acid	1.64	0.66	down
UMP	2.48	0.54	down
Sedoheptulose 1,7-bisphosphate	1.77	0.64	down
Uridine 5'-Diphospho-N-Acetylgalactosamine	1.69	0.65	down
Sorbitol	1.58	0.65	down
Serotonin	2.05	1.71	up
S-(5-Adenosy)-L-Homocysteine	1.88	1.65	up
(S)-2-Hydroxybutanoicacid	4.50	3.06	up
D-Alanyl-D-Alanine	3.30	2.24	up
S-Sulfo-L-Cysteine	3.22	2.49	up
Dl-Lanthionine	3.83	2.43	up

Figure 7. Oxidative stress (induced by tBHP) changed the serum metabolites. (**A**) The principal component analysis (PCA) of the serum metabolites. (**B**) Heatmap of the different serum metabolites. (**C**) Ovarian different metabolites in OS group compared to controls. $p < 0.05$.

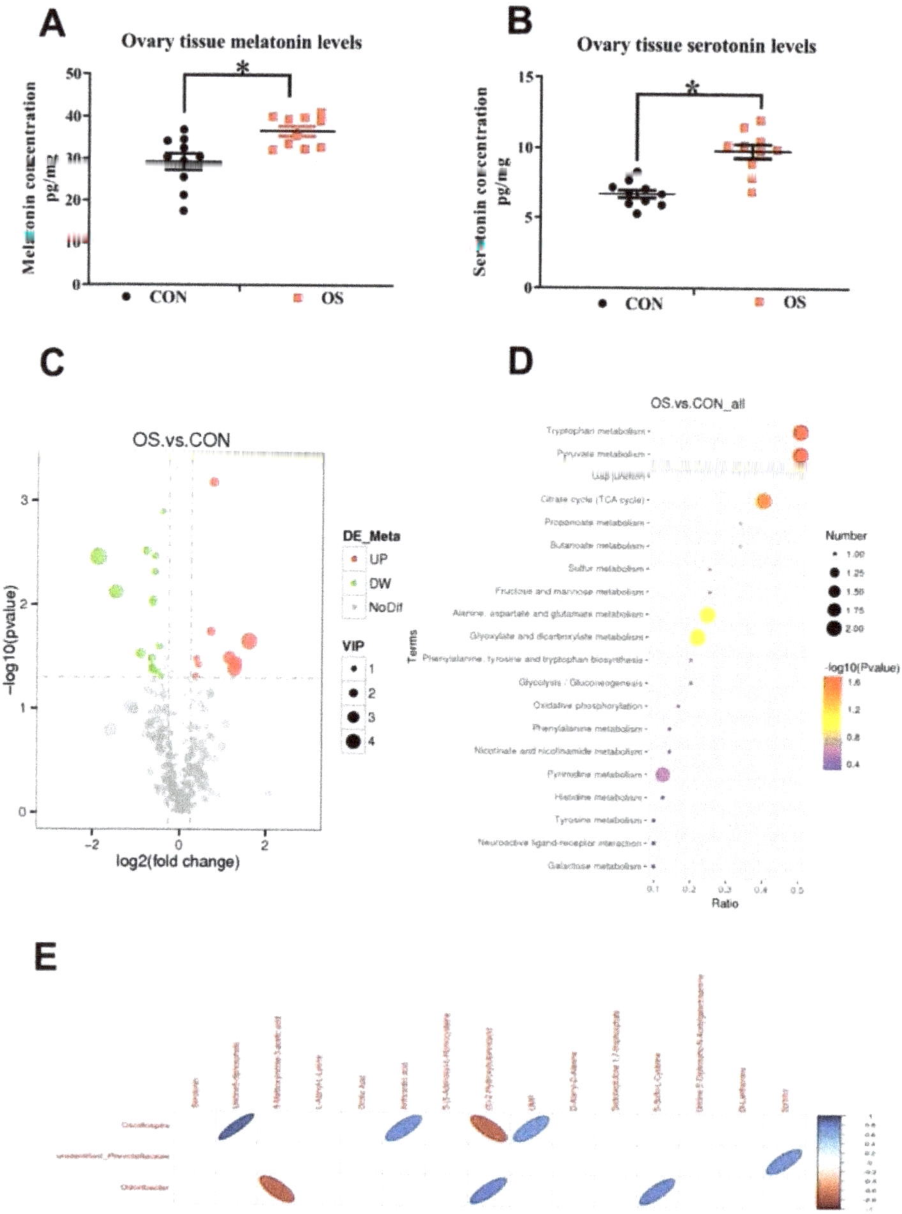

Figure 8. Oxidative stress (induced by tBHP) reduced increased serotonin and melatonin levels in the ovary and changed body amino acid biosynthesis and metabolism. (**A**) Ovarian tissue melatonin levels. (**B**) Ovarian tissue serotonin levels. (**C**) Description of the different metabolites between OS and CON group by volcano plot. (**D**) KEGG pathway enrichment of target metabolites. (**E**) Spearsman Correlationship between metabolites and microbiota. Statistical significance was evaluated by the Student's *t*-test, * $p < 0.05$.

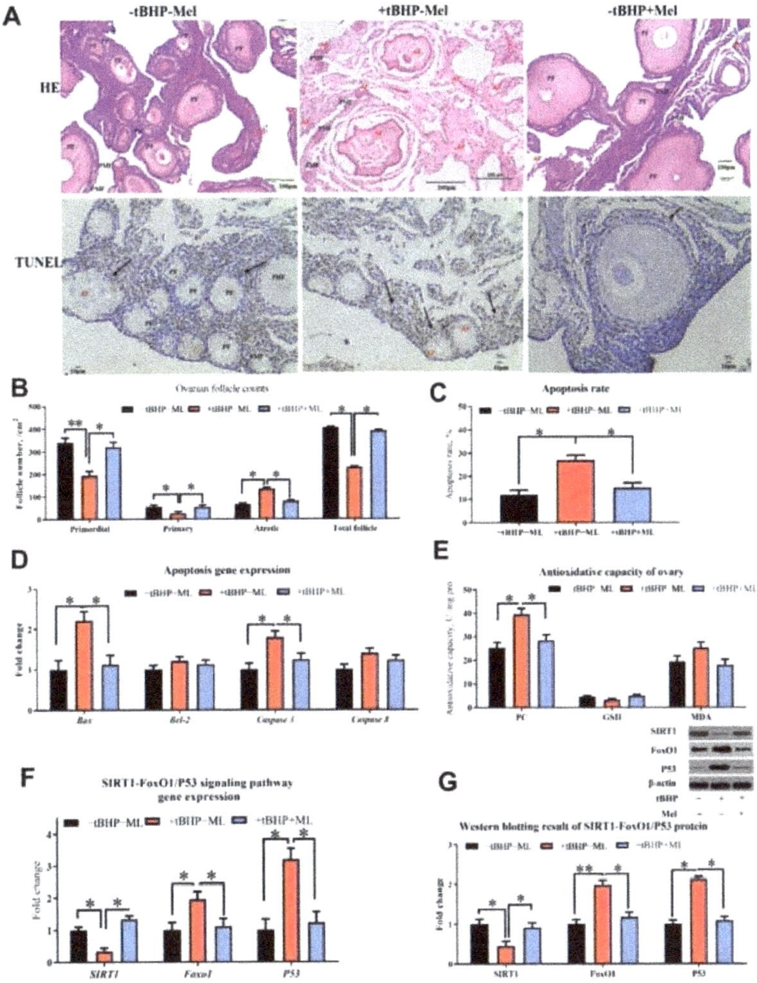

Figure 9. Melatonin ameliorates oxidative stress (tBHP) induced ovarian dysfunction in vitro (Experiment 2). (**A**,**B**) Ovarian histology and TUNEL analysis of layers after three days of treatment and follicle counts at each developmental stage (PMF = primordial follicle; PF = primary follicle; AF = atretic follicle). TUNEL analysis for cell apoptosis in ovary (**A**,**C**) showed the TUNEL results with the brown color presents the positive cells. (**D**) RT-PCR analysis for mRNA expression levels related to ovarian apoptosis gene expression. (**E**) Antioxidant capacity analysis for ovary with oxidation product. (**F**,**G**) the western blotting result of apoptosis associate protein (Bax, Bcl-2, caspase 3 and cleaved-caspase 3). Data are shown as averages and error bars represent SEM± SEM; statistically significant differences are shown with different letters ($p < 0.05$) with one-way ANOVA with Tukey's test. −tBHP − Mel = 0 tBHP + 0 melatonin, +tBHP − Mel = 50 μmol/mL tBHP, +tBHP + Mel = 50 μmol/mL tBHP + 100 ng/mL melatonin, Bcl-2 = B-cell lymphoma-2, SIRT1 = silent information regulator 1, FoxO1 = forkhead box O1, PC = protein carbonyl, GSH = glutathione, MDA = malondialdehyde. Statistical significance was evaluated by *t*-test, * $p < 0.05$, ** $p < 0.01$.

4. Discussion

Proper functioning of the ovary is critical to maintaining fertility and overall health, and ovarian function depends on the maintenance and normal development of ovarian follicles [39,40]. Accelerated metabolism occurs in rapidly proliferating granulosa cells (GCs) within developing follicles, leading to increased ROS production. Moreover, accumulating evidence demonstrates that excessive ROS are the key signals in the initiation of apoptosis in granulosa cells of both primordial and growing follicles by diverse stimuli, such as alcohol, radiation, and smoking, as well as malnutrition and obesity [14,41,42]. Melatonin, an endogenous component of follicular fluid, has been suggested to improve GCs resistance to oxidative stress in vitro cell model [43–45]. In this study, we investigated the role of Mel in protecting ovarian follicles from atresia via the SIRT1-FoxO1/P53 pathway; we also examined the role of the gut microbiota in this process.

Oxidative stress induces apoptosis and affects cellular homeostasis, which results from dysregulation between antioxidant and pro-oxidant availability [42]. PC and MDA are produced during protein and lipid peroxidation, and they can be indicative of OS [46]. The markedly lower levels of T-SOD, Mn-SOD, T-AOC activities, along with significantly elevated MDA and PC levels within the ovary of tBHP challenged layers (OS group) suggest the presence of oxidative damage and also indicates that our model was successfully established in the current study.

We observed that OS decreased the layers' reproductive performance as indicated by the lower egg-laying rate, decreased hormone levels (lower estradiol, FSH, IGF-1), and also by the smaller primordial follicle reserve and increased atresia in the ovaries of the tBHP challenged layers. In our previous study, we also found that oxidative stress (induced by high levels of molybdenum and vanadium) decreased egg production in layers and the addition of antioxidants (tea polyphenols) was able to reverse this effect by improving the antioxidant capacity and gut microbiota balance [9,47]. Studies in mammals have found that oxidative stress can reduce the number of follicles in each stage of the ovarian cycle and impair ovarian function [48–52]. Additionally, previous studies have also reported that oxidative stress decreases hormone secretion (including lower estradiol, FSH, LH and IGF-1) and impairs the glutathione redox cycle [50,53]. Recent animal studies indicate that IGF-1 exerts antioxidant effects and anti-inflammatory effects in animals [54,55]. Leptin was demonstrated to exert an important role in the regulation of ovarian folliculogenesis indirectly via control of luteinizing hormone and FSH secretion [52]; however, the OS did not affect leptin levels, but decreased LH and FSH levels in the current study.

Interestingly, both melatonin and serotonin serum levels were decreased after the OS challenge, but the levels of melatonin and serotonin levels were enhanced in the ovary, while the addition of melatonin in the in vitro experiment was able to reverse these adverse effects induced by the tBHP challenge. To date, there is a paucity of data elucidating the mechanism contributing to the gut–ovary axis in the literature. In the present study, we aimed to establish a link between the gastrointestinal microbiome and ovarian function. In the current study, we found that the microbial diversity was reduced by the tBHP challenge. In agreement with our observations, another study also reported that mice fed a high-fat diet resulted in oxidative stress and the increased ROS levels disrupted the intestinal microenvironment, ultimately resulting in dysbiosis [56,57].

In this study, we observed that Firmicutes and Bacteroidetes were the most abundant at the phylum level, and Bacteroides and Lactobacillus were the dominant genera in all dietary treatments. The Firmicutes to Bacteroidetes ratio is an important biomarker of gastrointestinal functionality and can be used as an indicator of eubiosis conditions in the gastrointestinal tract [58]. Previous studies have also demonstrated that the gut microbiota of obese subjects (humans) is characterized by a lower abundance of Firmicutes and a higher abundance of Bacteroidetes compared to their lean counterparts [51]. At the same time, it has been reported that oxidative stress has significantly increased the estimators of richness and community diversity of the gut microbiota of sows [57]. In addition, other studies have demonstrated that during conditions of intestinal dysbiosis, an excessive bioavailability of ROS molecules can contribute to an increase in oxidative stress [59,60]. It could be argued that these discrepancies may be related to the form of the type of animal or the level of oxidative stress.

In this study, although the cecal microbiota of the OS group was slightly similar to that of the CON one, the abundance of *Marinifilaceae, Odoribacter* and *Bacteroides_plebeius* in the OS group was significantly increased. Bacteroidetes are plant polysaccharide degraders and propionate producers that can improve intestinal barrier function and reduce inflammation and oxidative stress [46,61]. It has been reported that the relative abundance of Bacteroides is enriched in the gut of the host treated with antioxidants [62]. Our observation is in agreement with that of Wang [63], who found that oxidative stress significantly increases the relative abundance of *Bacteroides_f_Bacteroidaceae* in the gut of sows.

In this study, we observed that tryptophan metabolism, pyruvate metabolism, and the TCA cycle were disrupted by the tBHP challenge. While serotonin was upregulated, anthranilic acid, succinic acid, and oxaloacetate were downregulated. Tryptophan is an essential amino acid and is generally considered as the second-limiting amino acid in the most based diets of layers. Tryptophan metabolized in animals comes from two sources: one is an endogenous amino acid that is broken down by tissue protein, and the other is the exogenous amino acid that is digested and absorbed from the diet [64]. It has been reported that oxidative stress significantly decreased tryptophan/large neutral amino acids and serotonin concentrations in pigs, suggesting that oxidative stress might increase tryptophan metabolism [65]. This is consistent with our research results.

The tryptophan metabolism has been associated with various nutrients such as carbohydrates, proteins during the metabolic process [66]. An increase in the metabolism of tryptophan during oxidative stress could affect other physiological processes. On the other hand, tryptophan is also the precursor of Mel, and which can also be synthesized in the ovary [22]. Therefore, the decreasing levels of tryptophan in serum may correspond to the higher Mel and serotonin in the ovary. Pyruvate can realize the mutual conversion of carbohydrates, lipids, and amino acids in the body through the acetyl CoA and tricarboxylic acid cycles [67]. Therefore, pyruvate plays an important pivotal role in the metabolic connection of the three major nutrients. Moreover, the increased metabolism of pyruvate in cases of oxidative stress may be related to the increased need for oxygen and energy during the production of ROS.

It has been shown that supplementing exogenous pyruvate can prevent oxidative tissue stress [68]; however, in this study, we observed that oxidative stress promoted pyruvate metabolism. This apparent discrepancy may be due to different experimental models and different tissues tested, which needs to be clarified in further investigations. Oxidative stress could affect the levels of metabolites from glycolysis and the TCA cycle and production of adenine nucleotides [69]. In addition, it has been shown that an increase in oxidative stress and a decrease in the TCA cycle enzymes activity may cause the distal peripheral nerve to rely on truncated TCA cycle metabolism in rats [70]. Tryptophan catabolism has been recognized as an important player in inflammation and immune response. In this study, we found that oxidative stress caused a decrease in succinic acid and oxaloacetate in the TCA, which is consistent with previous studies. Oxidative stress will hinder the TCA and pyruvate metabolism, and ultimately affect the metabolism of carbohydrates, lipids, and amino acids.

SIRT1 is an NAD+-dependent protein deacetylase, and it has been proved to be involved in the protective effects of melatonin [71]. In this study, we found that melatonin levels were decreased in oxidative stress and the exogenous addition of Mel could mitigate the negative impact of oxidative stress and improve ovarian functionality in tBHP challenged layers. It has been shown that SIRT1 exerts its beneficial effects via the reduction of oxidative stress and endoplasmic reticulum stress in mitochondria [28,72]. Our results also indicated that the SIRT1 was increased by melatonin addition in vitro and decreased FoxO1 and P53 expression. As reported previously, melatonin was able to protect mouse granulosa cells against oxidative damage by inhibiting FoxO1-mediated autophagy [72,73].

The animal model we used (ovaries from laying hens) may be different from the ovaries of productive mammalian animals. There are distinct histological and physiological differences according to the reproductive stage of the animal. Moreover, "a link between the gastrointestinal microbiome and ovarian function" could be speculated in the current animal model, but cannot be established for other species and humans especially. However, the literature about oxidative stress on the ovary function of poultry is not well elucidated, which needs to be explored in future studies.

5. Conclusions

In conclusion, we found that oxidative stress could decrease laying performance, ovarian function and induce gut microbiota dysbiosis and disrupt serum metabolites in tBHP challenged layers; melatonin was able to reverse the impact of oxidative stress at the ovarian level through the SIRT1-P53/FoxO1 pathway (Graphical Abstract).

Supplementary Materials: The following are available online at https://www.mdpi.com/article/10.3390/antiox10091422/s1, Table S1: Composition and nutrient levels of basal diet, Table S2: Related gene and primer information.

Author Contributions: J.W. and K.Z. conceived and designed the experiments; J.W., R.J., H.Y. and H.G. performed the experiments; R.J., and H.G. analyzed the data; J.W. and H.G. wrote the paper; P.C., Y.Z., X.D., S.B., Q.Z., H.Y., J.L., X.M. and S.X. helped revise the manuscript. All authors have read and agreed to the published version of the manuscript.

Funding: This project was fanatically supported by the National Natural Science Foundation of China (31872792) and the Sichuan Science and Technology Program (2019YFH0062, 2018NZ20009) for financial support.

Institutional Review Board Statement: Experimental procedures were approved by the Animal Care and Use Committee of Sichuan Agricultural University (SICAU-2015-034) and were in accordance with the National Research Council's Guide for Care and Use of Laboratory Animals.

Informed Consent Statement: Not applicable.

Data Availability Statement: The data presented in this study are available in this article and also Supplementary Material here.

Conflicts of Interest: The authors declare no conflict of interest.

References

1. Shkolnik, K.; Tadmor, A.; Ben-Dor, S.; Nevo, N.; Galiani, D.; Dekel, N. Reactive oxygen species are indispensable in ovulation. *Proc. Natl. Acad. Sci. USA* **2011**, *108*, 1462–1467. [CrossRef]
2. Talukder, S.; Kerrisk, K.L.; Gabai, G.; Celi, P. Role of oxidant–antioxidant balance in reproduction of domestic animals. *Anim. Prod. Sci.* **2017**, *57*, 1588–1597. [CrossRef]
3. Martínez-Álvarez, R.M.; Morales, A.E.; Sanz, A. Antioxidant defenses in fish: Biotic and abiotic factors. *Rev. Fish. Biol. Fish.* **2005**, *15*, 75–88. [CrossRef]
4. Ma, Y.; Zhu, M.; Miao, L.; Zhang, X.; Dong, X.; Zou, X. Mercuric chloride induced ovarian oxidative stress by suppressing Nrf2-Keap1 signal pathway and its downstream genes in laying hens. *Biol. Trace Elem. Res.* **2018**, *185*, 185–196. [CrossRef] [PubMed]
5. Yeh, J.; Bowman, M.J.; Browne, R.W.; Chen, N. Reproductive aging results in a reconfigured ovarian antioxidant defense profile in rats. *Fertil. Steril.* **2005**, *84*, 1109–1113. [CrossRef] [PubMed]
6. Yang, Y.Z.; Cheung, H.H.; Zhang, C.; Wu, J.; Chan, W.Y. Melatonin as potential targets for delaying ovarian aging. *Curr. Drug Targets* **2019**, *20*, 16–28. [CrossRef] [PubMed]
7. Celi, P.; Gabai, G. Oxidant/antioxidant balance in animal nutrition and health: The role of protein oxidation. *Front. Vet. Sci.* **2015**, *2*, 48. [CrossRef]
8. Liguori, I.; Russo, G.; Curcio, F.; Bulli, G.; Aran, L.; Della-Morte, D.; Gargiulo, G.; Testa, G.; Cacciatore, F.; Bonaduce, D.; et al. Oxidative stress, aging, and diseases. *Clin. Interv. Aging* **2018**, *13*, 757–772. [CrossRef]
9. Wang, J.; Yang, Z.Q.; Celi, P.; Yan, L.; Ding, X.M.; Bai, S.P.; Zeng, Q.F.; Mao, X.B.; Feng, B.; Xu, S.Y.; et al. Alteration of the antioxidant capacity and gut microbiota under high levels of molybdenum and green tea polyphenols in laying hens. *Antioxidants* **2019**, *8*, 503. [CrossRef]
10. Paithankar, J.G.; Saini, S.; Dwivedi, S.; Sharma, A.; Chowdhuri, D.K. Heavy metal associated health hazards: An interplay of oxidative stress and signal transduction. *Chemosphere* **2021**, *262*, 128350. [CrossRef]
11. Gupta, R.K.; Miller, K.P.; Babus, J.K.; Flaws, J.A. Methoxychlor inhibits growth and induces atresia of antral follicles through an oxidative stress pathway. *Toxicol. Sci.* **2006**, *93*, 382–392. [CrossRef] [PubMed]

12. Devine, P.J.; Perreault, S.D.; Luderer, U. Roles of reactive oxygen species and antioxidants in ovarian toxicity. *Biol. Reprod.* **2012**, *86*, 27. [CrossRef]
13. Shen, M.; Lin, F.; Zhang, J.; Tang, Y.; Chen, W.K.; Liu, H. Involvement of the up-regulated FoxO1 expression in follicular granulosa cell apoptosis induced by oxidative stress. *J. Biol. Chem.* **2012**, *287*, 25727–25740. [CrossRef]
14. Shen, M.; Jiang, Y.; Guan, Z.Q.; Cao, Y.; Li, L.C.; Liu, H.L.; Sun, S.C. Protective mechanism of FSH against oxidative damage in mouse ovarian granulosa cells by repressing autophagy. *Autophagy* **2017**, *13*, 1364–1385. [CrossRef] [PubMed]
15. Zhang, S.; Zheng, C.; Lanza, I.R.; Nair, K.S.; Raftery, D.; Vitek, O. Interdependence of signal processing and analysis of urine 1H NMR spectra for metabolic profiling. *Anal. Chem.* **2009**, *81*, 6080–6088. [CrossRef]
16. Li, Y.; Wang, P.; Yin, J.; Jin, S.; Su, W.; Tian, J.; Li, T.; Yao, K. Effects of ornithine α-ketoglutarate on growth performance and gut microbiota in a chronic oxidative stress pig model induced by D-galactose. *Food Funct.* **2020**, *11*, 472–482. [CrossRef]
17. Hu, Y.; Chen, D.; Zheng, P.; Yu, J.; He, J.; Mao, X.; Yu, B. The bidirectional interactions between resveratrol and gut microbiota: An insight into oxidative stress and inflammatory bowel disease therapy. *BioMed Res. Int.* **2019**, *2019*, 5403761. [CrossRef] [PubMed]
18. Zhu, D.; Ma, Y.; Ding, S.; Jiang, H.; Fang, J. Effects of melatonin on intestinal microbiota and oxidative stress in colitis mice. *BioMed Res. Int.* **2018**, *2018*, 2607679. [CrossRef] [PubMed]
19. Zhang, H.; Yan, A.; Liu, X.; Ma, Y.; Zhao, F.; Wang, M.; Loor, J.J.; Wang, H. Melatonin ameliorates ochratoxin A induced liver inflammation, oxidative stress and mitophagy in mice involving in intestinal microbiota and restoring the intestinal barrier function. *J. Hazard. Mater.* **2021**, *407*, 124489. [CrossRef] [PubMed]
20. Tamura, H.; Nakamura, Y.; Korkmaz, A.; Manchester, L.C.; Tan, D.X.; Sugino, N.; Reiter, R.J. Melatonin and the ovary: Physiological and pathophysiological implications. *Fertil. Steril.* **2009**, *92*, 328–343. [CrossRef]
21. Galano, A.; Tan, D.X.; Reiter, R.J. Melatonin as a natural ally against oxidative stress: A physicochemical examination. *J. Pineal Res.* **2011**, *51*, 1–16. [CrossRef]
22. Hao, E.Y.; Chen, H.; Wang, D.H.; Huang, C.X.; Tong, Y.G.; Chen, Y.F.; Zhou, R.Y.; Huang, R.L. Melatonin regulates the ovarian function and enhances follicle growth in aging laying hens via activating the mammalian target of rapamycin pathway. *Poult. Sci.* **2020**, *99*, 2185–2195. [CrossRef]
23. Jia, Y.; Yang, M.; Zhu, K.; Wang, L.; Song, Y.; Wang, J.; Qin, W.; Xu, Z.; Chen, Y.; Liu, G. Melatonin implantation improved the egg-laying rate and quality in hens past their peak egg-laying age. *Sci. Rep.* **2016**, *6*, 39799. [CrossRef] [PubMed]
24. Bordone, L.; Guarente, L. Calorie restriction, SIRT1 and metabolism: Understanding longevity. *Nat. Rev. Mol. Cell Biol.* **2005**, *6*, 298–305. [CrossRef] [PubMed]
25. Alam, F.; Syed, H.; Amjad, S.; Baig, M.; Khan, T.A.; Rehman, R. Interplay between oxidative stress, SIRT1, reproductive and metabolism functions. *Curr. Res. Physiol.* **2021**, *4*, 119–124. [CrossRef]
26. Singh, V.; Ubaid, S. Role of silent information regulator 1 (SRIT1) in regulation oxidative stress and inflammation. *Inflammation* **2020**, *43*, 1589–1598. [CrossRef]
27. Tao, X.; Zhang, X.; Ge, S.Q.; Zhang, E.H.; Zhang, B. Expression of SIRT1 in the ovaries of rats with polycystic ovary syndrome before and after therapeutic intervention with exenatide. *Int. J. Clin. Exp. Pathol.* **2015**, *8*, 8276–8283.
28. Ding, M.; Lei, J.; Han, H.; Li, W.; Qu, Y.; Fu, E.; Fu, F.; Wang, X. SIRT1 protects against myocardial ischemia-reperfusion injury via activating eNOS in diabetic rats. *Cardiovasc. Diabetol.* **2015**, *14*, 143. [CrossRef]
29. Salminen, A.; Kaarniranta, K.; Kauppinen, A. Crosstalk between oxidative stress and SIRT1: Impact on the aging process. *Int. J. Mol. Sci.* **2013**, *14*, 3834–3859. [CrossRef] [PubMed]
30. Feng, Y.L.; Jiang, X.T.; Ma, F.F.; Han, J.; Tang, X.L. Resveratrol prevents osteoporosis by upregulating FoxO1 transcriptional activity. *Int. J. Mol. Med.* **2018**, *41*, 202–212. [CrossRef] [PubMed]
31. Ren, Z.; Sun, W.; Liu, Y.; Li, Z.; Han, D.; Cheng, X.; Yan, J.; Yang, X. Dynamics of serum phosphorus, calcium, and hormones during egg laying cycle in Hy-Line Brown laying hens. *Poult. Sci.* **2019**, *98*, 2193–2200. [CrossRef] [PubMed]
32. Luan, Y.; Edmonds, W.E.; Woodruff, T.K.; Kim, S.Y. Inhibitors of apoptosis protect the ovarian reserve from cyclophosphamide. *J. Endocrinol.* **2019**, *240*, 243–256. [CrossRef] [PubMed]
33. Xu, S.; Wu, X.; Dong, Y.; Li, Z.; Lin, Y.; Che, L.; Li, J.; Feng, B.; Fang, Z.; Zhuo, Y.; et al. Glucose activates the primordial follicle through the AMPK/mTOR signaling pathway. *Clin. Transl. Med.* **2020**, *10*, e122. [CrossRef]
34. Johnson, A.L. Ovarian follicle selection and granulosa cell differentiation. *Poult. Sci.* **2015**, *94*, 2193–2200. [CrossRef]
35. Myers, M.; Britt, K.L.; Wreford, N.G.M.; Ebling, F.J.P.; Kerr, J.B. Methods for quantifying follicular number within the mouse ovary. *Reproduction* **2004**, *127*, 569–580. [CrossRef]
36. Zhuo, Y.; Hua, L.; Feng, B.; Jiang, X.M.; Jiang, L.; Jiang, D.D.; Huang, X.H.; Zhu, Y.G.; Li, Z.; Yan, L.J.; et al. Fibroblast growth factor 21 coordinates adiponectin to mediate the beneficial effect of low-protein diet on primordial follicle reserve. *EbioMedicine* **2019**, *41*, 623–635. [CrossRef] [PubMed]
37. Wang, J.; Wan, C.; Zhao, S.; Yang, Z.; Celi, P.; Ding, X.; Bai, S.; Zeng, Q.; Mao, X.; Xu, S.; et al. Differential analysis of gut microbiota and the effect of dietary *Enterococcus feacium* supplementation in broiler breeders with high or low laying performance. *Poult. Sci.* **2021**, *100*, 1109–1119. [CrossRef]
38. Zhuo, Y.; Feng, B.; Xuan, Y.; Che, L.; Feng, Z.; Lin, Y.; Xu, S.; Li, J.; Wu, D. Inclusion of purified dietary fiber during gestation improved the reproductive performance of sows. *J. Anim. Sci. Biotechnol.* **2020**, *11*, 47. [CrossRef] [PubMed]
39. Van den Hurk, R.; Zhao, J. Formation of mammalian oocytes and their growth, differentiation and maturation within ovarian follicles. *Theriogenology* **2005**, *63*, 1717–1751. [CrossRef] [PubMed]

40. Findlay, J.K.; Hutt, K.J.; Hickey, M.; Anderson, R.A. How is the number of primordial follicles in the ovarian reserve established? *Biol. Reprod.* **2015**, *93*, 111–117. [CrossRef]
41. Agarwal, A.; Aponte-Mellado, A.; Premkumar, B.J.; Shaman, A.; Gupta, S. The effects of oxidative stress on female reproduction: A review. *Reprod. Biol. Endocrinol.* **2012**, *10*, 49. [CrossRef]
42. Cao, Y.; Shen, M.; Jiang, Y.; Sun, S.C.; Liu, H. Melatonin reduces oxidative damage in mouse granulosa cells via restraining JNK-dependent autophagy. *Reproduction* **2018**, *155*, 307–319. [CrossRef]
43. Fu, Y.; He, C.J.; Ji, P.Y.; Zhuo, Z.Y.; Tian, X.Z.; Wang, F.; Tan, D.X.; Liu, G.S. Effects of melatonin on the proliferation and apoptosis of sheep granulosa cells under thermal stress. *Int. J. Mol. Sci.* **2014**, *15*, 21090–21104. [CrossRef]
44. Wang, H.; Ji, Y.; Yin, C.; Deng, M.; Tang, T.; Deng, B.; Ren, W.; Deng, J.; Yin, Y.; Tan, C. Differential analysis of gut microbiota correlated with oxidative stress in sows with high or low litter performance during lactation. *Front. Microbiol.* **2018**, *9*, 1665. [CrossRef]
45. He, Y.; Deng, H.; Jiang, Z.; Li, Q.; Shi, M.; Chen, H.; Han, Z. Effects of melatonin on follicular atresia and granulosa cell apoptosis in the porcine. *Mol. Reprod. Dev.* **2016**, *83*, 692–700. [CrossRef] [PubMed]
46. Zhang, J.; Zhao, X.; Jiang, Y.; Zhao, W.; Guo, T.; Cao, Y.; Teng, J.; Hao, X.; Zhao, J.; Yang, Z. Antioxidant status and gut microbiota change in an aging mouse model as influenced by exopolysaccharide produced by *Lactobacillus plantarum* YW11 isolated from Tibetan kefir. *J. Dairy Sci.* **2017**, *100*, 6025–6041. [CrossRef]
47. Yuan, Z.H.; Wang, J.P.; Zhang, K.Y.; Ding, X.M.; Bai, S.P.; Zeng, Q.F.; Xuan, Y.; Su, Z.W. Effect of vanadium and tea polyphenols on intestinal morphology, microflora and short-chain fatty acid profile of laying hens. *Poult. Sci.* **2017**, *174*, 419–427. [CrossRef] [PubMed]
48. Abou Ueil, M.A.; Youssef, A.A. Oxidative stress and male IGF-1, gonadotropin and related hormones in diabetic patients. *Clin. Chem. Lab. Med.* **2001**, *39*, 618–623. [CrossRef]
49. Tsai-Turton, M.; Luderer, U. Opposing effects of glutathione depletion and follicle-stimulating hormone on reactive oxygen species and apoptosis in cultured prevulatory rat follicles. *Endocrinology* **2006**, *147*, 1224–1236. [CrossRef] [PubMed]
50. Miyamoto, K.; Sato, F.E.; Kasahara, E.; Jikumaru, M.; Hiramoto, K.; Tabata, H.; Katsuragi, M.; Odo, S.; Utsumi, K.; Indoue, M. Effect of oxidative stress during repeated ovulation on the structure and functions of the ovary, oocytes, and their mitochondria. *Free Radic. Biol. Med.* **2010**, *49*, 674–681. [CrossRef]
51. Ley, R.E.; Turnbaugh, P.J.; Klein, S.; Gordon, J.I. Microbial ecology: Human gut microbes associated with obesity. *Nature* **2006**, *444*, 1022–1023. [CrossRef]
52. Brannian, J.D.; Hansen, K.A. Leptin and ovarian folliculogenesis: Implications for ovulation induction and ART outcomes. *Semin. Reprod. Med.* **2002**, *20*, 103–112. [CrossRef]
53. Yu, H.; Kuang, M.; Wang, Y.; Rodeni, S.; Wei, Q.; Wang, W.; Mao, D. Sodium arsenite injection induces ovarian oxidative stress and affects steroidogenesis in rats. *Biol. Trace Elem. Res.* **2019**, *189*, 186–193. [CrossRef]
54. Higashi, Y.; Sukhanov, S.; Anwar, A.; Shai, S.Y.; Delafontaine, P. IGF-1, oxidative stress and atheroprotection. *Trends Endocrinol. Metab.* **2010**, *21*, 245–254. [CrossRef] [PubMed]
55. Luo, X.; Jiang, X.; Li, J.; Bai, Y.; Li, Z.; Wei, P.; Sun, S.; Liang, Y.; Han, S.; Li, X.; et al. Insulin-like growth factor-1 attenuates oxidative stress-induced hepatocyte premature senescence in liver fibrogenesis via regulating nuclear p53-progerin interaction. *Cell Death Dis.* **2019**, *10*, 451. [CrossRef] [PubMed]
56. Cheng, M.; Zhang, X.; Miao, Y.J.; Cao, J.X.; Wu, Z.F.; Weng, P.F. The modulatory effect of (−)-epigallocatechin 3-O-(3-O-methyl) gallate (EGCG3″Me) on intestinal microbiota of high fat diet-induced obesity mice model. *Food Res. Int.* **2017**, *92*, 9–16. [CrossRef] [PubMed]
57. Dam, B.; Misra, A.; Banerjee, S. Role of gut microbiota in combating oxidative stress. In *Oxidative Stress in Microbial Diseases;* Chakraborti, S., Chakraborti, T., Chattopadhyay, D., Shaha, C., Eds.; Springer: Singapore, 2019; pp. 43–82. [CrossRef]
58. Pereira, T.M.C.; Pimenta, F.S.; Porto, M.L.; Baldo, M.P.; Campagnaro, B.P.; Gava, A.L.; Meyrelles, S.S.; Vasquez, E.C. Coadjuvants in the diabetic complications: Nutraceuticals and drugs with pleiotropic effects. *Int. J. Mol. Sci.* **2016**, *17*, 1273. [CrossRef] [PubMed]
59. Zhou, L.; Xiao, X.; Zhang, Q.; Zheng, J.; Deng, M.Q. Maternal genistein intake mitigates the deleterious effects of high-fat diet on glucose and lipid metabolism and modulates gut microbiota in adult life of male mice. *Front. Physiol.* **2019**, *10*, 985. [CrossRef] [PubMed]
60. Corrigan, A.; de Leeuw, M.; Penaud-Frezet, S.; Dimova, D.; Murphy, R.A. Phylogenetic and functional alterations in bacterial community compositions in broiler ceca as a result of mannan oligosaccharide supplementation. *Appl. Environ. Microbiol.* **2015**, *81*, 3460–3470. [CrossRef]
61. Tong, L.C.; Wang, Y.; Wang, Z.B.; Liu, W.Y.; Sun, S.; Li, L.; Su, D.F.; Zhang, L.C. Propionate ameliorates dextran sodium sulfate-induced colitis by improving intestinal barrier function and reducing inflammation and oxidative stress. *Front. Pharmacol.* **2016**, *7*, 253. [CrossRef]
62. Wang, Y.Z.; Zeng, S.M. Melatonin promotes ubiquitination of phosphorylated pro-apoptotic protein Bcl-2-interacting mediator of cell death-extra long (BimEL) in porcine granulosa cells. *Int. J. Mol. Sci.* **2018**, *19*, 3431. [CrossRef]
63. Allison, A.; Julien, P.; Harry, S. Gut microbiota regulation of tryptophan metabolism in health and disease. *Cell Host Microbe* **2018**, *23*, 716–724.

64. Lv, M.; Yu, B.; Mao, X.B.; Zheng, P.; He, J.; Chen, D.W. Responses of growth performance and tryptophan metabolism to oxidative stress induced by diquat in weaned pigs. *Animal* **2012**, *6*, 928–934. [CrossRef] [PubMed]
65. O'Mahony, S.M.; Clarke, G.; Borre, Y.E.; Dinan, T.G.; Cryan, J.F. Serotonin, tryptophan metabolism and the brain-gut-microbiome axis. *Behav. Brain Res.* **2015**, *15*, 32–48. [CrossRef]
66. Hegde, K.R.; Varma, S.D. Prevention of oxidative stress to the retina by pyruvate. A preliminary report. *Ophthalmologica* **2008**, *222*, 194–198. [CrossRef] [PubMed]
67. Varma, S.D.; Hegde, K.; Henein, M. Oxidative damage to mouse lens in culture. Protective effect of pyruvate. *Biochim. Biophys. Acta* **2003**, *1621*, 246–252. [CrossRef]
68. Tavsan, Z.; Kayali, H.A. Influence of the oxidative stress induced by the organophosphate pesticide bromopropylate on the mitochondrial respiratory chain in Trichoderma harzianum. *Process. Biochem.* **2014**, *49*, 745–750. [CrossRef]
69. Hinder, L.M.; Vivekanandan-Giri, A.; Mclean, L.L.; Pennathur, S.; Feldman, E.L. Decreased glycolytic and tricarboxylic acid cycle intermediates coincide with peripheral nervous system oxidative stress in a murine model of type 2 diabetes. *J. Endocrinol.* **2013**, *216*, 1–11. [CrossRef]
70. Ding, M.; Feng, N.; Tang, D.; Feng, J.; Li, Z.; Jia, M.; Liu, Z.; Gu, X.; Wang, Y.; Fu, F.; et al. Melatonin prevents Drp1-mediated mitochondrial fission in diabetic hearts through SIRT1-PGC1α pathway. *J. Pineal Res.* **2018**, *65*, e12491. [CrossRef]
71. Hsu, Y.J.; Hsu, S.C.; Hsu, C.P.; Chen, Y.H.; Chang, Y.L.; Sadoshima, J.C.; Huang, S.M.; Tsai, C.S.; Lin, C.Y. Sirtuin 1 protects the aging heart from contractile dysfundcion mediated through the inhibition of endoplasmic reticulum stress-mediated apoptosis in cardiac-specific Sirtuin 1 knockout mouse model. *Int. J. Cardiol.* **2017**, *228*, 543–552. [CrossRef]
72. Shen, M.; Cao, Y.; Jiang, Y.; Wei, Y.H.; Liu, H.L. Melatonin protects mouse granulosa cells against oxidative damage by inhibiting FOXO1-mediated autophagy: Implication of an antioxidation-independent mechanism. *Redox Biol.* **2018**, *18*, 138–157. [CrossRef] [PubMed]
73. Hao, E.Y.; Wang, D.H.; Chang, L.Y.; Huang, C.X.; Chen, H.; Yue, Q.X.; Zhou, R.Y.; Huang, R.L. Melatonin regulates chicken granulosa cell proliferation and apoptosis by activating the mTOR signaling pathway via its receptors. *Poult. Sci.* **2020**, *99*, 6147–6162. [CrossRef] [PubMed]

MDPI
St. Alban-Anlage 66
4052 Basel
Switzerland
Tel. +41 61 683 77 34
Fax +41 61 302 89 18
www.mdpi.com

Antioxidants Editorial Office
E-mail: antioxidants@mdpi.com
www.mdpi.com/journal/antioxidants

www.ingramcontent.com/pod-product-compliance
Lightning Source LLC
Chambersburg PA
CBHW061758210926
51850CB00006B/584
* 9 7 8 3 0 3 6 5 5 4 7 1 6 *